Developments in Mathematics

VOLUME 41

More information about this series at http://www.springer.com/series/5834

Sabir Umarov

Introduction to Fractional and Pseudo-Differential Equations with Singular Symbols

Springer

Sabir Umarov
Department of Mathematics
University of New Haven
West Haven, CT, USA

ISSN 1389-2177 ISSN 2197-795X (electronic)
Developments in Mathematics
ISBN 978-3-319-36846-7 ISBN 978-3-319-20771-1 (eBook)
DOI 10.1007/978-3-319-20771-1

Mathematics Subject Classification (2010): 46Bxx, 46Sxx, 35Sxx, 35Gxx, 32W25, 47G30, 26A33, 60H10, 60G50, 35Q84

Springer Cham Heidelberg New York Dordrecht London

Softcover reprint of the hardcover 1st edition 2015

Printed on acid-free paper

Springer International Publishing AG Switzerland is part of Springer Science+Business Media (www.springer.com)

To my parents,
to my family

Preface

The 20th century was rich with great scientific and mathematical discoveries. One of the most influential events in mathematics was the introduction of the Lebesgue integral (Lebesgue, 1901) followed soon after by Borel's development of measure theory (Borel, 1903). Combined with Cantor's set theory, whose axiomatic form appeared in the beginning of the century (Zermelo, 1907), these sources with their deep and remarkable ideas made possible the development of a large number of function spaces that are extremely important in modern analysis, such as L_p-spaces (F. Riesz, 1910), Sobolev spaces (Sobolev, 1938), Schwatz's distributions (Schwartz, 1951), and many more. Undoubtedly, modern functional analysis, including measure theory, the theory of topological vector spaces, and operator theory, with their various powerful techniques and methods, ultimately takes their origin from these remarkable sources.

Another extremely influential series of events was the emergence of the mathematical formalism of probability theory (Kolmogorov, 1933) in conjunction with the introduction of Brownian motion (Einstein, 1905; Wiener, 1927) and Lévy processes (Lévy, 1938). These led to a number of deep developments in the theory of stochastic processes, two of which are the Itô's stochastic calculus (Itô, 1948) and the theory of stochastic ordinary and partial differential equations. An important attribute in this chain is the Fokker-Planck-Kolmogorov equation (Fokker, 1913; Planck, 1917; Kolmogorov, 1931), which provides a deterministic way of describing stochastic processes.

By the 1950s the ideas behind Sobolev spaces were well understood, Schwartz distributions were introduced (justifying, in particular, the Dirac delta function intensively used in physics), mathematical theories of Brownian motion and Lévy processes were developed, and mathematical reasoning of stochastic integrals, Itô's stochastic calculus had emerged (justifying the Langevin stochastic differential equations introduced in 1908). This spawned the rapid development in the 1960s not only of new powerful methods and generalizations of the above theories, but also the initiation of a number of new theories. Just to mention a few, these are the theory of pseudo-differential operators (J.J. Kohn and L. Nirenberg 1965, and L. Hörmander,

1965) in its modern form, Fourier integral operators (Hörmander, 1968–71), continuous time random walks (Montroll and Weiss, 1965), and fractional order Fokker-Planck and diffusion equations. There was an abundance of applications of these theories in the natural and social sciences and engineering, including the filtering theory (Kalman, Bucy, 1961), etc. In the scope of these theories, a number of long-standing mathematical problems have found their solutions. The related theories of fractional order differential operators and equations have also led to a huge number of applications in science and engineering, as well as inside mathematics itself.

Many good books and papers have been written about the classical theories mentioned above, and continue to appear, since extensive investigations, discovering new concepts and developing new theories, are ongoing. With no pretense of completeness, we refer the reader to the books [Nik77, BIN75, Tri77, Tri83] on function spaces, [Hor83, Tre80, Tay81, Shu78, Won99] on pseudo-differential operators and Fourier integral operators, [IW81, Pro91, KSh91, Sit05, App09] on Itô calculus and stochastic differential equations, and [OS74, SKM87, Rub96, Pod99, KST06, Mai10] on fractional calculus and fractional order differential equations.

What is the present book about? This book is an introduction to the theory of pseudo-differential operators with symbols singular with respect to dual variables combined with fractional order differential equations and their applications to various applied fields. One of the essential requirements in the classical theory of pseudo-differential operators is the symbol must be smooth on the cotangent bundle, that is $a(x,\xi) \in C^{\infty}(\Omega,\mathbb{R}^n)$. However, in the solution of many problems of mathematical physics pseudo-differential operators with singular symbols arise. For instance, solution operators of two- or multi-point boundary value problems frequently appear to have symbols singular in dual variables. For example, the symbol of the solution operator of the following simple nonlocal boundary value problem

$$\frac{\partial^2 u(t,x)}{\partial t^2} + \frac{\partial^2 u(t,x)}{\partial x^2}(t,x) = 0, \ t \in (0,2), \ x \in \mathbb{R},$$
$$u(0,x) = \varphi(x), \ u(1,x) = u(2,x), \ x \in \mathbb{R},$$

is (see for details, Section 2.2)

$$s(t,\xi) = \frac{\cos(\xi t)[\sin\xi - \sin(2\xi)] - \sin(\xi t)[\cos\xi - \cos(2\xi)]}{\sin\xi - \sin(2\xi)},$$

which has irreducible singularities at the points $\pi k, k = \pm 1, \pm 2, \ldots,$ and $\pm\pi/3 + 2\pi m, m = 0, \pm 1, \ldots$. Nonlocal boundary value problems arise, for instance, in plasma physics, diffusion in porous media, etc. Pseudo-differential operators with symbols singular with respect to dual variables (ΨDOSS) allow us to revisit from a new angle many classic boundary value problems, as well. Examples include the Schrödinger operator $i\frac{\partial}{\partial t} - \frac{\omega}{c\hbar}\sqrt{I - \omega^2\Delta}$, arising in the theory of relativistically free particles, the theory of Bessel potentials with symbols $|\xi|^{-\alpha}$, where α is an arbitrary positive number. Many other examples, including fractional order pseudo-differential equations with singular symbols, will be considered throughout the book.

Pseudo-differential operators with symbols singular with respect to dual variables were first considered systematically by Cordes, Williams [CW77], Plamenevski [Pla86], and Ju. A. Dubinskii [Dub82], though episodically appeared earlier in works of other authors (e.g., [Kon67]). Cordes and Williams used methods of abstract algebra to define algebras of pseudo-differential operators with non-smooth symbols with respect to dual variables $\xi = (\xi_1, \dots, \xi_n)$, and applied these algebras to solution of singular elliptic pseudo-differential equations. Plamenevski used the Mellin's transform approach to the construction of singular pseudo-differential operators. Dubinskii defined his operators with the help of infinite order differential operators. The latter requires the analyticity of the symbols in domains not containing singularities. Hence, the corresponding algebras of pseudo-differential operators have locally analytic symbols. We also note that pseudo-differential operators with symbols singular with respect to the current variable were studied by Nagase in [Nag77].

Our approach is different from those mentioned above. In Chapter 2 of this book we introduce and study ΨDOSS in the form

$$Af(x) = \frac{1}{(2\pi)^n} \int_{\mathbb{R}^n} a(x,\xi) F[f](\xi) e^{-i(x,\xi)} d\xi,$$

where $F[f](\xi)$ is the Fourier transform of f and $(x,\xi) = x_1\xi_1 + \dots, + x_n\xi_n$. Due to singularity of the symbol $a(x,\xi)$, this operator is not well defined even on infinitely differentiable functions with compact support. Indeed, let $f_0 \in C_0^\infty(\mathbb{R}^n)$, such that $F[f_0](\xi_0) = 1$ in a neighborhood of a non-integrable singular point ξ_0 of the symbol $a(x,\xi)$. Then the integral in the above definition of the operator A diverges. Therefore, we introduce a special class of test functions and the corresponding space of distributions, for which the operator A with a singular symbol $a(x,\xi)$ is well defined. The space of test functions denoted by $\Psi_{G,p}(\mathbb{R}^n)$, where $G \subset \mathbb{R}^n$ and $p \in [1,\infty)$, and the corresponding space of distributions (we call them ψ-distributions) denoted by $\Psi'_{-G,p'}(\mathbb{R}^n)$, are relatively new and not adequately presented in the literature. These spaces are studied in detail in Chapter 1. The Fourier transform of any function in $\Psi_{G,p}(\mathbb{R}^n)$, by definition, has a compact support contained in G. The dependence of spaces $\Psi_{G,p}(\mathbb{R}^n)$ and $\Psi'_{-G,p'}(\mathbb{R}^n)$ on the parameter p is not formal (see details in Section 1.10). L. Hörmander in his book [Hor83] pointed out that the Fourier transform is not well adapted to L_p-spaces. However, in the above construction one cannot avoid working with Fourier transform in L_p-spaces, since $\Psi_{G,p}(\mathbb{R}^n)$ and $\Psi'_{-G,p'}(\mathbb{R}^n)$ depend in an essential way on p.

Chapter 4 discusses the existence and uniqueness problem of general nonlocal boundary value problems for ΨDOSS as well as their applications to various problems of analysis. Initial and other types of boundary value problems for fractional order ΨDOSS are important to include in our book due to their wide applications in science and engineering. For instance, the fractional order Fokker-Planck-Kolmogorov (FPK) equations are used as mathematical models of various processes in physics, biology, finance, hydrology, etc. In particular, it is well known that a deep triple relationship exists between the classical FPK equation, its associated Itô stochastic differential equation, and the corresponding driving process. The driving process in the classical case is a Brownian motion. If one changes the driving process, then its corresponding stochastic differential equation and FPK equation counterparts will also change. In this sense the driving process plays an important role. What is the driving process in the case of fractional order FPK equations? And what form does the associated stochastic differential equation assume? These are natural questions arising in the case of fractional FPK equations. Chapter 7 discusses these questions in detail. The reader will see that the driving process can be described as a scaling limit of continuous time random walks, which first appeared in the literature in 1965 in the paper [MW65] by Montroll and Weiss. The stochastic differential equation associated with fractional FPK equation will take a form that is driven by a time-changed Brownian motion if there are no jump components. Random walks, including continuous time random walks, approximating mixed and time-changed Lévy processes, which will serve as driving processes in SDEs arising in Chapter 7, will be discussed in Chapter 8.

Chapters 3, 5, and 6 provide fractional calculus background including recent developments in this area, such as distributed and variable fractional order differential operators, as well as boundary value problems for fractional order differential equations. Chapters 5 and 6, in particular, discuss fractional generalizations of the famous Duhamel principle to various forms of fractional order differential equations.

Finally, in Chapter 9 we develop a complex analogue of pseudo-differential operators with singular symbols. We note that in this theory many issues still remain open. In the earlier works, the complex theory was constructed by methods essentially different from the methods used in real analysis. The main reason for that was the absence of a suitable Fourier transform technique in the complex case. The existence of the Borel and the Fourier-Laplace transforms do not give appropriate results because of their narrow application. We will develop a new complex Fourier transform technique on fiber spaces of analytic and exponential functions and functionals. Based on this construction we will be able to study not only the new wide classes of differential and pseudo-differential equations but also to improve the existing results. This method allows us also to detect the deep connection between analytic and exponential solvability of initial and boundary value problems for pseudo-differential equations with analytic and meromorphic symbols.

Nowadays the number of applications of the theory of ΨDOSS and fractional order differential equations is rapidly increasing. The author hopes that the selected material reflects the current state and will serve as a good source for those who want to study the theory of ΨDOSS and fractional differential equations and use their methods in their own research. It seems as though this is the first attempt to present systematically the theory of ΨDOSS in the chosen format. Therefore, the style of the book is introductory. Each chapter supplies a section containing historical and additional notes on related topics for those readers who want further reading (Tashkent-Boston-New Haven).

Acknowledgment. I am indebted to my colleagues Shavkat Alimov, Erik Andries, Ravshan Ashurov, Frederick Daum, Yulii Dubinskii, Rudolf Gorenflo, Marjorie Hahn, Dinh Háo, Xinxin Jiang, Alimdjan Khalmukhamedov, Nikolai Kislov, Kei Kobayashi, Joseph Kolibal, Yury Luchko, Mark Meerschaert, Marc Mehlman, Makhmuda Nazarova, Kenric Nelson, Jelena Ryvkina, Erkin Saydamatov, Hans-Jürgen Schmießer, Stanly Steinberg, Hans Triebel, Constantino Tsallis, Batirkhan Turmetov, and many others, with whom I have collaborated and/or discussed various parts of the topics included in this book.

Special thanks go to Marjorie Hahn (Tufts University) and Kei Kobayashi (University of Tennessee, Knohville) for reading of some parts of the book and for their insightful comments. My gratitude also goes to Bruce Boghosian and Boris Hasselblatt (chairs, Tufts University), Joseph Kolibal (chair, University of New Haven), and Lourdes Alvares (dean, University of New Haven) for their encouragement and support during the work on the book.

West Haven, CT, USA Sabir Umarov
January 5, 2015

Contents

Chapter 1
Function spaces and distributions

1.1 Introduction

This chapter is devoted to function and distribution spaces. We first recall definitions of some well-known classical function and distribution spaces, simultaneously introducing the terminology and notations used in this book. Then we introduce (see Section 1.10) a new class of test functions and the corresponding space of distributions (generalized functions), which play an important role in the theory of pseudo-differential operators with singular symbols introduced in Chapter 2. By singular symbols we mean, if not otherwise assumed, symbols singular in dual variables. We will denote the space of test functions endowed with the strong topology by $\Psi_{G,p}(\mathbb{R}^n)$, where $G \subset \mathbb{R}^n$ and $p \in [1,\infty)$, while the corresponding space of distributions, called ψ-distributions, by $\Psi'_{-G,p'}(\mathbb{R}^n)$. The dependence of spaces $\Psi_{G,p}(\mathbb{R}^n)$ and $\Psi'_{-G,p'}(\mathbb{R}^n)$ on the parameter p is not formal (see examples in Section 1.10). For $p = 2$ these spaces coincide with the spaces $H^{\pm\infty}(\pm G)$ introduced and studied in [Dub82]. The general case of $p \geq 1$ was introduced in [Uma97, Uma98]. For further historical details see Section 1.13.

Pseudo-differential operators with singular symbols in the dual variable (ΨDOSS) are important in the modern theory of partial differential equations (see examples in Section 2.2). To solve boundary value problems for pseudo-differential equations with ΨDOSS in the Sobolev, Besov, BMO, Lizorkin, and other classical spaces one needs to know when ΨDOSS has a continuous closure to these spaces. The closure of ΨDOSS to classical function spaces requires the denseness of $\Psi_{G,p}(\mathbb{R}^n)$ in these spaces. The necessary and sufficient conditions of the denseness of $\Psi_{G,p}(\mathbb{R}^n)$ in $L_p(\mathbb{R}^n)$ and other classical function spaces will be studied in detail in Section 1.11. By definition, elements of $\Psi_{G,p}(\mathbb{R}^n)$, i.e., test functions, are L_p-functions such that their Fourier transforms have compact support. In accordance with the Paley-Wiener-Schwartz theorem elements of $\Psi_{G,p}(\mathbb{R}^n)$ are entire functions of a finite exponential type whose restrictions to $\mathbb{R}^n$ belong to $L_p(\mathbb{R}^n)$.

S. Umarov, *Introduction to Fractional and Pseudo-Differential Equations with Singular Symbols*, Developments in Mathematics 41,
DOI 10.1007/978-3-319-20771-1_1

The class of entire functions of a finite exponential type and its various subspaces are broadly used by many researchers in different fields of analysis. In our construction there is one subtlety: G in the definition of $\Psi_{G,p}(\mathbb{R}^n)$ allows to localize singularities of symbols of ΨDOSS. At the same time, $\Psi_{G,p}(\mathbb{R}^n)$ is a subclass of the class of exponential functions of a finite type. The smaller is G, the narrower is $\Psi_{G,p}(\mathbb{R}^n)$. It is not hard to indicate such a G so that the corresponding space $\Psi_{G,p}(\mathbb{R}^n)$ will not be dense in $L_p(R^n)$.

Lizorkin type spaces, introduced in Section 1.12, are exact spaces for which ΨDOSS' with symbols singular at the origin are well defined. The exactness used here means that the number of orthogonality conditions, under which ΨDOSS is well defined, is minimal.

In this book we use the following notations: $\mathbb{R}$, $\mathbb{C}$, $\mathbb{N}$, and $\mathbb{Z}$ denote the set of real, complex, natural, and integer numbers, respectively. $\mathbb{N}_0$ is the set of extended natural numbers: $\mathbb{N}_0 \equiv \mathbb{N} \cup \{0\}$. For $n \in \mathbb{N}$ by $\mathbb{R}^n$ we denote the n-dimensional Euclidean space of n-tuples $x = (x_1, \dots, x_n)$, $x_j \in \mathbb{R}$, $j = 1, \dots, n$. Let r be a positive real number and a be a point in $\mathbb{R}^n$. Then by $B_r(a)$ we denote the n-dimensional open ball with the radius r and center a, that is $B_r(a) = \{x \in \mathbb{R}^n : |x - a| < r\}$, where $|x| = \sqrt{x_1^2 + \dots + x_n^2}$. A nonempty set $A \subset \mathbb{R}^n$ is called *open* if $B_\varepsilon(a) \subset A$ for every $a \in A$ with some $\varepsilon = \varepsilon(a) > 0$. By a neighborhood of a point $a \in A$ we understand any open subset of A which contains a. For a given sequence $y_k \in A, k = 1, 2, \dots,$ its limit, in general, may not belong to A. The set A is called *closed* if it contains all its limit points. One can make a set closed adding its all the limit points, which is called a *a closure* of A and denoted by $\overline{A}$. For instance, $\overline{B_r(a)} = \{x \in \mathbb{R}^n : |x - a| \leq r\}$. A point of A is *interior* if it has a neighborhood contained in A. By the boundary of a given set A we understand $\overline{A} \setminus \overset{\circ}{A}$, where $\overset{\circ}{A}$ is the set of interior points of A, and the symbol "\" means, as usual, "set-minus." For the boundary of A we will use the notation ∂A. For instance, the boundary of $B_r(a)$ is the $n-1$-dimensional sphere $S_r(a) = \partial B_r(a) \equiv \{x \in \mathbb{R}^n : |x - a| = r\}$. There are sets with no boundary, i.e., $\partial A = \emptyset$. An example of such a set is $S_r(a)$. The set A is called *bounded* if there is a ball with a finite radius containing A. Any closed bounded set in $\mathbb{R}^n$ will be called a *compact*. If A is a compact subset of B, then we write $A \Subset B$. We will say that a property P holds *almost everywhere (a.e.)* in A if P breaks down only in a set of zero measure in A.

We assume that the reader is familiar with the theory of Borel sets, the Lebesgue integral, the Lebesgue measure, as well as some basic notions of normed/Banach and Hilbert spaces, including the notions such as norm, inner product, fundamental sequence, and completeness. If x is an element of a Banach space X, then the norm of x will be denoted by $\|x|X\|$ or $\|x\|_X$.

Finally, for the reader's convenience in Chapter 1 we provide the necessary auxiliary results, but not all of them with proofs. In case, when proof is not provided we refer the reader to other appropriate sources.

1.2 Hölder-Zygmund spaces

Let Ω be a bounded open set in $\mathbb{R}^n$ with a smooth boundary $\partial\Omega$. We denote by $C(\Omega)$ the set of functions f defined and continuous on Ω and by $C(\overline{\Omega})$ the set of functions defined on the closed set $\overline{\Omega}$ and continuous up to the boundary of Ω. $C(\overline{\Omega})$ is endowed with the norm

$$\|f\|_{C(\overline{\Omega})} = \max_{x\in\bar{\Omega}} |f(x)|.$$

Further, let $C^m(\Omega)$ be the set of functions differentiable in Ω up to order m:

$$C^m(\Omega) = \{f \in C(\Omega) : D^\gamma f \in C(\Omega), \quad |\gamma| \leq m\}.$$

And finally,

$$C^m(\overline{\Omega}) = \{f \in C(\overline{\Omega}) : D^\gamma f \in C(\overline{\Omega}), \quad |\gamma| \leq m\}$$

is the set of functions, whose derivatives up to order m are continuous on $\overline{\Omega}$. We define the norm in $C^m(\overline{\Omega})$ by

$$\|f\|_{C^m(\overline{\Omega})} = \sum_{|\gamma|\leq m} \|D^\gamma f\|_{C(\overline{\Omega})}. \tag{1.1}$$

The space $C^m(\overline{\Omega}), m \in \mathbb{N}_0$, is a Banach space with respect to the norm (1.1). For a continuous function f the set $\overline{\{x : f(x) \neq 0\}}$ is called *a support* of f and denoted supp$f$. Further, we denote by $C_0^m(\Omega)$ the subset of $C^m(\overline{\Omega})$ consisting of functions f for which $D^\gamma f(x) = 0, x \in \partial\Omega, |\gamma| \leq m$.

Definition 1.1. Let $0 < \lambda < 1$. We denote by $C^\lambda(\overline{\Omega})$ the set of functions $f \in C(\overline{\Omega})$ such that

$$|f(x) - f(y)| \leq C|x-y|^\lambda \tag{1.2}$$

holds for arbitrary $x, y \in \overline{\Omega}$ with a positive constant C. $C^\lambda(\overline{\Omega})$ is a Banach space with respect to the norm

$$\|f\|_{C^\lambda(\overline{\Omega})} = \|f\|_{C(\overline{\Omega})} + \sup_{x\neq y} \frac{|f(x) - f(y)|}{|x-y|^\lambda}.$$

If $\lambda = 1$ then condition (1.2) is called a Lipschitz condition, and the corresponding set of functions is called *a Lipschitz class*. The Lipschitz class will be denoted by $Lip(\overline{\Omega})$. We notice that if $\lambda > 1$ then a function satisfying condition (1.2) is a constant. For $\lambda > 1$ Hölder spaces can be defined as follows. Let $\lambda = m + \mu$, where $m \in \mathbb{N}$, and $0 < \mu < 1$. Then we denote by $C^\lambda(\overline{\Omega})$ the set of functions $f \in C^m(\overline{\Omega})$ such that

$$|D^\alpha f(x) - D^\alpha f(y)| \leq C|x-y|^\mu$$

holds for all $\alpha, |\alpha| = m$, and $x, y \in \overline{\Omega}$ with a positive constant C, not depending on x and y. Here $\alpha = (\alpha_1, \dots, \alpha_n)$, $\alpha_j \in \mathbb{N}_0$, $j = 1, \dots, n$, is a multi-index, $|\alpha| = \alpha_1 + \dots + \alpha_n$, and

$$D^\alpha f(x) = \frac{\partial^{|\alpha|} f(x)}{\partial x_1^{\alpha_1} \dots \partial x_n^{\alpha_n}}.$$

The Hölder space $C^\lambda(\overline{\Omega})$ with $\lambda = m + \mu, 0 < \mu < 1$, is also denoted by $C^{m,\mu}(\overline{\Omega})$, emphasizing that $m < \lambda < m+1$. In this notation if $0 < \lambda < 1$, one has $C^\lambda(\overline{\Omega}) = C^{0,\lambda}(\overline{\Omega})$ and $Lip(\overline{\Omega}) = C^{0,1}(\overline{\Omega})$. The space $C^\lambda(\overline{\Omega}) = C^{m,\mu}(\overline{\Omega})$ is a Banach space with respect to the norm

$$\|f\|_{C^\lambda(\overline{\Omega})} = \|f\|_{C^m(\overline{\Omega})} + \sup_{x \neq y} \frac{|D^\alpha f(x) - D^\alpha f(y)|}{|x-y|^\mu}.$$

We also introduce the space $C^\lambda(\Omega)$ of functions f such that $f \in C^\lambda(K)$ for arbitrary compact $K \Subset \Omega$. The spaces $C^\lambda(\overline{\Omega})$ and $C^\lambda(\Omega)$, where $0 < \lambda \neq 1, 2, \dots$, are called a *Hölder spaces of order* λ. A function in $C^\lambda(\overline{\Omega})$ is called uniformly Hölder continuous, while a function in $C^\lambda(\Omega)$ is called locally Hölder continuous. We notice that the class of Lipschitz continuous functions does not coincide with the class of continuously differentiable functions, that is $C^1(\overline{\Omega}) \neq C^{0,1}(\overline{\Omega})$. Indeed, a simple example $f(x) = |x|$ shows that f is Lipschitz continuous on $[-1, 1]$, but not continuously differentiable on this interval.

Definition 1.2. Let $0 < \lambda \leq 1$. We denote by $\mathscr{C}^\lambda(\overline{\Omega})$ the set of functions $f \in C(\overline{\Omega})$ such that

$$|f(x-h) + f(x+h) - 2f(x)| \leq C|h|^\lambda$$

holds for arbitrary $x, x \pm h \in \overline{\Omega}$ and with a positive constant C not depending on h and $x \in \overline{\Omega}$. The expression $\Delta_h^2 f(x) = f(x-h) + f(x+h) - 2f(x)$ is called a second finite difference of f at the point x with the step vector $h \in \mathbb{R}^n$. $\mathscr{C}^\lambda(\overline{\Omega})$ is a Banach space with respect to the norm

$$\|f\|_{\mathscr{C}^\lambda(\overline{\Omega})} = \|f\|_{C(\overline{\Omega})} + \sup_{x, x \pm h \in \overline{\Omega}, h \neq 0} \frac{|\Delta_h^2 f(x)|}{|h|^\lambda}.$$

The space $\mathscr{C}^\lambda(\overline{\Omega})$ is called a *Zygmund space of order* λ. Obviously, any Lipschitz continuous function is in $\mathscr{C}^1(\overline{\Omega})$. On the other hand, Zygmund [Zyg45] showed that there is a function in $\mathscr{C}^1(\overline{\Omega})$ and nowhere differentiable, and hence, is not Lipschitz continuous. Thus the inclusions $C^1(\overline{\Omega}) \subset C^{0,1}(\overline{\Omega}) \subset \mathscr{C}^1(\overline{\Omega})$ are strict.

For $\lambda > 1$ Zygmund spaces can be defined as follows. Let $\lambda = m + \mu$, where $m \in \mathbb{N}$, and $0 < \mu \leq 1$. Then we denote by $\mathscr{C}^\lambda(\overline{\Omega})$ the set of functions $f \in C^m(\overline{\Omega})$ such that

$$|\Delta_h^2 D^\alpha f(x)| \leq C|h|^\mu$$

holds for all $\alpha, |\alpha| = m$, and $x, x \pm h \in \overline{\Omega}$ with a positive constant C not depending on h and x. The space $\mathscr{C}^\lambda(\overline{\Omega})$ is a Banach space with respect to the norm

$$\|f\|_{C^\lambda(\overline{\Omega})} = \|f\|_{C^m(\overline{\Omega})} + \sup_{x,x\pm h\in\overline{\Omega},h\neq 0} \frac{|\Delta_h^2 D^\alpha f(x)|}{|h|^\lambda}.$$

Introduce also a Zygmund space $\mathscr{C}^\lambda(\Omega)$ of functions f such that $f \in \mathscr{C}^\lambda(K)$ for an arbitrary compact $K \Subset \Omega$. Thus, the Zygmund spaces $\mathscr{C}^\lambda(\overline{\Omega})$ and $\mathscr{C}^\lambda(\Omega)$ are defined for all $\lambda \geq 0$, including integers. If $\lambda = m$, then the spaces $C^m(\overline{\Omega})$ and $C^m(\Omega)$ are strict subspaces of $\mathscr{C}^m(\overline{\Omega})$ and $\mathscr{C}^m(\Omega)$, respectively. The following proposition says that Hölder and Zygmund spaces coincide if λ is not integer.

Proposition 1.1. *[Tri77] Let $\lambda > 0$ and $\lambda \notin \mathbb{N}$. Then $C^\lambda(\Omega) = \mathscr{C}^\lambda(\Omega)$.*

If $\Omega = \mathbb{R}^n$, or unbounded open set, then one needs to take into account the behavior of a function near infinity. For instance, the set of functions $f : \mathbb{R}^n \to \mathbb{C}$ with continuous and bounded derivatives $D^\alpha f$ for all α, $|\alpha| \leq m$, endowed by the norm

$$\|f\|_{C^m(\mathbb{R}^n)} = \sum_{|\alpha|\leq m} \sup_{x\in\mathbb{R}^n} |D^\alpha f(x)|,$$

is a Banach space. This space is denoted by $C_b^m(\mathbb{R}^n)$. In analysis the set $C^\infty(\mathbb{R}^n)$ of infinite differentiable functions defined on $\mathbb{R}^n$ play an important role. This set is not normalized; however the notion of convergence, or a topology, can be defined in it (see Section 1.5). An important subsets of $C^\infty(\mathbb{R}^n)$ are infinite differentiable functions with compact support, and functions with bounded derivatives of all orders, respectively defined by $C_0^\infty(\mathbb{R}^n)$ and $C_b^\infty(\mathbb{R}^n)$.

The following statement known as the Arzela-Ascoli Lemma (see, e.g., [Tri83]) plays an important role in the theory of function spaces.

Lemma 1.1. *(Arzela-Ascoli Lemma) Let $K \Subset \mathbb{R}^n$ and a sequence of functions $\{f_n\}_{n\in\mathbb{N}} \subset C(K)$ satisfy the following two conditions:*

1. *"Uniform boundedness": for all $n = 1, 2, \ldots$, there exists a number $M > 0$, such that $\|f_n|C(K)\| \leq M$;*
2. *"Equicontinuity": for an arbitrary $\varepsilon > 0$ there exists a number $\delta > 0$, such that if $x, y \in K$, and $|x - y| < \delta$, then $|f(x) - f(y)| < \varepsilon$ for all $n = 1, 2, \ldots$.*

Then there is a uniformly convergent subsequence of $\{f_n(x)\}_{n\in\mathbb{N}}$.

1.3 L_p-spaces

Definition 1.3. Let $1 \leq p < \infty$. Introduce $L_p(\Omega)$, the set of Lebesgue measurable functions $f : \Omega \to \mathbb{C}$ for which the Lebesgue integral

$$\int_\Omega |f(x)|^p dx$$

is finite; $L_\infty(\Omega)$ denotes the set of Lebesgue measurable functions bounded almost everywhere on Ω, that is $|f(x)| \le C < \infty$ for almost all $x \in \Omega$. For $f \in L_\infty(\Omega)$ the smallest number C such that $f(x) \le C$ a.e. in Ω, is called the essential supremum and denoted by $ess\sup f$. For instance, the function $f(x) = x$ if x rational, and $f(x) = 1/(1+x^2)$ otherwise, belongs to $L_\infty(R)$ with $ess\sup|f| = 1$ (even though $\sup|f| = \infty$). It follows from the definition of $L_p(\Omega)$ that two functions f and g are equal in $L_p(\Omega)$ if $f(x) = g(x)$ *a.e.* on Ω.

$L_p(\Omega)$ is a Banach space for all $1 \le p \le \infty$ with respect to the norm

$$\|f\|_{L_p} = \begin{cases} (\int\limits_\Omega |f(x)|^p dx)^{1/p}, & \text{if } 1 \le p < \infty; \\ ess\sup|f|, & \text{if } p = \infty. \end{cases}$$

Only the case $p = 2$, i.e., $L_2(\Omega)$ defines a Hilbert space with the inner product

$$(f,g) = \int\limits_\Omega f(x)\overline{g(x)}dx$$

for $f,g \in L_2(\Omega)$. Here $\bar{g}$ is the complex conjugate of g. Functions in $L_1(\Omega)$ are called "absolutely integrable."

Proposition 1.2. *The two following inequalities hold:*

1. *Minkowski's inequality:* $\|f+g\|_{L_p} \le \|f\|_{L_p} + \|g\|_{L_p}$ *is valid for arbitrary* $f,g \in L_p(\Omega)$, $p \ge 1$;
2. *Hölder's inequality:* $|(f,g)| \le \|f\|_{L_p}\|g\|_{L_q}$ *is valid for arbitrary* $f \in L_p(\Omega)$ *and* $g \in L_q(\Omega)$, *where* $p,q \ge 1$, $1/p + 1/q = 1$.

Two numbers $p,q \ge 1$ satisfying the condition $1/p + 1/q = 1$ are called a *conjugate pair.* The Minkowski's inequality generalizes immediately to a finite number of functions $f_j \in L_p(\Omega)$, $j = 1,\ldots,N$:

$$\|\sum_{j=1}^N f_j\|_{L_p} \le \sum_{j=1}^N \|f_j\|_{L_p}.$$

In general, if $f(x,a) \in L_p(\Omega)$ is a family of functions depending on a parameter $a \in A \subset \mathbb{R}^m$, such that $\|f(x,a)\|_{L_p} \in L_1(A)$, then the following generalized Minkowski's inequality

$$\left\|\int\limits_A f(x,a)da\right\|_{L_p} \le \int\limits_A \|f(x,a)\|_{L_p} da \qquad (1.3)$$

holds.

Proposition 1.3. *Let $\Omega \subset \mathbb{R}^n$ be bounded and $p_1 < p_2$. Then $L_{p_2}(\Omega) \subset L_{p_1}(\Omega)$.*

In other words this proposition states that for Ω bounded, $L_p(\Omega)$ decreases when p increases. The proof of this proposition can be easily obtained using the Hölder inequality. For unbounded Ω it is not so.

Example 1.1. Indeed, let $p_1 < p_2$. Then, for instance, the function

$$f_1(x) = (|x|^n + 1)^{-(1+\varepsilon_1)/p_2} \in L_{p_2}(\mathbb{R}^n)$$

for any positive real ε_1. But, $f_1(x) \notin L_{p_1}(\mathbb{R}^n)$, if $0 < \varepsilon_1 < p_2/p_1 - 1$. On the other hand, for the function

$$f_2(x) = \begin{cases} |x|^{-n(1-\varepsilon_2)/p_1}, & \text{if } |x| \le 1, \\ 0, & \text{if } |x| > 1, \end{cases}$$

one can easily verify that $f_2 \in L_{p_1}(\mathbb{R}^n)$ for any positive ε_2, but $f_2(x) \notin L_{p_2}(\mathbb{R}^n)$, if $0 < \varepsilon_2 < 1 - p_1/p_2$.

The proposition below provides some well-known properties of L_p-spaces.

Proposition 1.4. *Let $\Omega \subseteq \mathbb{R}^n$. The space $L_p(\Omega)$ possesses the following properties:*

1. *For all $1 \le p < \infty$ the space $L_p(\Omega)$ is separable;*
2. *If $1 < p < \infty$ and q is its conjugate then $L_p(\Omega)$ is reflexive, and its dual $(L_p(\Omega))^* = L_q(\Omega)$ in the sense of isometric isomorphism. For $p = 1$ one has $(L_1(\Omega))^* = L_\infty(\Omega)$. However, if $p = \infty$, then the dual $(L_\infty(\Omega))^*$ is not isomorphic to $L_1(\Omega)$;*[1]
3. *The set of step functions defined on Ω form a dense set in $L_p(\Omega)$ for all $p \in [1,\infty)$. Moreover, $C_0^\infty(\Omega)$ is also dense in $L_p(\Omega)$, $p \in [1,\infty)$. Here C_0^∞ is the set of infinitely differentiable functions vanishing outside a compact set in Ω. (These denseness statements are not valid if $p = \infty$.)*

The second statement in this Proposition is a part of the Riesz representation theorem on L_p-spaces. Namely, for any linear continuous functional defined on $L_p(\Omega)$, $p \in [1,\infty)$, there exists a unique function $g \in L_q(\Omega)$, where q is the conjugate of p, so that

$$L(f) = < f, g > = \int_\Omega f(x)g(x)dx,$$

and the norm of L is equal to $\|g\|_{L_q}$. Recalling the norm of linear continuous functionals, one has

$$\|L\| = \sup_{0 \ne f \in L_p(\Omega)} \frac{|L(f)|}{\|f\|_{L_p(\Omega)}} = \|g\|_{L_q(\Omega)}.$$

---

[1] In fact, $(L_\infty(\Omega))^*$ is isomorphic to the space of finite Borel measures with the total variation norm. The latter contains $L_1(\Omega)$ as a linear subspace, see, e.g., [Tri77].

The last equality in this chain shows also that for the norm of $g \in L_q(\Omega)$, $q \in (0,\infty]$, the relations

$$\|g\|_{L_q(\Omega)} = \sup_{0\neq f\in L_p(\Omega)} \frac{|<g,f>|}{\|f\|_{L_p(\Omega)}} = \sup_{f\in L_p(\Omega),\|f\|_{L_p(\Omega)}=1,} |<g,f>| \tag{1.4}$$

hold.

1.3.0.1 The Riesz-Thorin interpolation theorem

Interpolation theorems play an important role in modern analysis, see, for example, Bergh and Löfström [BL76], or Triebel [Tri77]. The Riesz-Thorin theorem on interpolation of L_p-spaces was the first theorem in this theory.

Theorem 1.1. *Let T be a linear mapping on a generic space containing all the spaces $L_p(\Omega)$, $1 \leq p \leq \infty$, $\Omega \subseteq \mathbb{R}^n$. Suppose T_1 and T_2 are restrictions of T, such that operators*

$$T_1 : L_{p_1}(\Omega) \to L_{q_1}(\Omega) \text{ and } T_2 : L_{p_2}(\Omega) \to L_{q_2}(\Omega)$$

where $1 \leq p_1, p_2, q_1, q_2 \leq \infty$, are bounded and have norms M_1 and M_2, respectively. Further, let p_θ and q_θ be defined by

$$\frac{1}{p_\theta} = \frac{1-\theta}{p_1} + \frac{\theta}{p_2} \text{ and } \frac{1}{q_\theta} = \frac{1-\theta}{q_1} + \frac{\theta}{q_2},$$

where $\theta \in [0,1]$. Then the restriction T_θ of T to $L_{p_\theta}(\Omega)$:

$$T_\theta : L_{p_\theta}(\Omega) \to L_{q_\theta}(\Omega) \tag{1.5}$$

is a bounded operator with the norm

$$\|T_\theta\| \equiv M \leq M_1^{1-\theta} M_2^{\theta}. \tag{1.6}$$

1.4 Euler's gamma- and beta-functions

The function

$$\Gamma(z) = \int_0^\infty e^{-t} t^{z-1} dt, \quad \Re(z) > 0,$$

is called *Euler's gamma-function.* $\Gamma(z)$ can be analytically extended to the whole complex plain $\mathbb{C}$ except points $z = 0, -1, -2, \ldots,$ which are simple poles of the gamma-function. Using the integration by parts, one can show that $\Gamma(1+z) = z\Gamma(z)$. Obviously, $\Gamma(1) = 1$. These two facts immediately imply $\Gamma(n+1) = n!$ for $n \in \mathbb{N}_0$ (with the convention $0! = 1$).

For $z = \frac{1}{2}$, using the substitution $t = s^2$, one has

$$\Gamma\left(\frac{1}{2}\right) = \int_0^\infty \frac{e^{-t}dt}{\sqrt{t}} = 2\int_0^\infty e^{-s^2}ds.$$

One can easily show that $A = \int_0^\infty e^{-s^2}ds = \pi/2$. Indeed, changing to the polar coordinates,

$$A^2 = \left(\int_0^\infty e^{-s^2}ds\right)\left(\int_0^\infty e^{-u^2}du\right) = \frac{1}{4}\int_{\mathbb{R}^2} e^{-(s^2+u^2)}dsdu = \frac{\pi}{2}\int_0^\infty e^{r^2}rdr = \frac{\pi}{4}.$$

Hence, $\Gamma(\frac{1}{2}) = \sqrt{\pi}$.

Further, *Euler's beta-function* $B(s,u)$ is defined by

$$B(s,u) = \int_0^1 x^{s-1}(1-x)^{u-1}dx, \quad \Re(s) > 0, \Re(u) > 0.$$

Taking the product $\Gamma(s)\Gamma(u) = \int_{\mathbb{R}_+^2} x^{s-1}y^{u-1}e^{-(x+y)}dxdy$, and using the substitution $x+y=z$, one obtains $\Gamma(s)\Gamma(u) = \Gamma(s+u)B(s,u)$. Hence, Euler's beta- and gamma-functions are connected through the formula

$$B(s,u) = \frac{\Gamma(s)\Gamma(u)}{\Gamma(s+u)}, \quad \Re(s) > 0, \Re(u) > 0.$$

We also note the following property of the gamma-function, which will be used in our further analysis [AS64]:

$$\Gamma(z)\Gamma(1-z) = \frac{\pi}{\sin \pi z}. \tag{1.7}$$

(We will prove this equality in Section 3.13).

1.4.0.2 The Fourier transform

Definition 1.4. Let $f \in L_1(\mathbb{R}^n)$. The Fourier transform of f denoted by $\hat{f}$ or $F[f]$, by definition, is

$$\hat{f}(\xi) = F[f](\xi) = \int_{\mathbb{R}^n} f(x)e^{ix\xi}dx, \quad \xi \in \mathbb{R}^n, \tag{1.8}$$

where $i = \sqrt{-1}$ and $x\xi = (x,\xi) = x_1\xi_1 + \cdots + x_n\xi_n$.

The Fourier transform of $f \in L_1(\mathbb{R}^n)$ is continuous. Namely, the following property holds (see, e.g., [RS80]):

Theorem 1.2. *Let $f \in L_1(\mathbb{R}^n)$. Then its Fourier transform $\hat{f}(\xi)$ is continuous on $\mathbb{R}^n$, and $\hat{f}(\xi) \to 0$, as $|\xi| \to \infty$.*

It follows from this theorem that F is a linear operator, mapping $L_1(\mathbb{R}^n)$ into the set of functions $C_0(\mathbb{R}^n)$ continuous on $\mathbb{R}^n$ and tending to zero at infinity. The inverse Fourier transform F^{-1} is given by the formula

$$f(x) = F^{-1}[\hat{f}](x) = \frac{1}{(2\pi)^n} \int_{\mathbb{R}^n} \hat{f}(\xi) e^{-ix\xi} d\xi. \tag{1.9}$$

We note that for $f \in L_1(\mathbb{R}^n)$ its Fourier transform $\hat{f}$ may not belong to $L_1(\mathbb{R}^n)$. In fact, the Fourier preimage of $C_0(\mathbb{R}^n)$ is not $L_1(\mathbb{R}^n)$. In fact, $F^{-1}(C_0(\mathbb{R}^n)) \supset L_1(\mathbb{R}^n)$. Therefore, one must be careful when using the inversion formula (1.9). We will discuss the question how to extend the Fourier transform to spaces of functions much larger than $L_1(\mathbb{R}^n)$ in Section 1.5.3.

The properties given in the following proposition can easily be verified by direct calculation.

Proposition 1.5. *Let $f \in L_1(\mathbb{R}^n)$. Then for $0 \neq a \in \mathbb{R}$ and $y \in \mathbb{R}^n$ the following formulas hold:*

1. $F[f(ax)](\xi) = \frac{1}{a^n} F[f](\frac{\xi}{a})$;
2. $\frac{1}{a^n} F[f(\frac{x}{a})](\xi) = F[f](a\xi)$;
3. $F[f(x+y)](\xi) = e^{-iy\xi} F[f](\xi)$;
4. $F[e^{iyx} f(x)](\xi) = F[f](\xi + y)$.

In the two propositions below x^α and ξ^α for a multi-index $\alpha = (\alpha_1, \dots, \alpha_n)$ mean $x_1^{\alpha_1} \dots x_n^{\alpha_n}$, and $\xi_1^{\alpha_1} \dots \xi_n^{\alpha_n}$. These properties of the Fourier transform can be proved by integration by parts.

Proposition 1.6. *Let $x^\alpha f \in L_1(\mathbb{R}^n)$. Then $\widehat{(ix)^\alpha f}(\xi) = D^\alpha \hat{f}(\xi)$.*

Proposition 1.7. *Let $D^\alpha f \in L_1(\mathbb{R}^n)$. Then $\widehat{D^\alpha f}(\xi) = \xi^\alpha \hat{f}(\xi)$.*

Consider some examples which will be exploited later.

Example 1.2. 1. Let $f(x) = e^{-|x|}$, $x \in \mathbb{R}$. This function belongs to $L_1(\mathbb{R})$. We have

$$F[e^{-|x|}](\xi) = \int_{-\infty}^{\infty} e^{-|x|} e^{ix\xi} dx = \int_{-\infty}^{0} e^{x(1+i\xi)} dx + \int_{0}^{\infty} e^{-x(1-i\xi)} dx$$

$$= \frac{1}{1+i\xi} + \frac{1}{1-i\xi} = \frac{2}{1+\xi^2}.$$

Using the first formula in Proposition 1.5 with $a > 0$, one obtains

$$F[e^{-a|x|}](\xi) = \frac{2a}{a^2 + \xi^2}. \tag{1.10}$$

One can easily see that in this fortunate case the Fourier transform $\hat{f}(\xi) = \frac{2}{1+\xi^2}$ of f is also in $L_1(\mathbb{R})$. Therefore one can use the inverse Fourier transform formula (1.9) to equation (1.10), to obtain

$$F^{-1}\left[\frac{2}{1+\xi^2}\right](x) = e^{-|x|}.$$

2. It follows from the previous example that

$$F\left[\frac{1}{\pi(1+x^2)}\right](\xi) = e^{-|\xi|}. \tag{1.11}$$

Further, using the second formula in Proposition 1.5 and (1.11), one has

$$F\left[\frac{t}{\pi(t^2+x^2)}\right](\xi) = e^{-t|\xi|}, \quad t > 0. \tag{1.12}$$

3. Now we find the Fourier transform of the function $f(x) = e^{-x^2}, x \in \mathbb{R}$. This function belongs to $L_1(\mathbb{R})$. Differentiating $\hat{f}(\xi) = \int_{\mathbb{R}} e^{-x^2+ix\xi}dx$, and using the relation

$$ixe^{-x^2+ix\xi} = -\frac{i}{2}d(e^{-x^2+ix\xi}) - \frac{\xi}{2}e^{-x^2+ix\xi},$$

one can see that $\hat{f}$ satisfies the ordinary differential equation $\frac{d\hat{f}(\xi)}{d\xi} = -\frac{\xi}{2}\hat{f}(\xi)$. The solution to this equation is $\hat{f}(\xi) = Ce^{-\frac{\xi^2}{4}}$, where $C = \hat{f}(0) = \int_{\mathbb{R}} e^{-x^2}dx = \pi$. Hence,

$$F[e^{-x^2}](\xi) = \sqrt{\pi}e^{-\frac{\xi^2}{4}}. \tag{1.13}$$

Using again the second formula in Proposition 1.5 with $b = \sqrt{4t}, t > 0$, we obtain

$$F\left[\frac{1}{\sqrt{4\pi t}}e^{-\frac{x^2}{4t}}\right](\xi) = e^{-t\xi^2}. \tag{1.14}$$

4. Let $f(x) = e^{-|x|^2}, x \in \mathbb{R}^n$. Then it follows immediately from formula (1.13) that

$$F\left[\frac{1}{(\sqrt{\pi})^n}e^{-|x|^2}\right](\xi) = e^{-\frac{|\xi|^2}{4}}, \quad \xi \in \mathbb{R}^n. \tag{1.15}$$

Now using this formula and proposition 1.5, one can derive the multidimensional case of (1.14):

$$F\left[\frac{1}{(\sqrt{4\pi t})^n}e^{-\frac{|x|^2}{4t}}\right](\xi) = e^{-t|\xi|^2}, \quad \xi \in \mathbb{R}^n. \tag{1.16}$$

5. The n-dimensional analog of formula (1.12) is

$$F\left[\frac{\Gamma(\frac{n+1}{2})}{\pi^{\frac{n+1}{2}}}\frac{t}{(|x|^2+t^2)^{\frac{n+1}{2}}}\right](\xi) = e^{-t|\xi|}, \quad t>0,\ \xi \in \mathbb{R}^n. \tag{1.17}$$

where $\Gamma(s)$ is Euler's gamma-function. See Section 1.13 "Additional notes" for the proof of (1.17).

1.4.0.3 The Laplace transform. Watson's lemma

Let a function $f(t)$, defined on $[0,\infty)$, be piece-wise continuous and satisfy the condition $|f(t)| \le Ce^{\sigma t}$ for some $\sigma > 0$. We denote the set of such functions by M_σ. The Laplace transform of $f \in M_\sigma$ is defined by

$$L[f](s) = \int_0^\infty f(t)e^{-st}dt, \quad \Re(s) > \sigma,$$

For the Laplace transform of f we also use the notation $\tilde{f}(s)$. The Laplace transform $\tilde{f}(s)$ is an analytic function of $s = p + i\eta$ in the half-plane $\Re(s) = p > \sigma$.

Here are some well-known properties of the Laplace transform, which follow directly from the definition.

Proposition 1.8. *Let $f,g \in M_\sigma$. Then for s with $\Re(s) > \sigma$,*

1. $L[af+bg] = aL[f](s) + bL[g](s),\ a,b \in \mathbb{C}$;
2. $L[f*g](s) = L[f](s)\cdot L[g](s)$;
3. $L[e^{\beta t}f](s) = L[f](s-\beta),\ \Re(s) > \sigma+\beta$;
4. $L[f'](s) = sL[f](s) - f(0)$;
5. $\frac{d}{ds}L[f](s) = -L[tf](s)$.

In property 2) $(f*g)(t)$ is a *convolution* of functions f and g defined by

$$(f*g)(t) = \int_0^t f(\tau)g(t-\tau)d\tau.$$

One can easily verify that if $f,\ g \in M_\sigma$, then $f*g \in M_\sigma$, as well.

Example 1.3. Let $\beta > 0$ be a real number. Then

$$L[t^{\beta-1}](s) = \frac{\Gamma(\beta)}{s^\beta}, \quad s>0. \tag{1.18}$$

Indeed, using the substitution $ts = u$ in the integral

$$L[t^{\beta-1}](s) = \int_0^\infty e^{-st} t^{\beta-1} dt$$

we have

$$L[t^{\beta-1}](s) = s^{-\beta} \int_0^\infty e^{-u} u^{\beta-1} du = \frac{\Gamma(\beta)}{s^\beta}.$$

If $\beta = n$, a positive integer, then the latter reduces to the well-known formula

$$L[t^{n-1}](s) = \frac{(n-1)!}{s^n}.$$

In particular, if $n = 1$, then $L[1](s) = \frac{1}{s}$. Obviously, the formula (1.18) extends for all complex s with $\Re(s) > 0$.

If $0 < \beta < 1$, then (1.18) extends also for all s with $\Re(s) = 0$, except $s = 0$, as well, that is $s = i\eta$, $\eta \in \mathbb{R}$, $\eta \neq 0$. In fact, in this case the left-hand side of (1.18) takes the form

$$L[t^{\beta-1}](i\eta) = \int_0^\infty \frac{e^{-i\eta t}}{t^{1-\beta}} dt. \tag{1.19}$$

Using the known formulas (see, for instance, [AS64], formulas 6.5.7,8,20)

$$\int_0^\infty \frac{\cos \eta t}{t^{1-\beta}} dt = \frac{\Gamma(\beta) \cos \frac{\pi\beta}{2}}{\eta^\beta} \quad \text{and} \quad \int_0^\infty \frac{\sin \eta t}{t^{1-\beta}} dt = \frac{\Gamma(\beta) \sin \frac{\pi\beta}{2}}{\eta^\beta},$$

one obtains

$$\int_0^\infty \frac{e^{-i\eta t}}{t^{1-\beta}} dt = \frac{\Gamma(\beta) e^{i\frac{\pi\beta}{2}}}{\eta^\beta} = \frac{\Gamma(\beta)}{(i\eta)^\beta}. \tag{1.20}$$

Hence, (1.19) and (1.20) imply the formula

$$L[t^{\beta-1}](i\eta) = \frac{\Gamma(\beta)}{(i\eta)^\beta}, \quad \eta \neq 0. \tag{1.21}$$

Below are two other properties of the Laplace transform important for our further considerations. The first one is the *differentiation formula* for the Laplace transform, which generalizes Property 4 in Proposition 1.8 for arbitrary integer order $m \geq 1$.

Proposition 1.9. *Let $f \in C^m(0,\infty) \cap M_\sigma$ has finite values $f^{(k)}(0)$, $k = 0,\ldots,m-1$. Then the formula*

$$L[f^{(m)}](s) = s^m L[f](s) - \sum_{k=0}^{m-1} f^{(k)}(0) s^{m-1-k} \tag{1.22}$$

holds.

The second property is known as Watson's lemma.

Proposition 1.10. *(Watson's Lemma) Let $f(t) = t^\gamma g(t)$, $g(0) \neq 0$, where $\gamma > -1$, and $g \in M_\sigma$. Suppose the function $g(t)$ has the expansion*

$$g(t) = \sum_{k=0}^{n} a_k t^k + R_n(t), \quad 0 < t < t_0,$$

with $|R_n(t)| \leq Ct^{n+1}, t \in (0, t_0)$. Then

$$L[f](s) = \sum_{k=0}^{n} \frac{a_k \Gamma(k+\gamma+1)}{s^{k+\gamma+1}} + O\left(\frac{1}{s^{n+\gamma+2}}\right), \quad s \to \infty.$$

1.5 Distribution spaces

1.5.1 Schwartz distributions

Definition 1.5. Let $C_0^\infty(\Omega)$ be the set of infinitely differentiable functions with compact support in Ω. For a sequence of functions $\varphi_m \in C_0^\infty(\Omega), m = 1, 2, \ldots$, we introduce the following convergence: φ_m converges to an element $\varphi_0 \in C_0^\infty(\Omega)$ if the following two conditions are fulfilled:

1. there exists a compact $K \subset \Omega$ such that $\operatorname{supp} \varphi_m \subset K$ for all $m = 1, 2, \ldots$;
2. $D^\gamma \varphi_m(x) \rightrightarrows D^\gamma \varphi_0(x)$ uniformly on K for all $|\gamma| = 0, 1, \ldots$.

$C_0^\infty(\Omega)$ with the introduced convergence is denoted by $\mathfrak{D}(\Omega)$, and called a space of *test functions.*

$\mathfrak{D}(\Omega)$ is a linear space. Obviously, if $\varphi_1, \varphi_2 \in \mathfrak{D}(\Omega)$, then for arbitrary complex numbers $c_1, c_2 \in \mathbb{C}$, one has $c_1\varphi_1 + c_2\varphi_2 \in \mathfrak{D}(\Omega)$.

Let f be a linear and continuous functional defined on $\mathfrak{D}(\Omega)$. This means that for $f : \mathfrak{D}(\Omega) \to \mathbb{C}$ the following two conditions, namely the linearity condition:

$$f(c_1\varphi_1 + c_2\varphi_2) = c_1 f(\varphi_1) + c_2 f(\varphi_2), \quad \forall c_1, c_2 \in \mathbb{C},$$

and the continuity condition:

$$\varphi_m \to \varphi_0 \Rightarrow f(\varphi_m) \to f(\varphi_0), \quad m \to \infty,$$

are fulfilled. Here $f(\varphi)$ stands for the value of f on φ. For instance, the functional defined as $\delta_a(\varphi) = \varphi(a), \varphi \in \mathfrak{D}(\Omega)$, $a \in \Omega$, satisfies both conditions.

Definition 1.6. Denote by $\mathfrak{D}'(\Omega)$ the set of all linear and continuous functionals defined on $\mathfrak{D}(\Omega)$. A sequence $f_m \in \mathfrak{D}'(\Omega)$ is said to converge weakly to $f_0 \in \mathfrak{D}'(\Omega)$ if for arbitrary $\varphi \in \mathfrak{D}(\Omega)$ the sequence of numbers $f_m(\varphi)$ converges to $f_0(\varphi)$. $\mathfrak{D}'(\Omega)$ with this convergence is called the space of *Schwartz distributions.*

The functional $\delta_a(\varphi) = \varphi(a)$ introduced in the example above is a Schwartz distribution. This distribution is called the *Dirac delta function* with mass on a.

The space $\mathfrak{D}'(\Omega)$ is a vector space. The sum $f_1 + f_2$ of two distributions $f_1, f_2 \in \mathfrak{D}'(\Omega)$ is defined as $(f_1+f_2)(\varphi) = f_1(\varphi)+f_2(\varphi)$ for arbitrary $\varphi \in \mathfrak{D}(\Omega)$. Similarly, the scalar multiplication cf, where $c \in \mathbb{C}$ and $f \in \mathfrak{D}'(\Omega)$, is defined as $(cf)(\varphi) = cf(\varphi)$ for arbitrary $\varphi \in \mathfrak{D}(\Omega)$. Thus, the space of Schwartz distributions, as well as the space of corresponding test functions, have both structures, the vector structure and the topological structure (defined through convergence). Therefore, both spaces $\mathfrak{D}(\Omega)$ and $\mathfrak{D}'(\Omega)$ are topological-vector spaces. Their deeper properties, such as the completeness, local convexity, and inductive and projective limit structures, will follow from general assertions presented in Section 1.5.4 in the abstract case.

The following statement provides a criterion for a linear functional defined on $\mathfrak{D}(\Omega)$ to be a Schwartz distribution [Sch51]:

Proposition 1.11. *A linear functional f defined on $\mathfrak{D}(\Omega)$ is in $\mathfrak{D}'(\Omega)$ if and only if for arbitrary open set $\Omega_0 \Subset \Omega$ there exist an integer $m = m(\Omega_0) \geq 0$ and a constant $C = C(\Omega_0) > 0$, such that for all $\varphi \in \mathfrak{D}(\Omega)$ the estimate*

$$|f(\varphi)| \leq C\|\varphi\|_{C^m(\Omega)}$$

holds.

If m in this Proposition does not depend on Ω_0, then it is called *an order of distribution f*. The Dirac delta function satisfies the estimate

$$|\delta_a(\varphi)| = |\varphi(a)| \leq \|\varphi\|_{C(\Omega)}.$$

Therefore, δ_a is a distribution of order 0.

Definition 1.7. The derivative of order α of a distribution $f \in \mathfrak{D}'(\Omega)$ is defined by

$$D^\alpha f(\varphi) = (-1)^{|\alpha|} f(D^\alpha \varphi), \quad \forall \varphi \in \mathfrak{D}(\Omega).$$

Let $a(x) \in C^\infty(\Omega)$ and $f \in \mathfrak{D}'(\Omega)$. Then the multiplication of the distribution f by $a(x)$ is defined by

$$af(\varphi) = f(a\varphi), \quad \forall \varphi \in \mathfrak{D}(\Omega).$$

It follows from Proposition 1.11 that for a Schwartz distribution f both $D^\alpha f$ for any multi-index α and af for any $a \in C^\infty(\Omega)$ are Schwartz distributions again. Moreover, the mappings

$$D^\alpha : \mathfrak{D}'(\Omega) \to \mathfrak{D}'(\Omega) \quad \text{and} \quad a(x)\cdot : \mathfrak{D}'(\Omega) \to \mathfrak{D}'(\Omega)$$

are continuous. This statement is a direct implication of Definition 1.7.

The notion of support can be extended to distributions as well. We say that a distribution $f \in \mathfrak{D}'(\Omega)$ is zero on an open set $\Omega' \subset \Omega$, if for all $\varphi \in \mathfrak{D}(\Omega')$, one has $f(\varphi) = 0$. The union of all open sets where f is zero is called a *null-set* of the distribution f. By definition, the support of $f \in \mathfrak{D}'(\Omega)$, denoted by $\mathrm{supp} f$, is the closure of $\Omega \setminus \Omega_0(f)$, where $\Omega_0(f)$ is the null-set of f, namely, $\mathrm{supp} f = \overline{\Omega \setminus \Omega_0(f)}$.

For example, the support of the Dirac delta function δ_a is $\{a\}$. Indeed, if Ω' is an open set not containing a, then for all $\varphi \in \mathfrak{D}(\Omega')$, one has $\delta_a(\varphi) = 0$. Therefore, the null-set of δ_a is $\Omega \setminus \{a\}$, implying $\mathrm{supp}\,\delta_a = \{a\}$.

It is not hard to see that if $\mathrm{supp}\, f \Subset \Omega$, then m in Proposition 1.11 can be chosen independent of all open sets $\Omega_0 \Subset \Omega$. Hence, any distribution with compact support has a finite order.

Example 1.4. 1. Let a function g be locally integrable on Ω, that is for arbitrary compact set $A \Subset \Omega$ the integral $\int\limits_A |g(x)|dx$ is finite. Then, g defines a linear functional G on $\mathfrak{D}(\Omega)$ by the expression

$$G(\varphi) = \int\limits_{\Omega} g(x)\varphi(x)dx.$$

Moreover, for each $\Omega_0 \Subset \Omega$ the estimate $|f(\varphi)| \leq C\|\varphi\|_{C(\Omega)}$, with $C = C(\Omega_0) = \int_{\Omega_0} |g(x)|dx$, holds. Due to Proposition 1.11, G is a Schwartz distribution of order 0. A distribution defined by a locally integrable function is called *regular*, otherwise it is called *singular.* For instance, the Dirac delta function is singular (try to prove this. It is a good exercise!).

The Heaviside function

$$\theta(x) = \begin{cases} 1, & \text{if } x > 0, \\ 0, & \text{otherwise,} \end{cases}$$

is locally integrable on $\mathbb{R} = (-\infty, \infty)$, and therefore, is a regular distribution. For the derivative (in the sense of distributions) of $\theta(x)$ one has $D\theta = \delta_0$. Indeed, for an arbitrary $\varphi \in \mathfrak{D}(\mathbb{R})$,

$$D\theta(\varphi) = -\theta(D\varphi) = -\int\limits_0^\infty \varphi'(x)dx = \varphi(0) = \delta_0(\varphi).$$

Similarly, one can easily verify that $\Theta(x-a) = \theta(x_1 - a_1) \cdot \dots \cdot \theta(x_n - a_n)$, where $a = (a_1, \dots, a_n) \in \mathbb{R}^n$ is a fixed point, is a regular distribution, and

$$D_1 \dots D_n \Theta(x-a) = \delta_a(x). \tag{1.23}$$

2. Introduce the distribution $P.v.\frac{1}{x}$, where $P.v.$ stands for *principal value* (in the sense of Cauchy), and defined by

$$< P.v.\frac{1}{x}, \varphi >= \lim_{\varepsilon \to 0} \left(\int\limits_{-\infty}^{-\varepsilon} \frac{\varphi(x)}{x} dx + \int\limits_{\varepsilon}^{\infty} \frac{\varphi(x)}{x} dx \right).$$

One can easily verify that for any $d > 0$ the integral $P.v. \int\limits_{-d}^{d} \frac{dx}{x} = 0$. Taking this into account, and assuming $\mathrm{supp}\,\varphi \subset [-d, d]$, we have

$$|< P.v.\frac{1}{x}, \varphi >| = \left| \int_{-\infty}^{\infty} \frac{\varphi(x) - \varphi(0)}{x} dx \right| \leq C \|\varphi\|_{C^1(\mathbb{R})},$$

where $C = 2d$. Hence, $P.v.\frac{1}{x} \in \mathfrak{D}'(\mathbb{R})$, singular, and of order 1.

Proposition 1.12. *(Sokhotski-Plemelj formulas) The following formulas hold:*

$$\frac{1}{x+i0} = -i\pi\delta_0(x) + V.p.\frac{1}{x}, \tag{1.24}$$

$$\frac{1}{x-i0} = i\pi\delta_0(x) + V.p.\frac{1}{x}. \tag{1.25}$$

Proof. To prove this statement we calculate the limit $\lim_{\varepsilon\to 0} \frac{1}{x+i\varepsilon}$ in $\mathfrak{D}'(\mathbb{R})$. Let $\varphi \in \mathfrak{D}(\mathbb{R})$ with the support $\operatorname{supp}\varphi \subset [-d,d]$. We have

$$\begin{aligned}
\lim_{\varepsilon\to 0} < \frac{1}{x+i\varepsilon}, \varphi(x) > &= \lim_{\varepsilon\to 0} \int_{\mathbb{R}} \frac{\varphi(x)}{x+i\varepsilon} dx = \lim_{\varepsilon\to 0} \int_{-d}^{d} \frac{x-i\varepsilon}{x^2+\varepsilon^2} \left[\varphi(0) + \Big(\varphi(x) - \varphi(0)\Big)\right] dx \\
&= -2i\varphi(0) \lim_{\varepsilon\to 0} \tan^{-1}\left(\frac{d}{\varepsilon}\right) + \int_{\mathbb{R}} \frac{\varphi(x)-\varphi(0)}{x} dx \\
&= -i\pi\varphi(0) + \int_{\mathbb{R}} \frac{\varphi(x)-\varphi(0)}{x} dx \\
&= < -i\pi\delta_0(x) + P.v.\frac{1}{x}, \varphi(x) >,
\end{aligned}$$

obtaining (1.24). Formula (1.25) follows from (1.24) replacing i by $-i$.

1.5.2 Distributions with compact support

Definition 1.8. Denote by $\mathscr{E}(\Omega)$, $\Omega \subset \mathbb{R}^n$, the set of functions $\varphi \in C^\infty(\Omega)$ with the following convergence: a sequence $\varphi_k \in \mathscr{E}(\Omega), k = 1,2,\ldots$ is said to converge to $\varphi_0 \in \mathscr{E}(\Omega)$ in $\mathscr{E}(\Omega)$ if for every multi-index γ and any compact set $K \Subset \Omega$

$$\sup_{x\in K} |D^\gamma \varphi_k(x) - D^\gamma \varphi_0(x)| \to 0, \tag{1.26}$$

as $k \to \infty$.

Definition 1.9. Denote by $\mathscr{E}'(\Omega)$ the set of all linear and continuous functionals defined on $\mathscr{E}(\Omega)$. A sequence $f_m \in \mathscr{E}'(\Omega)$ is said to converge weakly to $f_0 \in \mathscr{E}'(\Omega)$, if for arbitrary $\varphi \in \mathscr{E}$ the sequence $f_m(\varphi)$ converges to $f_0(\varphi)$. With this convergence the set $\mathscr{E}'(\Omega)$ is said to be the space of *distributions with compact support.*

The name of the space $\mathscr{E}'(\Omega)$ is not a game. In fact, any distribution in $\mathscr{E}'(\Omega)$ has a compact support. This fact can be proved by contradiction. Assume the support of $f \in \mathscr{E}'(\Omega)$ is not compact. Then there exist a sequence of sets $\omega_k, k = 1,2,\ldots$, and a sequence of test functions $\varphi_k \in \mathscr{E}, k = 1,2,\ldots$, with the following requirements: the sets ω_k, $k = 1,2,\ldots$, are compact in Ω, $\omega_k \subset \omega_{k+1}$ for all $k = 1,2,\ldots$, and $\cup\omega_k = \Omega$; the function $\varphi_k \in \mathscr{E}(\Omega)$ has compact support $\operatorname{supp}\varphi_k \subset \Omega \setminus \omega_k$, and $f(\varphi_k) = 1$ for all $k = 1,2,\ldots$. By construction, obviously, $\varphi_k \to 0$ as $k \to \infty$ in $\mathscr{E}(\Omega)$. Hence, $f(\varphi_k) \to 0$, as $k \to 0$. The letter contradicts to $f(\varphi_k) = 1 \nrightarrow 0$.

Example 1.5. As we have seen, the Dirac delta function δ_a with mass on $a \in \Omega$ is a distribution with compact support, $\operatorname{supp}(\delta_a) = \{a\}$. Hence, $\delta_a \in \mathscr{E}'(\Omega)$.

The following two propositions (see, e.g., [Vla79]) describe the structure of distributions with compact support and distributions concentrated at a point $a \in \mathbb{R}^n$, respectively.

Proposition 1.13. *Let $f \in \mathscr{E}'(\Omega)$. Then f is a distribution of a finite order m and there exists a function $h \in L_\infty(\Omega)$, such that*

$$f(x) = \sum_{|\alpha| \le m} D^\alpha h(x).$$

An illustration of this proposition is (1.23) for $f(x) = \delta_a(x)$.

Proposition 1.14. *Let a distribution $f \in \mathscr{E}'(\Omega)$ has the support $\operatorname{supp} f = \{a\}$. Then there exist an integer m and numbers b_α, $|\alpha| \le m$, such that*

$$f(x) = \sum_{|\alpha| \le m} b_\alpha D^\alpha \delta_a(x).$$

Proposition 1.15. *The general solution of the equation*

$$(x-a)^\beta u(x) = 0 \tag{1.27}$$

in the space $\mathscr{E}'(\mathbb{R}^n)$ is

$$u(x) = \sum_{\substack{\alpha_j \le \beta_j - 1 \\ j = 1,\ldots,n}} C_\alpha D^\alpha \delta_a(x),$$

where C_α are arbitrary constants.

Proof. First we notice that since $(x-a)^\beta \in C^\infty(\mathbb{R}^n)$, the left side of equation (1.27) is meaningful in $\mathscr{E}'(\Omega)$. Moreover, equation (1.27) immediately implies $\operatorname{supp} u = \{a\}$. Hence, due to Proposition 1.14,

$$u(x) = \sum_{|\alpha| \le m} C_\alpha D^\alpha \delta_a(x),$$

for some integer m and constants C_α. Substituting the latter to equation (1.27), we have

$$\begin{aligned} 0 &=< (x-a)^\beta u(x), \varphi(x) >=< u(x), (x-a)^\beta \varphi(x) > \\ &= \sum_{|\alpha|\le m} C_\alpha < D^\alpha \delta_a(x), (x-a)^\beta \varphi(x) > \\ &= \sum_{|\alpha|\le m} (-1)^{|\alpha|} C_\alpha D^\alpha \left[(x-a)^\beta \varphi(x) \right]_{|x=a} . \end{aligned}$$

which implies $C_\alpha = 0$ if $\alpha_j \ge \beta_j$, $j = 1, \ldots, n$.

1.5.3 Tempered distributions

Definition 1.10. Denote by $\mathscr{G}$ the set of functions in $C^\infty(\mathbb{R}^n)$ satisfying the following condition: for every multi-index γ and $m \in \mathbb{N}_0$,

$$p_{\gamma,m}(\varphi) = \max_{x\in\mathbb{R}^n} \{ (1+|x|)^m |D^\gamma \varphi(x)| \} \tag{1.28}$$

is finite. We say that a sequence $\varphi_k \in \mathscr{G}, k = 1, 2, \ldots$ converges to $\varphi_0 \in \mathscr{G}$ in $\mathscr{G}$ if for all multi-indices γ and $m \in \mathbb{N}_0$,

$$p_{\gamma,m}(\varphi_k - \varphi_0) \to 0, \quad k \to \infty.$$

Definition 1.11. Denote by $\mathscr{G}'$ the set of linear and continuous functionals defined on $\mathscr{G}$. A sequence $f_m \in \mathscr{G}'$ is said to converge weakly to $f_0 \in \mathscr{G}'$ if for arbitrary $\varphi \in \mathscr{G}$ the sequence $f_m(\varphi)$ converges to $f_0(\varphi)$. $\mathscr{G}'$ endowed with this convergence is called the space of *tempered distributions*. For convenience we use the notation $< f, \varphi >$ for $f(\varphi)$.

The inclusions $\mathfrak{D}(\mathbb{R}^n) \subset \mathscr{G} \subset \mathscr{E}(\mathbb{R}^n)$ imply $\mathscr{E}'(\mathbb{R}^n) \subset \mathscr{G}' \subset \mathfrak{D}'(\mathbb{R}^n)$. Hence, a distribution with compact support is also a tempered distribution. The distribution $P.v.\frac{1}{x}$ is an example of a tempered distribution with non-compact support.

Using Propositions 1.6 and 1.7 one can easily see that if φ belongs to $\mathscr{G}$, then the same does its Fourier transform, $F[\varphi] \in \mathscr{G}$. In other words $\mathscr{G}$ is invariant with respect to the Fourier transform F, and the mapping

$$F : \mathscr{G} \to \mathscr{G} \tag{1.29}$$

is continuous. Moreover, the mapping (1.29) is onto, and the inverse F^{-1} is given by the formula

$$F^{-1} g(x) = \frac{1}{(2\pi)^n} \int_{\mathbb{R}^n} g(\xi) e^{-ix\xi} d\xi, \quad g \in \mathscr{G}. \tag{1.30}$$

The continuity of F in $\mathscr{G}$ implies the following assertion:

Proposition 1.16. *A sequence $\varphi_n \to 0$ as $n \to \infty$ in $\mathscr{G}$ if and only if $F[\varphi_n] \to 0$ as $n \to \infty$ in $\mathscr{G}$.*

Let S and F^* be the operators on $\mathscr{G}$ defined as $S\varphi(x) = \varphi(-x)$ and $F^* = SF$, i.e.

$$F^*\varphi(\xi) = \int_{\mathbb{R}^n} \varphi(x)e^{-ix\xi}dx, \quad \xi \in \mathbb{R}^n.$$

Obviously, the mapping $S: \mathscr{G} \to \mathscr{G}$ is onto with the inverse $S^{-1} = S$. This implies that $F^* = SF$ is also onto with the inverse operator $F^{*-1} = F^{-1}S$. Notice that (1.30) and definitions of S and F^* yield

$$F^{-1} = \frac{1}{(2\pi)^n}F^* = \frac{1}{(2\pi)^n}SF. \tag{1.31}$$

Now assume that $\varphi \in \mathscr{G}$ and f is a regular distribution in $\mathscr{G}'$, that is f is identified with a locally integrable function. In other words, $f \in L_1(K)$ for any compact $K \subset \mathbb{R}^n$ and grows at infinity at a polynomial rate. Then changing order of integration, which is legitimate under our assumptions, one has

$$<\hat{f},\varphi> = \int_{\mathbb{R}^n_\xi}\int_{\mathbb{R}^n_x} f(x)\varphi(\xi)e^{ix\xi}dxd\xi =<f,\hat{\varphi}>.$$

Making use of this fact and the continuity of the Fourier transform F in $\mathscr{G}$ one can extend the Fourier transform to the space of tempered distributions. Namely, by definition, if $f \in \mathscr{G}'$ then its Fourier transform $\hat{f}$ is defined by

$$<\hat{f},\varphi> = <f,\hat{\varphi}> \tag{1.32}$$

for all $\varphi \in \mathscr{G}$. As a direct implication of (1.29) we obtain that the mapping

$$F: \mathscr{G}' \to \mathscr{G}'$$

is also continuous. This fact due to Proposition 1.16 immediately implies

Proposition 1.17. *A sequence $f_n \to 0$ as $n \to \infty$ in $\mathscr{G}'$ if and only if $F[f_n] \to 0$ as $n \to \infty$ in $\mathscr{G}'$.*

Moreover, the properties of F indicated in Propositions 1.6 and 1.7 are extended to distributions in $\mathscr{G}'$:

Proposition 1.18. *Let $f \in \mathscr{G}'$. Then*

(1) $\widehat{(ix)^\alpha f}(\xi) = D^\alpha \hat{f}(\xi)$.
(2) $\widehat{D^\alpha f}(\xi) = \xi^\alpha \hat{f}(\xi)$.

Since any function $f \in L_p(\mathbb{R}^n)$, $p \geq 1$, is also a tempered distribution, the Fourier transform for these functions is well defined in the sense of distributions. In particular, $L_2(\mathbb{R}^n)$ is invariant with respect to the Fourier transform. This fact follows from celebrated Parseval's equality (see, e.g., [RS80]).

Theorem 1.3. *(Parseval) Let $f,g \in L_2(\mathbb{R}^n)$. Then*

$$(f,g) = (2\pi)^{-n}(\hat{f},\hat{g}). \tag{1.33}$$

Corollary 1.1. *Let $f \in L_2(\mathbb{R}^n)$. Then*

$$\|\hat{f}\|_{L_2}^2 = (2\pi)^n \|f\|_{L_2}^2. \tag{1.34}$$

Corollary 1.2. *The transform $F : L_2(\mathbb{R}^n) \to L_2(\mathbb{R}^n)$ is continuous.*

L_p-spaces are not invariant with respect to the Fourier transform if $p \neq 2$. In particular, for $1 \leq p < 2$ the following statement holds.

Theorem 1.4. *(Hausdorff-Young) If $f \in L_p(\mathbb{R}^n)$, $p \in [1,2)$, then $\hat{f} \in L_q(\mathbb{R}^n)$, where $q = p/(p-1)$ and $\|\hat{f}\|_{L_q} \leq C_p \|f\|_{L_p}$, with a constant $C_p > 0$ not depending on f.*

The proof immediately follows from the Riesz-Thorin theorem (Theorem 1.1) taking $p_2 = q_2 = 2$ due to Corollary 1.2, and $p_1 = 1$, $q_1 = \infty$ due to Theorem 1.2. Moreover, inequality (1.6) implies that $C_p \leq (2\pi)^{n(1-1/p)}$.

Definition 1.12. For $f,g \in \mathcal{G}$ define *the convolution* $f * g$ by the integral

$$(f * g)(x) = \int\limits_{\mathbb{R}^n} f(y)g(x-y)dy. \tag{1.35}$$

It is easy to verify that if $f,g \in \mathcal{G}$, then (1.35) exists and $f * g \in \mathcal{G}$. The convolution exists for any pair of functions $f \in L_1(\mathbb{R}^n)$ and $g \in L_\infty(\mathbb{R}^n)$, as well. Moreover, in this case

$$\|f * g\|_{L_\infty} \leq \|f\|_{L_1} \|g\|_{L_\infty}. \tag{1.36}$$

For functions $f \in L_p(\mathbb{R}^n)$, $p \in [1,\infty]$, and $g \in L_q(\mathbb{R}^n)$ with $q = p/(p-1)$, the conjugate of p, due to Hölder's inequality, one has

$$\|f * g\|_{L_\infty} \leq \|f\|_{L_p} \|g\|_{L_q}. \tag{1.37}$$

Further, for functions $f,g \in L_1(\mathbb{R}^n)$ using Minkowski's inequality in the integral form (1.3), we obtain

$$\begin{aligned} \|f * g\|_{L_1} &= \left\| \int\limits_{\mathbb{R}^n} f(y)g(x-y)dy \right\|_{L_1} \leq \int\limits_{\mathbb{R}^n} \|f(y)g(x-y)\|_{L_1} dy \\ &\leq \|g\|_{L_1} \int\limits_{\mathbb{R}^n} |f(y)|dy = \|f\|_{L_1} \|g\|_{L_1}. \end{aligned} \tag{1.38}$$

The following theorem represents the general case.

Theorem 1.5. *(Young) Let $1 \leq p, q \leq \infty$, and $f \in L_p(\mathbb{R}^n)$, $g \in L_q(\mathbb{R}^n)$. Suppose*

$$1 + \frac{1}{r} = \frac{1}{p} + \frac{1}{q}. \tag{1.39}$$

Then $f * g \in L_r(\mathbb{R}^n)$ and the inequality

$$\|f * g\|_{L_r} \leq \|f\|_{L_p} \|g\|_{L_q} \tag{1.40}$$

holds.

Proof. We sketch a brief proof based on the Riesz-Thorin theorem (Theorem 1.1). Let $f \in L_1(\mathbb{R}^n)$ be fixed. Then the operator $T_f(g) = f * g$ is bounded as a mapping: (a) $T_f : L_1(\mathbb{R}^n) \to L_1(\mathbb{R}^n)$ with the norm $\|f\|_{L_1}$, due to (1.38); (b) $T_f : L_\infty(\mathbb{R}^n) \to L_\infty(\mathbb{R}^n)$ with the same norm $\|f\|_{L_1}$, due to (1.36). Hence, the Riesz-Thorin theorem yields that the mapping $T_f : L_q(\mathbb{R}^n) \to L_q(\mathbb{R}^n)$ is bounded for any $q \in [1, \infty]$ with the norm $\leq \|f\|_{L_1}$. This conclusion for fixed $g \in L_q(\mathbb{R}^n)$ can also be interpreted as follows: (c) the mapping $T_g : L_1(\mathbb{R}^n) \to L_q(\mathbb{R}^n)$ is bounded with the norm $\|g\|_{L_q}$. At the same time inequality (1.37) implies that (d) $T_g : L_{q'}(\mathbb{R}^n) \to L_\infty(\mathbb{R}^n)$ is bounded with the norm $\|g\|_{L_q}$. Here q' is the conjugate of q. Now again using the Riesz-Thorin theorem for T_g as mappings in (c) and (d), we obtain that the operator $T_g : L_p(\mathbb{R}^n) \to L_r(\mathbb{R}^n)$ is bounded with the norm $\leq \|g\|_{L_q}$, and r satisfying the condition (1.39). This is equivalent to desired inequality (1.40).

Remark 1.1. The inequality (1.40) is called *Young's inequality.* To feel it better, it is useful to look at some particular cases. One particular case is $p = 1$ and $q = \infty$ in equation (1.39). In this case one has $r = \infty$. Hence, (1.40) recovers inequality (1.36). In general, for any conjugate pare p and q, i.e., $1/p + 1/q = 1$, one obtains $r = \infty$, and (1.40) takes the form $\|f * g\|_{L_\infty} \leq \|f\|_{L_p} \|g\|_{L_q}$, recovering (1.37). Another particular case is $q = 1$. In this case $r = p$ and Young's inequality (1.40) reduces to

$$\|f * g\|_{L_p} \leq \|g\|_{L_1} \|f\|_{L_p}. \tag{1.41}$$

The latter is valid for $p = 1$ as well, obtaining (1.38). Inequality (1.38) shows that $L_1(\mathbb{R}^n)$ is closed with respect to the convolution operation "$*$."

Proposition 1.19. *Let $f, g \in L_1(\mathbb{R}^n)$. Then*

$$F[f * g](\xi) = F[f](\xi) \cdot F[g](\xi). \tag{1.42}$$

Formula (1.42) is valid, in particular, for any functions $f, g \in \mathscr{G}$. Hence, it can be extended by continuity to any tempered distribution $f \in \mathscr{G}'$ and $g \in \mathscr{G}$.

Consider some examples of the Fourier transform of tempered distributions.

Example 1.6. 1. The Fourier transform of the Dirac delta function is

$$F[\delta_0] = 1. \tag{1.43}$$

Indeed, for any $\varphi \in \mathscr{G}$,

$$< F[\delta_0], \varphi >=< \delta_0, F\varphi >= F\varphi(0) = \int_{\mathbb{R}^n} \varphi(x)dx =< 1, \varphi >,$$

which means $F[\delta_0] = 1$ in the distributional sense. Further, applying the inverse Fourier transform F^{-1} to the letter and using (1.31), one has $\delta_0 = (2\pi)^{-n} SF[1]$. Now inverting the operator S and taking into account the symmetry of the Dirac delta function δ_0, we obtain $F[1] = (2\pi)^n \delta_0$.

2. We show that

$$F\left[\frac{1}{\pi}\frac{\sin x}{x}\right](\xi) = I_{[-1,1]}(\xi). \tag{1.44}$$

The function $\frac{1}{\pi}\frac{\sin x}{x}$ does not belong to $L_1(\mathbb{R})$, but it is in $L_p(\mathbb{R})$ for any $p > 1$, and hence, a tempered distribution. Using formula (1.30), one has

$$< F^{-1}[I_{[-1,1]}(\xi)], \varphi > =< I_{[-1,1]}(\xi), F^{-1}\varphi >= \frac{1}{2\pi} \int_{-1}^{1} \int_{\mathbb{R}} \varphi(x) e^{-ix\xi} dx d\xi$$

$$= \frac{1}{2\pi} \int_{\mathbb{R}} \varphi(x) \Big(\int_{-1}^{1} e^{-ix\xi} d\xi \Big) dx = \frac{1}{2\pi} \int_{\mathbb{R}} \varphi(x) \frac{\sin x}{\pi x} dx$$

$$=< \frac{\sin x}{\pi x}, \varphi > .$$

Hence, $F^{-1}[I_{[-1,1]}(\xi)] = \frac{1}{\pi}\frac{\sin x}{x}$ in the sense of $\mathscr{G}'$. Now applying the operator F to both sides we obtain (1.44). Using Proposition 1.5, for arbitrary $t > 0$, we have

$$F\left[\frac{1}{\pi}\frac{\sin tx}{tx}\right](\xi) = \frac{1}{t} I_{[-1,1]}\left(\frac{\xi}{t}\right) = \frac{I_{[-t,t]}(\xi)}{t},$$

or, canceling t in the denominators,

$$F\left[\frac{1}{\pi}\frac{\sin tx}{x}\right](\xi) = I_{[-t,t]}(\xi). \tag{1.45}$$

3. One can easily verify that the right-hand side of (1.45) converges to 1 in $\mathscr{G}'$, when $t \to \infty$. Hence, due to Proposition 1.17 and equation (1.43),

$$\frac{1}{\pi}\frac{\sin tx}{x} \to \delta_0(x), \quad t \to \infty.$$

Similarly, formulas (1.16) and (1.17) imply

$$\frac{1}{(\sqrt{4\pi t})^n} e^{-\frac{|x|^2}{4t}} \to \delta_0(x), \quad t \to 0 \ (x \in \mathbb{R}^n),$$

and

$$\frac{\Gamma(\frac{n+1}{2})}{\pi^{\frac{n+1}{2}}} \frac{t}{(|x|^2+t^2)^{\frac{n+1}{2}}} \to \delta_0(x), \quad t \to 0 \ (x \in \mathbb{R}^n),$$

respectively, in the topology of $\mathscr{S}'$.

4. The Heaviside function $\theta(x)$ does not belong to $L_p(\mathbb{R})$ for $p \in [1,\infty)$, but $\theta \in L_\infty(\mathbb{R})$. Hence, it is a tempered distribution. Since $\theta(x) = \lim_{\varepsilon\to 0} \theta(x)e^{-\varepsilon x}$ in $\mathscr{S}'$, one has

$$F[\theta](\xi) = \lim_{\varepsilon\to 0} F[\theta(x)e^{-\varepsilon x}](\xi) = \lim_{\varepsilon\to 0} \int_0^\infty e^{-\varepsilon x+ix\xi} dx$$
$$= \lim_{\varepsilon\to 0} \frac{-1}{-\varepsilon+i\xi} = \frac{i}{\xi+i0}.$$

Due to Sokhotski-Plemelj formula (1.24), the latter takes the form

$$F[\theta(x)](\xi) = \pi\delta_0(\xi) + iV.p.\frac{1}{\xi}. \tag{1.46}$$

Similarly,

$$F[\theta(-x)](\xi) = \pi\delta_0(\xi) - iV.p.\frac{1}{\xi}. \tag{1.47}$$

5. Next, we find the Fourier transform of sign(x). Using the obvious equality sign(x)=$\theta(x) - \theta(-x)$, and formulas (1.46) and (1.47), one obtains

$$F[\text{sign}(x)](\xi) = F[\theta(x)](\xi) - F[\theta(-x)](\xi) = 2iV.p.\frac{1}{\xi}. \tag{1.48}$$

6. The Fourier transform of the distribution $P.v.\frac{1}{x}$ is

$$F[P.v.\frac{1}{x}](\xi) = i\pi\text{sign}(\xi) = i\pi \begin{cases} 1, & \text{if } \xi > 0; \\ 0, & \text{if } \xi = 0; \\ -1, & \text{if } \xi < 0. \end{cases} \tag{1.49}$$

Indeed, using the relationship $F[f](\xi) = 2\pi F^{-1}[f](-\xi)$ (formula (1.31) in the one-dimensional case), and (1.48), we have

$$F[P.v.\frac{1}{x}](\xi) = 2\pi F^{-1}[P.v.\frac{1}{x}](-\xi) = \frac{\pi}{i}\text{sign}(-\xi) = i\pi\text{sign}(\xi).$$

7. Consider the function $f(x) = e^{i|x|^2}$, $x \in \mathbb{R}^n$. This function is in $L_\infty(\mathbb{R}^n)$, and hence, a tempered distribution. We find the Fourier transform of this function. It suffices to find the Fourier transform for $n = 1$. First, consider the functions

$$C(x) = \sqrt{\frac{2}{\pi}} \int_0^x \cos t^2 dt \quad \text{and} \quad S(x) = \sqrt{\frac{2}{\pi}} \int_0^x \sin t^2 dt$$

called Fresnel's cosine and sine integrals. It is known (see [AS64], formula 7.3.20), that

$$\lim_{x\to\infty} C(x) = \lim_{x\to\infty} S(x) = \frac{1}{2}.$$

Using this fact, one has

$$\int_{-\infty}^{\infty} e^{ix^2} dx = 2\int_0^{\infty} (\cos x^2 + i \sin x^2) dx$$

$$= 2\sqrt{\frac{\pi}{2}}\Big(C(\infty) + iS(\infty)\Big) = \sqrt{\pi} e^{i\frac{\pi}{4}}. \tag{1.50}$$

Now, it is easy to compute the Fourier transform of e^{ix^2}. Indeed, exploiting (1.50),

$$F[e^{ix^2}](\xi) = \int_{-\infty}^{\infty} e^{ix\xi + ix^2} dx = e^{-i\xi^2/4} \int_{-\infty}^{\infty} e^{iz^2} dz = \sqrt{\pi} e^{-i\frac{\xi^2 - \pi}{4}}.$$

It follows in the n-dimensional case that

$$F[e^{i|x|^2}](\xi) = (\pi)^{n/2} e^{-\frac{i}{4}|\xi|^2 + \frac{i}{4}n\pi}. \tag{1.51}$$

The latter can be rewritten in the form

$$F\left[\frac{1}{(i\pi)^{n/2}} e^{i|x|^2}\right](\xi) = e^{-\frac{i}{4}|\xi|^2}, \quad \xi \in \mathbb{R}^n. \tag{1.52}$$

8. Let $f(x) = \frac{1}{|x|^{1-\alpha}}$, where $x \in \mathbb{R}$ and $0 < \alpha < 1$. Obviously, f is a tempered (regular) distribution, but $f \notin L_p(\mathbb{R})$ for all $p \in [1, \infty]$. The Fourier transform of this distribution exists in $\mathscr{G}'$. In fact, the following relation

$$F[f](\xi) = b_\alpha |\xi|^{-\alpha}, \tag{1.53}$$

where $b_\alpha = 2\Gamma(\alpha)\cos\frac{\alpha\pi}{2}$, holds. In order to compute the Fourier transform of f, it suffices to compute the oscillatory integral

$$\int_{\mathbb{R}} \frac{e^{ix\xi}}{|x|^{1-\alpha}} dx. \tag{1.54}$$

This is seen from the following relation

$$< F\left[\frac{1}{|x|^{1-\alpha}}\right], \varphi >= \int_{\mathbb{R}} \varphi(\xi) \Big(\int_{\mathbb{R}} \frac{e^{ix\xi}}{|x|^{1-\alpha}} dx \Big) d\xi$$

between the Fourier transform of $f(x) = |x|^{-(1-\alpha)}$ and the integral in (1.54). The integral (1.54) is (conditionally) convergent. In the integral in (1.54) setting $\xi = |\xi|\mu$, where $\mu = sign(\xi)$, and substituting $y = |\xi|x$, one has

$$\int_{\mathbb{R}} \frac{e^{ix\xi}}{|x|^{1-\alpha}} dx = |\xi|^{-\alpha} \int_{\mathbb{R}} \frac{e^{i\mu y}}{|y|^{1-\alpha}} dy.$$

In fact, the integral on the right-hand side does not depend on $\mu = \pm 1$, and hence, we get (1.53) with constant

$$b_\alpha = \int_{\mathbb{R}} \frac{e^{i\mu y}}{|y|^{1-\alpha}} dy = 2\int_0^\infty \frac{\cos y}{y^{1-\alpha}} dy = 2\Gamma(\alpha)\cos\frac{\alpha\pi}{2}.$$

In Section 3.7 we demonstrate a different method of calculation of b_α.

9. Let $f(x) = \frac{1}{|x|^\sigma}, x \in \mathbb{R}^n$, where $0 < \sigma < n$. This function is locally integrable, hence is a regular tempered distribution. We will show that the Fourier transform in the sense of $\mathscr{G}'$ of this function is

$$F\left[\frac{1}{|x|^\sigma}\right](\xi) = \frac{b_{\sigma,n}}{|\xi|^{n-\sigma}}, \quad \xi \in \mathbb{R}^n, \tag{1.55}$$

where

$$b_{\sigma,n} = \frac{2^{n-\sigma}\pi^{n/2}\Gamma(\frac{n-\sigma}{2})}{\Gamma(\frac{\sigma}{2})}. \tag{1.56}$$

In order to show this fact we use formula (1.16):

$$F\left[\frac{1}{(\sqrt{4\pi t})^n} e^{-\frac{|x|^2}{4t}}\right](\xi) = e^{-t|\xi|^2}, \xi \in \mathbb{R}^n,$$

in the sense of $\mathscr{G}'$ (see (1.32)). Namely, for arbitrary function $\varphi \in \mathscr{G}$,

$$\langle \frac{1}{(\sqrt{4\pi t})^n} e^{-\frac{|x|^2}{4t}}, F[\varphi](x)\rangle = \langle e^{-t|\xi|^2}, \varphi(\xi)\rangle. \tag{1.57}$$

Multiplying both sides of (1.57) by $t^{(n-\sigma)/2-1}$ and integrating over the interval $(0,\infty)$, we obtain

$$\frac{1}{2^n\pi^{n/2}}\langle \int_0^\infty t^{-\frac{\sigma}{2}-1} e^{-\frac{|x|^2}{4t}} dt, F[\varphi]\rangle = \langle \int_0^\infty t^{\frac{n-\sigma}{2}-1} e^{-t|\xi|^2} dt, \varphi\rangle.$$

Changing the order of integration performed above is valid. Now using the substitution $\frac{|x|^2}{4t} = s$ on the left integral, and $t|\xi|^2 = u$ on the right integral, one gets

$$\langle F\left[\frac{1}{|x|^\sigma}\right](\xi),\varphi(\xi)\rangle = \langle \frac{2^{n-\sigma}\pi^{n/2}\Gamma(\frac{n-\sigma}{2})}{\Gamma(\frac{\sigma}{2})}\frac{1}{|\xi|^{n-\sigma}},\varphi(\xi)\rangle,$$

proving (1.55), (1.56). We note that the Fourier transform $F[\frac{1}{|x|^{n-\alpha}}](\xi)$ serves as a symbol of, so-called, Riesz potential, considered in Section 3.7. See also Section 1.13 for additional notes.

Finally, we notice that the Fourier transform in the last example allows the analytic continuation to all $\sigma \in \mathbb{C}$, except those which satisfy the equations:

$$\frac{n-\sigma}{2} = -1,-2,\ldots, \quad \text{and} \quad \frac{\sigma}{2} = -1,-2,\ldots.$$

Thus, the following statement holds:

Proposition 1.20. *Let $\sigma \in \mathbb{C}$ such that $\sigma \neq -2m$ and $\sigma \neq n+2m$, for all $m \in \mathbb{N}$. Then*

$$F\left[\frac{1}{|x|^\sigma}\right](\xi) = \frac{2^{n-\sigma}\pi^{n/2}\Gamma(\frac{n-\sigma}{2})}{\Gamma(\frac{\sigma}{2})}\frac{1}{|\xi|^{n-\sigma}}, \quad \xi \neq 0. \tag{1.58}$$

Let $f \in M_\sigma$ and $s = p+i\eta$. We have

$$L[f](s) = L[f](p+i\eta) = \int_0^\infty f(t)e^{-(p+i\eta)t}dt = \int_{\mathbb{R}}[f(t)e^{-pt}]e^{-it\eta}dt$$
$$= F_\eta[f(t)e^{-pt}](-\eta).$$

This relation can be taken as the base for the Laplace transform of distributions. Namely, let $f \in \mathscr{D}'(\mathbb{R})$ with the null-set $\Omega_0(f) = (-\infty,0)$, and such that $f(t)e^{-st} \in \mathscr{G}'$. Then the Laplace transform $L[f](s)$ is defined by

$$L[f](s) = F_\eta[f(t)e^{-pt}](-\eta) \tag{1.59}$$

The reader can easily verify that $L[\delta_0(t)](s) = 1$ and $L[\theta(t)](s) = 1/s$, where $\theta(t)$ is the Heaviside function.

Since the Fourier transform is a continuous mapping in $\mathscr{G}'$, the Laplace transform $L[f](p+i\eta)$ is a tempered distribution in the variable η for each fixed $p > \sigma$. All the properties of the Laplace transform, mentioned in Section 1.4.0.3, are valid for distributions as well. Let us briefly discuss the property

$$L[f'](s) = sL[f](s) - f(0). \tag{1.60}$$

For distributions this property takes the form

$$L[Df](s) = sL[f](s), \tag{1.61}$$

where $D = d/dt$ in the sense of distributions. Indeed, using the equality $D[f(t)e^{-pt}] = Df(t)e^{-pt} - pf(t)e^{-pt}$, one has

$$L[Df](s) = F_\eta[Dfe^{-pt}](-\eta) = F_\eta\left[D[f(t)e^{-pt}]\right](-\eta) + pF_\eta[f(t)e^{-pt}](-\eta)$$
$$= (i\eta)L[f](s) + pL[f](s) = sL[f](s),$$

obtaining (1.61). This is consistent with (1.60). To see this we use the relationship $Dg(t) = f'(t) + \delta_0(t)f(0+)$ between generalized and usual derivatives of a function f differentiable on $\mathbb{R} \setminus \{0\}$, and having a jump $f(0+) - f(0)$ at $t = 0$, to obtain

$$L[f'](s) = L[Df - \delta_0(t)f(0+)](s) = L[Df](s) - f(0+)L[\delta_0](s)$$
$$= sL[f](s) - f(0+).$$

1.5.4 Some basic principles of distribution spaces

In Sections 1.5.1–1.5.3 we introduced three different distribution spaces. In Section 1.10 we will introduce a new space of distributions appeared in the literature relatively recently. In construction of distribution spaces one should follow some basic principles common for all distribution spaces. Below we briefly discuss these principles in abstract case referring the reader for details, for instance, to [Hor83, GS53].

In this context Fréchet type locally convex topological vector spaces play an important role. They generalize Banach spaces and can be defined with the help of a family of seminorms. By definition, a function $p : \mathscr{X} \to R_+$ defines a seminorm in a vector space $\mathscr{X}$, if for arbitrary $x, y \in \mathscr{X}$ and $\lambda \in \mathbb{C}$,

(1) $p(x+y) \leq p(x) + p(y)$,
(2) $p(\lambda x) = |\lambda| p(x)$.

If additionally, $p(x) = 0$ implies $x = 0$, then $p : \mathscr{X} \to R_+$ defines a norm in $\mathscr{X}$. An example of a seminorm is $p(x) = max_{[0,1]}|x'(t)|$ for functions $x \in C^{(1)}[0,1]$. Let $U = \{x \in \mathscr{X} : p(x) < \varepsilon\}$, where p is a seminorm and $\varepsilon > 0$. Then, conditions (1) and (2) imply that for arbitrary $x, y \in U$ and $\alpha \geq 0, \beta \geq 0$ such that $\alpha + \beta = 1$, one has $\alpha x + \beta y \in U$. In other words, U is a convex subset of $\mathscr{X}$.

Since topology of a topological vector space is translation-invariant, one can assume that its topology consists of a family of neighborhoods of zero and their translations. Let $\mathscr{X}$ be a topological vector space with a family of neighborhoods τ of zero of $\mathscr{X}$. We say that $\mathscr{X}$ is locally convex if it is Hausdorff (that is for any x and y there exists neighborhoods U_x and U_y such that $U_x \cap U_y = \emptyset$), and members of τ are convex. Locally convex topological vector spaces can be defined with the help of a family of seminorms. Let p_j, $j \in \mathbb{J}$, where $\mathbb{J}$ is an index set, be a family of seminorms in $\mathscr{X}$. Then the set of convex neighborhoods of zero $U_{j,\varepsilon} = \{x \in \mathscr{X} : p_j(x) < \varepsilon,\ j \in \mathbb{J},\ \varepsilon > 0\}$ form a base topology of zero of $\mathscr{X}$. Hence, locally convex topological vector space $\mathscr{X}$ has a fundamental base of convex neighborhoods of every point $x \in \mathscr{X}$. If the family of seminorms p_j is separating, that is $p_j(x) = 0$ for all $j \in \mathbb{J}$ implies $x = 0$, and $\mathscr{X}$ is complete, then $\mathscr{X}$ is called a Fréchet type space. Fréchet type spaces are metrizable, and a metric in $\mathscr{X}$ can be introduced by

$$d(x,y) = \sum_{j \in \mathbb{J}} \frac{1}{2^j} \frac{p_j(x-y)}{1 + p_j(x-y)}.$$

Distribution spaces can be constructed as a strict inductive or projective limits of sequences of locally convex topological vector spaces. Let $\mathscr{X}_n$, $n = 1, 2, \ldots$, be a sequence of locally convex spaces, such that the inclusion $\mathscr{X}_n \subset \mathscr{X}_{n+1}$ is continuous, and let

$$\mathscr{X} = \cup_{n=1}^{\infty} \mathscr{X}_n.$$

The set $\mathscr{X}$ equipped with the finest topology such that $\mathscr{X}_n \subset \mathscr{X}$ is continuous for each $n \in \mathbb{N}$ is called *a strict inductive limit* of the sequence $\mathscr{X}_n$, and denoted by

$$\mathscr{X} = \operatorname{ind} \lim_{n \to \infty} \mathscr{X}_n.$$

Proposition 1.21. *([R64, SW66]) Let $\mathscr{X}_n$ be a sequence of locally convex topological vector spaces, $\mathscr{X}_n$ be a closed subspace of $\mathscr{X}_{n+1}$ for each $n \in \mathbb{N}$, and $\mathscr{X} = ind \lim_{n \to \infty} \mathscr{X}_n$. Then*

1. *a sequence $x_k \in \mathscr{X}$ converges to $x_0 \in \mathscr{X}$ if and only if there exists some $n \in \mathbb{N}$ such that all x_k are elements of $\mathscr{X}_n$ and $x_k \to x_0$ in the topology of $\mathscr{X}_n$;*
2. *a subset A of $\mathscr{X}$ is bounded if and only if there exists some $n \in \mathbb{N}$ such that $A \subset \mathscr{X}_n$ and bounded in the topology of $\mathscr{X}_n$;*
3. *a set K in $\mathscr{X}$ is compact in $\mathscr{X}$ if and only if there is some $n \in \mathbb{N}$ such that K is compact in $\mathscr{X}_n$;*
4. *if each $\mathscr{X}_n$ is complete, then $\mathscr{X}$ is also complete.*

Let $\mathscr{Y}_n$, $n = 1, 2, \ldots$, be a sequence of locally convex spaces, such that the inclusion $\mathscr{Y}_{n+1} \subset \mathscr{Y}_n$ is continuous, and let

$$\mathscr{Y} = \cap_{n=1}^{\infty} \mathscr{Y}_n.$$

The set $\mathscr{Y}$ equipped with the coarsest topology such that $\mathscr{Y} \subset \mathscr{Y}_n$ is continuous for each $n \in \mathbb{N}$ is called *a strict projective limit* of the sequence $\mathscr{Y}_n$, and denoted by

$$\mathscr{Y} = \operatorname{pr} \lim_{n \to \infty} \mathscr{Y}_n.$$

Proposition 1.22. *([R64, SW66]) Let $\mathscr{Y}_n$ be a sequence of locally convex topological vector spaces, $\mathscr{Y}_{n+1}$ be a closed subspace of $\mathscr{X}_n$ for each $n \in \mathbb{N}$, and $\mathscr{Y} = pr \lim_{n \to \infty} \mathscr{Y}_n$. Then*

1. *$\mathscr{Y}$ is a locally convex topological vector space;*
2. *a sequence $y_k \in \mathscr{Y}$ converges to $y_0 \in \mathscr{Y}$ if and only if $y_k \to y_0$ in the topology of $\mathscr{Y}_n$ for all $n \in \mathbb{N}$;*
3. *a subset A of $\mathscr{Y}$ is bounded if and only if $A \subset \mathscr{Y}_n$ and bounded in the topology of $\mathscr{Y}_n$ for all $n \in \mathbb{N}$;*
4. *if each $\mathscr{Y}_n$ is complete, then $\mathscr{Y}$ is also complete.*

As an example consider the space of test functions $\mathscr{E}(\Omega)$ introduced in Section 1.5.2. Let $K \subset \Omega$ be a compact set and $E(K)$ be the set of functions infinitely differentiable in a neighborhood of K. Introduce in $E(K)$ a family of seminorms as follows:

$$p_j(\varphi) = \sup_{\substack{x \in K \\ |\alpha| = j}} \{|D^\alpha \varphi(x)|\}.$$

The set $E(K)$ is metrizable locally convex space and the metric is

$$d(\varphi, \psi) = \sum_{j=0}^{\infty} \frac{1}{2^j} \frac{p_j(\varphi - \psi)}{1 + p_j(\varphi - \psi)}, \quad \varphi, \psi \in E(K).$$

Using the classical theorem on uniform convergence of uniformly continuous functions on a compact set one can easily verify that $E(K)$ is complete. Further, let $K_n, n \in \mathbb{N}$, be a sequence of compact sets in Ω, such that $K_n \subset K_{n+1}$ and $\cup_{n=1}^{\infty} K_n = \Omega$. Consider the sequence of locally convex spaces $\mathscr{E}_n = E(K_n), n \in \mathbb{N}$. Obviously, if $\varphi \in \mathscr{E}_{n+1}$ then $\varphi \in \mathscr{E}_n$. Moreover, since $K_n \subset K_{n+1}$ the topology of $\mathscr{E}_n$ is coarser than the topology of $\mathscr{E}_{n+1}$, implying continuity of the inclusion $\mathscr{E}_{n+1} \subset \mathscr{E}_n$. Therefore, we can define a strict projective limit

$$\mathscr{E}(\Omega) = \text{pr} \lim_{n \to \infty} \mathscr{E}_n.$$

The convergence of φ_m to φ_0 associated with the strict projective limit topology of $\mathscr{E}(\Omega)$ means that for every compact $K_n \subset \Omega$ and for every multi-index α with $|\alpha| = j$,

$$p_{j,n}(\varphi_m - \varphi_0) = \sup_{x \in K_n} |D^\alpha \varphi_m(x) - D^\alpha \varphi_0(x)| \to 0, \, m \to \infty,$$

which coincides with the convergence (1.26) of the space $\mathscr{E}(\Omega)$. Completeness of each $\mathscr{E}_n$ implies, due to Proposition 1.22, completeness of $\mathscr{E}(\Omega)$. Hence, $\mathscr{E}(\Omega)$ is a locally convex Fréchet type topological vector space.

Now consider the space of test functions $\mathfrak{D}(\Omega)$. Let K_n again be a sequence of compact sets in Ω, such that $K_n \subset K_{n+1}$ and $\cup_{n=1}^{\infty} K_n = \Omega$. Let $\mathfrak{D}_n$ be the set of infinite differentiable functions φ with the support $\operatorname{supp} \varphi \subset K_n$. We equip $\mathfrak{D}_n$ with the topology of $\mathscr{E}(K_n)$, which makes $\mathfrak{D}_n$ a locally convex topological vector space. Obviously, $\mathfrak{D}_n \subset \mathfrak{D}_{n+1}$ and since $K_n \subset K_{n+1}$, the topology of $\mathfrak{D}_{n+1}$ is finer than the topology of $\mathfrak{D}_n$. Therefore each inclusion $\mathfrak{D}_n \subset \mathfrak{D}_{n+1}$ is continuous. Moreover,

$$\cup_{n=1}^{\infty} \mathfrak{D}_n = \mathfrak{D}(\Omega).$$

Hence,

$$\mathfrak{D}(\Omega) = \text{ind} \lim_{n \to \infty} \mathfrak{D}_n,$$

with the finest topology for which each inclusion $\mathfrak{D}_n \subset \mathfrak{D}(\Omega)$ is continuous. Since each $\mathfrak{D}_n$ is complete, due to Proposition 1.21, $\mathfrak{D}(\Omega)$ is complete. The convergence of φ_m to φ_0 in $\mathfrak{D}(\Omega)$ means, in accordance with Proposition 1.21, that there exists a natural number n_0 such that $\varphi_m \to \varphi_0$ in the topology of $\mathfrak{D}_{n_0}$, which in turn, means that there is a compact $K \subset \Omega$, such that supp$\varphi_m \subset K$ for all m, and $D^\alpha \varphi_m(x) \to D^\alpha \varphi_0$ uniformly on K for all α. This convergence is exactly the convergence introduced in the definition of $\mathfrak{D}(\Omega)$ (see Definition 1.5).

The topology of the space of test functions $\mathscr{G}$ also can be defined with the help of a family of seminorms. Let $\varphi \in C^\infty(\mathbb{R}^n)$ and

$$p_n(\varphi) = \max_{m+|\alpha|=n} \{p_{m,\alpha}(\varphi)\}, \quad n = 0, 1, \ldots, \tag{1.62}$$

where (see (1.28))

$$p_{m,\alpha}(\varphi) = \sup_{x \in \mathbb{R}^n} \{(1+|x|)^m |D^\alpha \varphi(x)|\}. \tag{1.63}$$

One can easily verify that p_n defined in (1.62) is indeed a norm. Let $p_n(\varphi) = 0$. Then it follows from (1.62) and (1.63) that $D^\alpha \varphi(x) \equiv 0$, $|\alpha| = n$, that is $\varphi(x)$ is a polynomial of order less than n. The only polynomial for which the expression in (1.28) is finite, is zero-polynomial. Therefore, $\varphi(x) \equiv 0$. This immediately implies that $\mathscr{G}$ is a Fréchet type locally convex topological vector space. The reader can verify as an exercise what is the metric in $\mathscr{G}$ and that the convergence associated with this metric coincides with the convergence introduced in the definition of $\mathscr{G}$ (Definition 1.10).

Thus all the three spaces of test functions $\mathfrak{D}(\Omega)$, $\mathscr{E}(\Omega)$, and $\mathscr{G}$ are *complete*. Moreover, they are *dense in L_2-spaces*. Namely, $\mathfrak{D}(\Omega)$ is densely embedded into $L_2(\Omega)$ (see Proposition 1.4), $\mathscr{G}$ is densely embedded into $L_2(\mathbb{R}^n)$, and $\mathscr{E}(\Omega)$ is densely embedded into $L_{2,loc}(\Omega)$, where $L_{2,loc}$ is the set of locally square-integrable functions. These two properties, completeness and denseness, are important in the construction of corresponding distribution spaces. The denseness of spaces of test functions in the L_p-spaces is important in applications. For instance, a solution space of a differential equation found in the frame of test functions can be extended to wider classes of functions (Sobolev, Besov, Triebel-Lizorkin, etc.) if the denseness in these classes of functions holds. Thus, we assume that

(A) $\mathscr{X}$ is a complete metrizable locally convex topological vector space, and
(B) $\mathscr{X}$ is densely embedded into a Banach space $\mathbb{X}$ in the sense of the norm of $\mathbb{X}$.

Further, let $\mathscr{X}'$ be the dual space to $\mathscr{X}$ with respect to $\mathbb{X}$, i.e., the space of linear continuous functionals endowed with the weak topology. Namely,

(i) if $F \in \mathscr{X}'$ and $\varphi_1, \varphi_2 \in \mathscr{X}$, then

$$F(\lambda_1 \varphi_1 + \lambda_2 \varphi_2) = \lambda_1 F(\varphi_1) + \lambda_2 F(\varphi_2),$$

for all $\lambda_1, \lambda_2 \in \mathbb{C}$.
(ii) if $\varphi_m \to \varphi_0$ in $\mathscr{X}$, then $\lim_{m\to\infty} F(\varphi_m) = F(\varphi_0)$.

It follows from properties (A) and (B) that

(C) $\mathbb{X}^* \subset \mathscr{X}'$,

which shows that indeed the space of distributions generalizes elements of the Banach space $\mathbb{X}^*$, containing them as a particular case. In some cases the duality between the strict inductive and strict projective topologies helps to study a structure of distribution spaces.

Proposition 1.23. *Let a sequence of complete locally convex topological vector spaces $\mathscr{X}_n$ form a sequence of embeddings*

$$\mathscr{X}_1 \subset \dots \mathscr{X}_n \subset \mathscr{X}_{n+1} \subset \dots \subset \mathscr{X} \subset \mathbb{X},$$

where $\mathscr{X} = ind\lim_{n\to\infty} \mathscr{X}_n$, which densely embedded into a Banach space $\mathbb{X}$. Then the duals $\mathscr{X}_n'$ with the weak topologies form the following sequence of embeddings

$$\mathbb{X}' \subset \mathscr{X}' \subset \dots \mathscr{X}_{n+1}' \subset \mathscr{X}_n' \subset \dots \subset \mathscr{X}_1',$$

where $\mathscr{X}' = pr\lim_{n\to\infty} \mathscr{X}_n'$ and $\mathbb{X}'$ is the dual of $\mathbb{X}$.

Another important principle in construction of the spaces of distributions is that all the derivatives of distributions should exist in some weaker sense. Of course, speaking about the derivatives, we assume that elements of $\mathbb{X}$ are functions. Since by definition the (generalized) derivative of order α of a distribution $F \in \mathscr{X}'$ is

$$D^\alpha F(\varphi) = (-1)^\alpha F(D^\alpha \varphi) \tag{1.64}$$

for all $\varphi \in \mathscr{X}$, then the corresponding test functions must have all the derivatives. This leads to the following property of the space of test functions:

(D) $\mathscr{X}$ is invariant with respect to differentiation operator D^α.

This property together with (1.64) immediately implies that

(E) $F \in \mathscr{X}'$ has all the derivatives $D^\alpha F$ in the sense of (1.64).

1.6 Fourier multipliers

Now we briefly discuss Fourier L_p-multipliers, which play an important role in our further considerations. Suppose $\varphi \in \mathscr{G}$ and $m(\xi)$ s a bounded function. Then the operator T defined as $Tf = F^{-1}[mF[f]]$ performs a mapping $T : \mathscr{G} \to \mathscr{G}'$. In the multiplier problem we are interested in functions $m(\xi)$ for which the operator T extends to a continuous mapping $T : L_p(\mathbb{R}^n) \to L_q(\mathbb{R}^n)$ for some $p,\ q \in [0,\infty]$.

Definition 1.13. A bounded function $m(\xi),\ \xi \in \mathbb{R}^n$, is called *a Fourier multiplier of type* (p,q) if for all $f \in L_p(\mathbb{R}^n)$ the inequality

$$\|Tf\| = \|F^{-1}[mF[f]]\|_{L_q} \leq C\|f\|_{L_p} \tag{1.65}$$

holds. Here the constant $C > 0$ does not depend on f. The operator T is called a *Fourier multiplier operator.* The set of Fourier multipliers of type (p,q) is denoted by M_p^q. If $q = p$, then m is called *an L_p-multiplier.* Correspondingly, in this case we write M_p instead of M_p^p.

Proposition 1.24. *The following assertions hold:*

(a) If $m_1, m_2 \in M_p^q$, and $\lambda_1, \lambda_2 \in \mathbb{C}$, then $\lambda_1 m_1 + \lambda_2 m_2 \in M_p^q$;
(b) If $m_1 \in M_r^q$ and $m_2 \in M_p^r$, then their product $m_1 \cdot m_2 \in M_p^q$;
(c) If $m \in M_p^q$, then $m(\cdot + c) \in M_p^q$ for any $c \in \mathbb{R}^n$;
(d) $M_2 = L_\infty(\mathbb{R}^n)$;
(e) $M_p \subset L_\infty(\mathbb{R}^n)$, $1 < p < \infty$, $p \neq 2$;
(f) $M_p^q = M_{q'}^{p'}$, where $1 < p,q < \infty$ and (p,p') and (q,q') are conjugate pairs;
(g) $M_1 = M_\infty = F[\mathscr{B}(\mathbb{R}^n)]$, where $F[\mathscr{B}(\mathbb{R}^n)]$ is the Fourier image of the set of bounded Borel measures $\mathscr{B}(\mathbb{R}^n)$.

Proof. Part (a) immediately follows from Definition 1.13. To show (b) we assume that $m_1 \in M_r^q$ and $m_2 \in M_p^r$. Then it follows from Definition 1.13 that

$$\begin{aligned} \|F^{-1}[m_1 m_2 F[f]]\|_{L_q} &= \|F^{-1}[m_1 F F^{-1}[m_2 F[f]]]\|_{L_q} \\ &\leq C_1 \|F^{-1}[m_2 F[f]]\|_{L_r} \leq C_1 C_2 \|f\|_{L_p}. \end{aligned}$$

Since $F^{-1}[m(\xi + c)F[f]] = \exp(icx)F^{-1}[m(\xi)F[f]](\xi - c)$, every Fourier multiplier of type (p,q) is translation invariant, that is if $m \in M_p^q$, then $m(\cdot + c) \in M_p^q$ for any $c \in \mathbb{R}^n$, yielding (c). Part (d) follows easily from Parseval's equality, see Lemma 1.3. Part (e) follows from Mikhlin's theorem (see Theorem 1.7 below) when $\alpha = 0$. Mikhlin's theorem provides a description of the class M_p. To show (f), suppose $m \in M_p^q$ and $\varphi \in L_{q'}(\mathbb{R}^n)$. Then, using relations $F^{-1} = (2\pi)^{-n} FS$ and $F = (2\pi)^n F^{-1} S$, where S is a reflection operator acting as $Sg(x) = g(-x)$, one has

$$\langle F^{-1} m F f, \varphi \rangle = \langle f, F^{-1} m F \varphi \rangle.$$

Therefore,

$$\begin{aligned} |\langle f, F^{-1}[mF[\varphi]] \rangle| = |\langle F^{-1}[mF[f]], \varphi \rangle| &\leq \|F^{-1}[mF[f]]\|_{L_q} \|\varphi\|_{L_{q'}} \\ &\leq C \|f\|_{L_p} \|\varphi\|_{L_{q'}}. \end{aligned}$$

Due to (1.4) this implies

$$\|F^{-1}[mF[\varphi]]\|_{L_{p'}} = \sup_{0 \neq f \in L_p(\mathbb{R}^n)} \frac{|\langle f, F^{-1}[mF[\varphi]] \rangle|}{\|f\|_{L_p}} \leq C \|\varphi\|_{L_{q'}},$$

yielding $m \in M_{q'}^{p'}$. For the proof of the fact that $M_\infty = F[\mathscr{B}(\mathbb{R}^n)]$ see [Ste70].

Let $K(x) \in \mathcal{G}'$ denote the inverse Fourier transform of $m \in M_p^q$, i.e., $K = F^{-1}[m]$. Assume that $f \in \mathcal{G}$. Then, taking into account the equation $F^{-1}[m(\xi)F[f]] = K * f$, inequality (1.65) can be rewritten in the form

$$\|K * f\|_{L_q} \le C\|f\|_{L_p} \tag{1.66}$$

which can be extended to functions $f \in L_p$ by continuity. Moreover, (1.40) implies that if $K = F^{-1}[m] \in L_r(\mathbb{R}^n)$, where $r \in [1,\infty)$ and satisfies the condition

$$1 + \frac{1}{q} = \frac{1}{r} + \frac{1}{p},$$

then (1.66) holds with $C = \|K\|_{L_r}$. In particular, if $p = q$, and consequently, $K = F^{-1}[m] \in L_1(\mathbb{R}^n)$, then m is a L_p-multiplier with $C = \|K\|_{L_1}$ in inequality (1.66). However, the condition $F^{-1}[m] \in L_1(\mathbb{R}^n)$ is not necessary for m to be an L_p-multiplier. Consider an example. Let $m(\xi) = I_{[-1,1]}(\xi)$, where $I_{[-1,1]}(\xi)$ is the indicator function of the interval $[-1,1]$. Then $F^{-1}[m](x) = \frac{\sin x}{\pi x} \notin L_1(\mathbb{R})$. However, $F^{-1}[m] \in L_r(\mathbb{R})$ for any $r > 1$. Therefore, $m \in M_p^q$, where $1 < p, q < \infty$, and (p,q) is a conjugate pair. In particular, $m \in M_2$. In the one-dimensional case, in fact, $m \in M_p$ for all $p > 1$.

In dimensions $n \ge 2$, surprisingly, the function $m = I_{|\xi| \le 1}(\xi)$ is an L_p-multiplier if and only if $p = 2$. This fact was proved by Fefferman [Fef71] in 1971.

Theorem 1.6. *(Fefferman [Fef71]) Let $\mathbb{D}$ be the unit disc in $\mathbb{R}^n$, $n \ge 2$. Then the Fourier multiplier operator $T_{\mathbb{D}} = F^{-1}I_{\mathbb{D}}F$ is unbounded in $L_p(\mathbb{R}^n)$ for every $p \ne 2$.*

The theorem below is due to Mikhlin [Mih56]. This theorem describes a class of L_p-multipliers in the whole scale $1 < p < \infty$.

Theorem 1.7. *(Mikhlin) Let $m(\xi) \in C^{1+[\frac{n}{2}]}(\mathbb{R}^n \setminus 0)$, where $[a]$ designates the integer part of a, and there exists a positive constant C such that*

$$|\xi|^{|\alpha|}|D^\alpha m(\xi)| \le C, \quad \xi \in \mathbb{R}^n \setminus \{0\}, \tag{1.67}$$

for all $|\alpha| \le 1 + [\frac{n}{2}]$. Then $m \in M_p$, $1 < p < \infty$.

Example 1.7. The function $m(\xi) = i\,\mathrm{sign}(\xi)$, $\xi \in \mathbb{R}^1$, satisfies the Mikhlin condition. Therefore, this function belongs to M_p for all $1 < p < \infty$. The corresponding multiplier operator $Tf = F^{-1}[m(\xi)F[f]]$, due to formula (1.49), has the form

$$Tf(x) = \frac{1}{\pi}(p.v.\frac{1}{x} * f)(x) = \frac{1}{\pi}\int_{-\infty}^{\infty} \frac{f(y)}{x-y}dy,$$

and is called a *Hilbert transform*.

Lizorkin [Liz67] proved the following theorem under a weaker condition than (1.67).

Theorem 1.8. *(Lizorkin [Liz67]) Let $m(\xi)$ be a function differentiable out of hyperplanes $\xi_j = 0,\ j = 1,\dots,n$, and satisfy the condition*

$$\sup_{\xi\in\mathbb{R}^n} |\xi^\alpha D^\alpha m(\xi)| \leq C, \tag{1.68}$$

for all $\alpha = (\alpha_1,\dots,\alpha_n)$ with components α_j, $j = 1,\dots,n$, equal either 0 or 1. Then $m \in M_p$, $1 < p < \infty$.

Example 1.8. 1. It is easy to see that the characteristic function $m_{\mathbb{R}^n_+}(\xi)$ of the set $\mathbb{R}^n_+ = \{\xi \in \mathbb{R}^n : \xi_1 > 0,\dots,\xi_n > 0\}$ satisfies the condition (1.68). Therefore, $m_{\mathbb{R}^n_+}(\xi) \in M_p$, $1 < p < \infty$.

2. Consider the following radial function for $\gamma > 0$:

$$m(\xi) = \rho(|\xi|) = \frac{|\xi|^\gamma}{(1+|\xi|)^\gamma}, \quad \xi \in \mathbb{R}^n.$$

Taking the derivative of $\rho(\tau) = \Big(\tau/(1+\tau)\Big)^\gamma$, $\tau > 0$, we have

$$\rho'(\tau) = \frac{\gamma}{\tau}\left(\frac{\tau}{1+\tau}\right)^\gamma \frac{1}{1+\tau} \leq \frac{\gamma}{\tau}, \quad \tau > 0.$$

Similarly,

$$|\tau^m \rho^{(m)}(\tau)| \leq C, \quad \tau > 0,$$

for all $m = 2,3,\dots$. Now it is readily seen that these estimates imply that $m(\xi)$ satisfies condition (1.68), and hence $m \in M_p$, $1 < p < \infty$.

The theorem below with an integral condition instead of (1.67) is due to Hörmander [Hor83].

Theorem 1.9. *([Hor83]) Let $m(\xi)$ for some integer $s > \frac{n}{2}$ satisfy the condition*

$$\sum_{|\alpha|\leq s} \int_{R/2<|\xi|<2R} |R^\alpha D^\alpha m(\xi)| d\xi \leq C < \infty, \quad \forall R > 0. \tag{1.69}$$

Then $m \in M_p$, $1 < p < \infty$.

Remark 1.2. Condition (1.69) in the Hörmander's multiplier theorem implies continuity of $m(\xi)$. The indicator function of the unit ball obviously does not satisfy this condition, as well as Lizorkin's condition (1.68).

1.7 Sobolev spaces and Bessel potentials: case p=2

Let m be a nonnegative integer number. Let a function $f \in L_2(\Omega)$, $\Omega \subseteq \mathbb{R}^n$, be such that its all the derivatives $D^\alpha f$, $|\alpha| \leq m$, in the sense of distributions, belong to $L_2(\Omega)$. The set of such functions endowed with the norm

$$\|f|W_2^m(\Omega)\| = \sum_{|\alpha|\leq m} \|D^\alpha f\|_{L_2(\Omega)}, \tag{1.70}$$

and denoted by $W_2^m(\Omega)$, is called a *Sobolev space of order m*. Another norm, equivalent to (1.70), is given by

$$\|f|W_2^m(\Omega)\| = \Big(\sum_{|\alpha|\leq m} \|D^\alpha f\|^2_{L_2(\Omega)} \Big)^{\frac{1}{2}}.$$

For Sobolev spaces as a direct implication of the definition, the following inclusions hold:

$$L_2(\Omega) \supset W_2^1(\Omega) \supset \cdots \supset W_2^m(\Omega) \supset W_2^{m+1}(\Omega) \supset \cdots. \tag{1.71}$$

Define by $W_2^{-m}(\Omega)$ the dual space to $W_2^m(\Omega)$. In other words $W_2^{-m}(\Omega)$ is the set of linear continuous functionals defined on $W_2^m(\Omega)$. The norm in $W_2^{-m}(\Omega)$ is defined by

$$\|f|W_2^{-m}(\Omega)\| = \sup_{0\neq\varphi\in W_2^m(\Omega)} \frac{|<f,\varphi>|}{\|\varphi|W_2^m(\Omega)\|}.$$

$W_2^{-m}(\Omega)$ is called *a negative order Sobolev space.* With this definition one can extend (1.71) to the full scale of Sobolev spaces

$$\begin{aligned} \cdots \subset W_2^{(m+1)}(\Omega) \subset W_2^m(\Omega) \subset \cdots \subset W_2^1(\Omega) \subset L_2(\Omega) \\ \subset W_2^{-1}(\Omega) \subset \cdots \subset W_2^{-m}(\Omega) \subset W_2^{-(m+1)}(\Omega) \subset \cdots. \end{aligned} \tag{1.72}$$

Obviously, the norms satisfy the inequalities

$$\begin{aligned} \cdots \leq \|f|W_2^{-(m+1)}(\Omega)\| \leq \|f|W_2^{-m}(\Omega)\| \leq \cdots \leq \|f\|_{L_2} \\ \leq \cdots \leq \|f|W_2^m(\Omega)\| \leq \|f|W_2^{m+1}(\Omega)\| \leq \cdots \end{aligned} \tag{1.73}$$

Hence, if a sequence $f_k \in W_2^{m+1}(\Omega)$, where $m \in \mathbb{Z}$, converges to $f_0 \in W_2^{m+1}(\Omega)$ in the norm of $W_2^{m+1}(\Omega)$, then $f_k \to f_0$ in the norm of $W_2^m(\Omega)$, too. This implies that each of the embeddings in (1.72) is continuous. We use the symbol $\hookrightarrow$ for continuous embeddings. Thus, unifying (1.72) and (1.73) one can write

$$\begin{aligned} \cdots \hookrightarrow W_2^{(m+1)}(\Omega) \hookrightarrow W_2^m(\Omega) \hookrightarrow \cdots \hookrightarrow W_2^1(\Omega) \hookrightarrow L_2(\Omega) \\ \hookrightarrow W_2^{-1}(\Omega) \hookrightarrow \cdots \hookrightarrow W_2^{-m}(\Omega) \hookrightarrow W_2^{-(m+1)}(\Omega) \hookrightarrow \cdots. \end{aligned}$$

Further, each of these embeddings is dense due to Proposition 1.4, Part (3). Applying Arzela-Ascoli Lemma (Lemma 1.1) and the denseness of $C_0^\infty(\Omega)$ in Sobolev spaces one can verify that each of the embeddings in the scale (1.72) is compact.

Due to Lemma 1.18 and Parseval's equality $D^\alpha f \in L_2(\mathbb{R}^n)$ is equivalent to $\xi^\alpha F[f] \in L_2(\mathbb{R}^n)$. Therefore, in the case $\Omega = \mathbb{R}^n$ Sobolev spaces can easily be extended to fractional order Sobolev spaces. These spaces serve also as local elements in construction of Sobolev spaces on manifolds $\mathcal{M} \subset \mathbb{R}^n$ without boundary.

Definition 1.14. Let $s \geq 0$ be a real number. Introduce the space

$$H^s \equiv H^s(\mathbb{R}^n) = \{f \in L_2(\mathbb{R}^n) : (1+|\xi|^2)^{\frac{s}{2}} F[f] \in L_2(\mathbb{R}^n)\}. \tag{1.74}$$

H^s is called a space of *Bessel potentials,* or *Liouville space.* Similar to the integer order Sobolev spaces one can introduce $H^{-s}(\mathbb{R}^n)$ as the space of linear continuous functionals defined on $H^s(\mathbb{R}^n)$. The norm of $f \in H^{-s}(\mathbb{R}^n)$ we denote by $\|f\|_{-s}$. Both $H^s(\mathbb{R}^n)$ $(s > 0)$ and $H^{-s}(\mathbb{R}^n)$ are Hilbert spaces. The inner product and the norm in $H^s(\mathbb{R}^n)$ are defined

$$(\varphi, \psi)_s = \Big((1+|\xi|^2)^{s/2} F[\varphi], (1+|\xi|^2)^{s/2} F[\psi]\Big)_{L_2},$$

and

$$\|\varphi\|_s = \|(1+|\xi|^2)^{s/2} F[\varphi]\|_{L_2},$$

respectively. Obviously, $H^s(\mathbb{R}^n)$ is equivalent to $W_2^m(\mathbb{R}^n)$ if $s = m \in \mathbb{Z}$. If $s_2 > s_1$, then $H^{s_2}(\mathbb{R}^n) \subset H^{s_1}(\mathbb{R}^n)$, and this inclusion is continuous, dense, and compact. It is also useful to note that continuous and dense inclusions $\mathcal{G} \subset H^s(\mathbb{R}^n) \subset L_2(\mathbb{R}^n)$ imply continuous and dense inclusions $L_2(\mathbb{R}^n) \subset H^{-s}(\mathbb{R}^n) \subset \mathcal{G}'$.

The definitions of the spaces $H^s(\mathbb{R}^n)$ for positive and negative numbers s can be unified for any real s. Namely, a distribution f in $\mathcal{G}'$ is in the space $H^s(\mathbb{R}^n)$, where s is a real number (not necessarily positive), if

$$\|f\|_s = \|(1+|\xi|^2)^{s/2} F[f](\xi)\|_{L_2} < \infty.$$

Assume $K_0 = \{\xi \in \mathbb{R}^n : |\xi| \leq 1\}$ and $K_j = \{2^{j-1} \leq |\xi| \leq 2^j\}$, $j = 1, 2, \ldots$. Obviously, $\cup_{j=0}^\infty K_j = \mathbb{R}^n$. Further, for a given set A denote by $I_A(\xi)$ the indicator function of A, i.e.

$$I_A(\xi) = \begin{cases} 1, & \text{if } \xi \in A; \\ 0, & \text{if } \xi \neq A. \end{cases}$$

Then for a $f \in H^s(\mathbb{R}^n)$ one has

$$C_1 \sum_{j=0}^\infty 2^{2js} \|f_j\|^2_{L_2} \leq \|f\|_s^2 \leq C_2 \sum_{j=0}^\infty 2^{2js} \|f_j\|^2_{L_2}, \tag{1.75}$$

where $C_1 < C_2$ are positive constants and $f_j(x) = F^{-1}(I_j(\xi) F[f](\xi)) \in L_2(\mathbb{R}^n)$ due to Proposition 1.24, part (d), since the indicator function $I_A(\xi)$ is a L_2-multiplier for any bounded set $A \subset \mathbb{R}^n$. Here $I_j(\xi) = I_{K_j}(\xi)$, and therefore supp $F[f_j] \subset K_j$ for each $j = 0, 1, \ldots$. Moreover, $\sum_{j=0}^\infty I_j(\xi) \equiv 1$, yielding $f(x) = \sum_{j=0}^\infty f_j(x)$. These facts and the Parseval equality imply (1.75). This technique of characterization of the spaces H^s is called a *dyadic (or spectral) decomposition method.* A modification of this idea will be used in the next section for introduction of the Besov and Lizorkin-Triebel spaces.

We have noted above that $L_2(\mathbb{R}^n)$ is invariant with respect to the Fourier transform F. The Hausdorff-Young Theorem (see Theorem 1.4) provides the range of the Fourier transform F acting on $L_p(\mathbb{R}^n)$ in the case $p \in [1,2)$. If $p > 2$, then the spaces $H^s(\mathbb{R}^n)$, $s \in \mathbb{R}$, can be used for description of the range $F[L_p(\mathbb{R}^n)]$.

Theorem 1.10. *(Hörmander) Let $p > 2$. Then the Fourier transform F maps the space $L_p(\mathbb{R}^n)$ into H^{-s} where $s > n\left(\frac{1}{2} - \frac{1}{p}\right)$, and the inequality $\|Ff\|_{-s} \leq C\|f\|_{L_p}$ holds with a constant $C > 0$, which does not depend on f.*

In Example 1.6 we obtained that $F[1] = (2\pi)^n \delta_0$. The function $f(x) \equiv 1$ belongs to $L_\infty(\mathbb{R}^n)$, but does not belong to $L_p(\mathbb{R}^n)$ for any $p \in [1,\infty)$. Hence, letting $p \to \infty$ in Theorem 1.10 one has that the Dirac delta function $\delta \in H^{-s}$, only if $s > n/2$.

The classic Paley-Wiener theorem characterizes the Fourier transform of functions $f \in C_0^\infty(\mathbb{R}^n)$.

Theorem 1.11. *(Paley-Wiener [Hor83]) Let $\varphi \in C_0^\infty(\mathbb{R}^n)$ with a support contained in the ball $B_R = \{x : |x| \leq R\}$. Then its Fourier transform $F(\xi) = F[\varphi](\xi)$ can be extended analytically to the entire complex space $\mathbb{C}^n$ and for any $m \in \mathbb{N}_0$ there exists a number $C_m > 0$, such that*

$$|F(\xi + i\eta)| \leq C_m(1 + |\xi|)^{-m} e^{R|\eta|}, \quad \xi + i\eta \in \mathbb{C}^n. \tag{1.76}$$

Conversely, if an entire function $F(\zeta)$, $\zeta = \xi + i\eta \in \mathbb{C}^n$, satisfies the inequality (1.76), then there exists a function $\varphi \in C_0^\infty(\mathbb{R}^n)$ with a support $supp\,\varphi \subset B_R$, such that $F[\varphi](\xi) = F(\xi)$.

By construction, the sets of test functions introduced in Sections 1.5.1–1.5.3 are in the following relation: $\mathfrak{D}(\mathbb{R}^n) \subset \mathscr{S}(\mathbb{R}^n) \subset \mathscr{E}(\mathbb{R}^n)$. Moreover, the topology of a larger set in this series of inclusions is finer making each of these inclusions continuous. Therefore, their duals endowed with the corresponding weak topologies are related as $\mathscr{E}'(\mathbb{R}^n) \subset \mathscr{S}'(\mathbb{R}^n) \subset \mathfrak{D}'(\mathbb{R}^n)$, where each inclusion is continuous. In Section 1.5.3 we established that every distribution f in $\mathscr{S}'(\mathbb{R}^n)$ has the Fourier transform $F[f]$, which is again a distribution in $\mathscr{S}'(\mathbb{R}^n)$. The Paley-Wiener-Schwartz theorem, proved by Laurent Schwartz [Sch51], shows that if a distribution f has a compact support, that is if $f \in \mathscr{E}'(\mathbb{R}^n)$, then its Fourier transform is actually a function $F[f](\xi) = F(\xi) = < f, e^{ix\xi} >$ called a *Fourier-Laplace transform,* and this function can be characterized as an entire function on $\mathbb{C}^n$.

Theorem 1.12. *(Paley-Wiener-Schwartz [Sch51, Hor83]) Let $f \in \mathscr{E}'(\mathbb{R}^n)$ with a support contained in the ball $B_R = \{x : |x| \leq R\}$. Then its Fourier transform $F(\xi) = F[\varphi](\xi)$ can be extended analytically to the entire complex space $\mathbb{C}^n$ and satisfies the estimate*

$$|F(\xi + i\eta)| \leq C_m(1 + |\xi|)^m e^{R|\eta|}, \quad \xi + i\eta \in \mathbb{C}^n. \tag{1.77}$$

for some $m \in \mathbb{N}_0$ *and* $C_m > 0$. *Conversely, if an entire function* $F(\zeta)$, $\zeta = \xi + i\eta \in \mathbb{C}^n$, *satisfies the inequality* (1.77) *for some* $m \in \mathbb{N}_0$ *and* $C_m > 0$, *then there exists a distribution* $f \in \mathscr{E}'(\mathbb{R}^n)$ *with a support* $\operatorname{supp}\varphi \subset B_R$, *such that*

$$F(\xi) = < f, e^{ix\xi} >, \quad \xi \in \mathbb{R}^n. \tag{1.78}$$

Remark 1.3. The Fourier transform of a tempered distribution $f \in \mathscr{S}'$ was defined as $F[f](\varphi) = f(F[\varphi])$ for all $\varphi \in \mathscr{S}$. This definition works well for a Schwartz distribution $f \in \mathfrak{D}'(\mathbb{R}^n)$, as well: $F[f](\varphi) = f(F[\varphi])$ for all $\varphi \in \mathfrak{D}(\mathbb{R}^n)$. This definition does not work for distributions with compact support $f \in \mathscr{E}'(\mathbb{R}^n)$, since $F[\varphi]$ may not exist for $\varphi \in \mathscr{E}(\mathbb{R}^n)$. However, it follows from Theorem 1.12 that the Fourier transform of $f \in \mathscr{E}'(\mathbb{R}^n)$ can be defined by the Fourier-Laplace transform (1.78). The Fourier-Laplace transform is valid for functions of complex variables also and widely used in complex analysis. We will exploit it, in particular, in Chapter 7 in the context of complex ΨDOSS'.

1.8 Sobolev, Sobolev-Slobodecki, and Besov spaces: general case

Sobolev spaces and spaces of Bessel potentials for arbitrary $1 \le p \le \infty$ are defined as follows.

Definition 1.15. Let $m \in \mathbb{N}$ and $1 \le p \le \infty$. Then

$$W_p^m(\Omega) \equiv \{f \in \mathfrak{D}'(\Omega) : \|f|W_p^m(\Omega)\| = \sum_{|\alpha| \le m} \|D^\alpha f\|_{L_p(\Omega)} < \infty\}. \tag{1.79}$$

Definition 1.16. Let $s \in \mathbb{R}$ and $1 \le p \le \infty$. Then

$$H_p^s \equiv H_p^s(\mathbb{R}^n) = \{f \in \mathscr{S}' : F^{-1}\left[(1+|\xi|^2)^{\frac{s}{2}} F[f]\right] \in L_p(\mathbb{R}^n)\}. \tag{1.80}$$

H_p^s is a Banach space with respect to the norm

$$\|f\|_{H_p^s} = \left\| F^{-1}\left[(1+|\xi|^2)^{s/2} F[f]\right] \right\|_{L_p}. \tag{1.81}$$

Elements of H_p^s are also called Bessel potentials. Sobolev-Slobodecki spaces interpolate Sobolev spaces to noninteger order. Let $\mu \in (0,1)$. Then for $1 \le p < \infty$ the Sobolev-Slobodecki space of order μ is

$$W_p^\mu(\Omega) \equiv \{f \in L_p(\Omega) : \|f|W_p^\mu(\Omega)\| = \Big(\int_\Omega \int_\Omega \frac{|f(x)-f(y)|^p}{|x-y|^{\mu p+n}} dxdy\Big)^{1/p}$$

$$= \Big(\int_{|h| \le h_0} \|\Delta_h f\|_{L_p(\Omega)}^p \frac{dh}{|h|^{n+\mu p}}\Big)^{1/p} < \infty\}.$$

Here $\Delta f(x) = f(x+h) - f(x)$, the first order finite difference, and h_0 is a positive number, $0 < h_0 < 1$. Note that norms in the above definition for all h_0 are equivalent, and hence, $W_p^{\mu}(\Omega)$ does not depend on h_0. In a similar manner, if $\lambda = m + \mu$, where $m \in \mathbb{N}$ and $0 < \mu < 1$, then

$$W_p^{\lambda}(\Omega) \equiv \{f \in W_p^m(\Omega) : \|f|W_p^{\lambda}(\Omega)\| = \|f|W_p^m(\Omega)\| + \sum_{|\alpha|=m} \Big(\int_{|h|\leq h_0} \|\Delta_h D^{\alpha} f\|^p_{L_p(\Omega)} \frac{dh}{|h|^{n+\mu p}} \Big)^{1/p} < \infty\}.$$

It is known (see, e.g., [Tri83]) that for integer $\mu = m$ the Sobolev-Slobodecki space does not coincide with the Sobolev space $W_p^m(\Omega)$. Sobolev-Slobodecki spaces are important in the study of traces of functions from Sobolev spaces. Namely, the theorem below holds. Denote by γ the trace operator $\gamma f = f_{|\partial\Omega}$, where $\partial\Omega$ is the boundary of Ω.

Theorem 1.13. *Let $1 < p < \infty$, $m \geq 1$, and Ω be a bounded domain with a Lipschitz boundary $\partial\Omega$. Then the trace operator*

$$\gamma : W_p^m(\Omega) \to W_p^{m-\frac{1}{p}}(\partial\Omega)$$

is continuous.

Definition 1.17. Let $s = m + \mu$, $0 < \mu \leq 1$, $1 \leq p, q \leq \infty$, $0 < h_0 < 1$, and $\Omega_h = \{x \in \Omega : x + h \in \Omega\}$. The Besov space is a normed space of functions

$$B^s_{p,q}(\Omega) \equiv \{f \in W_p^m(\Omega) : \sum_{|\alpha|=m} \int_{|h|<h_0} \|\Delta_h^2 D^{\alpha} f\|^q_{L_p(\Omega_h)} \frac{dh}{|h|^{n+\mu q}} < \infty\},$$

relative to the norm

$$\|f|B^s_{p,q}(\Omega)\| = \|f|W_p^m(\Omega)\| + \sum_{|\alpha|=m} \Big(\int_{|h|<h_0} \|\Delta_h^2 D^{\alpha} f\|^q_{L_p(\Omega_h)} \frac{dh}{|h|^{n+\mu q}} \Big)^{1/q},$$

if $q < \infty$, and

$$\|f|B^s_{p,q}(\Omega)\| = \|f|W_p^m(\Omega)\| + \sum_{|\alpha|=m} \sup_{|h|<h_0} \frac{\|\Delta_h^2 D^{\alpha} f\|_{L_p(\Omega_h)}}{|h|^{\mu}},$$

if $q = \infty$.

The Besov space is a Banach space and does not depend on h_0. If $\Omega = \mathbb{R}^n$, then obviously, $\Omega_h = \Omega$. In the particular case $q = \infty$, $B^s_{p,\infty}(\Omega)$ is called *Nikol'skii space*, and in the case $p = q = \infty$, $B^s_{\infty,\infty}(\Omega)$ coincides with the Hölder-Zygmund space $\mathscr{C}^s(\Omega)$. If $p = q = 2$, then $B^s_{2,2}(\Omega) = W_2^s(\Omega)$ (Sobolev-Slobodecki space). Thus, the Besov space represents a wide generalization of the spaces introduced above and in previous sections. The Besov and Lizorkin-Triebel spaces on the base of spectral decomposition method will be introduced in the next section.

Remark 1.4. 1. We note that the spectral decomposition method of characterization of the space of Bessel potentials, we used above in the case $p = 2$, does not work in the case $p \neq 2$. The reason is indicator functions of the sets $K_0 = \{\xi \in \mathbb{R}^n : |\xi| \leq 1\}$ and $K_j = \{2^{j-1} \leq |\xi| \leq 2^j\}$, $j = 1, 2, \ldots$, are not Fourier multipliers for $p \neq 2$, $n \geq 2$, due to Fefferman's theorem, see Theorem (1.6). We will see in the next section that a suitable modification of the method of spectral decomposition works for characterization of not only H_p^s, but also wider spaces, the Besov and Lizorkin-Triebel spaces.

2. We also note that one can define the Sobolev space $H_p^s(\Omega)$ for bounded domain $\Omega \subset \mathbb{R}^n$ with a smooth boundary, as the set

$$H_p^s(\Omega) = \{g = f_{|\Omega} : f \in H_p^s(\mathbb{R}^n)\},$$

where $f_{|\Omega}$ is a restriction of f onto Ω. The space $H_p^s(\Omega)$ relative to the norm

$$\|g|H_p^s(\Omega)\| = \inf_{f_{|\Omega}=g} \|f|H_p^s\|,$$

is a Banach space. This space is important in the modern theory of boundary value problems for linear and nonlinear differential and pseudo-differential equations (see, e.g., [Tri83, Tre80, Tay81, BL81]).

The relations of the space $W_p^\mu(\Omega)$ to the spaces $L_p(\Omega)$ and $C(\Omega)$ are given in the theorem below, which is known as the *Sobolev embedding theorem.*

Theorem 1.14. *1. Let $1 \leq p \leq \infty$ and $1 \leq q < \infty$. Assume that $\mu \geq n(1/p - 1/q)$. Then the embedding*

$$W_p^\mu(\Omega) \subset L_q(\Omega) \tag{1.82}$$

is continuous.

2. Let $1 \leq p \leq \infty$. Assume that $\mu > n/p$. Then the embedding

$$W_p^\mu(\Omega) \subset C(\Omega) \tag{1.83}$$

is continuous.

The particular case $p = 2$ will be used frequently in our further analysis. Namely, if $s \geq n(1/2 - 1/q)$, then the embedding

$$H_2^s(\Omega) \subset L_q(\Omega) \tag{1.84}$$

is continuous, and if $s > n/2$, then the embedding

$$H_2^s(\Omega) \subset C(\Omega) \tag{1.85}$$

is continuous.

Finally, we introduce two topological vector spaces important for our further analysis. Let $\Omega \subseteq \mathbb{R}^n$ be a domain, $p \geq 1$, and $s \in \mathbb{R}^n$. Introduce the following spaces:

$$H^s_{p,comp}(\Omega) = \{f \in H^s_p(\mathbb{R}^n) : \operatorname{supp} f \Subset \Omega\};$$
$$H^s_{p,loc}(\Omega) = \{f \in \mathscr{G}' : \|f|H^s_p(K)\| < \infty\} \text{ for all compact sets } K \subset \Omega.$$

If $s = 0$, then we write $H^0_{p,comp}(\Omega) = L_{p,comp}(\Omega)$ and $H^0_{p,loc}(\Omega) = L_{p,loc}(\Omega)$. The spaces $H^s_{p,comp}(\Omega)$, $H^s_{p,loc}(\Omega)$ are not normed. We say that a sequence $f_k \in H^s_{p,comp}(\Omega)$ converges to $f_0 \in H^s_{p,comp}(\Omega)$, if there is a compact set $K \subset \Omega$ such that $\operatorname{supp} f_k \subseteq K$ for all $k \in \mathbb{N}$, and $\|f_k - f_0|H^s_p(\mathbb{R}^n)\| \to 0$, when $k \to \infty$. Further, we say that a sequence $f_k \in H^s_{p,loc}(\Omega)$ converges to $f_0 \in H^s_{p,loc}(\Omega)$, if $\|f_k - f_0|H^s_p(K)\| \to 0$, as $k \to \infty$, for arbitrary compact set $K \subset \Omega$.

Proposition 1.25. *Let $1 < p < \infty$ and p' be its conjugate. Then the spaces $H^s_{p,comp}(\Omega)$ and $H^s_{p,loc}(\Omega)$ satisfy the following duality relations:*

1. $[H^s_{p,comp}(\Omega)]^* = H^{-s}_{p',loc}(\Omega)$;
2. $[H^s_{p,loc}(\Omega)]^* = H^{-s}_{p',comp}(\Omega)$.

1.9 Besov and Lizorkin-Triebel spaces

Besov and Lizorkin-Triebel type spaces widely generalize the spaces like L_p, Hölder, Sobolev, Bessel, and other spaces. In Definitions (1.18) and (1.19) below we assume that $1 \leq p, q < \infty$ and $s \in \mathbb{R}$.

Definition 1.18. A function f is said to belong to the *Besov space* $B^s_{pq}(\mathbb{R}^n)$ if f has the representation

$$f = \sum_{j=0}^{\infty} a_j(x)$$

in the sense of $\mathscr{G}'(\mathbb{R}^n)$, where $a_j \in L_p(\mathbb{R}^n)$,

$$\operatorname{supp} F[a_j] \subset M_j = \{2^{j-1} \leq |\xi| \leq 2^{j+1}\}, \quad j = 0, 1, \ldots,$$

and

$$\|f|B^s_{pq}\|_* = \Big(\sum_{j=0}^{\infty} 2^{sjq} \|a_j\|^q_{L_p}\Big)^{\frac{1}{q}} \tag{1.86}$$

is finite.

In fact, the expression $\|f|B^s_{pq}\|_*$ in equation (1.86) defines a norm. This can be easily verified. The triangle inequality follows due to Minkowski's inequality.

Definition 1.19. A function f is said to belong to the *Lizorkin-Triebel space* $F^s_{pq}(\mathbb{R}^n)$ if f has the representation

$$f = \sum_{j=0}^{\infty} a_j(x)$$

in the sense of $\mathscr{G}'(\mathbb{R}^n)$, where $a_j \in L_p(\mathbb{R}^n)$, supp $F[a_j] \subset M_j$, $j = 0, 1, \ldots$, and

$$\|f|F_{pq}^s\|_* = \|(\sum_{j=0}^{\infty} 2^{sjq}|a_j(x)|^q)^{\frac{1}{q}}\|_{L_p} \tag{1.87}$$

is finite.

Notice that these two function spaces differ by the form of their norms (1.86) and (1.87). Namely, the norm of $B_{pq}^s(\mathbb{R}^n)$ can be interpreted as l_q-norm of the sequence $\|2^{sj}a_j\|_{L_p}$, while the norm of $F_{pq}^s(\mathbb{R}^n)$ as L_p-norm of the function $\|2^{sj}a_j(x)\|_{l_q}$. Now, due to this fact, it is straightforward to extend the definitions of Besov and Lizorkin-Triebel spaces to the cases $p = \infty$ or $q = \infty$. For instance, $B_{p\infty}^s(\mathbb{R}^n)$ is defined by the finite norm $\sup_j \|2^{sj}a_j\|_{L_p}$. Similarly, $F_{p\infty}^s(\mathbb{R}^n)$ is defined by the finite norm $\left(\int_{\mathbb{R}^n}(\sup_j |2^{sj}a_j|)^p dx\right)^{1/p}$. We refer the reader for details to [Tri77, Tri83].

There is a wide class of norms equivalent to (1.86) (or (1.87)) and defined via the family Φ of functional sequences defined below, see [Tri83].

Definition 1.20. A sequence $\{\psi_j(\xi)\}_{j=0}^{\infty}$ is said to belong to Φ_N, $N \in \mathbb{N}$, if the following conditions hold:

1) $\psi_j \in \mathscr{G}(\mathbb{R}^n)$, $F[\psi_j] \geq 0$, $j \geq 0$;
2) supp $F[\psi_j] \subset \{\xi \in \mathbb{R}^n : 2^{j-N} \leq |\xi| \leq 2^{N+j}\}$, $j \geq 1$, and supp $F[\psi_0] \subset M_0 = \{|\xi| \leq 2^N\}$;
3) there exists a positive number $\eta > 0$ such that

$$F[\psi_0 + \psi_1 + \ldots](\xi) \geq \eta;$$

4) for an arbitrary α there exists $C(\alpha) > 0$ such that

$$|\xi|^{|\alpha|}|D^{\alpha}F[\psi_j](\xi)| \leq C(\alpha), \quad j = 1, 2, \ldots.$$

Definition 1.21. $\Phi = \cup_{N=1}^{\infty} \Phi_N$.

If $\{\psi_j(\xi)\}_{j=0}^{\infty} \in \Phi$, then the norm equivalent to (1.86) is defined as

$$\|f|B_{pq}^s\| = (\sum_{j=0}^{\infty} 2^{sjq}\|F^{-1}[\varphi_j F[f]]\|_{L_p}^q)^{\frac{1}{q}}, \tag{1.88}$$

where $\varphi_j = F[\psi_j]$, $j \geq 0$. Similarly, the norm equivalent to (1.87) is defined as

$$\|f|F_{pq}^s\| = \|(\sum_{j=0}^{\infty} 2^{sjq}|F^{-1}[\varphi_j F[f]]|^q)^{\frac{1}{q}}\|_{L_p}.$$

Hence, the definition of the Besov and Lizorkin-Triebel spaces can be given in the following equivalent forms:

$$B^s_{pq} = \{f \in \mathscr{G}' : \|f|B^s_{pq}\| = (\sum_{j=0}^{\infty} 2^{sjq} \|F^{-1}[\varphi_j F[f]]\|^q_{L_p})^{\frac{1}{q}} < \infty\}, \{F^{-1}[\varphi_j] = \psi_j\} \in \Phi,$$

and

$$F^s_{pq}(\mathbb{R}^n) = \{f \in \mathscr{G}' : \|f|F^s_{pq}\| = \|(\sum_{j=0}^{\infty} 2^{sjq} |F^{-1}[\varphi_j F[f]]|^q)^{\frac{1}{q}}\|_{L_p} < \infty\}, \{F^{-1}[\varphi_j] = \psi_j\} \in \Phi.$$

For the proof of the equivalence of corresponding norms we refer the reader to [Tri83].

Example 1.9. The example below shows that Φ_N, and hence, Φ are not empty. We construct a sequence of functions $\{\psi_k\}_{k=0}^{\infty}$, which belongs to Φ_N for $N = 1$. Consider two functions $\varphi_0(\xi)$ and $\phi(\xi)$ with the following properties:

1. $\varphi \in C_0^{\infty}(|\xi| < 2)$, $\phi \in C_0^{\infty}(\frac{1}{2} < |\xi| < 2)$;
2. $\varphi \geq 0$, $\phi \geq 0$, and $\varphi \equiv 1$ on $\{\xi : |\xi| \leq \sqrt{2}\}$, and $\phi = 1$ on $\{\xi : \frac{1}{\sqrt{2}} \leq |\xi| \leq \sqrt{2}\}$.

We set

$$\psi_0(x) = F^{-1}[\varphi(\xi)],$$
$$\psi_k(x) = F^{-1}[\phi(2^{-k}\xi)], \quad k = 1, 2, \ldots.$$

It is not hard to verify that $\{\psi_k\}_{k=0}^{\infty} \in \Phi_1$. By definition, the functions φ and ϕ are infinite differentiable with compact support, and therefore belong to $\mathscr{G}(\mathbb{R}^n)$. This implies that $\psi_k \in \mathscr{G}(\mathbb{R}^n)$ for each $k \in \mathbb{N}_0$. Moreover, by construction, $F[\psi_k](\xi) = \phi(2^{-k}\xi) \geq 0$ for all $k \in \mathbb{N}$, and $F[\psi_0](\xi) = \varphi(\xi) \geq 0$. Hence, condition 1) in Definition 1.20 is fulfilled. Condition 2) is immediate by construction of the system $\{\psi_k(x)\}$. Further, by construction, $F[\psi_0] \equiv 1$ on the set $|\xi| \leq \sqrt{2}$, $F[\psi_1] \equiv 1$ on the set $\sqrt{2} \leq |\xi| \leq 2\sqrt{2}$, $F[\psi_2] \equiv 1$ on the set $2\sqrt{2} \leq |\xi| \leq 2^2\sqrt{2}$, etc. These imply that $F[\psi_0 + \psi_1 + \ldots] \geq 1$, yielding Condition 3). Condition 4) is also verified, since $\varphi, \phi \in C_0^{\infty}(\mathbb{R}^n)$. In a similar manner one can easily construct a system $\{\psi_k(x)\} \in \Phi_N$ for any $N > 1$.

Besov and Lizorkin-Triebel spaces represent a wide class of function spaces containing Hölder-Zigmund, Sobolev, Slobodecki, Liouville, and other spaces as particular cases. For instance, for $s \in \mathbb{R}$ one has $F^s_{p,2}(\mathbb{R}^n) = H^s_p(\mathbb{R}^n)$, $1 < p < \infty$, (Sobolev spaces), or $B^s_{\infty,\infty}(\mathbb{R}^n) = \mathscr{C}^s(\mathbb{R}^n)$ (Hölder-Zigmund spaces) for $s > 0$. The proposition below contains some important properties of the Besov space. Similar properties hold for the Lizorkin-Triebel space. See details in [Tri83].

Proposition 1.26. *Let $s \in \mathbb{R}$ and $1 < p < \infty$.*

1. The spaces $\mathscr{G}(\mathbb{R}^n)$ and $C_0^{\infty}(\mathbb{R}^n)$ are dense in $B^s_{p,q}(\mathbb{R}^n)$ if $1 \leq q < \infty$, and are not dense if $q = \infty$.

2. *If* $1 \leq q_1 \leq q_2 \leq \infty$, *then* $B^s_{p,1}(\mathbb{R}^n) \hookrightarrow B^s_{p,q_1}(\mathbb{R}^n) \hookrightarrow B^s_{p,q_2}(\mathbb{R}^n) \hookrightarrow B^s_{p,\infty}(\mathbb{R}^n)$;
3. *For* $\varepsilon > 0$ *and any* $1 \leq q_1, q_2 \leq \infty$ *the embedding* $B^{s+\varepsilon}_{p,q_1}(\mathbb{R}^n) \hookrightarrow B^s_{p,q_2}(\mathbb{R}^n)$ *is compact;*
4. *If* $s > \frac{n}{p}$, *then the continuous embedding* $B^s_{p,q}(\mathbb{R}^n) \hookrightarrow C(\mathbb{R}^n)$, *holds for every* $1 \leq q \leq \infty$;
5. *If* $1 < q \leq p < \infty$, *then* $B^s_{p,q}(\mathbb{R}^n) \hookrightarrow F^s_{p,q}(\mathbb{R}^n) \hookrightarrow B^s_{p,p}(\mathbb{R}^n)$, *and if* $1 < p \leq q < \infty$, *then* $B^s_{p,p}(\mathbb{R}^n) \hookrightarrow F^s_{p,q}(\mathbb{R}^n) \hookrightarrow B^s_{p,q}(\mathbb{R}^n)$;
6. *If* $2 \leq p < \infty$, *then* $B^s_{p,2}(\mathbb{R}^n) \hookrightarrow H^s_p(\mathbb{R}^n) \hookrightarrow B^s_{p,p}(\mathbb{R}^n)$, *and if* $1 < p \leq 2$, *then* $B^s_{p,p}(\mathbb{R}^n) \hookrightarrow H^s_p(\mathbb{R}^n) \hookrightarrow B^s_{p,2}(\mathbb{R}^n)$. *Moreover, If* $1 < p = q < \infty$, *then* $B^s_{p,p}(\mathbb{R}^n) = F^s_{p,p}(\mathbb{R}^n)$ *and if* $p = q = 2$, *then* $B^s_{2,2}(\mathbb{R}^n) = H^s(\mathbb{R}^n)$.

Remark 1.5. 1. Statements 4) and 6) of the above proposition imply the Sobolev embedding theorem, Theorem 1.14. Namely, taking $q = p$, for $s > n/p$ one has $H^s_p(\mathbb{R}^n) \hookrightarrow B^s_{p,p}(\mathbb{R}^n) \hookrightarrow C(\mathbb{R}^n)$, recovering Theorem 1.14.

2. Besov spaces $B^s_{p,q}(\mathbb{R}^n)$ are reflexive Banach spaces for all $1 < p < \infty$, $1 \leq q < \infty$. For the conjugate space $(B^s_{p,q}(\mathbb{R}^n))^*$ to $B^s_{p,q}(\mathbb{R}^n)$ the following duality relation holds:
$$(B^s_{p,q})^* = B^{-s}_{p',q'}(\mathbb{R}^n),$$
where (p, p') and (q, q') are conjugate pairs.

3. Similar to Sobolev spaces (see Remark 1.4), if $\Omega \subset \mathbb{R}^n$ is a bounded domain with a smooth boundary, then the space
$$B^s_{p,q}(\Omega) = \{g = f_{|\Omega} : f \in B^s_{p,q}(\mathbb{R}^n)\},$$
is a Banach space relative to the norm
$$\|g|B^s_{p,q}(\Omega)\| = \inf_{f_{|\Omega}=g} \|f|B^s_{p,q}(\mathbb{R}^n)\|.$$

1.10 ψ-distributions

We have introduced three types of distributions above. Now we introduce a fourth type of distributions space appeared relatively recently and therefore not well represented in the literature. Therefore, we will study its properties in detail. We will show that this space possesses all the basic principles discussed in Section 1.5.4. Moreover, this space is often convenient and preferable for the study of various initial and boundary value problems for pseudo-differential equations with symbols depending only on dual variables and with possible singularities.

Let $1 \leq p \leq \infty$ and let G be an open set in $\mathbb{R}^n$. Suppose that $\{g_k\}_{k=0}^\infty$ is a locally finite covering of G, i.e., $G = \cup_{k=0}^\infty g_k$, $g_k \subset G$ is compact and every compact set $K \subset G$ has a nonempty intersection with a finite number of sets g_k. For a set K compact in G we write $K \Subset G$. We denote by $\{\phi_k\}_{k=0}^\infty$ the smooth partition of unity for G corresponding to the covering $\{g_k\}_{k=0}^\infty$.

Definition 1.22. Denote by $\Psi_{G,p}(\mathbb{R}^n)$ the set of functions $\varphi \in L_p(\mathbb{R}^n)$ such that $\operatorname{supp} F[\varphi] \Subset G$, i.e., the Fourier transform of φ has a compact support in G. We say that a sequence $\varphi_m \in \Psi_{G,p}(\mathbb{R}^n)$, $m = 1, 2, \ldots$, converges to $\varphi_0 \in \Psi_{G,p}(\mathbb{R}^n)$ in $\Psi_{G,p}(\mathbb{R}^n)$ if the following two conditions are fulfilled:

(i) there exists a compact set $K \subset G$ such that $\operatorname{supp} F\varphi_m \subseteq K$, $\forall m \in \mathbb{N}$;
(ii) $\varphi_m \to \varphi_0$ in the norm of $L_p(\mathbb{R}^n)$, as $m \to \infty$.

It follows from the Paley-Wiener-Schwartz Theorem (Theorem 1.12) that $\varphi \in \Psi_{G,p}(\mathbb{R}^n)$ if and only if it has the analytic extension $\varphi(x+iy)$ to $\mathbb{C}^n$, which is an entire function of a finite exponential type, i.e., there exist constants $C > 0$, $r > 0$, and $\gamma \in \mathbb{R}$, such that

$$|\varphi(x+iy)| \leq C(1+|x|)^{\gamma} e^{r|y|}.$$

Since $\varphi \in L_p(\mathbb{R}^n)$, obviously $\gamma < -n/p$. This observation supplies examples of functions of $\Psi_{G,p}(\mathbb{R}^n)$. Namely, any entire function of a finite exponential type whose restriction to $\mathbb{R}^n$ belongs to $L_p(\mathbb{R}^n)$, gives an example of $\Psi_{G,p}(\mathbb{R}^n)$.

Example 1.10. 1. Consider the function $\varphi_1(z) = \frac{\sin rz}{\pi z}$, $z \in \mathbb{C}$, $r > 0$, which is an entire function of type r. The restriction of this function to the real axis $\varphi_1(x) = \frac{\sin rx}{\pi x}$, belongs to $L_p(\mathbb{R})$ for any $p > 1$. Therefore, $\varphi_1(x) \in \Psi_{G,p}(\mathbb{R}^n)$, for all $p > 1$, where G is any interval containing $[-r, r]$. The Fourier transform of φ_1 is $F[\varphi_1](\xi) = I_{[-r,r]}(\xi)$, where $I_{[-r,r]}(\xi)$ is the indicator function of the interval $[-r, r]$, yielding $\operatorname{supp} F[\varphi_1] = [-r, r]$. This again confirms that $\varphi_1(x)$ is a function in $\Psi_{G,p}(\mathbb{R}^n)$. Similarly, one has

$$\varphi_2(x) = \frac{e^{irx} - 1}{x} \in \Psi_{G,p}(\mathbb{R}^n), \quad p > 1, \text{and}$$
$$\varphi_3(x) = \frac{\sin^2 rx}{x^2} \in \Psi_{G,p}(\mathbb{R}^n), \quad p \geq 1.$$

2. The function $\varphi_1(x)$ in the first example gives rise to n-dimensional case. Let

$$\varphi(x) = \prod_{j=1}^{n} \frac{\sin x_j}{\pi x_j}.$$

Then the n-dimensional Fourier transform of this function is $F[\varphi](\xi) = I_K(\xi)$, where $K = \{-1 \leq \xi_1 \leq 1, \ldots, -1 \leq \xi_n \leq 1\}$, is an n-dimensional cub. Applying Fubini's theorem we have that $\varphi \in L_p(\mathbb{R}^n)$ for any $p > 1$. Therefore, $\varphi \in \Psi_{G,p}(\mathbb{R}^n)$ for any $p > 1$ and any domain $G \subseteq \mathbb{R}^n$ containing K.
3. The Bessel function of the first kind $J_\ell(x)$ with nonnegative integer ℓ admits the analytic extension to the whole complex plane $\mathbb{C}$. This extension $J_\ell(z)$ is an entire function of finite exponential type and of order one. Moreover, $J_\ell(x) \in L_p(\mathbb{R})$ for any $p > 2$, since it behaves like $1/\sqrt{x}$ when $|x| \to \infty$. Therefore, $J_\ell \in \Psi_{G,p}(\mathbb{R}^n)$, $p > 2$.

4. Consider the function

$$\phi(x) = \frac{1}{(2\pi|x|)^{\frac{n}{2}}} J_{\frac{n}{2}}(|x|), \quad x \in \mathbb{R}^n,$$

where again $J_{\frac{n}{2}}$ is the Bessel function of the first kind of order $n/2$. Passing on to the spherical coordinates one can show that the Fourier transform of ϕ is $F[\phi](\xi) = I_B(\xi)$, with $B = \{\xi \in \mathbb{R}^n : |x| \leq 1\}$, the unit ball. Remembering the asymptotic behavior $J_{n/2}(x) \sim 1/\sqrt{x}$, when $|x| \to \infty$, one can see that $\phi \in L_p(\mathbb{R}^n)$, if $p > \frac{2n}{n+1}$. Hence,

$$\phi \in \Psi_{G,p}(\mathbb{R}^n) \text{ for all } p > \frac{2n}{n+1}, \text{ and } G \supset B.$$

Remark 1.6. 1. The above examples show that the parameter p in the definition of $\Psi_{p,G}$ is essential. The space $\Psi_{p,G}$ in the particular case $p = 2$ was introduced in 1981 by Dubinski [Dub82] (Dubinski used the notation $H^\infty(G)$ for test functions and $H^{-\infty}(G)$ for distributions). For arbitrary $p \in [1,\infty]$ the space $\Psi_{G,p}(\mathbb{R}^n)$ and its dual were introduced in 1997 in [Uma97, Uma98].

2. How $\Psi_{G,p}(\mathbb{R}^n)$ is related to other spaces of test functions? First, it is obvious by construction, that $\Psi_{G,p}(\mathbb{R}^n) \cap \mathfrak{D}(\mathbb{R}^n) = \emptyset$ for any $p \geq 1$ and any $G \subseteq \mathbb{R}^n$. A function with compact support cannot have Fourier transform with compact support. Secondly, the space $\Psi_{G,p}(\mathbb{R}^n)$ is a topological subspace of $\mathscr{E}(\mathbb{R}^n)$. This follows from the fact that any entire function is infinitely differentiable, and the convergence in $\Psi_{G,p}(\mathbb{R}^n)$ implies the convergence in $\mathscr{E}(\mathbb{R}^n)$. To show the latter we notice that if $\varphi_n \in \Psi_{G,p}(\mathbb{R}^n)$ and $\|\varphi_n\|_{L_p} \to 0$, then $\varphi_n(x) \to 0$ for every point $x \in \mathbb{R}^n$. Therefore $\varphi_n \to 0$ uniformly on every compact $K \subset \mathbb{R}^n$. Since $\varphi_n, n \in \mathbb{N}$, are entire functions, it follows that $D^\alpha \varphi_n \to 0$ locally uniformly for all multi-indices α. However, the embedding $\Psi_{G,p}(\mathbb{R}^n) \subset \mathscr{E}(\mathbb{R}^n)$ may not be dense. The denseness of this and other embeddings will be discussed in Section 1.11. Finally, one can see that there is a function $\varphi \in \mathscr{G}$ such that $\varphi \notin \Psi_{G,p}(\mathbb{R}^n)$, and vise versa, there is a function $\psi \in \Psi_{G,p}(\mathbb{R}^n)$ such that $\psi \notin \mathscr{G}$. Indeed, a function $\varphi \in C_0^\infty(\mathbb{R}^n)$, $\varphi(0) = 1$, is also a function in $\mathscr{G}$. On the other hand, $\varphi \notin \Psi_{G,p}(\mathbb{R}^n)$, since any analytic function with compact support is identically zero. Conversely, as we have seen above, the function $\psi(x) = \sin(x)/x \in \Psi_{G,p}(\mathbb{R})$ for all $p > 1$. However, obviously, $\psi \notin \mathscr{G}$.

Now we turn our attention to the topological structure of $\Psi_{G,p}(\mathbb{R}^n)$. This space of test functions can be represented as an inductive limit of a sequence of Banach spaces $\Psi_{N,p}$, for instance, defined as follows. Let

$$G_N = \cup_{k=0}^N g_k, \ \kappa_N(\xi) = \sum_{k=0}^N \phi_k(\xi). \tag{1.89}$$

It is clear that $G_N \subset G_{N+1}, N = 0,1,\ldots$, and $G_N \to G$ when $N \to \infty$.

Definition 1.23. Denote by $\Psi_{N,p}$ the set of functions $\varphi \in L_p(\mathbb{R}^n)$, satisfying the following conditions:

1. $\operatorname{supp} F\varphi \subset G_N$;
2. $\operatorname{supp} F\varphi \cap \operatorname{supp} \phi_j = \emptyset, j > N$;
3. $p_N(\varphi) = \|F^{-1}[\kappa_N F[\varphi]]\|_{L_p} < \infty$.

One can easily verify that $\Psi_{N,p}$ is a Banach space for each $N \in \mathbb{N}$ with norm $\|\varphi|\Psi_{N,p}\| = p_N(\varphi)$. We want to show that $\Psi_{G,p}(\mathbb{R}^n)$ can be represented as the inductive limit of the sequence of Banach spaces $\Psi_{N,p}$. First we prove the following assertion.

Proposition 1.27. *The following continuous embeddings hold:*

1. $\Psi_{N,p} \subset \Psi_{N+1,p}$, $\forall N \geq 0$;
2. $\Psi_{N,p} \subset L_p(\mathbb{R}^n)$, $N \geq 0$.

Proof. Let $\varphi \in \Psi_{N,p}$, which implies $\operatorname{supp} F[\varphi] \subset G_N$. Hence, $\operatorname{supp} F[\varphi] \subset G_{N+1}$. Since $\operatorname{supp} \phi_{N+1,p} \cap \operatorname{supp} F[\varphi] = \emptyset$, we obtain

$$\begin{aligned} p_{N+1}(\varphi) &= \|F^{-1}[\kappa_{N+1} F[\varphi]]\|_{L_p} \\ &\leq \|F^{-1}[\kappa_N F[\varphi]]\|_{L_p} + \|F^{-1}[\phi_{N+1} F[\varphi]]\|_{L_p} \\ &= p_N(\varphi). \end{aligned}$$

Since $\varphi \in \Psi_{N,p}$, then its Fourier transform has a compact support. Furthermore, $\operatorname{supp} F\varphi$ has empty intersection with each g_k, $k > N$. Therefore, one has

$$\begin{aligned} \|\varphi\|_{L_p} &= \left\| F^{-1}\left[\sum_{k=0}^{\infty} \phi_k F[\varphi]\right]\right\|_{L_p} = \left\| F^{-1}\left[\sum_{k=0}^{N} \phi_k F[\varphi]\right]\right\|_{L_p} \\ &= \|F^{-1}[\kappa_N[\varphi]]\|_{L_p} = p_N(\varphi) < \infty. \end{aligned}$$

Proposition 1.28. $\Psi_{G,p}(\mathbb{R}^n) = ind \lim_{N\to\infty} \Psi_{N,p}$.

Proof. We show that the topology of the inductive limit agrees with the topology of $\Psi_{G,p}(\mathbb{R}^n)$. In other words, we show that the inductive limit agrees with the limit in the sense of the convergence introduced in Definition 1.22. In fact, let $\varphi_m \to \varphi_0$ in $\Psi_{G,p}(\mathbb{R}^n)$. Then there exists a compact set $K \subset G$ such that $\operatorname{supp} \hat{\varphi}_m \subset K$. Since K has a finite number of nonempty intersections with the system $\{g_k\}$, there is a number N such that $K \subset G_N$. Moreover, κ_N is a multiplier in L_p for arbitrary $N \geq 0$. Therefore,

$$\begin{aligned} p_N(\varphi_m - \varphi_0) &= \|F^{-1}[\kappa_N F[\varphi_m - \varphi_0]]\|_{L_p} \\ &\leq C_{N,p}\|\varphi_m - \varphi_0\|_{L_p} \to 0, \text{ as } m \to \infty, \ \forall N \in \mathbb{N}. \end{aligned}$$

Vice versa, let $p_N(\varphi_m - \varphi_0) \to 0$, as $m \to \infty$, for all $N \geq N_0$. Then $\operatorname{supp} F[\varphi_m] \subset G_{N_0}$. Hence, there exists a compact set $K \subset G_{N_0} \subset G$ such that $\operatorname{supp} F[\varphi_m] \subset K$ for all $m \in \mathbb{N}$. Moreover, due to Theorem 1.27, one has

$$\|\varphi_m - \varphi_0\|_{L_p} \leq p_N(\varphi_m - \varphi_0) \to 0, m \to \infty, N \geq N_0.$$

Theorem 1.15. *$\Psi_{G,p}(\mathbb{R}^n)$ is complete.*

Proof. Let f_m, $m = 1,2,\dots$, be a fundamental sequence, i.e., for all $m = 1,2,\dots$ there exists a compact set $K \Subset G$ such that

i) $\mathrm{supp} f_m \subset K$, $m = 1,2,\dots$;
ii) $\|f_m - f_n\|_{L_p} \to 0, m,n \to \infty$.

Since $L_p(\mathbb{R}^n)$ is complete, then there exists $f_0 \in L_p(\mathbb{R}^n)$ such that $f_m \to f_0$ in the norm of $L_p(\mathbb{R}^n)$. Besides, it is easy to see that $\mathrm{supp} F[f_0] \subset K$.

Theorem 1.16. *The Fourier transform F is continuous as the mapping:*

$$F : \Psi_{G,p}(\mathbb{R}^n) \to \begin{cases} L_{p',comp}(G), & \text{if } 1 \leq p \leq 2; \\ H^{-s}_{2,comp}(G), & \text{if } 2 < p < \infty. \end{cases} \tag{1.90}$$

where s is any number satisfying the inequality $s > n\left(\frac{1}{2} - \frac{1}{p}\right)$.

Proof. The inclusion $f \in \Psi_{G,p}(\mathbb{R}^n)$ implies immediately that supp $F[f] \Subset G$. Besides, in accordance with Theorem 1.27, $\Psi_{G,p}(\mathbb{R}^n) \subset L_p(\mathbb{R}^n)$. If $p \in [1,2]$, then Young's inequality implies $F[f] \in L_{p'}(\mathbb{R}^n)$, where $p' = p/(p-1)$. Hence, in this case $F[f] \in L_{p',comp}(G)$. If $p > 2$, then in accordance with Theorem 1.10, one has

$$F[L_p(\mathbb{R}^n)] \hookrightarrow H^{-s}(\mathbb{R}^n),\ s > n(\frac{1}{2} - \frac{1}{p}).$$

Further, let $f_m \to 0$ in $\Psi_{G,p}(\mathbb{R}^n)$. Then there exists a total compact set $K \Subset G$ with supp $F[f_m] \subset K$ for every $m \in \mathbb{N}$. If $1 \leq p \leq 2$, then using Young's inequality again, we obtain

$$\|F[f_m]\|_{L_{p'}} \leq C_p \|f_m\|_{L_p} \leq C_p\, p_N(f_m) \to 0, \quad m \to \infty,$$

for all $N \geq 0$. This implies continuity of mapping (1.90) in the case $1 \leq p \leq 2$. Similarly, one can readily see that F is continuous in the case $p > 2$, as well. Indeed, in this case due to Theorem, we have the estimate

$$\|F f_m\|_{H^{-s}(\mathbb{R}^n)} \leq C_p \|f_m\|_{L_p} \leq C_p\, p_N(f_m),$$

which implies the continuity of the mapping (1.90) in the case $p > 2$, as well.

Denote the right side of (1.90) by $F[\Psi_{G,p}](G)$. Theorem 1.16 implies the following assertion (cf. Proposition 1.16):

Proposition 1.29. *A sequence $\varphi_n \to 0$ as $n \to \infty$ in $\Psi_{G,p}(\mathbb{R}^n)$ if and only if $F[\varphi_n] \to 0$ as $n \to \infty$ in $F[\Psi_{G,p}](G)$.*

Due to the relation $F^{-1} = (2\pi)^{-n}SF$ (see (1.31)), or the same $F^{-1}[\varphi](x) = (2\pi)^{-n}F[\varphi](-x)$, Theorem 1.16 implies also

$$F^{-1} : \Psi_{G,p} \to \begin{cases} L_{p',comp}(-G), & \text{if } 1 \le p \le 2; \\ H^{-s}_{2,comp}(-G), & \text{if } 2 < p < \infty. \end{cases} \tag{1.91}$$

Theorem 1.16 shows that unlike $\mathscr{G}$ the space $\Psi_{G,p}(\mathbb{R}^n)$ is not invariant with respect to the Fourier transform F. It follows from the statement below that, in general, $\Psi_{G,p}(\mathbb{R}^n)$ is not closed with respect to the convolution operation as well.

Theorem 1.17. *Let $\varphi_1 \in \Psi_{G,p_1}(\mathbb{R}^n)$ and $\varphi_2 \in \Psi_{G,p_2}(\mathbb{R}^n)$. Then $\varphi_1 * \varphi_2 \in \Psi_{G,p}(\mathbb{R}^n)$, where p satisfies the equation $1 + 1/p = 1/p_1 + 1/p_2$.*

Proof. The equality $F[\varphi_1 * \varphi_2](\xi) = F[\varphi_1](\xi) \cdot F[\varphi_2](\xi)$ immediately implies that $supp F[\varphi_1 * \varphi_2] \Subset G$. The fact that p satisfies the equation $1 + 1/p = 1/p_1 + 1/p_2$, follows from Theorem 1.5.

We note that the space $\Psi_{G,p}(\mathbb{R}^n)$ is closed with respect to the convolution operation only if $p = 1$.

Definition 1.24. Let $1 < p < \infty$. We denote by $\Psi'_{-G,p'}(\mathbb{R}^n), p' = p(p-1)^{-1}$, the space of linear continuous functionals defined on $\Psi_{G,p}(\mathbb{R}^n)$. A sequence $f_m \in \Psi'_{-G,p'}(\mathbb{R}^n)$ is said to converge weakly to $f_0 \in \Psi'_{-G,p'}(\mathbb{R}^n)$ if for arbitrary $\varphi \in \Psi_{G,p}(\mathbb{R}^n)$ the sequence $f_m(\varphi)$ converges to $f_0(\varphi)$. The elements of the space $\Psi'_{-G,p'}(\mathbb{R}^n)$ will be called *ψ-distributions.* The duality of a pair of elements $f \in \Psi'_{-G,p'}(\mathbb{R}^n)$ and $\varphi \in \Psi_{G,p}(\mathbb{R}^n)$ we will denote by $< f, \varphi >$. Two elements f and g in $\Psi'_{-G,p'}(\mathbb{R}^n)$ coincide if and only if $< f, \varphi > = < g, \varphi >$ for all $\varphi \in \Psi_{G,p}(\mathbb{R}^n)$.

It follows from Propositions 1.23 and 1.28 that

$$\Psi'_{-G,p'}(\mathbb{R}^n) = pr \lim_{N \to \infty} \Psi'_{N,p'},$$

where $\Psi'_{N,p}$ is dual to the Banach space $\Psi_{N,p}$ introduced in Definition 1.23. Proposition 1.22 implies that $\Psi'_{-G,p'}(\mathbb{R}^n)$ is a locally convex topological vector space.

Definition 1.25. The differentiation operator D^α (with multi-index α) on $\Psi'_{-G,p'}(\mathbb{R}^n)$ is defined by

$$< D^\alpha f, \varphi > \equiv (-1)^{|\alpha|} < f, D^\alpha \varphi >, \tag{1.92}$$

where $f \in \Psi'_{-G,p'}(\mathbb{R}^n)$ and φ is an arbitrary function in $\Psi_{G,p}(\mathbb{R}^n)$.

Proposition 1.30. *The differentiation operator D^α maps continuously the space $\Psi'_{-G,p'}(\mathbb{R}^n)$ into itself.*

Proof. Let $\varphi \in \Psi_{G,p}(\mathbb{R}^n)$. Then, obviously, $D^\alpha \varphi \in \Psi_{G,p}(\mathbb{R}^n)$ for each multi-index α. Therefore, the right-hand side of (1.92), and consequently its left-hand side define a linear continuous functional on $\Psi_{G,p}(\mathbb{R}^n)$. This yields $D^\alpha f \in \Psi'_{-G,p'}(\mathbb{R}^n)$.

Definition 1.26. Let $f \in \Psi'_{-G,p'}(\mathbb{R}^n)$ and $\psi \in \Psi_{G,q}$. Then the convolution $f * \psi$ is defined by $< f * \psi, \varphi > = < f, \psi * \varphi >$ for an arbitrary $\varphi \in \Psi_{G,r}$. Here p, q, and r are connected through the relation $1 + 1/p = 1/q + 1/r$. Due to Theorem 1.17 $\psi * \varphi \in \Psi_{G,p}(\mathbb{R}^n)$. Therefore, the convolution $f * \psi$ is well-defined.

Theorem 1.18. *Let $f \in \Psi'_{-G,p'}(\mathbb{R}^n)$ and $\psi \in \Psi_{G,q}(\mathbb{R}^n)$. Then $f * \psi \in \Psi'_{-G,r'}(\mathbb{R}^n)$, where r' is the conjugate of r, which satisfies the equation $1/r = 1 + 1/p - 1/q$.*

Proof. By the above definition the convolution of $f \in \Psi'_{-G,p'}(\mathbb{R}^n)$ and $\psi \in \Psi_{G,q}(\mathbb{R}^n)$ is

$$< f * \psi, \varphi > = < f, \psi * \varphi >, \quad \forall \varphi \in \Psi_{G,r}(\mathbb{R}^n).$$

Since $\psi * \varphi \in \Psi_{G,p}(\mathbb{R}^n)$ (Theorem 1.17), the right-hand side is well defined for all $\varphi \in \Psi_{G,r}(\mathbb{R}^n)$ and ψ fixed. This means that the left-hand side is well defined for all $\varphi \in \Psi_{G,r}(\mathbb{R}^n)$, implying $f * \psi \in \Psi'_{-G,r'}(\mathbb{R}^n)$.

Since $\Psi_{G,p}(\mathbb{R}^n)$ is not invariant with respect to the Fourier transform, one has to be careful in extending the Fourier transform F to $\Psi'_{-G,p'}(\mathbb{R}^n)$. First, if $f, \varphi \in L_2(\mathbb{R}^n)$, then the following Parseval relation

$$(F[f], \varphi) = \int_{\mathbb{R}^n} \int_{\mathbb{R}^n} f(x)\varphi(\xi)e^{ix\xi}dxd\xi = (f, F[\varphi]) \tag{1.93}$$

holds. Here $(\cdot, \cdot)$ is the inner product of $L_2(\mathbb{R}^n)$.

Now suppose that $f \in \Psi'_{-G,p'}(\mathbb{R}^n)$. Then, in order to have a well-defined quantity on the right-hand side of (1.93), the Fourier transform of φ must be in $\Psi_{G,p}(\mathbb{R}^n)$. Denoting $\phi = F[\varphi]$, one can write $< F[f], \varphi > = < f, \phi >$, where $\phi \in \Psi_{G,p}(\mathbb{R}^n)$ is an arbitrary function, and $\varphi = F^{-1}[\phi]$. By virtue of relations (1.91) we have $\varphi \in X_{p,comp}(G)$, where

$$X_{p,comp}(G) = \begin{cases} L_{p',comp}(-G), & \text{if } 1 \le p \le 2; \\ H^{-s}_{2,comp}(-G), & \text{if } 2 < p < \infty. \end{cases} \tag{1.94}$$

These facts lead to the following definition of the Fourier transform in the space of ψ-distributions.

Definition 1.27. The Fourier transform of a ψ-distribution $f \in \Psi'_{-G,p'}(\mathbb{R}^n)$ is an element $F[f]$ defined by the relation

$$< F[f], \varphi > = < f, F[\varphi] >, \tag{1.95}$$

for all $\varphi \in X_{p,comp}(G)$.

Sometimes we will use the following form of the Fourier transform for a ψ-distribution f, which is convenient from the calculations point of view:

$$< F[f](\xi), \varphi(\xi) > = < f(x), \phi(x) >,$$

where, as was mentioned above, $\phi = F[\varphi] \in \Psi_{G,p}(\mathbb{R}^n)$, or equivalently,

$$< F[f], F[\phi] > = (2\pi)^n < f, \phi >, \quad \forall \phi \in \Psi_{G,p}(\mathbb{R}^n). \tag{1.96}$$

Theorem 1.19. *Let $1 \leq p < \infty$ and (p, p') be a conjugate pair. Then the Fourier transform F acts continuously as the mapping*

$$F : \Psi'_{-G,p'}(\mathbb{R}^n) \to L_{p,loc}(G) \tag{1.97}$$

in the weak topology.

Proof. Let $1 \leq p \leq 2$. Then the proof of this assertion follows immediately from Theorem 1.16 and Proposition 1.25. Suppose that $p > 2$ and $f \in \Psi'_{-G,p'}(\mathbb{R}^n)$. Then again using Theorem 1.16 and the duality relationship stated in Proposition 1.25, one obtains that $F[f] \in H^s_{2,loc}(G)$, where $s > n\left(\frac{1}{2} - \frac{1}{p}\right)$. Let K_m, $m \in \mathbb{N}$, be a sequence of expanding compact sets such that $\lim_{m\to\infty} K_m = \cup_{m=1}^{\infty} K_m = G$. We have $F[f] \in H^s_2(K_m)$ for each m. Therefore, due to Theorem 1.14 it follows that $F[f] \in L_p(K_m)$ for each m, that is $F[f] \in L_{p,loc}(G)$.

To prove continuity of mapping (1.97), assume that the sequence $f_m \in \Psi'_{-G,p'}(\mathbb{R}^n)$ weakly converges to zero in $\Psi'_{-G,p'}(\mathbb{R}^n)$, that is for an arbitrary element $\phi \in \Psi_{G,p}(\mathbb{R}^n)$, we have $< f_m, \phi > \to 0$, as $m \to \infty$. Then relation (1.95) implies that $< F[f_m], \varphi > \to 0$ for all $\varphi \in X_{p,comp}(G)$. In the case $1 \leq p \leq 2$, due to definition (1.94) of $X_{p,comp}$, the latter immediately implies the weak convergence of the sequence $F[f_m]$ to zero in $L_{p,loc}(G)$. If $p > 2$, then again due to (1.94) we have the convergence $< F[f_m], \varphi > \to 0$ as $m \to \infty$ for an arbitrary element $\varphi \in H^{-s}_{2,comp}(-G)$, where $s > n(1/2 - 1/p)$. Further, Theorem 1.14 by duality implies the inclusion $L_{p',comp}(G) \subset H^{-s}_{2,comp}(G)$. Hence, $< F[f_m], \varphi > \to 0$ for all $\varphi \in L_{p',comp}(G)$, yielding $F[f_m] \to 0$, as $m \to \infty$, in the weak topology of $L_{p,loc}(G)$.

As a corollary we obtain the following assertion:

Proposition 1.31. *A sequence $f_n \to 0$ as $n \to \infty$ in the space $\Psi'_{-G,p'}(\mathbb{R}^n)$ if and only if $F[f_n] \to 0$ as $n \to \infty$ in $L_{p,loc}(G)$.*

Proposition 1.32. *Let $f_n \in \Psi'_{-G,p'}(\mathbb{R}^n)$ and $\psi \in \Psi_{G,q}(\mathbb{R}^n)$. Further, let r satisfy the equation $1/r = 1 + 1/p - 1/q$ and r' be the conjugate number to r. Then the sequence $f_n * \psi$ converges to $f * \psi$ in $\Psi'_{-G,r'}(\mathbb{R}^n)$, if and only if the sequence $F[f_n](\xi) \cdot F[\psi](\xi)$ converges to $F[f](\xi) \cdot F[\psi](\xi)$ as $n \to \infty$ in $L_{r,loc}(G)$.*

Proof. Due to Theorem 1.18, $f_n * \psi \in \Psi'_{-G,r'}(\mathbb{R}^n)$. Now the statement of the proposition follows from Proposition 1.31.

Now we prove the structure theorem for ψ-distributions.

Theorem 1.20. *Let $1 < p < \infty$ and p' be its conjugate. For each distribution $f \in \Psi'_{-G,p'}(\mathbb{R}^n)$ there exist a function $A(\xi)$ analytic in G and a function f_0 with the Fourier transform $F[f_0] \in L_p(G)$, such that the representation*

$$f(x) = \sum_{|\alpha|=0}^{\infty} \frac{D^\alpha A(\xi_0)}{\alpha!}(D-\xi_0)^\alpha f_0(x), \quad x \in \mathbb{R}^n, \tag{1.98}$$

holds for any fixed $\xi_0 \in G$.

Proof. Let $f \in \Psi'_{-G,p'}(\mathbb{R}^n)$. Then, in accordance with Theorem 1.19, for the Fourier transform of f one has $F[f] \in L_{p,loc}(G)$. The Fourier transform of the distribution f in (1.98) has the form

$$F[f](\xi) = \sum_{|\alpha|=0}^{\infty} \frac{D^\alpha A(\xi_0)}{\alpha!}(\xi-\xi_0)^\alpha F[f_0](\xi) = A(\xi)F[f_0](\xi), \tag{1.99}$$

for all ξ in a neighborhood of any point $\xi_0 \in G$. More precisely, equation (1.99) holds for all $\xi \in \{\xi : |\xi - \xi_0| < dist(\xi_0, \partial G)\}$, where $dist(\xi_0, \partial G)$ is the distance between ξ_0 and the boundary of G. Therefore, the theorem will be proved, if one proves the following statement:

Lemma 1.2. *Every function $g \in L_{p,loc}(G)$ can be represented in the form $g(\xi) = A(\xi)g_0(\xi)$, where $A(\xi)$ is an analytic function in G and $g_0(\xi) \in L_p(G)$.*

Proof of Lemma. We recall that for G there exists a locally finite covering $\{g_k\}_{k=1}^{\infty}$, that is $G = lim_{k\to\infty} g_k = \cup_{k=1}^{\infty} g_k$, with compact sets $g_k \Subset G$. Let $h \in L_{p,loc}(G)$ be fixed. We will construct functions $A(\xi)$ and $h_0(\xi)$, which will satisfy all the requirements of the statement. We choose g_k and $A(\xi)$ so that the following condition is satisfied:

$$\sup_{\xi \in g_k} \frac{1}{|A(\xi)|^p} \leq \frac{C}{M_k k^2},$$

where

$$M_k = \int_{g_k} |g(\xi)|^p d\xi.$$

Such a function $A(\xi)$ exists since analytic functions may grow arbitrarily fast when it approaches to the boundary of domain. Further, defining the function

$$g_0(\xi) = \frac{g(\xi)}{A(\xi)}, \quad \xi \in G, \tag{1.100}$$

one obtains

$$\int_G |g_0(\xi)|^p d\xi = \sum_{k=1}^{\infty} \int_{g_k} \left| \frac{g(\xi)}{A(\xi)} \right|^p d\xi \leq \sum_{k=1}^{\infty} \sup_{\xi \in g_k} \frac{1}{|A(\xi)|^p} \int_{g_k} |g(\xi)|^p d\xi$$
$$\leq C \sum_{k=1}^{\infty} \frac{M_k}{M_k k^2} = C \sum_{k=1}^{\infty} \frac{1}{k^2} < \infty.$$

From equation (1.100) the desired result follows.

Hence, for $F[f] \in L_{p,loc}(G)$ there exist an analytic function $A(\xi)$ defined on G and $h_0 \in L_p(G)$, such that

$$F[f](\xi) = A(\xi)h_0(\xi) = \sum_{|\alpha|=0}^{\infty} \frac{D^\alpha A(\xi_0)}{\alpha!}(\xi - \xi_0)^\alpha h_0(\xi), \tag{1.101}$$

for $\xi \in B_r(\xi_0)$, $r < dist(\xi_0, \partial G)$, $\xi_0 \in G$. The proof of the theorem is complete.

Remark 1.7. 1. Theorem 1.20 requires that $F[f_0] \in L_p(G)$. Extending $F[f_0]$ by zero outside of G, one can assume $F[f_0] \in L_p(\mathbb{R}^n)$ (the extension is again denoted by $F[f_0]$). If $1 \leq p \leq 2$, due to Hausdorff-Young's Theorem (Theorem 1.4), for $F[f_0] \in L_p(\mathbb{R}^n)$ one can find the inverse Fourier transform $F^{-1}[F[f_0]] = f_0 \in L_{p'}(\mathbb{R}^n)$, where $p' = p/(p-1)$ is the conjugate number to p. Since $1 \leq p \leq 2$, its conjugate satisfies $p' \geq 2$. Thus, if $1 \leq p \leq 2$, then for a distribution $f \in \Psi'_{-G,p'}(\mathbb{R}^n)$ there exist an analytic function $A(\xi)$ defined on G and a function $f_0 \in L_{p'}(\mathbb{R}^n)$ with $\operatorname{supp} F[f_0] \subset G$ (note that $\operatorname{supp} F[f_0]$ is not necessarily compact), such that the representation

$$f(x) = \sum_{|\alpha|=0}^{\infty} \frac{D^\alpha A(\xi)}{\alpha!}(D - \xi)^\alpha f_0(x) \tag{1.102}$$

holds. Thus, for $1 \leq p \leq 2$, in Theorem 1.20 one can assume that $f_0 \in L_{p'}(\mathbb{R}^n)$ with $\operatorname{supp} F[f_0] \subset G$.
As the following example shows this argument, in general, is not valid if $p > 2$. Indeed, consider a ψ-distribution

$$f(x) = \sum_{|\alpha|=0}^{\infty} \frac{1}{\alpha!} D^\alpha \delta_0(x).$$

This distribution belongs to $\Psi'_{-G,p'}(\mathbb{R}^n)$ with $G \ni 0$. However, $F[f](\xi) = e^\xi 1(\xi)$, and $F[f_0](\xi) = 1(\xi) \notin L_p(\mathbb{R}^n)$.
2. In representation (1.102) the order of the differential operator is not bounded from above. Therefore, in general, ψ-distributions *may have infinite order*, unlike tempered distributions and distributions with compact support.
3. In Section 2.3 we reformulate Theorem 1.20 in terms of pseudo-differential operators with singular symbols.

1.11 Denseness of $\Psi_{G,p}(\mathbb{R}^n)$ in Besov and Lizorkin-Triebel spaces

In this section we prove the denseness theorem of $\Psi_{G,p}(\mathbb{R}^n)$ in the Besov and Lizorkin-Triebel type spaces, and in particular in $L_p(\mathbb{R}^n)$. For a set $A \subset \mathbb{R}^n$ we denote by $\mu_n(A)$ its n-dimensional Lebesgue measure. First we prove the following denseness lemma.

Lemma 1.3. *(Denseness lemma) For the embedding*

$$\Psi_{G,p}(\mathbb{R}^n) \cap \mathscr{G} \hookrightarrow \mathscr{G} \tag{1.103}$$

to be dense the condition $\mu_n(\mathbb{R}^n \setminus G) = 0$ is necessary if $1 < p < \infty$, sufficient if $2 \le p < \infty$.

Proof. Sufficiency: Let the complement of G in $\mathbb{R}^n$ have the n-dimensional measure zero, i.e., $m_n(\mathbb{R}^n \setminus G) = 0$. Let $f \in \mathscr{G}$ and the collection $\{\psi_j(\xi)\} \in \Phi_1$ (see Definition 1.20 for Φ_N). Namely, let the functions $\psi_0,\ \psi \in \mathscr{G}$ satisfy the conditions:

1. $0 \le \psi_0,\ \psi \le 1$;
2. $\text{supp}\ \psi_0 \subset \{|\xi| \le 2\}$, $\text{supp}\ \psi \subset \{1 \le |\xi| \le 4\}$;
3. $\sum_{j=0}^{\infty} \psi_j(\xi) = 1,\ \psi_j(\xi) = \psi(2^{1-j}\xi),\ j = 0,1,2\dots$

Then due to (1.88) the sequence of functions

$$f_m(x) = \sum_{j=0}^{m} F^{-1}\left[\psi_j F[f]\right](x), \quad m = 0,1,\dots$$

tends to f in the norm of $L_p(\mathbb{R}^n)$ as $m \to \infty$. Obviously,

$$\text{supp}\ F[f_m] \subset \{|\xi| \le 2^{m+1}\} = B_m.$$

So we can assume $\text{supp} F[f] \subset B_m$. As an approximating sequence we take the sequence of functions

$$f_N(x) = F^{-1}\left[\kappa_N F[f]\right](x), \quad N = 0,1,\dots,$$

where functions $\kappa_N(\xi)$ are defined in (1.89). By construction, $f_N \in \Psi_{G,p}(R^n)$ for all $N \ge 0$. We have

$$f(x) - f_N(x) = F^{-1}\Big[(1-\kappa_N)F[f]\Big](x) = F^{-1}\left[\gamma_N F[f]\right](x),$$

where $\gamma_N(\xi) = 1 - \kappa_N(\xi)$. Putting $w_{m,N} = \text{supp}\ (\gamma_N \cdot \chi_{B_m})$, where $\chi_{B_m}(\xi)$ is the characteristic function of the ball B_m, we see that the set $w_{m,N}$ is a compact subset of $(\mathbb{R}^n \setminus G_N) \cap B_m$. Evidently, $\mu_n(w_{m,N}) \to 0$ as $N \to \infty$ for each fixed $m = 0,1,\dots,$ since $\mu_n((\mathbb{R}^n \setminus G_N) \cap B_m) \to 0$ for $N \to \infty$.

To prove the lemma we have to show that the sequence of functions $F^{-1}\left[\gamma_N F[f]\right]$ converges to zero in the norm of $L_p(\mathbb{R}^n)$ for all $p \in [2,\infty)$ when $N \to \infty$. Since the functions $\gamma_N \in C^\infty(G)$ for all $N = 0,1,\ldots,$ and $|\gamma_N(\xi)| \le 1,\ \xi \in G$, we have $\gamma_N F f \in L_{p'}(\mathbb{R}^n)$, where $p' = p/(p-1)$, the conjugate to p. Using the Housdorff-Young inequality (Theorem 1.4), we obtain

$$\|F^{-1}[\gamma_N F[f]]\|_p^{p'} \le C(p')\|\gamma_N F[f]\|_{p'}^{p'} = C_{p'} \int_{\mathbb{R}^n} \left|\gamma_N(\xi)F[f](\xi)\right|^{p'} d\xi$$

$$\le C_{p'} \int_{w_{m,N}} \left|F[f](\xi)\right|^{p'} d\xi. \tag{1.104}$$

The integral on the right-hand side tends to 0 as $N \to \infty$, since $F[f] \in \mathscr{G}$, and therefore,

$$\int_{w_{m,N}} \left|F[f](\xi)\right|^{p'} d\xi \le C\mu(w_{m,N}) \to 0, \quad N \to \infty. \tag{1.105}$$

Estimates (1.104) and (1.105) imply $F^{-1}[\gamma_N F[f]] \to 0$ as $N \to \infty$ in the norm of $L_p(\mathbb{R}^n)$.

Necessity: Suppose that $m_n(\mathbb{R}^n \setminus G) > 0$. Then there exists a compact $V \subset \mathbb{R}^n \setminus G$ such that $m_n(V) > 0$. Let us take a function $f \in L_p,\ f \ne 0$ and let $\operatorname{supp} Ff = V$. Then for any $v \in \Psi_{G,p}(\mathbb{R}^n)$ we have $\operatorname{supp} Ff \cap \operatorname{supp} Fv = \emptyset$. For $p = 2$ this yields

$$\|f - v\|_2^2 = (f - v, f - v)_{L_2} = \|f\|_2^2 + \|v\|_2^2 \ge \|f\|_2^2 > 0.$$

For $1 < p < 2$ we obtain the analogous estimate using the Hausdorff-Young lemma. In fact,

$$(2\pi)^{\frac{n}{q}}\|f - v\|_p \ge \|Ff - Fv\|_q = (\|Ff\|_q^q + \|Fv\|_q^q)^{1/q} \ge \|Ff\|_q > 0.$$

For $2 < p < \infty$ and for an arbitrary fixed $g \in L_q(\mathbb{R}^n),\ \|g\|_q = 1,\ q = p/(p-1)$, we have

$$\|f - v\|_p = \sup_{h \in L_p, h \ne 0} \frac{|(f - v, h)|}{\|h\|_q} \ge |(f - v, g)|.$$

We choose $g \in L_q(\mathbb{R}^n)$ (note that $q \in (1,2)$) satisfying the following conditions: (i) $\operatorname{supp} Fg = V$; (ii) $Fg \in H_2^\beta(\mathbb{R}^n),\ \beta > n(\frac{1}{2} - \frac{1}{p})$; (iii) $\int_{\mathbb{R}^n} f(x)g(x)dx \ne 0$. It is not difficult to verify that such function exists. According to the Hausdorff-Young lemma the Fourier transform of g belongs to $L_p(\mathbb{R}^n)$ and $\|Fg\|_p \le (2\pi)^{\frac{n}{p}}\|g\|_q$. For any $v \in \Psi_{G,p}(\mathbb{R}^n)$ its Fourier transform has a compact support in G by the definition of the space $\Psi_{G,p}(\mathbb{R}^n)$. Moreover, the Fourier transform of v is an element of a Sobolev space $H_2^{-\beta}(\mathbb{R}^n)$ for some $\beta > n(\frac{1}{2} - \frac{1}{p})$, because $v \in L_p(\mathbb{R}^n)$ (see Theorem 1.10). Hence

$$(g, v) = \int_{R^n} g(x)v(x)dx = \frac{1}{(2\pi)^n} < Fv, Fg >,$$

where $< Fv, Fg >$ means the value of the functional $Fv \in H_2^{-\beta}(\mathbb{R}^n)$ on Fg. It is well defined due to $Fg \in H_2^{\beta}(\mathbb{R}^n)$. Moreover, $< Fv, Fg > = 0$, since $\text{supp}\, Fv \cap \text{supp}\, Fg \subset G \cap V = \emptyset$. Hence, $\|f - v\|_p \geq |(f,g)| > 0$.

Since $\mathscr{G}$ is dense in the spaces $B^s_{pq}(\mathbb{R}^n)$ and $F^s_{pq}(\mathbb{R}^n)$ if $q < \infty$, Lemma 1.3 immediately implies the following theorem.

Theorem 1.21. *Let* $1 < p, q < \infty$, $-\infty < s < +\infty$.

1. *Then the following embeddings*

$$\Psi_{G,p}(\mathbb{R}^n) \hookrightarrow B^s_{pq}(\mathbb{R}^n) \hookrightarrow \Psi'_{-G,p'}(\mathbb{R}^n), \tag{1.106}$$

$$\Psi_{G,p}(\mathbb{R}^n) \hookrightarrow F^s_{pq}(\mathbb{R}^n) \hookrightarrow \Psi'_{-G,p'}(\mathbb{R}^n) \tag{1.107}$$

are continuous.

2. *Moreover, if* $\mathbb{R}^n \setminus G$ *has the n-dimensional zero measure and* $2 \leq p < \infty$, *then the left embeddings in (1.106) and (1.107) are dense. Conversely, the denseness of the left embeddings in (1.106) and (1.107) implies that* $\mathbb{R}^n \setminus G$ *has the n-dimensional zero measure, that is* $\mu_n(\mathbb{R}^n \setminus G) = 0$.

Proof. We prove only Part 1. Part 2 of this theorem, as was noted above, is the direct implication of the denseness lemma (Lemma 1.3). Let $f \in \Psi_{G,p}(\mathbb{R}^n)$ and $s \geq 0$. Assume that $\text{supp}\, F[f] \subset G_N \subset \{|\xi| \leq 2^{N_1}\}$, where N and N_1 are positive integers. Since the functions $\varphi_j(\xi)$, $j = 0, \ldots, N_1$, are multipliers in $L_p(\mathbb{R}^n)$, $1 < p < \infty$, there exist constants $C_{j,p} > 0$ such that $\|F^{-1}[\varphi_j F[f]]\|_{L_p} \leq C_{j,p}\|f\|_{L_p}$, $j = 0, \ldots, N_1$. These estimates and Proposition 1.27 imply

$$\|f|B^s_{pq}\| = \left(\sum_{j=0}^{N_1} 2^{sjq} \|F^{-1}[\varphi_j F[f]]\|^q_{L_p} \right)^{\frac{1}{q}} \leq C p_N(f).$$

Thus, the embedding $\Psi_{G,p}(\mathbb{R}^n) \hookrightarrow B^s_{pq}(\mathbb{R}^n)$ is continuous. The latter also implies the continuity of the embedding $\Psi_{G,p}(\mathbb{R}^n) \hookrightarrow F^s_{pq}(\mathbb{R}^n)$, since according to Proposition 1.26 the embedding $B^s_{pq}(\mathbb{R}^n) \hookrightarrow F^s_{pq}(\mathbb{R}^n)$ is continuous. Now the continuity of the right embeddings in (1.106) and (1.107) follows by duality.

Remark 1.8. What concerns the case $1 < p < 2$, then in the particular case, when $\mathbb{R}^n \setminus G$ is a quasi-polygonal set the space $\Psi_{G,p}(\mathbb{R}^n)$ is dense in the spaces $B^s_{pq}(\mathbb{R}^n)$ and $F^s_{pq}(\mathbb{R}^n)$ for all $s \in \mathbb{R}$ and $1 < q < \infty$. For details see Section "Additional notes."

1.12 Lizorkin type spaces

It is known that the classical Neumann boundary value problem for the Laplace equation in a bounded domain has a unique solution if and only if the boundary function is orthogonal to 1 (a polynomial of order 0). This is natural if the boundary operator is a (pseudo) differential operator of order ≥ 1 with the homogeneous symbol. The spaces of functions orthogonal to polynomials provide solution spaces for boundary value problems with such boundary operators. Lizorkin [Liz63, Liz69] studied some spaces of functions orthogonal to monomials of selected variables with the induced topology of $\mathcal{G}$ and their dual spaces. Therefore, the spaces of functions orthogonal to a collection of monomials with an appropriate topology will be called *Lizorkin type spaces*. In this section we construct Lizorkin type spaces, originating from Besov and Lizorkin-Triebel spaces endowed with topologies of strict inductive or projective limits of sequences of Banach spaces. These spaces will be exploited in Chapter 5 in the context of boundary value problems with special fractional order boundary operators.

Lizorkin spaces $\Psi(\mathbb{R}^n)$ and $\Phi(\mathbb{R}^n)$ are defined as follows. Let $\Psi(\mathbb{R}^n)$ be the set of functions $v \in \mathcal{G}$ satisfying the conditions:

$$D^\gamma v(0) = 0, \quad |\gamma| = 0, 1, \ldots, \tag{1.108}$$

and $\Phi(\mathbb{R}^n)$ is the Fourier pre-image of the space $\Psi(\mathbb{R}^n)$. We assume that both spaces have topologies induced from $\mathcal{G}$. Hence, $\Phi(\mathbb{R}^n) = \{\varphi \in \mathcal{G} : F[\varphi] \in \Psi(\mathbb{R}^n)\}$ and it follows from (1.108) that any function $\varphi \in \Phi(\mathbb{R}^n)$ satisfies the orthogonality conditions

$$\int_{\mathbb{R}^n} x^\gamma \varphi(x) dx = 0, \quad |\gamma| = 0, 1, \ldots.$$

We need also the spaces $\Phi^{'}(\mathbb{R}^n)$ and $\Psi^{'}(\mathbb{R}^n)$, which are topologically dual to $\Phi(\mathbb{R}^n)$ and $\Psi(\mathbb{R}^n)$, respectively. For $\Psi^{'}(\mathbb{R}^n)$ the factorization $\Psi^{'}(\mathbb{R}^n) = \mathcal{G}^{'}/\Psi_0^{'}$ holds, where $\Psi_0^{'}$ is the set of all functionals of the form

$$v(x) = \sum_{|\gamma|=0}^{M_v} a_\gamma D^\gamma \delta(x).$$

Here δ is the Dirac distribution and M_v is a finite number depending on v. Analogously, $\Phi^{'}(\mathbb{R}^n)$ can be represented as the quotient space $\mathcal{G}^{'}/P$, where P is the set of all polynomials on $\mathbb{R}^n$. The Lizorkin spaces were studied in [Liz63, Liz69], and more general Lizorkin type spaces in [Sam77, Sam82]. See more in Section "Additional Notes."

Now we introduce other variations of Lizorkin type spaces, which will be defined on the base of Besov and Lizorkin-Triebel spaces. In this section we assume that $s \geq 0$, $1 < p < \infty$, and $1 < q < \infty$.

Definition 1.28. Let a function $f \in B^s_{pq}(\mathbb{R}^n)$ has a compact support $\text{supp} f \subset \mathbb{R}^n$ and satisfies the following orthogonality conditions

$$(f, x^\beta) = \int_{\mathbb{R}^n} f(x) x^\beta dx = 0, \quad \text{for all } \beta : |\beta| \leq m-1.$$

We shall denote the class of such functions by $\overset{\circ}{B}{}^{s}_{pq,m}(\mathbb{R}^n)$. The topology of this space is defined below as a strict inductive limit of a sequence of Banach spaces.

Definition 1.29. Similarly, we introduce the class $\overset{\circ}{B}{}^{s',s_n}_{pq,m}(\mathbb{R}^n)$ of functions f in non-isotropic Besov space $B^{s',s_n}_{pq}(\mathbb{R}^n)$, $s' = (s_1,\dots,s_{n-1})$, with a support in the strip $\{|x_n| \le N\}$, $N > 0$, and satisfying the orthogonality conditions

$$\int_{-\infty}^{+\infty} x_n^k f(x) dx_n = 0, k = 0,\dots,m-1.$$

Let R_N, $N = 1,2,\dots$, be a sequence, such that $R_N \to \infty$. We introduce a sequence of Banach spaces X_N :

$$X_N = \{f \in \overset{\circ}{B}{}^{s}_{pq,m}(\mathbb{R}^n) : \operatorname{supp} f \subseteq B_{R_N}(0)\},$$

with the induced norm $\|f|X_N\| = \|f|B^s_{p,q}\|$. Obviously,

$$X_1 \subset X_2 \subset \cdots \subset X_N \subset \dots,$$

with the norms

$$\|f|X_1\| \ge \|f|X_2\| \cdots \ge \|f|X_N\| \ge \dots. \tag{1.109}$$

It is not hard to see that $\overset{\circ}{B}{}^{s}_{pq,m}(\mathbb{R}^n) = \cup_{N=1}^{\infty} X_N$ in the theoretical-set sense. Now taking into account the norm relations (1.109) we can define the inductive limit space

$$X_\infty = \operatorname{ind} \lim_{N\to\infty} X_N,$$

The space X_∞ coincides with $\overset{\circ}{B}{}^{s}_{pq,m}(\mathbb{R}^n)$, if the latter is endowed with the strict inductive topology of X_∞. Hence, $\overset{\circ}{B}{}^{s}_{pq,m}(\mathbb{R}^n)$ is a locally convex topological vector space. In a similar manner $\overset{\circ}{B}{}^{s',s_n}_{pq,m}(\mathbb{R}^n)$ also can be defined as a strict inductive limit of Banach spaces, thus being a locally convex topological vector space.

Theorem 1.22. *Let $f \in B^s_{pq}(\mathbb{R}^n)$, $s \ge 0$, and $\operatorname{supp} f \subset Q_{\overline{N}} \equiv \{|x_j| \le N_j\}, N_j > 0$. Then the inequality*

$$\|F^{\pm 1}[f]|C(B_1)\| = \sup_{\xi \in B_1} \left|F^{\pm 1}[f](\xi)\right| \le C_{\overline{N},p}\|f|B^s_{pq}\| \tag{1.110}$$

holds. Here B_1 is the unit ball in R^n_ξ with the center at the origin, and $C_{\overline{N},p}$ is a positive constant independent on f.

Proof. First suppose that $f \in L_p(\mathbb{R}^n)$ and has a compact support in $Q_{\overline{N}}$. Then $f \in \mathscr{E}(\mathbb{R}^n)$, that is a distribution with compact support. In accordance with the

Paley-Wiener-Schwartz theorem the Fourier transform $F[f]$ is an entire function. This function is bounded on $\mathbb{R}^n$. Indeed,

$$\left|F^{\pm 1}[f](\xi)\right| \leq \int\limits_{Q_{\overline{N}}} |f(x)|dx \leq |Q_{\overline{N}}|^{\frac{1}{p'}} \|f|L_p\|, \tag{1.111}$$

where $|Q_{\overline{N}}|$ is the volume of $Q_{\overline{N}}$, and p' is the conjugate number to p. Now estimate (1.110) follows from (1.111) taking into account the embedding $B^r_{pq}(\mathbb{R}^n) \hookrightarrow B^s_{pq}(\mathbb{R}^n)$ valid for any $r > s$ and $1 < p, q < \infty$.

Remark 1.9. Analogously, one can show that for the derivatives of order $\beta = (\beta_1, \ldots, \beta_n)$ of $F[f]$ the inequality

$$\|D^\beta F[f]|C(B_1)\| \leq C \prod_{j=1}^{n} N_j^{\beta_j + n/p} (p'\beta_j + 1)^{-1/p'} \|f|B^s_{pq}\| \tag{1.112}$$

holds with a conjugate pair (p, p').

Theorem 1.23. *Let $f \in B^{\circ s}_{pq,m}(\mathbb{R}^n)$. Then there exists a function $v \in B^{s+m}_{pq}(\mathbb{R}^n)$, such that*

(i) $F[f](\xi) = |\xi|^m F v(\xi)$;
(ii) $\|v|B^{s+m}_{pq}\| \leq C_1 \|f|B^s_{pq}\|$;
(iii) $\|F[v]|C(B_1)\| \leq C_2 \|f|B^s_{pq}\|$,

where C_1, C_2 are positive constants depending on the size of supp f.

Proof. Since $f \in B^{\circ s}_{pq,m}(\mathbb{R}^n)$ the orthogonality conditions $(f, x^\beta) = 0$, $|\beta| \leq m-1$, hold. These conditions imply that

$$D^\beta F[f](0) = 0, \quad |\beta| \leq m-1. \tag{1.113}$$

Further, since f has a compact support, due to the Paley-Wiener-Schwartz theorem $F[f]$ is an entire function. Therefore, taking into account (1.113), we have

$$\begin{aligned} F[f](\xi) &= \sum_{|\beta|=0}^{\infty} \frac{D^\beta F[f](0)}{\beta!} \xi^\beta = \sum_{|\beta| \geq m} \frac{D^\beta F[f](0)}{\beta!} |\xi|^{|\beta|} \theta^\beta \\ &= |\xi|^m \sum_{|\beta| \geq m} |\xi|^{|\beta|-m} \frac{1}{\beta!} D^\beta F[f](0) \theta^\beta, \end{aligned} \tag{1.114}$$

where we have used the substitution $\xi = |\xi|\theta$, $\theta = \xi/|\xi| \in S$, and S is the unit sphere in R^n with the center at the origin. Hence, $F[f](\xi) = |\xi|^m V(\xi)$, where

$$V(\xi) = \sum_{|\beta| \geq m} |\xi|^{|\beta|-m} \frac{1}{\beta!} D^\beta F[f](0) \theta^\beta.$$

We set $v = F^{-1}[V]$, and show that, in fact, $v \in B^{s+m}_{pq}(\mathbb{R}^n)$. Indeed, using the representation (1.114) and conditions $D^\beta F[f](0) = 0$, $|\beta| \le m-1$, we obtain

$$
\begin{aligned}
\|v|B^{s+m}_{pq}\|^q &= \sum_{j=0}^{\infty} 2^{(s+m)jq} \left\| F^{-1}\left[\varphi_j \frac{1}{|\xi|^m} \sum_{|\beta|\ge m} |\xi|^{|\beta|} \frac{1}{\beta!} D^\beta F f(0)\theta^\beta\right] |L_p\right\|^q \\
&\le C \sum_{j=0}^{\infty} 2^{sjq} \left\| F^{-1}\left[\varphi_j \sum_{|\beta|=m}^{\infty} \frac{1}{\beta!} D^\beta F[f](0)\xi^\beta\right] |L_p\right\|^q \\
&= \sum_{j=0}^{\infty} 2^{sjq} \|F^{-1}[\varphi_j F[f]]\,|L_p\|^q = C\|f|B^s_{pq}\|^q.
\end{aligned}
$$

In the second inequality of this chain we used the fact that $\operatorname{supp}\varphi_j \subset M_j \equiv \{2^{j-1} \le |\xi| \le 2^{j+1}\}$, and consequently, in the j-th term of the sum, $1/|\xi|^m \le 2^m 2^{-jm}$.

Now let us prove Part (*iii*). Suppose $|\xi| \le 1$. Then it follows from (1.112) that

$$
\begin{aligned}
|Fv(\xi)| &\le \sum_{|\beta|\ge m} \frac{1}{\beta!} |D^\beta F[f](0)| \le C N^{n/p'} \|f|B^s_{pq}\| \sum_{|\beta|\ge m} \frac{N^\beta}{(p'\beta+1)^{1/p'}} \\
&\le C\Big(\frac{N^n}{(p'm+1)}\Big)^{1/p'} (e^{nN}-1)\|f|B^s_{pq}\|.
\end{aligned}
$$

This inequality immediately implies the desired estimate (*iii*).

Theorem 1.24. *Let $1 < p < \infty$, p' the conjugate of p, and α a positive number satisfying the condition $\alpha p' < n$. Further, let $\varphi_0 \in C_0^\infty(B_1)$ and $f \in L_p(\mathbb{R}^n)$ has a compact support. Then the following estimates hold:*

(*i*) $\left\| F^{-1}\left[\frac{\varphi_0}{|\xi|^\alpha} F[f]\right] |L_p\right\| \le C_1 \|f|L_p\|$,

(*ii*) $\left\| F\left[\frac{\varphi_0}{|\xi|^\alpha} F^{-1}[f]\right] |L_p\right\| \le C_2 \|f|L_p\|$,

where C_1 and C_2 are positive constants depending on the size of $\operatorname{supp} f$.

Proof. First assume that $p \in [2,\infty)$. Let $f \in L_p(\mathbb{R}^n)$ has a compact support. The Paley-Wiener-Schwartz theorem implies that $F^{\pm}[f]$ is a smooth function on $\mathbb{R}^n$. Therefore $\sup_{\xi\in B_1} |F^{\pm}[f](\xi)| \le C_{1,1}$, with some constant $C_{1,1} > 0$. Using this fact and $\operatorname{supp}(\varphi_0 F[f]) \subseteq B_1$, one can verify that

$$
\frac{\varphi_0}{|\xi|^\alpha} F^{\pm 1}[f] \in L_{p'}, \tag{1.115}
$$

if α satisfies the condition $\alpha p' < n$. Indeed, making use of Theorem 1.22,

$$\|\frac{\varphi_0}{|\xi|^\alpha}F^{\pm 1}[f]|L_{p'}\| \le C\Big(\int\limits_{B_1}\frac{|F^{\pm 1}[f]|^{p'}}{|\xi|^{\alpha p'}}d\xi\Big)^{1/p'}$$
$$\le C\sup_{B_1}|F^{\pm 1}[f]|\int\limits_{B_1}\frac{d\xi}{|\xi|^{\alpha p'}} \le C_{1,2}\|f|L_p\|. \tag{1.116}$$

As the conjugate of p, the number p' belongs to the interval $(1,2]$. Therefore, Hausdorff-Young's inequality is applicable implying the estimate in part (i) in the case $p \in [2,\infty)$:

$$\left\|F\left[\frac{\varphi_0}{|\xi|^\alpha}F^{-1}[f]\right]|L_p\right\| \le C_{1,3}\|\frac{\varphi_0}{|\xi|^\alpha}F^{-1}[f]|L_{p'}\| \le C_1\|f|L_p\|. \tag{1.117}$$

Now we show that this estimate is valid for $p \in (1,2]$, too. Let g be an arbitrary function in $L_{p',comp}$, where p' is conjugate of p, and hence, $p' \in [2,\infty)$. Then, in accordance with estimate (1.117), we have $F\left[\frac{\varphi_0}{|\xi|^\alpha}F^{-1}[g]\right](x) \in L_{p'}(\mathbb{R}^n)$. Moreover,

$$\|F\frac{\varphi_0}{|\xi|^\alpha}F^{-1}g|L_{p'}\| \le C_1\|g|L_{p'}\|. \tag{1.118}$$

Further, using the form of the L_p-norm given by (1.4) and estimate (1.118), we obtain

$$\left\|F^{-1}\left[\frac{\varphi_0}{|\xi|^\alpha}F[f]\right]|L_p\right\| = \sup_{g\neq 0}\frac{\left|\left(F^{-1}\left[\frac{\varphi_0}{|\xi|^\alpha}F[f]\right],g\right)\right|}{\|g|L_{p'}\|} = \sup_{g\neq 0}\frac{\left|\left(f,F\left[\frac{\varphi_0}{|\xi|^\alpha}F^{-1}[g]\right]\right)\right|}{\|g|L_{p'}\|}$$
$$\le \sup_{g\neq 0}\frac{\|f|L_p\|\left\|F\left[\frac{\varphi_0}{|\xi|^\alpha}F^{-1}[g]\right]|L_{p'}\right\|}{\|g|L_{p'}\|} \le C_1\|f|L_p\|.$$

In the L_p-norm given by relation (1.4) the sup is taken over all functions $g \in L_{p'}(\mathbb{R}^n)$. It is not hard to see that the space $L_{p',comp}(\mathbb{R}^n)$ is dense in $L_{p'}(\mathbb{R}^n)$. Therefore, it suffices to take sup over all functions $g \in L_{p',comp}(\mathbb{R}^n)$ in the definition of the L_p-norm. The proof of part (i) is complete. The proof of part (ii) is similar.

Remark 1.10. 1. Obviously, Theorem 1.24 remains valid if one changes $|\xi|^\alpha$ in conditions 1) and 2) to a function $h(\xi)$ satisfying the condition $c|\xi|^\alpha \le h(\xi) \le C|\xi|^\alpha$, where c, C are positive constants.

2. We note that constants in all the estimates obtained in this section (Theorems 1.22–1.24) depend only on the size of the supp f, but not on the function f itself. This is important, since such estimates are coherent with the inductive topologies of spaces $\overset{\circ}{B}{}^s_{pq,m}(\mathbb{R}^n)$ and $\overset{\circ}{B}{}^{s',s_n}_{pq,m}(\mathbb{R}^n)$. In the sequel we will use this fact for the study of the uniqueness of a solution of some initial-boundary value problems, as well as the continuous dependence on boundary and initial functions.

1.13 Additional notes

1. *The Lebesgue integral and the Lebesgue space.* The Lebesgue measure and Lebesgue integral were introduced in 1901 by H. Lebesgue in his seminal paper [Leb01]. The first paper on L_p-spaces seems was [Ri10] by Friedrich Riesz published in 1910. Let (Ω,Σ,μ) be a measure space, where $\Omega \subseteq \mathbb{R}^n$, Σ is a σ-algebra of subsets of Ω, and μ is a measure defined on the measurable space (Ω,Σ). Let $f:\Omega\to\mathbb{R}$ be a Σ-measurable function: for any open interval $A\subset\mathbb{R}$ the inclusion $f^{-1}(A)\in\Sigma$ holds. Then the Lebesgue integral $\int_\Omega f(x)dx$ and Lebesgue space $L_p(\Omega)$, discussed in Section 1.3, can be generalized to the integral $\int_\Omega f(x)d\mu$ with respect to the measure μ, and to the space

$$L_p(\Omega;\mu) = \{f \text{ is } \Sigma\text{-measurable} : \int_\Omega |f|^p d\mu < \infty\},\ p\geq 1,$$

respectively. $L_p(\Omega;\mu)$ is a Banach space with the norm $\|f|L_p(\Omega;\mu)\|$. The Hölder and Minkowski inequalities in the general form take the following forms, respectively:

$$\left|\int_\Omega f(x)g(x)d\mu\right| \leq \left(\int_\Omega |f(x)|^p d\mu\right)^{\frac{1}{p}}\left(\int_\Omega |g(x)|^q d\mu\right)^{\frac{1}{p}},$$

where $f\in L_p(\Omega;\mu)$, $g\in L_q(\Omega;\mu)$, and $p,q\geq 1$ and $1/p+1/q=1$;

$$\left(\int_\Omega \left|\int_A f(x,a)da\right|^p d\mu\right)^{\frac{1}{p}} \leq \int_A \left(\int_\Omega |f(x,a)|^p d\mu\right)^{\frac{1}{p}} da$$

where $f(x,a)\in L_p(\Omega;\mu)$, $p\geq 1$, is a family of functions depending on a parameter $a\in A\subset\mathbb{R}^m$, such that $\|f(x,a)|L_p(\Omega;\mu)\|\in L_1(A)$.

Theorem 1.25. *(Lebesgue's dominated convergence theorem [RS80]) Let a sequence $f_n(x)$ be defined on a measure space (Ω,Σ,μ), such that $f_n(x)\to f(x)$ as $n\to\infty$ for μ-a.e. $x\in\Omega$, and $|f_n(x)|\leq g(x)$, where $g\in L_1(\Omega;\mu)$. Then $f\in L_1(\Omega;\mu)$ and $f_n\to f$ as $n\to\infty$ in the norm of $L_1(\Omega;\mu)$. Moreover,*

$$\lim_{n\to\infty}\int_\Omega f_n(x)d\mu = \int_\Omega f(x)d\mu.$$

2. *Hölder-Zygmund spaces.* Hölder-Zygmund spaces appear naturally in the context of solution of boundary value problems for Poisson equation $\Delta u = f$ in an open domain Ω with a smooth boundary $\partial\Omega$; see, e.g., [GT83]. In fact, a solution of the Poisson equation with the Dirichlet boundary condition $u_{|\partial\Omega}=\varphi$, can be represented as

$$u(x) = \int_\Omega G f dx + \int_{\partial\Omega}\frac{\partial G}{\partial \mathbf{n}}\varphi ds,$$

where $G(x)$ is the Green function, and $\mathbf{n}$ is the outside normal. The continuity of f is not sufficient for the solution $u(x)$ to be in class $C^2(\Omega)$. However, if $f\in C^\lambda(\overline{\Omega})$, $0<\lambda<1$, then the twice differentiable unique classic solution exists for any continuous boundary function φ; see details in [GT83].

If $m<\lambda<\mu<m+1$, and Ω is bounded, then $C^\mu(\overline{\Omega})\subset C^\lambda(\overline{\Omega})$. This fact is seen from the inequality

$$\frac{f(x)-f(y)}{|x-y|^\lambda} = \frac{f(x)-f(y)}{|x-y|^\mu}|x-y|^{\lambda-\mu} \leq d^{\lambda-\mu}\frac{f(x)-f(y)}{|x-y|^\mu},$$

where $d=diam(\Omega)$. The above inequality implies $\|f|C^\lambda(\overline{\Omega})\|\leq C\|f|C^\mu(\overline{\Omega})\|$. However, in general, the inclusion $C^\mu(\overline{\Omega})\subset C^\lambda(\overline{\Omega})$ for $\lambda<\mu$ may not be valid. Here is an example [GT83]:

Let $\Omega = \{(x,y) \in \mathbb{R}^2 : x^2 + y^2 < 1, y \leq \sqrt{|x|}\}$. Then the function

$$f(x,y) = \begin{cases} sign(x) y^\alpha, & \text{if } y > 0, \\ 0, & \text{if } y \leq 0, \end{cases}$$

where $1 < \alpha < 2$, defined on Ω, belongs to $C^1(\overline{\Omega})$, and hence, is Lipschitz continuous. At the same time $f \notin C^\lambda(\overline{\Omega})$ for all λ satisfying the inequality $\alpha/2 < \lambda < 1$.

3. *Function spaces and distributions.* Sobolev spaces $W_p^m(\Omega)$ were introduced by S.L. Sobolev in his papers [Sob35, Sob36, Sob38] in 1935–38, and with comprehensive presentation in his book [Sob50] published in 1950. His monograph [Sob74] uses completion procedure to describe $W_p^m(\Omega)$. The dyadic approach for description of Besov and, in particular, Sobolev spaces was used by Peetre in [Pee76]. Various results related to function spaces are provided in monographs by Nikolskii [Nik77], Besov, Il'in, Nikolskii [BIN75], and Triebel [Tri77, Tri83], which now became classic books in the theory of function spaces. Methods used in these spaces differ. Nikolskii [Nik77] used a technique based on the approximation by exponential functions. Anisotropic Sobolev, Besov, and other spaces are studied in [BIN75]. Triebel [Tri77, Tri83] uses modifications of the dyadic approach and Fourier multiplier theorems to describe Besov and Lizorkin-Triebel spaces. The theory of distributions was published in 1951 by L. Schwartz [Sch51]. The fundamental spaces (test functions) in building of Schwartz distributions are C^∞-functions. Wider classes of distributions, called *ultra-distributions*, use as a fundamental spaces versions of non-quasi-analytic classes. Fundamental spaces of ψ-distributions are entire functions of finite exponential type belonging to L_p-spaces. Examples in Section 1.10 show that the dependence of these spaces on p is not formal.
4. *Lizorkin type spaces.* The Lizorkin spaces $V(\mathbb{R}^n)$ and $\Phi(\mathbb{R}^n)$ and their duals were introduced in [Liz63]. Samko [Sam77] introduced more general spaces $\Psi_V(\mathbb{R}^n)$ and $\Phi_V(\mathbb{R}^n)$, where V is a closed subset of $\mathbb{R}^n$. The space $\Psi_V(\mathbb{R}^n)$ consists of functions $\varphi \in \mathscr{G}$ such that $D^\alpha \varphi(x) = 0, x \in V$, for all $|\alpha| = 0, 1, \ldots$. The space $\Phi_V(\mathbb{R}^n)$ is the Fourier pre-image of $\Psi_V(\mathbb{R}^n)$. The topologies of both spaces are induced from the topology of $\mathscr{G}$. If $V = \{0\}$, then the latter spaces coincide with Lizorkin spaces $\Psi(\mathbb{R}^n)$ and $\Phi(\mathbb{R}^n)$. In the paper [Sam82] Samko studied the denseness of these spaces in $L_p(\mathbb{R}^n)$. He established that if the n-dimensional measure of V is zero, then: (a) $\Psi_V(\mathbb{R}^n)$ is dense in $L_p(\mathbb{R}^n)$ for all $1 \leq p < \infty$; (b) $\Phi_V(\mathbb{R}^n)$ is dense in $L_p(\mathbb{R}^n)$ for all $2 \leq p < \infty$. He also proved that if V is a quasi-polygonal set, then $\Phi_V(\mathbb{R}^n)$ is dense in $L_p(\mathbb{R}^n)$ in the case $1 < p < 2$, as well. A set in $\mathbb{R}^n$ is called quasi-polygonal if in each finite ball it can be embedded in the union of a finite number of hypersubspaces of dimension $\leq n-1$. Samko [Sam95] announced a conjecture on the density of $\Phi_V(\mathbb{R}^n)$ for any V satisfying $\mu_n(V) = 0$, which still remains an open problem. Finally, we note that the fundamental space $\Psi_{G,p}(\mathbb{R}^n)$ of ψ-distributions as a set of functions is isomorphic to $\Phi_V(\mathbb{R}^n)$ (with different topologies), where $V = \mathbb{R}^n \setminus G$. This fact immediately implies the denseness of $\Psi_{G,p}(\mathbb{R}^n)$ in $B_{p,q}^s(\mathbb{R}^n)$, $s \in \mathbb{R}$, $1 < q < \infty$, for all $1 < p < 2$, if $\mathbb{R}^n \setminus G$ is quasi-polygonal.
5. *The Fourier transform.*

   (a) Proof of formula (1.17). In order to prove formula (1.17) we first find the Fourier transform of the function
   $$f(x) = \frac{\Gamma(\frac{n+1}{2})}{\pi^{\frac{n+1}{2}}} \frac{1}{(1+|x|^2)^{\frac{n+1}{2}}}, \quad x \in \mathbb{R}^n. \tag{1.119}$$
   We have
   $$F[f](\xi) = \frac{\Gamma(\frac{n+1}{2})}{\pi^{\frac{n+1}{2}}} \int_{\mathbb{R}^n} \frac{e^{ix\xi} dx}{(1+|x|^2)^{\frac{n+1}{2}}}. \tag{1.120}$$
   Using the transformation $x = Ty$, where T is an orthogonal matrix with entries $t_{k,j}$:
   $$t_{k,1} = \frac{\xi_k}{|\xi|}, \quad \sum_{k=1}^n t_{k,j} \xi_k = 0, \quad j = 2, \ldots, n,$$

one can reduce (1.120) to

$$F[f](\xi) = \frac{\Gamma(\frac{n+1}{2})}{\pi^{\frac{n+1}{2}}} \int_{-\infty}^{\infty} e^{iy_1|\xi|} \left[\int_{\mathbb{R}^{n-1}} \frac{dy_2 \dots dy_n}{(1+y_1^2+\dots+y_n^2)^{\frac{n+1}{2}}} \right] dy_1$$
$$= \frac{\Gamma(\frac{n+1}{2})}{\pi^{\frac{n+1}{2}}} \int_{-\infty}^{\infty} e^{iy_1|\xi|} H(y_1) dy_1. \tag{1.121}$$

Setting $\omega^2 = 1 + y_1^2 + \dots + y_{n-1}^2$ and integrating the inner integral with respect to y_n, we have

$$H(y_1) = \int_{\mathbb{R}^{n-2}} \left(\int_{-\infty}^{\infty} \frac{dy_n}{(\omega^2+y_n^2)^{\frac{n+1}{2}}} \right) dy_2 \dots dy_{n-1}$$
$$= \int_{\mathbb{R}^{n-2}} \frac{1}{\omega^{\frac{n}{2}}} \left(\int_{-\infty}^{\infty} \frac{d(\frac{y_n}{\omega})}{(1+(\frac{y_n}{\omega})^2)^{\frac{n+1}{2}}} \right) dy_2 \dots dy_{n-1}$$
$$= C_n \int_{\mathbb{R}^{n-2}} \frac{1}{(1+y_1^2+\dots+y_{n-1}^2)^{n/2}} dy_2 \dots dy_{n-1},$$

where $C_n = 2\int_0^\infty \frac{dz}{(1+z^2)^{(n+1)/2}}$. Repeating this with respect to $y_{n-1}, \dots, y_2$, one obtains

$$H(y_1) = C_n C_{n-1} \dots C_2 \frac{1}{1+y_1^2}, \quad C_j = 2\int_0^\infty \frac{dz}{(1+z^2)^{(j+1)/2}}, \quad j = 2, \dots, n.$$

Further, using substitution $s = 1/(1+z^2)$, one can verify that $C_j = \frac{\sqrt{\pi}\Gamma(j/2)}{\Gamma((j+1)/2)}$. Hence,

$$H(y_1) = (\sqrt{\pi})^{n-1} \frac{\Gamma(\frac{2}{2})}{\Gamma(\frac{3}{2})} \frac{\Gamma(\frac{3}{2})}{\Gamma(\frac{4}{2})} \cdots \frac{\Gamma(\frac{n}{2})}{\Gamma(\frac{n+1}{2})} \frac{1}{1+y_1^2} = \frac{\pi^{\frac{n-1}{2}}}{\Gamma(\frac{n+1}{2})} \frac{1}{1+y_1^2}.$$

Therefore, substituting the latter to (1.121) and using relation (1.11), one obtains

$$F[f](\xi) = \frac{1}{\pi} \int_{-\infty}^{\infty} \frac{e^{iy_1|\xi|} dy_1}{1+y_1^2} = e^{-|\xi|}.$$

Finally, due to Proposition 1.5, formula (1.17) follows.

(b) Let

$$f(A,\mu,x) = \frac{\Gamma(\frac{n+1}{2})}{\sqrt{det(A)}\pi^{\frac{n+1}{2}}} \frac{1}{\Big(1+(x-\mu)^T A^{-1}(x-\mu)\Big)^{\frac{n+1}{2}}}, \quad x \in \mathbb{R}^n,$$

where A is a positive definite $n \times n$-matrix and $\mu = (\mu_1, \dots, \mu_n)$ is a fixed vector. The function $f(A,\mu,x)$ generalizes (1.119), coinciding with it if $A = I$, the identity matrix, and $\mu = 0$, zero vector. The Fourier transform of $f(A,\mu,x)$ is

$$F[f(A,\mu,x)](\xi) = e^{i\mu\xi - \sqrt{\xi^T A \xi}}, \quad \xi \in \mathbb{R}^n.$$

The function $f(A,\mu,x)$ is the Cauchy-Poisson density function, a particular case of the multivariate Student's t-distribution with one degree of freedom. See [Sut86] for the Fourier transform (characteristic function) of Student's t-distribution with an arbitrary degrees of freedom.

(c) Let

$$g(A,\mu,x) = \frac{\sqrt{det(A)}}{\pi^{n/2}} e^{(x-\mu)^T A(x-\mu)}, \quad x \in \mathbb{R}^n,$$

where again A is a positive definite matrix and $\mu \in \mathbb{R}^n$ is a fixed vector. Then,

$$F[g(A,\mu,x)](\xi) = e^{i\mu\xi - \frac{1}{4}\xi^T A\xi}, \quad \xi \in \mathbb{R}^n.$$

The latter generalizes formula (1.15). The function $g(A,\mu,x)$ represents the density of the multivariate Gaussian variable with a position μ and a covariance matrix A.

(d) Similarly, one can generalize formula (1.51), as well. Let A be a positive definite $n \times n$-matrix. Then

$$F[e^{ix^T Ax}](\xi) = \frac{\pi^{n/2}}{\sqrt{det(A)}} e^{-\frac{i}{4}(\xi^T A^{-1}\xi - n\pi)}.$$

This formula is important in the study of solution properties of the Cauchy problem for the Schrödinger equation.

6. *The Bochner-Schwartz theorem.* A continuous function $f(x)$ defined on $\mathbb{R}^n$ is called positive definite, if for arbitrary $x_1,\ldots,x_N \in \mathbb{R}^n$ and complex numbers $z_1,\ldots,z_n$, one has

$$\sum_{i,j=1}^{N} f(x_i - x_j) z_i \bar{z}_j \geq 0.$$

One can reduce the latter to

$$\int_{\mathbb{R}^n} f(x-y)\varphi(x)\bar{\varphi}(x)dx \geq 0, \tag{1.122}$$

where $\varphi \in \mathscr{D}(\mathbb{R}^n)$. Using (1.122) one can extend the definition of positive definite functions to any distribution $f \in \mathscr{D}'(\mathbb{R}^n)$. Namely, a distribution f is positive definite if $< f, \varphi * \bar{\varphi} > \geq 0$ for all $\varphi \in \mathscr{D}(\mathbb{R}^n)$. The Bochner-Schwartz theorem describes positive definite distributions through the Fourier transform.

Theorem 1.26. *A Schwartz distribution ϕ is positive definite if and only if it is the Fourier transform of a tempered positive measure μ.*

7. *Sato's hyperfunctions.* The statement of Proposition 1.15 can be generalized to equations of the form

$$(x-a)^{\beta} u(x) = f(x), \tag{1.123}$$

where $f \in \mathscr{E}'(\mathbb{R}^n)$. If $u_0(x)$ is a particular solution to equation (1.123), then the general solution has the representation

$$u(x) = u_0(x) + \sum_{\substack{\alpha_j \leq \beta_j - 1 \\ j=1,\ldots,n}} C_\alpha D^\alpha \delta_a(x),$$

where C_α are arbitrary constants. In fact, this is valid for arbitrary hyperfunction f, as well (see, e.g., [Gra10]). The space of hyperfunctions was introduced by M. Sato in papers [Sat59, Sat60], in 1959. By definition, a hyperfunction defined on an interval $I \subset \mathbb{R}$ is an equivalence class of differences $f(x) = F_1(x+i0) - F_2(x-i0)$, where $F_1(z)$ and $F_2(z)$ are analytic functions on upper and lower complex neighborhoods of the interval I, respectively. Sato showed that the functions F_1, F_2, called defining functions of the hyperfunction f, can also be selected harmonic. Any Schwartz distribution is also a hyperfunction. Moreover, there is a hyperfunction, which is not a Schwartz distribution. Hence, the space of hyperfunctions is wider than the space of Schwartz distributions. The following statement on the structure of hyperfunctions is due to Kaneko [Kan72]:

Theorem 1.27. *Any hyperfunction f is globally represented as*

$$f = J(D)g, \tag{1.124}$$

where $J(D)$ is a local operator with constant coefficients, that is, $J(D)$ is an infinite order differential operator $J(D) = \sum_\alpha a_\alpha D^\alpha$ with the coefficients satisfying

$$\lim_{|\alpha|\to\infty} \sqrt[|\alpha|]{|a_\alpha|\alpha!} = 0,$$

and g is an infinitely differentiable function.

Thus, hyperfunctions may have the infinite order singularity like ψ-distributions (see Remark 1.7.2.). This fact and Kaneko's structure theorem in conjunction with the structure theorem for ψ-distributions (Theorem 1.20) shows that ψ-distributions are in close relationship with hyperfunctions. However, what is the exact relationship between them is currently an open question.

Chapter 2
Pseudo-differential operators with singular symbols (ΨDOSS)

2.1 Introduction

We begin Chapter 2 with simple examples of initial and boundary value problems, solution operators of which have singularities of one or another type in the dual variable. The presence of a singularity often causes a failure of well posedness of the problem in the sense of Hadamard. Let A be a linear differential operator mapping a function space X into another function space F. The differential equation $Ax = f$ is (X,F) well posed in the sense of Hadamard,[1] if

(1) for any $f \in F$ a solution $x \in X$ exists,
(2) the solution is unique, and
(3) the solution continuously depends on the data in terms of norms of F and X.

A classical example of the boundary value problem, which can be found in textbooks and is not well posed in the sense of Hadamard, is the initial value problem for the Laplace equation

$$\Delta u(t,x) \equiv \frac{\partial^2 u(t,x)}{\partial t^2} + \frac{\partial^2 u(t,x)}{\partial x^2} = 0, \quad t > 0, x \in \mathbb{R},$$
$$u(0,x) = 0, \ \frac{\partial u(0,x)}{\partial t} = 0, \quad x \in \mathbb{R}.$$

The operator A in this example is the pair $A = (\Delta, B)$, where B is the boundary operator $B[u(t,x)] = (u(0,x), \frac{\partial u(0,x)}{\partial t})$; $X = C^2(\mathbb{R}^{n+1}_+)$, and $F = C(\mathbb{R}^{n+1}_+) \times C(\mathbb{R}^n) \times C(\mathbb{R}^n)$. Obviously, $u(t,x) \equiv 0$ solves this problem. If one changes the initial conditions to

$$u(0,x) = 0, \ \frac{\partial u(0,x)}{\partial t} = \varepsilon \sin\frac{x}{\varepsilon},$$

[1] Jacques Hadamard introduced the notion of well posedness in his 1902 paper [Had02].

S. Umarov, *Introduction to Fractional and Pseudo-Differential Equations with Singular Symbols*, Developments in Mathematics 41,
DOI 10.1007/978-3-319-20771-1_2

where $\varepsilon > 0$ is a small number, then the corresponding solution is $v(t,x) = \varepsilon^2 \sin\frac{x}{\varepsilon}$ $\sinh\frac{t}{\varepsilon}$. Taking, for instance $x = \varepsilon\pi/2$ and $t = 1$, one can see that a small change of the data affected to an arbitrarily large (exponential) change of the solution. Textbooks usually ignore mathematical explanation of this ill-posedness phenomenon, just giving a physical explanation, that the elliptic equations describe a stationary processes, and therefore, initial-value problems for them are not physical. However, mathematical reason can easily be explained in terms of singularities of the solution operators.

As we will see in Section 2.2, in the above example the symbols of the solution operators have a singularity at infinity causing ill-posedness. Examples discussed there show that depending on the problem formulation different types of singularities may arise for the solution operators. In Section 2.3 we introduce an algebra of pseudo-differential operators with singular symbols (ΨDOSS) and study their properties. In the subsequent sections we develop a periodic and an abstract form of ΨDOSS, namely an operator calculus with symbols which have singularities on the spectrum of the generic operator.

2.2 Some examples of boundary value problems leading to ΨDOSS

2.2.0.1 The Cauchy Problem

The first example is the Cauchy problem for the one-parameter family of differential equations

$$\frac{\partial^2 u(t,x)}{\partial t^2} + \lambda^2 D^2 u(t,x) = 0,\ t > 0, x \in \mathbb{R}, \tag{2.1}$$

$$u(0,x) = \varphi(x),\ \frac{\partial u(0,x)}{\partial t} = \psi(x), x \in \mathbb{R}, \tag{2.2}$$

where $D = -i\partial/\partial x,\ \lambda = \sigma + i\tau \in \mathbb{C}$. As is well known, (2.1),(2.2) represents the Cauchy problem for the wave equation if $\lambda = 1$, and it is well posed in the classical sense of Hadamard. However, if $\lambda = i$, then (2.1),(2.2) represents the Cauchy problem for the Laplace equation, which, as discussed in Introduction, is not well posed in the sense of Hadamard. Now we will take an attempt to understand why it is so in terms of symbols of solution operators.

Let us temporarily replace in (2.1) D by a parameter ξ, assuming that ξ takes values in $\mathbb{R}$. Then we have a linear ordinary differential equation depending on parameters $\xi \in \mathbb{R}$ and $\lambda \in \mathbb{C}$

$$u''(t) + \lambda^2 \xi^2 u(t) = 0,\ t > 0.$$

Solving this equation, we obtain

$$u(t) = C_1 \cos(\xi\lambda t) + C_2 \sin(\xi\lambda t), t > 0,$$

where C_1, C_2 do not depend on t. Further, taking into account conditions (2.2), we have

$$u(t,\xi) = \Phi(\xi)\cos(\xi\lambda t) + \Psi(\xi)\frac{\sin(\xi\lambda t)}{\lambda\xi},$$

where Φ and Ψ are the Fourier transforms of φ and ψ, respectively.

The well posedness of problem (2.1),(2.2) essentially depends on the behavior of two functions, namely

$$\cos(\xi\lambda t) \quad \text{and} \quad \frac{\sin(\xi\lambda t)}{\lambda\xi}. \tag{2.3}$$

In fact, a solution, as we will see later, can be written in the form (returning back to D)

$$u(t,x) = [\cos\, \lambda tD]\varphi(x) + \left[\frac{\sin\, \lambda tD}{\lambda D}\right]\psi(x), \tag{2.4}$$

where

$$\cos\, \lambda tD \quad \text{and} \quad \frac{\sin\, \lambda tD}{\lambda D}$$

are pseudo-differential operators with symbols in equation (2.3), respectively. So, these two important functions in (2.3) are symbols of solution operators. Now, assume $\lambda = 1$ (in fact, λ may be any real number). Then, both symbols $\cos(\xi t)$ and $\frac{\sin(\xi t)}{\xi}$ are well defined and bounded on $\mathbb{R}$:

$$|\cos t\xi| \leq 1, \; |\frac{\sin\, t\xi}{\xi}| \leq \frac{C_t}{1+|\xi|}, \quad \xi \in \mathbb{R}.$$

These estimates imply well posedness in the sense of Hadamard (see Chapter 4) in this case. If $\lambda = i$ (or $\lambda \in \mathbb{C} \setminus \mathbb{R}$), then the symbols of solution operators become $\cosh(\xi t)$ and $\frac{\sinh(\xi t)}{\xi}$. These functions are not bounded on $\mathbb{R}$, exponentially increasing at infinity. That is, the symbols of the solution operators now have asymptotic behavior

$$\cosh(t\xi) = O(e^{t\xi}), \; \frac{\sinh\,(t\xi)}{\xi} = O(e^{t\xi}),\; t > 0,\; |\xi| \to \infty, \tag{2.5}$$

resulting the failure of the problem to be well posed in the sense of Hadamard. We keep in mind that in this particular case a *singularity appears at infinity*.

The solution operators in this example are examples of ΨDOSS. We will see in Section 2.3 that such operators are well defined on the spaces $\Psi_{G,p}(\mathbb{R}^n)$, where G is determined by singularities of the symbol. Therefore, depending on singularities of the symbols of solution operators the problem is well posed in certain space. This space we will call *a well-posedness space.* In our example the well-posedness space corresponding to $\lambda = i$ is the most narrow space in terms of $\Psi_{G,p}(\mathbb{R}^n)$ and it is not closable up to classical Sobolev, Besov, or Lizorkin-Triebel spaces, thus is not well posed in the sense of Hadamard. Vice versa, in the case $\lambda = 1$ the well-posedness space is the widest among the spaces $\Psi_{G,p}(\mathbb{R}^n)$, and in this case it is closable up to Sobolev and other classical function spaces of finite order, making the problem well posed in the sense of Hadamard.

Finally, one can notice that the representation (2.4) is just a different form of the well-known D'Alembert's formula. Taking into account that the operator $[\exp t\frac{\partial}{\partial t}]$ acts as a translation operator

$$[\exp t\frac{\partial}{\partial x}] : \varphi(\cdot) \to \varphi(\cdot + t),$$

and accepting

$$(\frac{\partial}{\partial x})^{-1}\varphi(x) = \int_0^x \varphi(\xi)d\xi + c,$$

one can see that (2.4) can be reduced to the usual form of D'Alembert's formula

$$u(t,x) = \frac{\varphi(x+\lambda t) + \varphi(x-\lambda t)}{2} + \frac{1}{2\lambda}\int_{x-\lambda t}^{x+\lambda t} \psi(\xi)d\xi. \tag{2.6}$$

Remark 2.1. Instead of differential equation (2.1) one can consider the equation

$$\frac{\partial^2 u(t,x)}{\partial t^2} + \lambda^2 A u(t,x) = 0,\ t > 0, x \in \Omega,$$

where A is an elliptic operator with an appropriate domain $\mathscr{D}(A)$ containing functions defined on $\Omega \subseteq \mathbb{R}^n$. However, the essence of the question concerning the singularity of the symbols of solution operators remains the same.

2.2.0.2 The Dirichlet problem

The second problem we want to consider is the Dirichlet problem in the infinite strip $\{(t,x) : 0 \le t \le 1, x \in \mathbb{R}\}$ for the same one-parameter differential equation as in the previous example. A peculiarity of this problem is now we observe totally different type of singularities of the solution operators. Namely, in this case for $\lambda = 1$ the symbols of the solution operators have a pole type singularities. Ergo, consider the following boundary value problem:

$$\frac{\partial^2 u(t,x)}{\partial t^2} + \lambda^2 D^2 u(t,x) = 0, \tag{2.7}$$

$$u(0,x) = \varphi(x),\ u(1,x) = \psi(x), \tag{2.8}$$

where again $\lambda = \sigma + i\tau \in \mathbb{C}$. It is well known that if $\lambda = i$, then this problem represents the Dirichlet problem for the Laplace equation, which is well posed. In contrary, if $\lambda = 1$, then we have the Dirichlet problem for the wave equation, which is not well posed in the sense of Hadamard. Similar to the previous example, applying the operator method we obtain the following representation for a solution

$$u(t,x) = \left[\frac{\sin\, \lambda(1-t)D}{\sin\, \lambda D}\right] \varphi(x) + \left[\frac{\sin\, \lambda t D}{\sin\, \lambda D}\right] \psi(x).$$

The symbols of solution operators in this case are

$$\frac{\sin\,(1-t)\lambda\xi}{\sin\, \lambda\xi}, \quad \frac{\sin\, t\lambda\xi}{\sin \lambda\xi}.$$

They are bounded if $\lambda = i$:

$$\left|\frac{\sinh\,(1-t)\xi}{\sinh\, \xi}\right| \leq 1, \quad \left|\frac{\sinh\, t\xi}{\sinh \xi}\right| \leq 1, \quad \xi \in \mathbb{R},\, 0 < t < 1.$$

In accordance with results obtained in Chapter 4 this fact implies a possibility of closure of the well posedness space up to classical spaces. At the same time if $\lambda = 1$ the symbols of the solution operators

$$\frac{\sin\,(1-t)\xi}{\sin\, \xi}, \quad \frac{\sin\, t\xi}{\sin \xi}$$

have *pole type singularities* at points $\xi_k = k\pi,\, k \neq 0$. Again from the results of next sections it follows that the well-posedness space is not closable up to classical spaces. This fact implies ill-posedness in the sense of Hadamard of the considering problem in the case when λ is a real number, i.e., the Dirichlet problem for the wave equation.

We note that in the multi-dimensional case, that is $x \in \mathbb{R}^n$ in (2.7), (2.8), the symbols of solution operators when $\lambda = 1$, are

$$\frac{\sin\,(1-t)|\xi|}{\sin\, |\xi|}, \quad \frac{\sin\, t|\xi|}{\sin |\xi|}.$$

In this case singularities occur on concentric hyperspheres $\{\xi \in \mathbb{R}^n : |\xi| = k\pi,\, k = 1,2,\ldots\}$.

2.2.0.3 Diffusion equation

The classical diffusion process without drift can be described by the Cauchy problem

$$\frac{\partial u(t,x)}{\partial t} = \kappa \Delta u(t,x),\; t > 0,\; x \in \mathbb{R}^n, \tag{2.9}$$

$$u(0,x) = f(x), \tag{2.10}$$

where $\Delta = -(D_1^2 + \ldots + D_n^2)$, $D_j = -i\frac{\partial}{\partial x_j}, j = 1, \ldots, n$, is the Laplace operator, and $\kappa > 0$ is a (constant) diffusion coefficient. The operator method leads now to the following representation for a solution

$$u(t,x) = e^{\kappa t \Delta} f(x)$$
$$= \frac{1}{(2\pi)^n} \int_{\mathbb{R}^n} e^{-\kappa t |\xi|^2} \hat{f}(\xi) e^{-ix\xi} d\xi, \quad t > 0, \tag{2.11}$$

where $\hat{f}(\xi) = F[f](\xi)$, the Fourier transform of f. The symbol of the solution operator $e^{\kappa t \Delta}$ is $e^{-\kappa t |\xi|^2}$. This is a "nice" function in the sense that it belongs to $C^\infty(\mathbb{R}^n)$, bounded, and decreases exponentially when $|\xi| \to \infty$ with all its derivatives. These properties of the solution operator 'make' problem (2.9),(2.10) not only well posed in the sense of Hadamard in the classical spaces, but also provide hypo-ellipticity. By definition, a pseudo-differential operator A acting on distributions defined on $\Omega \subset \mathbb{R}^n$ is called *hypo-elliptic,* if for any distribution $u \in \mathscr{D}'(\Omega')$, $\Omega' \subset \Omega$, $Au \in C^\infty$ implies $u \in C^\infty(\Omega')$.

The solution in the variable x is the inverse Fourier transform of the product $e^{-\kappa t |\xi|^2} \hat{f}(\xi)$. Due to the convolution formula (1.42) and relation (1.16) the solution in the variable x is represented in the form

$$u(t,x) = \frac{1}{(\sqrt{4\pi\kappa t})^n} \int_{\mathbb{R}^n} e^{-\frac{|x-y|^2}{4\kappa t}} f(y) dy. \tag{2.12}$$

The Cauchy problem (2.9)–(2.10) and its solution (2.12) have a clear probabilistic interpretation. In fact, if the initial function f is the Dirac delta function, then the equation (2.9) is the Fokker-Planck equation (or forward Kolmogorov equation) for the density function of Brownian motion B_t. In this particular case the solution $u(t,x)$ in (2.12) is a Gaussian density[2] evolved in time:

$$u(t,x) = G_t(x) = \frac{1}{(\sqrt{4\pi\kappa t})^n} e^{-\frac{|x|^2}{4\kappa t}}. \tag{2.13}$$

If the initial function is the density function of a random variable Y, then the solution $u(t,x)$ in (2.12) represents the density function of the random process $X_t = Y + B_t$. In the theory of stochastic processes it is well known that Brownian motion has a continuous path. This implies that the random process X_t also has a continuous path. We will return to stochastic applications of differential and pseudo-differential equations in Chapter 7 where more detailed discussion will be provided.

In Chapter 7 we will discuss non-Gaussian random processes, sometimes called *anomalous diffusion* processes, the mathematical model of which is given by the equation

$$\frac{\partial u(t,x)}{\partial t} = \kappa_\alpha (-\Delta)^{\alpha/2} u(t,x),\ t > 0,\ x \in \mathbb{R}^n, \tag{2.14}$$

[2] With mean 0 and correlation matrix I

where α is a positive real number, $(-\Delta)^\alpha$ is the fractional power α of the positive definite operator $-\Delta$, see Section 3.9. This operator has the symbol $-|\xi|^\alpha$. In this case the solution operator has the symbol $e^{-t|\xi|^\alpha}$. Even though this function decreases at infinity at an exponential rate, but in contrast to the classical case ($\alpha = 2$) it is not so "nice." This symbol is continuous, but not differentiable at the origin. This type of singularity again effects in the well-posedness space, which as we will see later (Chapter 4), is narrower than the well-posedness space of the classic diffusion equation.

Note that for $\alpha < 2$ the continuity path discussed above breaks down. Let us check this for $\alpha = 1$. In this case the solution $u(t,x)$ of equation (2.14) satisfying the initial condition $u(0,x) = f(x)$, due to the convolution formula (1.42) and relation (1.12), is ($n = 1$)

$$u(t,x) = \frac{1}{\pi}\int_{-\infty}^{\infty} \frac{tf(y)}{t^2 + (x-y)^2} dy.$$

If the initial function $f(x)$ is the Dirac delta function, then the latter reduces to

$$u(t,x) = \frac{1}{\pi}\frac{t}{t^2 + x^2}.$$

This is an evolved Cauchy distribution, which is a representative of stable distributions with pure jumps. The next example discusses a model of the general jump processes which, in particular, covers all the values of $\alpha \in (0,2]$.

2.2.0.4 Equations associated with jump processes

Let $\rho(x), x \in \mathbb{R}^n$, be a function defined as

$$\rho(x) = \begin{cases} |x|^2, & \text{if } |x| \le 1; \\ 1, & \text{if } |x| > 1. \end{cases}$$

Definition 2.1. A measure ν defined on $\mathbb{R}^n$ is called a *Lévy measure* if it satisfies the conditions

$$\nu(\{0\}) = 0 \quad \text{and} \quad \int_{\mathbb{R}^n} \rho(x) d\nu < \infty.$$

Consider the following equation on $\{(t,y) \in \mathbb{R}^{n+1} : t > 0, y \in \mathbb{R}^n\}$

$$\frac{\partial u(t,x)}{\partial t} = \int_{\mathbb{R}^n} \Big(u(t,x+y) - u(t,x) - b(y)\sum_{j=1}^{n} y_j \frac{\partial u(t,x)}{\partial x_j}\Big) d\nu(y), \tag{2.15}$$

where $b(y)$ is a bounded function, so that the integral in (2.15) is finite. For instance, if $\int_{|y|\le 1} |y| d\nu(y) < \infty$, then b can be selected identically zero; if $\int_{|y|>1} |y| d\nu(y) < \infty$, then b can be the constant function $b(y) \equiv 1$. If both of these conditions are not fulfilled, then b possesses the following properties: $b(y) \to 1$ when $|y| \to 0$, and $b(y) = O(\frac{1}{|y|})$ when $|y| \to \infty$. Equation (2.15) arises in the study of random processes accompanied with "pure jumps." In order to see that, let us take $b = 0$ and $d\nu(y) = \delta_a(y)dx$, where δ_a is the Dirac delta function with mass on a, $0 \neq a \in \mathbb{R}^n$. Then the right-hand side of equation (2.15) takes the form $Ju(t,x) = u(t,x+a) - u(t,x)$. Using the formula $F[f(x+a)](\xi) = e^{ia\xi}F[f](\xi)$, where $a\xi = a_1\xi_1 + \cdots + a_n\xi_n$, it is easy to check that the Fourier transform of Ju is $F[Ju](t,\xi) = (e^{ia\xi} - 1)F[u](t,\xi)$. Therefore, the solution operator of the equation (2.15) with the initial condition $u(0,x) = \delta(x)$ has the symbol $e^{e^{ia\xi}-1}$. The latter is the characteristic function of the Poisson process with intensity parameter $\lambda = 1$ and values ak, $k \in \mathbb{N}$ (see, e.g., [Fel68]). It is well known in the probability theory that the Poisson process is a purely jump process. In particular, in our example the jump size is $|a|$. In general, the measure ν identifies all the possible jump sizes, and intensity of these jumps.

Now, denote the operator on the right-hand side of equation (2.15) by A, that is

$$Af(x) = \int_{\mathbb{R}^n} \Big(f(x+y) - f(x) - b(y)\sum_{j=1}^{n} y_j \frac{\partial f(x)}{\partial x_j}\Big) d\nu(y). \tag{2.16}$$

Computing the Fourier transform of Af, one has

$$F[Af](\xi) = F[f](\xi) \int_{\mathbb{R}^n} \Big(e^{i\xi y} - 1 - ib(y)\sum_{j=1}^{n} y_j\xi_j\Big) d\nu(y).$$

It follows that the symbol of operator A in (2.16) is

$$\sigma_A(\xi) = \int_{\mathbb{R}^n} \Big(e^{i\xi y} - 1 - ib(y)\sum_{j=1}^{n} y_j\xi_j\Big) d\nu(y). \tag{2.17}$$

Hence, due to the formula for the inverse Fourier transform the operator A can also be represented in the form

$$Af(x) = \frac{1}{(2\pi)^n} \int_{\mathbb{R}^n} \sigma_A(\xi) Ff(\xi) e^{-ix\xi} d\xi.$$

Suppose that there is a finite measure λ defined on the unit sphere $S = \{x \in \mathbb{R}^n : |x| = 1\}$, such that for any Borel set $E \subset \mathbb{R}^n$

$$\nu(E) = \int_S \Big(\int_0^\infty I_E(r\theta) \frac{dr}{r^{\alpha+1}}\Big) d\lambda(\theta),$$

where $\alpha \in (0,2]$. Then the symbol $\sigma_A(\xi)$ in (2.17) of the operator A can be written in the form

$$\sigma_A(\xi) = \int_S \Big(\int_0^\infty [e^{ir\xi\theta} - 1 - b(r\theta)r\xi\theta] \frac{dr}{r^{\alpha+1}}\Big) d\lambda(\theta),$$

with $\xi\theta = \xi_1\theta_1 \ldots \xi_n\theta_n$, $\xi \in \mathbb{R}^n, \theta \in S$.

Let $0 < \alpha < 1$. Then the condition $\int_{|y|\leq 1} |y| d\nu(y) < \infty$ is verified and we can set $b(y) \equiv 0$. Using the substitution $r\xi\theta = \eta$, we have

$$\sigma_A(\xi) = \int_S (\xi\theta)^\alpha \Big(\int_0^\infty [e^{i\eta} - 1] \frac{d\eta}{\eta^{\alpha+1}} \Big) d\lambda(\theta).$$

Obviously, the internal integral in this equation is finite and does not depend on ξ and θ. Let us denote it by j_α. Hence,

$$\sigma_A(\xi) = j_\alpha |\xi|^\alpha \int_S \left(\frac{\xi}{|\xi|} \theta \right)^\alpha d\lambda(\theta). \tag{2.18}$$

If we require additionally that the measure λ is uniform on S then the integral $\int_S \theta_1 \theta d\lambda(\theta)$ does not depend on θ_1, i.e., this integral is constant depending only on α. Since $\xi/|\xi| \in S$, the integral in (2.18) does not depend on ξ. Therefore, the symbol $\sigma_A(\xi)$ has the representation

$$\sigma_A(\xi) = -\kappa_\alpha |\xi|^\alpha, \quad \text{where} \quad \kappa_\alpha = -j_\alpha \int_S (\theta_1 \theta) d\lambda(\theta). \tag{2.19}$$

The constant κ_α is positive and corresponds to the diffusion coefficient κ in the classic diffusion equation (2.9).

Further, let $1 < \alpha < 2$. Now the condition $\int_{|y>1|} |y| d\nu(y) < \infty$ is verified, and therefore we can take $b(y) \equiv 1$. In this case performing similar calculations, as we did in the case $0 < \alpha < 1$, we arrive again at the same representation (2.19), where the constant j_α is defined as $j_\alpha = \int_0^\infty (e^{i\eta} - 1 - \eta) \frac{d\eta}{\eta^{\alpha+1}}$. Not going into details we note that representation (2.19) is valid for $\alpha = 1$ as well under the additional condition on the measure λ : $\int_S \theta d\lambda(\theta) = 0$.

Thus, the anomalous diffusion equation (2.14) is a particular case of equation (2.15) describing jump processes. This fact explains the nature of anomalous diffusion processes modeled by equation (2.14), which is very different from the continuous nature of the classic diffusion modeled by equation (2.9). The general case of random processes, in which both continuous and jump components are present, will be discussed in Chapter 7.

2.2.0.5 The Schrödinger equation

If one takes the parameter D in equation (2.9) equal to $\frac{\omega}{i\hbar}$, where ω is a positive constant depending on the light velocity, mass of a particle, and Planck's constant $\hbar$, then (2.9) becomes the Schrödinger equation

$$i\frac{\partial u(t,x)}{\partial t} = \frac{\omega}{\hbar} \Delta u(t,x), \quad t > 0, \, x \in \mathbb{R}^n, \tag{2.20}$$

which describes the probability of particle distribution in the quantum mechanics. Now the symbol of the solution operator takes the form $e^{i\frac{\omega}{\hbar}t|\xi|^2}$. This function is not so "nice" to compare to the symbol of the solution operator of the diffusion equation (2.9), or anomalous diffusion equation (2.14). Unlike the symbols of solution operators of (2.9) and (2.14), the symbol $e^{i\frac{\omega}{\hbar}t|\xi|^2}$ does not decrease at infinity. It does not increase at infinity either, unlike the symbols in (2.5). In fact, this symbol is an oscillatory function with the amplitude equal to one, and hence, is in $L_\infty(\mathbb{R}^n)$.

The solution of the Schrödinger equation is frequently interpreted as a density of dispersive waves. Therefore, their L_p-estimates are useful. The Riesz-Thorin theorem can be used to get such an estimates. Indeed, if the solution satisfies the initial condition $u(0,x) = f(x), \quad x \in \mathbb{R}^n$, then, we have

$$u(t,x) = \frac{1}{(2\pi)^n} \int_{\mathbb{R}^n} e^{i\frac{\omega}{\hbar}t|\xi|^2} \hat{f}(\xi) e^{-ix\xi} d\xi, \quad t > 0, x \in \mathbb{R}^n, \tag{2.21}$$

or, inverting the latter in the distributional sense (see formula (1.52)),

$$u(t,x) = S_t f(x) = \left(\frac{\hbar}{i4\pi\omega t}\right)^{n/2} \int_{\mathbb{R}^n} e^{i\frac{\hbar}{\omega}\frac{|y|^2}{4t}} f(x-y)dy, \quad t > 0, x \in \mathbb{R}^n. \tag{2.22}$$

It follows from (2.22) that

$$|u(t,x)| \leq \frac{1}{(4\pi c_0 t)^{\frac{n}{2}}} \|f\|_{L_1}.$$

This estimate shows that $S_t : L_1(\mathbb{R}^n) \to L_\infty(\mathbb{R}^n)$ with the norm

$$\|S_t\|_{L_1(\mathbb{R}^n)\to L_\infty(\mathbb{R}^n)} \leq (4\pi c_0 t)^{-\frac{n}{2}},$$

where $c_0 = \omega/\hbar$. On the other hand, for $f \in L_2(\mathbb{R}^n)$ equation (2.21) together with Parseval's equality yields $S_t : L_2(\mathbb{R}^n) \to L_2(\mathbb{R}^n)$ with the norm $\|S_t\|_{L_2(\mathbb{R}^n)\to L_2(\mathbb{R}^n)}=1$. Now applying the Riesz-Thorin theorem, we obtain

$$\|u(t,x)\|_{L_p} \leq \frac{1}{(4\pi c_0 t)^{n(1/2-1/p)}} \|f\|_{L_q}, t > 0, \tag{2.23}$$

where $p \geq 2$ and q is the conjugate of p. Estimate (2.23) shows the rate of decay when $t \to \infty$. Estimates when f is an element of Besov or Lizorkin-Triebel spaces will be discussed in Chapter 4, as a corollary of general well-posedness theorems.

For relativistically free particles J. Björken and S. Drell [BD64] (see also [Dub82, Sam83]) considered the Schrödinger equation in the following form

$$\frac{\partial u(t,x)}{\partial t} = D\sqrt{I-\Delta}u(t,x), \tag{2.24}$$

where again $D = \frac{\omega}{i\hbar}$. A peculiarity of equation (2.24) is that the equation itself is given as a pseudo-differential equation. The symbol of the corresponding solution operator now has the form $e^{i\frac{\omega}{\hbar}t\sqrt{1+|\xi|^2}}$, which either does not decrease at infinity.

2.2.0.6 Multi-point problems

Consider a problem

$$\frac{\partial^2 u(t,x)}{\partial t^2} - \Delta u(t,x) = 0, \; t \in (0,2), \; x \in \mathbb{R}^n, \tag{2.25}$$

$$u(0,x) = \varphi(x), \, u(1,x) = u(2,x), \tag{2.26}$$

in which the values of a solution at three time instants are involved. This is a typical example of multi-point nonlocal boundary value problems. Multi-point and other nonlocal boundary value problems arise in various fields, including the plasma physics, fluid flows in porous media, etc. Equation (2.25) is included to the family of equations (2.1) with $\lambda = i$. The same method used in the previous examples gives a representation $u(t,x) = S(t,D)\varphi(x)$ for a solution, where the solution operator $S(t,D)$ has the symbol

$$s(t,\xi) = \frac{\cos(|\xi|t)[\sin|\xi| - \sin(2|\xi|)] - \sin(|\xi|t)[\cos|\xi| - \cos(2|\xi|)]}{\sin|\xi| - \sin(2|\xi|)}. \tag{2.27}$$

This symbol has non-integrable strong singularities at concentric spheres $|\xi| = k\pi, k \in \mathbb{Z}\setminus\{0\}$, and $|\xi| = \pm\pi/3 + 2\pi m, m \in \mathbb{Z}$. Due to these singularities the expression $S(t,D)\varphi(x)$ loses its meaning even on infinitely differentiable functions with compact support. Hence, one cannot expect the well posedness of this problem in the classical function spaces.

Later, when we gather enough information about ΨDOSS, we will return to these examples and discuss their well-posedness spaces in detail. We will also see other examples and applications of ΨDOSS, such that boundary values (traces) of harmonic functions (cf. with Sato's hyperfunctions; see Section 1.13 "Additional notes" to Chapter 1), uniqueness of a solution of polyharmonic equation, etc.

2.3 ΨDOSS: constant symbols

As we have seen in the previous section, in many boundary value problems of mathematical physics symbols of solution operators have different type of singularities: strong singularities on a finite part of the space, singularities due to increase or non-sufficiently fast decrease at infinity, or singularities due to irregular points of symbols. These kind of operators have two distinction from the standard pseudo-differential ones. First, their symbols contain singularities with respect to dual variable, and second, their orders, generally speaking, are not bounded.

In this section we determine ΨDO whose symbols have singularities with respect to dual variables.

Definition 2.2. Let $A(\xi) \in C^\infty(G)$, $G \subset R^n$. We determine an operator $A(D)$ by the formula

$$A(D)f(x) = \frac{1}{(2\pi)^n} \int_G A(\xi)F[f](\xi)e^{-ix\xi}d\xi, \tag{2.28}$$

provided the integral on the right-hand side exists. The function $A(\xi)$ is called a symbol of $A(D)$.

In definition 2.2 the function $A(\xi)$ may have arbitrary type of singularities outside G or on its boundary. Generally speaking, operators $A(D)$ with symbols $A \in C^\infty(G)$, may not be meaningful even for functions in the space $C_0^\infty(\mathbb{R}^n)$. Indeed, let $\xi_0 \in \mathbb{R}^n$ be a non-integrable singular point of $A(\xi)$ and denote by $O(\xi_0)$ some neighborhood of ξ_0. Let us take a function $f_0 \in C_0^\infty(\mathbb{R}^n)$ with $F[f_0](\xi) > 0$ for $\xi \in O(\xi_0)$ and $F[f_0](\xi_0) = 1$. Then it is easy to verify that $A(D)f_0(x) = \infty$. However, for functions $f \in \Psi_{G,p}(\mathbb{R}^n)$ the integral in (2.28) is convergent due to the compactness of $\operatorname{supp} F[f] \subset G$, and therefore, $A(D)f$ is well defined. In this sense the space $\Psi_{G,p}(\mathbb{R}^n)$ serves as a domain of pseudo-differential operators with symbols singular in the dual variable. We use the abbreviation ΨDOSS for *pseudodifferential operators with singular symbols.*

Theorem 2.1. *The space $\Psi_{G,p}(\mathbb{R}^n)$, $1 \le p \le \infty$, is invariant with respect to any operator $A(D)$ with the symbol $A(\xi) \in C^\infty(G)$. Moreover, the mapping $A(D)$: $\Psi_{G,p}(\mathbb{R}^n) \to \Psi_{G,p}(\mathbb{R}^n)$ is continuous.*

Proof. Let $f \in \Psi_{G,p}(\mathbb{R}^n)$ and $supp\, f \subset G_N$ for some $N \in \mathbb{N}$. Then $f \in \Psi_{N,p}$ (see Definition 1.23). It is obvious that

$$supp\, F[A(D)f] = supp\,(A(\xi)F[f]) = supp\, F[f] \subset G_N.$$

Moreover, since $\kappa_N \in C_0^\infty(G_N)$ the product $m_N(\xi) = \kappa_N(\xi)A(\xi) \in C_0^\infty(G)$. Mikhlin's theorem implies that any infinitely differentiable function with compact support is an L_p-multiplier. Therefore,

$$p_N(A(D)f) = \|F^{-1}[\kappa_N(\xi)A(\xi)F[f]]\|_{L_p} \le C_{N,p}\|f\|_{L_p} \le C_{N,p}p_N(f) < \infty, \tag{2.29}$$

where $C_{N,p} > 0$ is a constant not depending on f. It follows that $A(D)f \in \Psi_{N,p}$. Since $\Psi_{N,p} \subset L_p(\mathbb{R}^n)$ for each $N \in \mathbb{N}$, we have $A(D)f \in \Psi_{G,p}$. Estimate (2.29) together with Proposition 1.28 implies the continuity of $A(D)$ in $\Psi_{G,p}$.

Theorem 2.2. *The set of operators $A(D)$ with symbols $A \in C^\infty(G)$ and defined on $\Psi_{G,p}$ forms an operator algebra which is isomorphic to the algebra of symbols $C^\infty(G)$. This isomorphism is given by the correspondence $A(D) \leftrightarrow A(\xi)$, i.e.*

$$\alpha A(D) + \beta B(D) \leftrightarrow \alpha A(\xi) + \beta B(\xi), \alpha, \beta \in C^1,$$

$$A(D) \cdot B(D) \leftrightarrow A(\xi) \cdot B(\xi).$$

If $1/A(\xi)$ is also in $C^\infty(G)$, then the operator $A^{-1}(D)$ corresponding to the symbol $1/A(\xi)$ is the operator inverse to $A(D)$.

Proof. The proof is clear.

In the definition of the operator $A(D)$ for a fixed $p \in [0,\infty]$, it is not necessary that the symbol $A(\xi)$ was in $C^\infty(G)$. For example, assume that $p = 1$. Then the Fourier transform of $f \in \Psi_{G,1}$ belongs to $L_{\infty,com}(G)$. Therefore, for any $A \in L_{1,loc}(G)$, one has

$$|A(D)f(x)| \leq \|A|L_1(K)\| \|F[f]|L_\infty\| < \infty,$$

where K is any compact subset of G containing supp f. The latter together with the Housdorf-Young inequality implies

$$\|A(D)f\|_{L_\infty} \leq C\|f\|_{L_\infty}.$$

If $p = 2$, then a similar result holds true for the class of symbols $A \in L_{\infty,loc}(G)$. Namely, for any symbol $A \in L_{\infty,loc}(G)$ the corresponding operator $A(D)$ in (2.28) is meaningful for any function $f \in \Psi_{G,2}$, and maps $\Psi_{G,2}(\mathbb{R}^n)$ into $\Psi_{G,2}(\mathbb{R}^n)$. These two examples show that classes of symbols, for which (2.28) is meaningful, depends on p.

Let $S_p(G)$ denote the class of symbols for which (2.28) is well defined on the space $\Psi_{G,p}(\mathbb{R}^n)$. Theorem 2.1 shows that $C^\infty(G) \subset S_p(G)$ for each fixed $p \geq 1$. For a class of symbols $X(G)$ defined on G we set

$$XS_p(G) = \{a \in S_p(G) \text{ for some } p \in [1,\infty], \text{ such that } a \in X(G)\}.$$

For example, $C^\infty S_p(G) = C^\infty(G)$. The class $CS_p(G)$ consists of continuous symbols in $S_p(G)$. Further, we denote by $OPS_p(G)$ and $OPXS_p(G)$ the classes of operators corresponding to the classes of symbols $S_p(G)$ and $XS_p(G)$, respectively.

Example 2.1. Consider the symbols of solution operators of boundary value problems discussed in Problems 2.2.0.1–2.2.0.6 of the previous section.

1. *Problem 2.2.0.1:* The symbols of solution operators in this problem given in equation (2.3), namely $a_1(\lambda,t,\xi) = \cos(\xi\lambda t)$ and $a_2(\lambda,t,\xi) = \frac{\sin(\xi\lambda t)}{\lambda\xi}$ both belong to the class $C^\infty S_p(\mathbb{R})$ for every $\lambda \in \mathbb{C}$ and fixed $t > 0$.
2. *Problem 2.2.0.2:* The symbols of solution operators in this problem $a_3(\lambda,t,\xi) = \frac{\sin\,(1-t)\lambda\xi}{\sin\,\lambda\xi}$, and $a_4(\lambda,t,\xi) = \frac{\sin\,t\lambda\xi}{\sin\lambda\xi}$ are in the class $C^\infty S_p(G)$, where $G = \mathbb{R} \setminus \{\xi_k = \frac{\pi k}{\lambda}, k = \pm 1, \pm 2, \ldots\}$ for every fixed $\lambda \in \mathbb{C}$ and $t \geq 0$.
3. *Problem 2.2.0.3:* The symbol of the solution operator of the Cauchy problem (2.9), (2.10) is $a_5(t,\xi) = \exp(-t|\xi|^2) \in C^\infty S_p(\mathbb{R}^n)$ for every $t \geq 0$.

4. *Problem 2.2.0.4:* The symbol of the solution operator of the Cauchy problem for equation (2.15) is $a_6(t,\xi) = \exp(t\sigma_A(\xi))$, where $\sigma_A(\xi)$ is defined in equation (2.17). In particular, under some conditions to the Lévy measure ν one has $\sigma_A(\xi) = -\kappa_\alpha|\xi|^\alpha, 0 < \alpha < 2$; see equation (2.19). This particular case shows that the symbol $\sigma_A(\xi)$ is not differentiable. This is true in the general case as well. Thus, we have $\sigma_A(\xi) \in CS_p(\mathbb{R}^n)$ and $a_6(t,\xi) \in CS_p(\mathbb{R}^n)$ for every fixed $t \geq 0$.
5. *Problem 2.2.0.5:* The symbol of the solution operator of the Cauchy problem for the Schrödinger equation (2.20) is $a_7(t,\xi) = \exp\left(i\frac{\omega}{\hbar}t|\xi|^2\right) \in C^\infty S_p(\mathbb{R}^n)$ for each fixed $t \geq 0$.
6. *Problem 2.2.0.6:* The symbol $a_8(t,\xi) = s(t,\xi)$ of the solution operator for the three-point problem (2.25), (2.26) is defined in equation (2.27), and belongs to the class of symbols $C^\infty S_p(G)$, where $G = \mathbb{R} \setminus \mathcal{M}$, and

$$\mathcal{M} = \cup_{k\in\mathbb{Z}\setminus\{0\}}\{|\xi| = k\pi\} \cup \cup_{m\in\mathbb{Z}}\{|\xi| = \pm\pi/3 + 2\pi m\}.$$

The following theorem provides an extension of Theorem 2.2 to the general case of algebras of pseudo-differential operators.

Theorem 2.3. *Let $X(G)$ be an algebra of functions defined on $G \subset \mathbb{R}^n$ with respect to operations $\oplus$ and $\odot$. Then $OPXS_p(G)$ forms an operator algebra which is isomorphic to the algebra of symbols $XS_p(G)$. This isomorphism is given by the correspondence $A(D) \leftrightarrow A(\xi)$, i.e.*

$$\alpha A(D) \oplus \beta B(D) \leftrightarrow \alpha A(\xi) \oplus \beta B(\xi), \alpha, \beta \in C^1,$$

$$A(D) \odot B(D) \leftrightarrow A(\xi) \odot B(\xi).$$

If $1/A(\xi)$ is also in $XS_p(G)$, then the operator $A^{-1}(D)$ corresponding to the symbol $1/A(\xi)$ is the operator inverse to $A(D)$.

Theorem 2.4. *The space $\Psi_{G,p}$ is invariant with respect to any operator $A(D)$ with a symbol $A(\xi) \in H_{loc}^{\frac{n}{2}+\varepsilon}(G)$ for some $\varepsilon > 0$. Moreover, the mapping $A(D) : \Psi_{G,p}(\mathbb{R}^n) \to \Psi_{G,p}(\mathbb{R}^n)$ is continuous.*

Proof. Let $f \in \Psi_{G,p}(\mathbb{R}^n)$ and $supp f \subset G_N$. It is obvious that

$$\text{supp}\,(F[A(D)f]) \subseteq \text{supp}\,F[f].$$

Hence, $A(D)f \in \Psi_{G,p}(\mathbb{R}^n)$. Moreover, since $\kappa_N \in C_0^\infty(G_N)$ the product

$$m_N(\xi) = \kappa_N(\xi)A(\xi) \in H_{com}^{\frac{n}{2}+\varepsilon}(G), \varepsilon > 0.$$

The function m_N for each fixed N satisfies the condition of Mikhlin-Hörmander's multiplier theorem 1.9, and therefore is an L_p-multiplier. Hence, for each $N \in \mathbb{N}$,

$$p_N(A(D)f) = \|F^{-1}\left[\kappa_N(\xi)A(\xi)F[f]\right]\|_{L_p} \leq C_{p,N}\|f\|_{L_p} \leq C_{p,N}p_N(f) < \infty, \quad (2.30)$$

where $C_{p,N} > 0$ is a constant not depending on f. Estimate (2.30) together with Proposition 1.28 implies the continuity of $A(D)$ in $\Psi_{G,p}(\mathbb{R}^n)$.

Let $A^w(-D)$ be a weak extension of $A(D)$, i.e.

$$< A^w(-D)g, f > = < g, A(D)f >, \quad f \in \Psi_{G,p}(\mathbb{R}^n), g \in \Psi'_{-G,p'}(\mathbb{R}^n). \tag{2.31}$$

Theorem 2.5. *The space $\Psi'_{-G,p'}(\mathbb{R}^n)$ is invariant with respect to any weakly extended operator $A^w(-D)$ with a symbol $A(\xi)$ analytic in $-G$. Moreover, the mapping $A^w(-D) : \Psi'_{-G,p'}(\mathbb{R}^n) \to \Psi'_{-G,p'}(\mathbb{R}^n)$ is continuous.*

Proof. Let $g \in \Psi'_{-G,p'}(\mathbb{R}^n)$ be an arbitrary element. We show that $A^w(-D)g \in \Psi'_{-G,p'}(\mathbb{R}^n)$, as well. It is obvious from definition (2.31) that $A^w(-D)g$ is a linear functional. Assume that $f_j \in \Psi_{G,p}(\mathbb{R}^n)$ is an arbitrary convergent sequence. We have

$$< A^w(-D)g, f_j > = < g, A(D)f_j > = < g, v_j >, \tag{2.32}$$

where $v_j = A(D)f_j$. It follows from Theorem 2.1 that $v_j \to 0$ in $\Psi_{G,p}(\mathbb{R}^n)$ if $f_j \to 0$ in $\Psi_{G,p}(\mathbb{R}^n)$. Since $g \in \Psi'_{-G,p'}(\mathbb{R}^n)$, it follows that $< g, v_j > \to 0$, as $j \to \infty$. Hence, $A^w(-D)g$ is a continuous linear functional, implying $A^w(-D)g \in \Psi'_{-G,p'}(\mathbb{R}^n)$. Now let $g_m \in \Psi'_{-G,p'}(\mathbb{R}^n)$ be a sequence convergent to 0 in $\Psi'_{-G,p'}(\mathbb{R}^n)$. Then evidently, $< A^w(-D)g_m, f > = < g_m, v > \to 0$, $m \to \infty$, since $v = A(D)f \in \Psi_{G,p}(\mathbb{R}^n)$, obtaining the continuity of the operator $A^w(-D)$ in $\Psi'_{-G,p'}(\mathbb{R}^n)$.

Theorem 2.6. *The set of operators $A^w(-D)$ with symbols $A(\xi) \in XS_p(G)$ and defined on $\Psi'_{-G,p'}(\mathbb{R}^n)$ forms an operator algebra which is isomorphic to the algebra of functions $XS_p(G)$. This isomorphism is given by the correspondence $A^w(-D) \leftrightarrow A(\xi)$, i.e.*

$$\alpha A^w(-D) + \beta B^w(-D) \leftrightarrow \alpha A(\xi) + \beta B(\xi), \alpha, \beta \in C^1,$$
$$A^w(-D) \cdot B^w(-D) \leftrightarrow A(\xi) \cdot B(\xi).$$

If $1/A(\xi)$ is also in $XS_p(G)$, then the operator $\left(A^w\right)^{-1}(-D)$ corresponding to the symbol $1/A(\xi)$ is the operator inverse to $A^w(-D)$.

Proof. Follows immediately from Theorem 2.2.

Theorem 2.7. *Let $1 < q < \infty$, $-\infty < s < +\infty$ and $\mu(\mathbb{R}^n \setminus G) = 0$. For a pseudo-differential operator*

$$A(D) : \Psi_{G,2}(\mathbb{R}^n) \to \Psi_{G,2}(\mathbb{R}^n)$$

there exists a closed extension

$$\hat{A}(D) : B^s_{2q}(\mathbb{R}^n) \to B^\ell_{2q}(\mathbb{R}^n),$$

if and only if the symbol $A(\xi)$ satisfies the estimate

$$|A(\xi)| \leq C(1+|\xi|)^{s-\ell}, \quad C>0, \xi \in R^n. \tag{2.33}$$

Proof. First we show that the estimate

$$\|A(D)f|B^{\ell}_{2q}(\mathbb{R}^n)\| \leq C\|f|B^{s}_{2q}(\mathbb{R}^n)\|, \tag{2.34}$$

where $\varphi_k = F[\psi_k]$, holds for any test function $f \in \Psi_{G,2}(\mathbb{R}^n)$. Using the Parseval equality (1.34), one has $\|F^{-1}[\varphi F[f]]|L_2\| = (2\pi)^{-n}\|\varphi F[f]|L_2\|$. This implies

$$\begin{aligned}
\|A(D)f|B^{\ell}_{2q}(\mathbb{R}^n)\| &= (\sum_{k=0}^{\infty} 2^{\ell kq}\|F^{-1}\Big[\varphi_k F[A(D)f]\Big]\|^q_{L_2})^{\frac{1}{q}} \\
&= (2\pi)^{-n}(\sum_{k=0}^{\infty} 2^{\ell kq}\|\varphi_k(\xi)A(\xi)F[f](\xi)\|^q_{L_2})^{\frac{1}{q}} \\
&\leq C_1(\sum_{k=0}^{\infty} 2^{skq}\|F^{-1}\Big[\varphi_k F[f]\Big]\|^q_{L_2})^{\frac{1}{q}} = C_1\|f|B^{s}_{2q}(\mathbb{R}^n)\|,
\end{aligned}$$

since for $\xi \in \operatorname{supp}\varphi_k$ the estimate $|A(\xi)| \leq C_1 2^{(s-\ell)k}$ holds due to condition (2.33).

Extension of inequality (2.34) to an arbitrary function $f \in B^s_{2q}(\mathbb{R}^n)$ one can proceed in a standard way. Indeed, since the space of test functions $\Psi_{G,2}(\mathbb{R}^n)$ is dense in B^s_{2q}, then $f \in B^s_{2q}(\mathbb{R}^n)$ can be approximated by a sequence $f_j \in \Psi_{G,2}(\mathbb{R}^n)$ in the norm of $B^s_{2q}(\mathbb{R}^n)$. Due to the invariance theorem (Theorem 2.3 for $p=2$) the sequence $h_j = A(D)f_j \in \Psi_{G,2}(\mathbb{R}^n)$, and since $\Psi_{G,2}(\mathbb{R}^n)$ is dense in $B^{\ell}_{2q}(\mathbb{R}^n)$ as well, one has $h_j \in B^{\ell}_{2q}(\mathbb{R}^n)$, $j = 1,2,\ldots$. Further, since $B^{\ell}_{2q}(\mathbb{R}^n)$ is a complete Banach space, the sequence h_j converges to a limit h in the norm $B^{\ell}_{2q}(\mathbb{R}^n)$. Define an extension $\hat{A}$ of the operator $A(D)$ to $B^{\ell}_{2q}(\mathbb{R}^n)$ by setting $\hat{A}(D)f = h = \lim_{j\to\infty} h_j = \lim_{j\to\infty} A(D)f_j$. The inequality (2.34) implies the estimate

$$\|\hat{A}(D)f|B^{\ell}_{2q}\| \leq C\|f|B^{s}_{2q}\|, \tag{2.35}$$

valid for all $f \in B^s_{2q}(\mathbb{R}^n)$. Indeed, one needs to replace f by its approximating sequence $f_j \in \Psi_{G,p}(\mathbb{R}^n)$ in (2.34) and let $j \to \infty$, obtaining the estimate (2.35), and proving first part of the theorem.

Necessity. Let in a neighborhood of a point $\xi_* \in \mathbb{R}^n$ the inequality $|A(\xi)| > N(1+|\xi|)^{s-\ell}$ holds. Here N is an arbitrary number. We shall show that there exists a function $w \in B^s_{2q}(\mathbb{R}^n)$ such that $\|A(D)w|B^{\ell}_{2q}\| > \|w|B^s_{2q}\|$. Let $0 \neq w \in B^s_{2q}(\mathbb{R}^n)$ and $\operatorname{supp} F[w] \subset U(\xi_*)$. Without loss of generality we can take $U(\xi_*) \subset \{2^L \leq |\xi| \leq 2^{L+1}\}$ with some L. It is evident that $w \in \Psi_{G,2}(\mathbb{R}^n)$. Using the Parseval equality we have

$$\|\hat{A}(D)w|B^s_{2q}\|^q = \sum_{j=0}^{\infty} 2^{\ell jq}\|F^{-1}\left[\varphi_j F[A(D)w]\right]|L_2\|^q$$
$$= \sum_{j=0}^{\infty} 2^{\ell jq}\|\varphi_j F[A(D)w]|L_2\|^q \geq 2^{\ell qL}\Big(\int_{U(\xi_*)} |A(\xi)|^2|F[w]|^2 d\xi\Big)^{\frac{q}{2}}$$
$$> N^q 2^{\ell sL}\|\varphi_L F w|L_2\|^q = N^q\|w|B^s_{2q}\|^q.$$

Theorem 2.8. *Let $1 < p,q < \infty$, $-\infty < s < +\infty$, $m_n(R^n\backslash G) = 0$ and the symbol $A(\xi)$ of an operator $A(D) : \Psi_{G,p}(\mathbb{R}^n) \to \Psi_{G,p}(\mathbb{R}^n)$ satisfies the condition*

$$|\xi|^{|\alpha|}|D^\alpha_\xi A(\xi)| \leq C_\alpha(1+|\xi|)^{s-\ell}, C_\alpha > 0, \quad \xi \in G, |\alpha| \leq [\frac{n}{2}]+1. \tag{2.36}$$

Then

(i) there exists a unique continuous extension $\hat{A}(D) : B^s_{pq}(\mathbb{R}^n) \to B^\ell_{pq}(\mathbb{R}^n)$ of the operator $A(D)$;
(ii) there exists a unique continuous closed restriction $\hat{A}_c(-D) : B^{-\ell}_{p'q'}(\mathbb{R}^n) \to B^{-s}_{p'q'}(\mathbb{R}^n)$ of the operator $A^w(-D)$;
(iii) the equality $\hat{A}^+_c(-D) = \hat{A}(D)$, where $\hat{A}^+_c(-D)$ is the conjugate to $\hat{A}_c(-D)$, is valid.

Proof. The sufficiency of condition (2.36) can be proved similar to the previous case, however, now we use the multiplier property instead of Parseval's equality. Let again $f \in \Psi_{G,p}(\mathbb{R}^n)$ and the symbol $A(\xi)$ of the pseudo-differential operator $A(D)$ satisfy condition (2.36). Let $\{\psi_k\}_{k=0}^\infty \in \Phi$. For test functions $f \in \Psi_{G,p}(\mathbb{R}^n)$ we show the estimate

$$\|A(D)f|B^\ell_{pq}\| = (\sum_{k=0}^{\infty} 2^{lkq}\|F^{-1}\left[\varphi_k F[A(D)f]\right]\|^q_{L_p})^{\frac{1}{q}} \leq C\|f|B^s_{pq}\|, \tag{2.37}$$

where $\varphi_k = F[\psi_k]$, $k \geq 0$. Due to Part 4) of Definition 1.20, $\varphi_k \in M_p$, $1 < p < \infty$, for each $k = 0,1,\ldots$. Since $f \in \Psi_{G,p}(\mathbb{R}^n)$, there exists a compact set $K \subset G$, such that $\operatorname{supp} F[f] \subset K$. Moreover, there exists a natural number m, such that $\varphi_k F[A(D)f](\xi) \equiv 0$ for all $k > m$. Therefore, inequality (2.37) for $f \in \Psi_{G,p}(\mathbb{R}^n)$ takes the form

$$\|A(D)f|B^\ell_{pq}\| = (\sum_{k=0}^{m} 2^{lkq}\|F^{-1}\left[\varphi_k F[A(D)f]\right]\|^q_{L_p})^{\frac{1}{q}} \leq C\|f|B^s_{pq}\|,$$

with some natural number m depending on f.

Now consider a collection of functions

$$\phi_k(x) = F^{-1}\Big(\varphi_k(\xi)\cdot\frac{2^{\nu k}}{(1+|\xi|^2)^{\nu/2}}\Big), \quad k = 0,1,2,\ldots,$$

with $\nu = s - l$. One can check that the functions $\phi_k(x)$ satisfy conditions (1) - (4) of Definition 1.20. Hence, $\{\phi_k\}_{k=0}^{\infty} \in \Phi$. Since, both collections $\{\psi_k\}_{k=0}^{\infty}$ and $\{\phi_k\}_{k=0}^{\infty}$ belong to Φ, the respective norms $\|f|B^s_{pq}\|_\psi$ and $\|f|B^s_{pq}\|_\phi$ are equivalent. Hence,

$$\|A(D)f|B^\ell_{pq}\| \leq C\|A(D)f|B^\ell_{pq}\|_\phi = C(\sum_{k=0}^{m} 2^{lkq}\|F^{-1}\Big(F[\phi_k]\cdot F[A(D)f]\Big)\|^q_{L_p})^{\frac{1}{q}}$$

$$= C(\sum_{k=0}^{m} 2^{lkq}\|\phi_k * A(D)f\|^q_{L_p})^{\frac{1}{q}}, \tag{2.38}$$

where $C > 0$ is a constant. Further, it is easy to see that

$$\phi_k * A(D)f = F^{-1}(F\phi_k \cdot FA(D)f) = F^{-1}\Big(\varphi_k \cdot \frac{2^{\nu k}}{(1+|\xi|^2)^{\nu/2}} A(\xi)Ff\Big).$$

Denote by $m(\xi)$ the function $A(\xi)\cdot(1+|\xi|^2)^{-\frac{\nu}{2}}$. It follows from the Michlin's Theorem (see Theorem 1.7) that $m(\xi)$ is a Fourier multiplier in the space L_p. We have

$$\phi_k * A(D)f = 2^{\nu k}(F^{-1}m * \varphi_k * f).$$

Using the multiplier property of $m(\xi)$ we obtain the estimate

$$\|\phi_k * A(D)f|L_p\| = 2^{\nu k}\|F^{-1}\big(m(\xi)F[\varphi_k(\xi)f(\xi)]\big)|L_p\|$$

$$\leq C_p 2^{\nu k}\|\varphi_k * f|L_p\|. \tag{2.39}$$

Estimates obtained in (2.39) and (2.38) imply that

$$\|A(D)f|B^\ell_{pq}\| \leq C(\sum_{k=0}^{m} 2^{lkq}\|\phi_k * A(D)f|L_p\|^q)^{\frac{1}{q}}$$

$$\leq CC_p(\sum_{k=0}^{m} 2^{(l+\nu)kq}\|\varphi_k * f|L_p\|^q)^{\frac{1}{q}}$$

$$= CC_p\|f|B^{\nu+\ell}_{pq}\| = CC_p\|f|B^s_{pq}\|.$$

Thus, estimate (2.37) is valid for any test function $f \in \Psi_{G,p}(\mathbb{R}^n)$. The extension of inequality (2.37) to an arbitrary function $f \in B^s_{pq}(\mathbb{R}^n)$ repeats exactly the extension process proceeded in the previous theorem. Hence, part (i) of the theorem is proved.

Now let $g \in B^{-\ell}_{p'q'}(\mathbb{R}^n)$ and $f \in B^s_{pq}(\mathbb{R}^n)$. Since $f = \lim_{j\to\infty} f_j$, $f_j \in \Psi_{G,p}(\mathbb{R}^n)$, the restriction $\hat{A}_c(-D)$ to $B^{-\ell}_{p'q'}(\mathbb{R}^n)$ of the operator $A^w(-D) : \Psi'_{-G,p'}(\mathbb{R}^n) \to \Psi'_{-G,p'}(\mathbb{R}^n)$ we determine by

$$< \hat{A}_c(-D)g, f >= \lim_{j\to\infty} < g, A(D)f_j >=< g, \hat{A}(D)f > . \tag{2.40}$$

Here $\hat{A}(D)$ is the extension of the operator $A(D) : \Psi_{G,p}(\mathbb{R}^n) \to \Psi_{G,p}(\mathbb{R}^n)$ constructed above. We have the estimate

$$| < \hat{A}_c(-D)g, f > | \leq \|g|B^{-\ell}_{p'q'}\| \cdot \|\hat{A}(D)f|B^\ell_{pq}\| \leq C\|g|B^{-\ell}_{p'q'}\| \cdot \|f|B^s_{pq}\|.$$

It follows that

$$\|\hat{A}_c(D)g|B^{-\ell}_{p'q'}\| = \sup_{f\neq 0} \frac{\| < \hat{A}_c(-D)g, f > \|}{\|f|B^s_{pq}\|}$$
$$\leq C\|g|B^{-\ell}_{p'q'}\|, \tag{2.41}$$

proving part (ii).

Finally, when both operators $\hat{A}(D)$ and $\hat{A}_c(-D)$ are determined, it follows from (2.40) that $< \hat{A}_c(-D)g, f > = < g, \hat{A}(D)f >$ valid for arbitrary $g \in B^{-\ell}_{p'q'}(\mathbb{R}^n)$ and $f \in B^s_{pq}(\mathbb{R}^n)$. The latter yields the equality $\hat{A}^+_c(-D) = \hat{A}(D)$, completing the proof.

Theorem 2.8 remains valid for Lizorkin-Triebel spaces $F^s_{pq}(\mathbb{R}^n)$ as well.

Theorem 2.9. *Let $1 < p, q < \infty$, $-\infty < s < +\infty$, $m_n(R^n \setminus G) = 0$ and the symbol $A(\xi)$ of an operator $A(D) : \Psi_{G,p}(\mathbb{R}^n) \to \Psi_{G,p}(\mathbb{R}^n)$ satisfies the condition*

$$|\xi|^{|\alpha|}|D^\alpha_\xi A(\xi)| \leq C_\alpha(1+|\xi|)^{s-\ell}, C_\alpha > 0, \xi \in G, |\alpha| \leq [\frac{n}{2}]+1. \tag{2.42}$$

Then

(i) there exists a unique continuous extension $\hat{A}(D) : F^s_{pq}(\mathbb{R}^n) \to F^\ell_{pq}(\mathbb{R}^n)$ of the operator $A(D)$;
(ii) there exists a unique continuous closed restriction $\hat{A}_c(-D) : F^{-\ell}_{p'q'}(\mathbb{R}^n) \to F^{-s}_{p'q'}(\mathbb{R}^n)$ of the operator $A^w(-D)$;
(iii) the equality $\hat{A}^+_c(-D) = \hat{A}(D)$, where $\hat{A}^+_c(-D)$ is the conjugate to $\hat{A}_c(-D)$, is valid.

Returning to properties of pseudo-differential operators with singular symbols defined on $\Psi'_{-G,p'}(\mathbb{R}^n)$ one can reformulate the representation Theorem 1.20 in terms of pseudo-differential operators.

Theorem 2.10. *Let $1 < p < \infty$ and $p' = p/(p-1)$, the conjugate number of p. For each distribution $f \in \Psi'_{-G,p'}(\mathbb{R}^n)$ there exist a pseudo-differential operator $A(D)$ with the symbol $A(\xi)$ analytic in G and a function f_0 with $Ff_0 \in L_p(G)$ such that the representation $f(x) = A(D)f_0(x)$ holds.*

Remark 2.2. This result has an independent interest for analysis. We will use it essentially in proofs of uniqueness theorems in the next chapter.

2.4 Pseudo-differential operators with continuous symbols and hypersingular integrals

In Example 2.2.0.4 of Section 2.2, related to jump processes, the solution operator appeared to be a pseudo-differential operators with a non-differentiable symbol.

Moreover, the symbol $\sigma_A(\xi) = -\kappa_\alpha|\xi|^\alpha, 0 < \alpha < 2$, in equation (2.19) is continuous on $\mathbb{R}^n$, but not differentiable at the origin. Pseudo-differential operators like this will play an important role in the various random walk models discussed in Chapter 8 and used for description of fractional order diffusion processes.

Assume that G is an open domain in $\mathbb{R}^n$. Let a function f be continuous and bounded on $\mathbb{R}^n$, i.e., $f \in C_b(\mathbb{R}^n)$, and have a Fourier transform $F[f](\xi)$ in the sense of distributions, which has a compact support in G. The set of all such functions endowed with the convergence in the following sense is denoted by $C\Psi_G(\mathbb{R}^n)$: a sequence of functions $f_m \in C\Psi_G(\mathbb{R}^n)$ is said to converge to a function $f_0 \in C\Psi_G(\mathbb{R}^n)$ if:

1. there exists a compact set $K \subset G$ such that $\operatorname{supp} F[f_m] \subset K$ for all $m = 1,2,\ldots$;
2. $\|f_m - f_0|C_b\| = \sup_{x\in\mathbb{R}^n} |f_m - f_0| \to 0$ as $m \to \infty$.

In the case $G = \mathbb{R}^n$ we write simply $C\Psi(\mathbb{R}^n)$ omitting $\mathbb{R}^n$ in the subindex of $C\Psi_G(\mathbb{R}^n)$. Note that according to the Paley-Wiener-Schwartz theorem (Theorem 1.12) functions in $C\Psi_G(\mathbb{R}^n)$ are entire functions of finite exponential type. In accordance with Theorem 1.10 a function $f \in L_p(\mathbb{R}^n)$ with $p > 2$ has the Fourier transform $F[f]$ belonging to $H^{-s}(\mathbb{R}^n), s > n(\frac{1}{2} - \frac{1}{p})$. Letting $p \to \infty$ we have $F[f] \in H^{-s}(\mathbb{R}^n), s > \frac{n}{2}$ for $f \in L_\infty(\mathbb{R}^n)$. Taking into account this fact and the Paley-Wiener-Schwartz theorem we have that the Fourier transform of $f \in \Psi_G(\mathbb{R}^n)$ belongs to the space

$$\bigcap_{s>\frac{n}{2}} H^{-s}_{comp}(G),$$

where $H^{-s}_{comp}(G)$ is a negative order Sobolev space of functionals with compact support on G. Moreover, since $\lim_{p\to\infty} \|f\|_{L_p} = \|f\|_{L_\infty}$ for $f \in L_\infty(\mathbb{R}^n)$ (see, for example, [RS80]), it follows from Theorem 1.10 that

$$\|F[f]\|_{-s} \le C\|f\|_{L_\infty}. \tag{2.43}$$

Thus, $F[f]$ is a distribution well defined on the space of continuous functions with the topology of locally uniform convergence.

Denote by $C\Psi'_{-G}(\mathbb{R}^n)$ the space of all linear bounded functionals defined on the space $C\Psi_G(\mathbb{R}^n)$ and endowed with the weak topology. Namely, we say that a sequence of functionals $g_m \in C\Psi'_{-G}(\mathbb{R}^n)$ converges to an element $g_0 \in C\Psi'_{-G}(\mathbb{R}^n)$ in the weak sense, if for all $f \in C\Psi_G(\mathbb{R}^n)$ the sequence of numbers $< g_m, f >$ converges to $< g_0, f >$ as $m \to \infty$. By $< g, f >$ we mean the value of $g \in C\Psi'_{-G}(\mathbb{R}^n)$ on an element $f \in C\Psi_G(\mathbb{R}^n)$.

Let $A(\xi)$ be a continuous function defined in $G \subset \mathbb{R}^n$. By definition, a pseudo-differential operator $A(D)$ with the symbol $A(\xi)$ is

$$A(D)\varphi(x) = \frac{1}{(2\pi)^n}\langle F[\varphi], A(\xi)e^{-ix\xi}\rangle, \tag{2.44}$$

which is well defined on functions of $C\Psi_G(\mathbb{R}^n)$. We recall that $x\xi$ in equation (2.44) is the dot product of vectors $x \in \mathbb{R}^n_x$, $\xi \in \mathbb{R}^n_\xi$, that is $x\xi = x_1\xi_1 + \cdots + x_n\xi_n$. If $F[\varphi]$

is an integrable function with $\operatorname{supp} F[\varphi] \subset G$, then (2.44) becomes the usual form of pseudo-differential operator

$$A(D)\varphi(x) = \frac{1}{(2\pi)^n} \int\limits_G A(\xi)F[\varphi](\xi)e^{-ix\xi}d\xi,$$

with the integral taken over G. Note that, in general, this may not be meaningful even for infinitely differentiable functions with finite support (see Remark 2.3).

We define the operator $A(-D)$ acting on the space $C\Psi'_{-G}(\mathbb{R}^n)$ by the extension formula

$$< A(-D)f, \varphi > = < f, A(D)\varphi >, \; f \in C\Psi'_{-G}(\mathbb{R}^n), \; \varphi \in C\Psi_G(\mathbb{R}^n). \tag{2.45}$$

Lemma 2.1. *The pseudo-differential operators $A(D)$ and $A(-D)$ with a continuous symbol $A(\xi)$ are continuous mappings:*

$$A(D) : \Psi_G(\mathbb{R}^n) \to \Psi_G(\mathbb{R}^n), \quad A(-D) : \Psi'_{-G}(\mathbb{R}^n) \to \Psi'_{-G}(\mathbb{R}^n).$$

Proof. Indeed, since $\operatorname{supp} F[A(D)f] \subset \operatorname{supp} A(\xi)F[f]$, one has $\operatorname{supp} F[A(D)f] \Subset G$ for any function $f \in C\Psi_G(\mathbb{R}^n)$. Further, suppose that $f_m \to 0$ in $\Psi_G(\mathbb{R}^n)$, that is there exists a compact $K \Subset G$ such that $\operatorname{supp} F[f_m] \subset K$ for all $m \geq 1$, and $\sup |f_m| \to 0, m \to \infty$. Since $F[f_m] \in H^{-s}(K)$ for some $s > n/2$, using (2.43) we have the estimate

$$\begin{aligned} |A(D)f_m(x)| &= \left| \frac{1}{(2\pi)^n} \langle F[f_m](\xi), A(\xi)e^{-ix\xi} \rangle_\xi \right| \\ &\leq C \| F[f_m] | H^{-s}(K) \| \, \| A(\xi)e^{-ix\xi} | H^s(K) \| \\ &\leq C_1 \| f_m \|_{L_\infty} \| A(\xi)e^{-ix\xi} | H^s(K) \|. \end{aligned} \tag{2.46}$$

Further, since $A(\xi)$ is a function uniformly bounded over any compact, one has

$$\| A(\xi)e^{-ix\xi} | H^s(K) \|^2 = \int\limits_K |A(\xi)|^2 (1 + |\xi|^2)^s d\xi = C_K < \infty.$$

Taking this and estimate (2.46) into account, we finally get

$$\sup_{x \in \mathbb{R}^n} |A(D)f_m(x)| \leq C C_K \| f_m \| \to 0 \text{ as } m \to \infty.$$

The continuity of the second mapping in the lemma now follows by duality.

Lemma 2.2. *Let $A(\xi)$ be a function continuous on $\mathbb{R}^n$. Then for $\xi \in \mathbb{R}^n$*

$$A(D)\{e^{-ix\xi}\} = A(\xi)e^{-ix\xi}, \; x \in \mathbb{R}^n. \tag{2.47}$$

Proof. For any fixed $\xi \in \mathbb{R}^n$ the function $e^{-ix\xi}$ is in $C\Psi(\mathbb{R}^n)$. By definition (2.44) of $A(D)$, we have

$$\begin{aligned} A(D)\{e^{-ix\xi}\} &= \frac{1}{(2\pi)^n}\langle F[e^{-ix\xi}](\eta), A(\eta)e^{-ix\eta}\rangle \\ &= \langle \delta(\eta-\xi), A(\eta)e^{-ix\eta}\rangle = A(\xi)e^{-ix\xi}. \end{aligned}$$

Corollary 2.11 *The following relations hold:*

i) $A(\xi) = (A(D)e^{-ix\xi})e^{ix\xi}$;
ii) $A(\xi) = (A(D)e^{-ix\xi})_{|x=0}$;
iii) $A(\xi) =< A(-D)\delta_0(x), e^{-ix\xi} >$, *where* δ_0 *is the Dirac distribution.*

Proof. The first two assertions immediately follow from (2.47). To prove (iii), we have

$$< A(-D)\delta_0(x), e^{-ix\xi} > = < \delta_0(x), A(D)e^{-x\xi} > = < \delta_0(x), A(\xi)e^{-ix\xi} > = A(\xi).$$

Remark 2.3. Equality (2.47) holds in the space $C\Psi(\mathbb{R}^n)$, and therefore, understood in the usual pointwise sense. It is a valid equality, as indicated in many sources (see, e.g., [Hor83, Tre80]), in the space $\mathscr{E}(\mathbb{R}^n)$ of test functions, as well. However, since the function $e^{-ix\xi}$ does not belong to $S(\mathbb{R}^n)$ and $\mathscr{D}(\mathbb{R}^n)$, the representations for the symbol obtained in Lemma 2.2 and Corollary 2.11 are not applicable in these spaces.

Example 2.2. Consider the operator

$$\mathbb{D}_0^\alpha f(x) = \frac{1}{d(\alpha,l)}\int_{\mathbb{R}^n} \frac{\Delta_y^l f(x)}{|y|^{n+\alpha}}dy \equiv \frac{1}{d(\alpha,l)}\lim_{N\to\infty}\int_{|y|\le N} \frac{\Delta_y^l f(x)}{|y|^{n+\alpha}}dy \tag{2.48}$$

where $0 < \alpha < l$, l is a positive integer, Δ_y^l is the finite difference of order l in the y direction, either centered or non-centered, and $d(\alpha,l)$ is a constant defined in dependence on what type of difference is taken (see for details [SKM87]). Due to a strong singularity in the integrand, this operator is also called a hypersingular integral. As we will show in Section 3.8 (see Theorem 3.4), this operator is a pseudo-differential operator with the continuous symbol $-|\xi|^\alpha$, and plays an important role in random walk constructions studied in Chapter 8.

Note that in this book we consider only the centered case of the finite difference Δ_y^l in the definition of D_0^α. In this case $d(\alpha,l)$ is defined as (see [SKM87])

$$d(\alpha,l) = \frac{\pi^{1+n/2}A_l(\alpha)}{2^\alpha\Gamma(1+\frac{\alpha}{2})\Gamma(\frac{n+\alpha}{2})\sin(\alpha\pi/2)}, \tag{2.49}$$

with $A_l(\alpha)$ determined by the formula

$$A_l(\alpha) = 2\sum_{k=0}^{[l/2]}(-1)^{k-1}\binom{l}{k}\left(\frac{l}{2}-k\right)^\alpha.$$

Moreover $d(\alpha,l) \neq 0$ for all $\alpha > 0$ and for even l, but $d(\alpha,l)$ is identically zero for odd orders l. Let l be a given positive integer. Denote by τ_y a shift operator with spatial vector-step y

$$(\tau_y f)(x) = f(x-y), \quad x,y \in \mathbb{R}^n.$$

Using this operator we determine the symmetric difference operator of order l

$$(\Delta_y^l f)(x) = (\tau_{-\frac{y}{2}} - \tau_{\frac{y}{2}})^l f(x) = \sum_{k=0}^{l} (-1)^k \binom{l}{k} f\left(x + (\frac{l}{2} - k)y\right).$$

For $l = 2$, we have

$$A_2(\alpha) = -2,$$
$$d(\alpha,2) = -\frac{2\pi^{1+n/2}}{2^{\alpha}\frac{\alpha}{2}\Gamma(\frac{\alpha}{2})\Gamma(\frac{n+\alpha}{2})\sin(\alpha\pi/2)},$$
$$(\Delta_y^2 f)(x) = f(x-y) - 2f(x) + f(x+y).$$

Hence, the operator D_0^α in the case $l = 2$ can be written in the form

$$\mathbb{D}_0^\alpha f(x) = B(n,\alpha) \int_{\mathbb{R}^n} \frac{f(x-y) - 2f(x) + f(x+y)}{|y|^{n+\alpha}} dy, \tag{2.50}$$

where

$$B(n,\alpha) = -\frac{\alpha\Gamma(\frac{\alpha}{2})\Gamma(\frac{n+\alpha}{2}) sin\frac{\alpha\pi}{2}}{2^{2-\alpha}\pi^{1+n/2}}. \tag{2.51}$$

It is seen from (2.51) that the value $\alpha = 2$ is degenerate. For $0 < \alpha < 2$, it follows from Lemma 2.2 (and some calculations provided in Section 3.8, see Theorem 3.4) that

$$\mathbb{D}_0^\alpha \left(e^{ix\xi}\right)_{|x=0} = B(n,\alpha) \left(\int_{\mathbb{R}^n} \frac{\Delta_y^2 e^{ix\xi}}{|y|^{n+\alpha}} dy\right)_{|x=0} = -|\xi|^\alpha, 0 < \alpha < 2.$$

Therefore, for $f \in H^s(\mathbb{R}^n)$, $s \in \mathbb{R}$, one has

$$\|\mathbb{D}_0^\alpha f|H^{s-\alpha}\|^2 = \int_{\mathbb{R}^n} (1+|\xi|^2)^{s-\alpha} |\xi|^{2\alpha} |F[f](\xi)|^2 d\xi \leq C\|f|H^s\|^2.$$

Thus, the mapping $\mathbb{D}_0^\alpha : H^s(\mathbb{R}^n) \to H^{s-\alpha}(\mathbb{R}^n)$ is continuous.

We note also that $\mathbb{D}_0^\alpha$ in (2.50) can be considered as a fractional power of the Laplace operator, namely $\mathbb{D}_0^\alpha = -(-\Delta)^{\alpha/2}$. Fractional powers of positive definite operators are discussed in Section 3.9.

2.5 ΨDOSS: variable symbols

In this section we briefly discuss algebras of pseudo-differential operators with symbols $a(x,\xi)$ depending on both variables x and ξ. J. Kohn and L. Nirenberg first constructed an algebra of pseudo-differential operators $OPS^m(\Omega)$ with smooth symbols $a(x,\xi) \in C^\infty(\Omega, \mathbb{R}^n \setminus \{0\})$, satisfying the condition

$$|D_x^\beta D_\xi^\alpha a(x,\xi)| \leq C(1+|\xi|)^{m-|\alpha|}, \quad x \in K, \xi \in \mathbb{R}^n, \tag{2.52}$$

for all multi-indices α and β. Here $\alpha = (\alpha_1, \dots, \alpha_n)$ and $\beta = (\beta_1, \dots, \beta_n)$ are multi-indices, $|\alpha| = \alpha_1 + \cdots + \alpha_n$ is the length of α; $D_x = (-i\partial/\partial x_1, \dots, -i\partial/\partial x_n)$, $D_\xi = (-i\partial/\partial \xi_1, \dots, -i\partial/\partial \xi_n)$; $K \subset \Omega$ is a compact subset, and $C = C(\alpha, \beta, K)$ is a positive constant. By definition, $A \in OPS^m(\Omega)$ with the symbol $a(x,\xi) \in S^m(\Omega)$, if

$$Af(x) = \frac{1}{(2\pi)^n} \int_{\mathbb{R}^n} a(x,\xi) F[f](\xi) e^{-ix\xi} d\xi, \quad x \in \Omega.$$

A differential operator

$$A(x,D) = \sum_{|\alpha| \leq m} a_\alpha(x) D^\alpha$$

with coefficients $a_\alpha \in \mathbb{C}^\infty(\Omega)$ is an example of the operator in $OPS^m(\Omega)$. The corresponding symbol is a polynomial in the variable ξ,

$$a(x,\xi) = \sum_{|\alpha| \leq m} a_\alpha(x) \xi^\alpha.$$

Thus, the algebra $OPS^m(\Omega)$ contains all the differential operators of order m with infinite differentiable in Ω coefficients. In the algebra $OPS^m(\Omega)$ the addition and multiplication (composition) operations are well defined, as well as the adjoint operator. The reader is referred to books [Tay81, Tre80, Hor83] for details. Below we briefly mention the main ideas laid behind the construction of the algebra $OPS^m(\Omega)$ and its generalizations. The algebra $OPS^m(\Omega)$ is constructed so that it contains the parametrices of all the elliptic operators of order m. A differential operator $A(x,D)$ with the symbol $a(x,\xi)$ is called elliptic, if its main symbol

$$a_m(x,\xi) = \sum_{|\alpha| = m} a_\alpha(x) \xi^\alpha$$

satisfies the condition

$$a_m(x,\xi) \geq C_0 |\xi|^m \tag{2.53}$$

for all $x \in \Omega$ and $\xi \in \mathbb{R}^n$. For the symbol $a(x,\xi)$ of an elliptic differential operator it is obvious that $|a(x,\xi) - a_m(x,\xi)| \leq (1+|\xi|)^{m-1}$ for all ξ such that $|\xi| \geq R$ for some $R > 0$. Therefore, using this fact and (2.53), one can see that there exists a constant $C > 0$, such that the estimate

$$|a(x,\xi)| = |a_m(x,\xi) + [a(x,\xi) - a_m(x,\xi)]|$$
$$\geq |a_m(x,\xi)| - |a(x,\xi) - a_m(x,\xi)|$$
$$\geq C(1+|\xi|)^m$$

holds for all $x \in \Omega$ and for all $|\xi| > R$, where R is sufficiently large. This property can be used for extension of the definition of ellipticity for operators in $OPS^m(\Omega)$. Namely, an operator $A \in OPS^m(\Omega)$ is called *elliptic*, if there exists a constant $C > 0$, such that the symbol of A satisfies the condition

$$|a(x,\xi) \geq C(1+|\xi|)^m|, \quad x \in \Omega, \ |\xi| \geq R,$$

with some number $R > 0$. Further, an operator $P \in OPS^m(\Omega)$ with the symbol $p(x,\xi)$ is called a *parametrix* for an elliptic operator $A \in OPS^m(\Omega)$ with the symbol $a(x,\xi)$, if the following relations

$$a(x,\xi)p(x,\xi) = 1 + b(x,\xi)$$
$$p(x,\xi)a(x,\xi) = 1 + c(x,\xi),$$

hold, in which the symbols $b(x,\xi), c(x,\xi) \in S^{-\infty}(\Omega) \equiv \cap_{m=1}^{\infty} S^m(\Omega)$. As it follows from Proposition 2.1, Part (3) below, that the operators with symbols in $S^{-\infty}(\Omega)$ possess the smoothing property: these operators transfer distributions to infinitely differentiable functions.

Though within $OPS^m(\Omega)$ one can describe parametrices of elliptic operators, the class $OPS^m(\Omega)$ is too restrictive to describe, so-called, hypo-ellipticity of (pseudo) differential operators. If for arbitrary $f \in C^\infty(\Omega)$ a distributional solution $u \in \mathfrak{D}(\Omega')$, $\Omega' \subset \Omega$, of the equation $Au = f$, is in $C^\infty(\Omega')$, then A is hypo-elliptic. Any elliptic pseudo-differential operator is hypo-elliptic [Hor83]. Another example of hypo-elliptic operators is the heat operator $\frac{\partial}{\partial t} - \Delta$, which is not an elliptic operator. The hypo-ellipticity of differential operators was studied in works by Hörmander [Hor61, Hor67], Egorov [Ego67], etc. The class of symbols $S^m_{\rho,\delta}(\Omega)$ depending on parameters $\rho \in (0,1]$ and $\delta \in [0,1]$ was introduced by Hörmander [Hor65]. By definition, a symbol $a(x,\xi) \in C^\infty(\Omega, \mathbb{R}^n \setminus \{0\})$ belongs to the class $S^m_{\rho,\delta}(\Omega)$ if $a(x,\xi)$ satisfies the condition

$$|D_x^\beta D_\xi^\alpha a(x,\xi)| \leq C(1+|\xi|)^{m-\rho|\alpha|+\delta|\beta|}, \quad x \in K, \ \xi \in \mathbb{R}^n, \tag{2.54}$$

for all multi-indices α and β, and compacts $K \subset \Omega$. The corresponding class of pseudo-differential operators $OPS^m_{\rho,\delta}(\Omega)$ is wider than $OPS^m(\Omega)$ and within this class one can describe the hypo-ellipticity property of (pseudo) differential operators. The class of operators $OPS^m_{\rho,\delta}(\Omega)$ coincides with $OPS^m(\Omega)$ if $\rho = 1, \ \delta = 0$. We write S^m, $S^m_{\rho,\delta}$, OPS^m, and $OPS^m_{\rho,\delta}$, if $\Omega = \mathbb{R}^n$.

Proposition 2.1._(1) Let $A \in OPS^m_{\rho,\delta}(\Omega)$. Then the mapping $A : \mathscr{D}(\Omega) \to \mathscr{E}(\Omega)$, and by duality the mapping $A : \mathscr{E}'(\Omega) \to \mathscr{D}'(\Omega)$ are continuous;_

(2) *Let* $A \in OPS^m_{\rho,\delta}$, $\delta < 1$. *Then the mapping* $A : \mathscr{G} \to \mathscr{G}$, *and by duality the mapping* $A : \mathscr{G}' \to \mathscr{G}'$ *are continuous;*
(3) *Let* $A \in OPS^m_{\rho,\delta}$, $0 \le \delta < \rho \le 1$. *Then the mapping* $A : H^s(\mathbb{R}^n) \to H^{s-m}(\mathbb{R}^n)$ *is continuous for all* $s \in \mathbb{R}$;
(4) *Let* $A \in OPS^m_{\rho,\delta}$. *Then the mapping* $A : \Psi_{p,G}(\mathbb{R}^n) \to \mathscr{E}(\mathbb{R}^n)$, *and by duality the mapping* $A : \mathscr{E}'(\mathbb{R}^n) \to \Psi'_{p',-G}(\mathbb{R}^n)$ *are continuous; If additionally the Fourier transform of the symbol* $a(x,\xi)$ *of the operator* A *with respect to the variable* x *has a compact support* K_a *for all* $\xi \in G$, *then the mapping* $A : \Psi_{p,G}(\mathbb{R}^n) \to \Psi_{p,G_a}(\mathbb{R}^n)$, *where* $G_a = G + K_a$, *and by duality the mapping* $A : \Psi'_{p',-G_a}(\mathbb{R}^n) \to \Psi'_{p',-G}(\mathbb{R}^n)$ *are continuous.*

Proof. Parts (1)–(3) are known (see, e.g., [Tay81, Hor83]). Part (4) is a particular case of the more general statement established in Theorems 2.12 and 2.13.

One can notice that the symbols of solution pseudo-differential operators obtained in Section 2.2, except some of them, do not belong to $S^m_{\rho,\delta}$ for any finite m. For instance, the symbols $a_1(1,t,\xi) = \cos(t\xi)$ and $a_2(1,t,\xi) = \xi^{-1}\sin(t\xi)$ (these symbols do not depend on x) emerged in Problem 2.2.0.1, do not satisfy estimate (2.54). Indeed, for $\alpha = 0$ one has

$$|a_1(1,t,\xi)| \le 1, \ \ |a_2(1,t,\xi)| \le C_t(1+|\xi|)^{-1}, \xi \in \mathbb{R},$$

One can easily verify that derivatives of $a_1(1,t,\xi)$ and $a_2(1,t,\xi)$ in the variable ξ satisfy the inequalities

$$|D^m_\xi a_1(1,t,\xi)| \le C_t, \text{ and } |D^m_\xi a_2(1,t,\xi)| \le C_t(1+|\xi|)^{-1},$$

for all $\xi \in \mathbb{R}$. These facts show that (2.54) does not hold for any finite m and positive ρ, $0 < \rho \le 1$.

Remark 2.4. Notice that if one includes $\rho = 0$ in the definition of $S^m_{\rho,\delta}$, then $a_1 \in S^0_{0,0}$ and $a_2 \in S^{-1}_{0,0}$. Calderon and Vaillancourt [CV71] showed that operators with symbols in $S^m_{\rho,\rho}$, where $0 \le \rho < 1$ are bounded in the Sobolev spaces $H^s(\mathbb{R}^n)$. This fact implies well posedness in the sense of Hadamard in spaces $H^s(\mathbb{R}^n)$ of the Cauchy problem in Example 2.2.0.1 in the case $\lambda = 1$.

On the other hand, both symbols $a_1(1,t,\xi)$ and $a_2(1,t,\xi)$ can be expanded to the Taylor series

$$a_1(1,t,\xi) = \sum_{j=0}^{\infty} \frac{t^{2j}}{(2j)!}\xi^{2j}, \quad a_2(1,t,\xi) = \sum_{j=0}^{\infty} \frac{t^{2j+1}}{(2j+1)!}\xi^{2j}. \tag{2.55}$$

In other words $a_1, a_2 \in S^\infty$. The class of corresponding pseudo-differential operators OPS^∞ represents differential operators of infinite order. The differential operators of infinite order play an important role in the modern theory of differential equations [BG76, Dub82, K73], complex analysis [Dub84, Leo76], and functional analysis [Sat60, Kan72, Gra10].

Similarly, the solution operator $\exp(i\frac{\omega}{\hbar}t\Delta)$ in Example 2.2.0.6 for the Schrödinger equation is also a differential operator of infinite order with the symbol $a_7(t,\xi) = \exp(i\frac{\omega}{\hbar}t|\xi|^2)$, which does not belong to $S^m_{\rho,0}$ for any finite m and $\rho \in (0,1]$. Unlike the previous example a_3 does not belong $S^m_{\rho,0}$ for any finite m even for $\rho = 0$. However, this symbol does belong to S^∞, with the power series representation

$$a_7(t,\xi) = \sum_{j=0}^{\infty} \frac{(i\omega t)^j}{\hbar^J j!}|\xi|^{2j}. \tag{2.56}$$

The reader can verify that symbols $a_3(\lambda,t,\xi), a_4(\lambda,t,\xi), a_6(t,\xi), a_8(t,\xi)$ also do not belong to $S^m_{\rho,0}$ for any finite m and positive ρ. All these symbols can be represented as differential operators of infinite order locally or globally. The symbol $a_5(t,\xi)$ is an exception. Due to infinite differentiability and exponential decay at infinity, this symbol belongs to $S^m_{\rho,0}$ for any finite m and $0<\rho\leq 1$. Hence, $a_5(t,\xi) \in S^{-\infty}_{\rho,0}$, confirming that the operator $\frac{\partial}{\partial t} - \Delta$ is hypoelliptic.

The power series for the symbols $a_1(1,t,\xi)$ and $a_2(1,t,\xi)$ in (2.55), and for $a_7(t,\xi)$ in (2.56) converge for all $\xi \in \mathbb{R}$ and for all $\xi \in \mathbb{R}^n$, respectively. However, power series representations of the symbols

$$a_3(1,t,\xi) = \frac{\sin\,(1-t)\xi}{\sin\,\xi}, \quad a_4(1,t,\xi) = \frac{\sin\,t\xi}{\sin\xi}$$

of the solution operators arising in Problem 2.2.0.2 converge locally in the open set $G = \mathbb{R}\setminus\{k\pi, k=\pm1,\pm2,\dots\}$, and are functions of $C^\infty(G)$.

Now we introduce a class of symbols which contains all the above symbols.

Definition 2.3. Let $G\subseteq\mathbb{R}^n_\xi$ and $\Omega\subseteq\mathbb{R}^n_x$ be open sets. We denote by $S^\infty_G(\Omega)$ the class of symbols $a(x,\xi)\in C^\infty(\Omega\times G)$. We do not require any conditions for the growth in the variable ξ like estimates in equations (2.52) and (2.54). Symbols in $S^\infty_G(\Omega)$ as functions of variables (x,ξ) are jointly infinite differentiable in $\Omega\times G$, or in general, functions in $C^\infty(G)$ on the cotangent bundle $T^*(\Omega)$, and may have any type of singularities on the boundary of G or outside of G. The class of corresponding pseudodifferential operators will be denoted by $OPS^\infty_G(\Omega)$. We write S^∞_G and OPS^∞_G if $\Omega=\mathbb{R}^n$.

Theorem 2.12. *Let $a(x,\xi)\in S^\infty_G$.*

1. *The mapping*

$$A : \Psi_{G,p}(\mathbb{R}^n) \to \mathscr{E}(\mathbb{R}^n) \tag{2.57}$$

is continuous;

2. *If $a(x,\xi)$ has compact support in x, then the mapping*

$$A : \Psi_{G,p}(\mathbb{R}^n) \to \mathscr{D}(\mathbb{R}^n) \tag{2.58}$$

is continuous;

3. *If the Fourier transform of the symbol $a(x,\xi)$ of the operator A with respect to the variable x has a compact support K_a for all $\xi \in G$, then the mapping*

$$A : \Psi_{p,G}(\mathbb{R}^n) \to \Psi_{p,G_a}(\mathbb{R}^n), \tag{2.59}$$

where $G_a = G + K_a$, is continuous.

Proof. Let $a(x,\xi) \in S_G^\infty$ be the symbol of the operator A. Then for $f \in \Psi_{p,G}(\mathbb{R}^n)$, one has

$$Af(x) = \frac{1}{(2\pi)^n} \int_G a(x,\xi) F[f](\xi) e^{-ix\xi} d\xi. \tag{2.60}$$

Moreover, since $F[f]$ has a compact support in G, the integral in (2.60) is taken over a compact set. This implies that $Af \in C^\infty(\mathbb{R}^n)$. Now we show the continuity of the mapping (2.57). Let $f_m \in \Psi_{G,p}(\mathbb{R}^n)$ be a sequence converging to 0 in the topology of $\Psi_{G,p}(\mathbb{R}^n)$. Recall that $\Psi_{G,p}(\mathbb{R}^n)$ is the inductive limit of the Banach spaces $\Psi_{N,p}$ (see Section 1.10). Therefore, one can assume that $f_m \in \Psi_{N,p}$ for all $m \in \mathbb{N}$, and $f_m \to 0$ in the norm of $\Psi_{N,p}$. Moreover, since $f_m \in \Psi_{N,p}$, then $\operatorname{supp} F[f_m] \subset G_n$ and $\|f_m|\Psi_{N,p}\| = \|F^{-1}[\kappa_n(\xi)F[f]]|L_p\|$ (see Definition 1.23). We show that $Af_m \to 0$ in the topology of $\mathscr{E}(\mathbb{R}^n)$. Let $K \subset \mathbb{R}^n$ be an arbitrary compact. For the derivative of order α of Af_m, we have

$$\begin{aligned} D_x^\alpha Af_m(x) &= \frac{1}{(2\pi)^n} \sum_{\beta \preceq \alpha} \binom{\alpha}{\beta} \int_G D_x^\beta a(x,\xi)(i\xi)^{\alpha-\beta} \kappa_N(\xi) F[f_m](\xi) e^{-ix\xi} d\xi \\ &= \sum_{\beta \preceq \alpha} \binom{\alpha}{\beta} \int_{\mathbb{R}^n} b(x,y) f_m(y) dy, \quad x \in K, \end{aligned} \tag{2.61}$$

where

$$b(x,y) = \frac{1}{(2\pi)^n} \int_G D_x^\beta a(x,\xi)(i\xi)^{\alpha-\beta} \kappa_N(\xi) e^{-i(x-y)\xi} d\xi.$$

The function $b(x,y)$ for every fixed $x \in K$ is a function of $L_q(\mathbb{R}_y^n)$, where q is the conjugate of p, that is $q^{-1} + p^{-1} = 1$. Indeed, putting for convenience

$$c_N(x,\xi) = \frac{1}{(2\pi)^n} D_x^\beta a(x,\xi)(i\xi)^{\alpha-\beta} \kappa_N(\xi),$$

one has

$$\begin{aligned} \|b(x,y)\|_{L_q(\mathbb{R}_y^n)}^q &= \int_{\mathbb{R}^n} \left| \int_G c_N(x,\xi) e^{-i(x-y)\xi} d\xi \right|^q dy = \int_{|y-x|\le 1} \left| \int_G c_N(x,\xi) e^{-i(x-y)\xi} d\xi \right|^q dy \\ &+ \int_{|y-x|\ge 1} \left| \int_G c_N(x,\xi) e^{-i(x-y)\xi} d\xi \right|^q dy. \end{aligned} \tag{2.62}$$

The first integral on the right of (2.62) is finite, since the integral is taken over a compact set $\{|y-x| \leq 1\} \times g \subset \mathbb{R}^{2n}$. For the second integral using integration by parts and taking into account that $c_N(x,\xi)$ has a compact support $\operatorname{supp} c_N(x,\cdot) \subset G_N$ for every fixed $x \in K$, one has

$$\int\limits_{|y-x|\geq 1} \Big|\int\limits_{G} c_N(x,\xi)e^{-i(x-y)\xi}d\xi\Big|^q dy = \int\limits_{|y-x|\geq 1} \frac{1}{|x-y|^{qM}}\Big|\int\limits_{G} c_N(x,\xi)D_\xi^\gamma e^{-i(x-y)\xi}d\xi\Big|^q dy$$
$$= \int\limits_{|y-x|\geq 1} \frac{1}{|x-y|^{qM}}\Big|\int\limits_{G} (-D_\xi)^\gamma c_N(x,\xi)e^{-i(x-y)\xi}d\xi\Big|^q dy$$
$$\leq C_N \int\limits_{|y-x|\geq 1} \frac{1}{|x-y|^{qM}}dy < \infty,$$

where $|\gamma| = M$ and $Mq > n$, and C_N is a positive constant. Applying the Hölder inequality to (2.61), one obtains the desired estimate

$$\sup_{x\in K}|D^\alpha Af_m(x)| \leq C_N\|f_m|\Psi_{N,p}\|.$$

The proof of Part 1 is complete. The proof of Part 2 is similar to the proof of Part 1. Therefore, we leave it to the reader. To prove Part 3 we notice that the Fourier transform of $Af(x)$ is

$$F[Af](\eta) = \int\limits_{\mathbb{R}^n} \hat{a}(\eta-\xi,\xi)F[f](\xi)d\xi, \quad \eta \in \mathbb{R}^n, \tag{2.63}$$

where

$$\hat{a}(\eta,\xi) = \frac{1}{(2\pi)^n}\int\limits_{\mathbb{R}^n} a(x,\xi)e^{-ix\eta}dx. \tag{2.64}$$

It follows from (2.63) that $F[Af](\eta) = 0$ if $\eta \notin G + K_a$, implying

$$\operatorname{supp} F[Af] = \operatorname{supp} F[f] + K_a \subset G + K_a.$$

Further, let $\Psi_{G+K_a,p} = ind \lim_{L\to\infty} \Phi_{L,p}$, where $\Phi_{L,p}$ is a sequence of Banach spaces, corresponding to a locally finite covering $\{h_L\}_{L=0}^\infty$ of $G + K_a$, and the smooth partition of unity $\{v_L\}_{L=0}^\infty$. Since, this inductive limit does not depend on the partition of unity, one can construct it in the form

$$v_j(\eta) = \phi_k(\xi)w_\ell(\eta-\xi), \tag{2.65}$$

where $j = k+\ell$, $\eta \in G+K_a$, $\xi \in G$, and ϕ_k and w_ℓ are smooth partition of unities of G and K_a, respectively. Let

$$\overline{\kappa}_L(\eta) = \sum_{j=1}^{L} v(\eta), \quad \tilde{\kappa}_M(\zeta) = \sum_{\ell=1}^{M} w(\zeta), \quad \text{and} \quad \kappa_N(\xi) = \sum_{k=1}^{N} \phi_k(\xi).$$

The numbers L, N, and M are such that $L = N + M$. Using (2.65), one can easily verify that $\overline{\kappa}_L(\eta) = \kappa_N(\xi)\tilde{\kappa}_M(\eta - \xi)$. Exploiting this and (2.63), we have

$$\begin{aligned}
\|F^{-1}\Big[\overline{\kappa}_L(\eta)F[Af]\Big]|L_p\| &= \|\frac{1}{(2\pi)^n}\int\limits_{\mathbb{R}^n}\overline{\kappa}_L(\eta)\Big(\int\limits_{\mathbb{R}^n}\hat{a}(\eta-\xi,\xi)F[f](\xi)d\xi\Big)e^{-i\eta x}d\eta\,|L_p\| \\
&= \|\frac{1}{(2\pi)^n}\int\limits_{G}F[f](\xi)\Big(\int\limits_{\mathbb{R}^n}\overline{\kappa}_L(\eta)\hat{a}(\eta-\xi,\xi)e^{-i\eta x}d\eta\Big)d\xi\,|L_p\| \\
&= \|\frac{1}{(2\pi)^n}\int\limits_{G}\kappa_N(\xi)F[f](\xi)\Big(\int\limits_{\mathbb{R}^n}\tilde{\kappa}_M(\zeta)\hat{a}(\zeta,\xi)e^{-i\zeta x}d\zeta\Big)e^{-i\xi x}d\xi\,|L_p\| \\
&= \|(2\pi)^n F^{-1}\Big[\kappa_N(\xi)F^{-1}\Big(\tilde{\kappa}_M(\zeta)\hat{a}(\zeta,\xi)\Big)F[f](\xi)\Big]|L_p\|. \qquad (2.66)
\end{aligned}$$

In the third equality of this chain we used the change of variable $\eta = \xi + \zeta$. Now, due to definition of the symbol $a(x,\xi)$ the function

$$p_N(x,\xi) = \kappa_N(\xi)F^{-1}\Big[\tilde{\kappa}_M(\zeta)\hat{a}(\zeta,\xi)\Big] = \frac{\kappa_N(\xi)}{(2\pi)^n}\int\limits_{\mathbb{R}^n}\tilde{\kappa}_M(\zeta)\hat{a}(\zeta,\xi)e^{-ix\zeta}d\zeta \qquad (2.67)$$

is infinite differentiable and has a compact support as a function of ξ for every fixed x, and therefore can be considered as a symbol in $S^{-\infty}(\mathbb{R}^n)$. It is known (see [Hor83, Tay81]) that if the order of the pseudodifferential operator with a symbol in $S^m(\mathbb{R}^n)$ is negative, then this operator maps $L_p(\mathbb{R}^n)$ to itself. Using this fact, one obtains from (2.66) that

$$\begin{aligned}
\|Af|\Psi_{L,p}\| = \|F^{-1}\Big[\overline{\kappa}_L(\eta)F[Af]\Big]|L_p\| &= \|P_N(x,D)f(x)|L_p\| \\
&\leq C_N\|f|\Psi_{N,p}\|, \qquad (2.68)
\end{aligned}$$

where $P_N(x,D)$ is the pseudo-differential operator with the symbol in (2.67). The estimate obtained in (2.68) implies the continuity of mapping (2.59).

Theorem 2.13. *Let $a(x,\xi) \in S^{\infty}_{G}$.*

1. *The mapping*

$$A : \mathscr{E}'(\mathbb{R}^n) \to \Psi'_{-G,p'}(\mathbb{R}^n)$$

is continuous;

2. *If $a(x,\xi)$ has compact support in x, then the mapping*

$$A : \mathscr{D}'(\mathbb{R}^n) \to \Psi'_{-G,p'}(\mathbb{R}^n)$$

is continuous;

3. *If the Fourier transform of the symbol $a(x,\xi)$ of the operator A with respect to the variable x has a compact support K_a for all $\xi \in G$, then the mapping*

$$A : \Psi'_{p',-G_a}(\mathbb{R}^n) \to \Psi'_{p',-G}(\mathbb{R}^n),$$

where $G_a = G + K_a$, is continuous.

Proof. The proof of this theorem follows from Theorem 2.12 by duality.

Theorem 2.14. *Let $1 < p,q < \infty$, $-\infty < s < +\infty$, $G \subset \mathbb{R}^n$, and the symbol $a(x,\xi)$ of a pseudo-differential operator $A(x,D)$ belong to S^∞_G. Assume the following conditions:*

(i) $m_n(R^n \backslash G) = 0$;

(ii) the Fourier transform of the symbol $a(x,\xi)$ of the operator $A(x,D)$ with respect to the variable x has a compact support containing in a compact K_a for all $\xi \in G$;

(iii) there are a function $k \in L_1(\mathbb{R}^n)$ and a number $m \in \mathbb{R}$ such that the inequality

$$|a(x,\xi)| \leq k(x)(1+|\xi|^2)^{m/2}, \quad x \in \mathbb{R}^n,\ \xi \in G, \tag{2.69}$$

holds.

Then the mapping $A(x,D) : H^s(\mathbb{R}^n) \to H^{s-m}(\mathbb{R}^n)$ is continuous.

Proof. Due to Theorem 2.12 condition (ii) implies that

$$A(x,D) : \Psi_{G,p}(\mathbb{R}^n) \to \Psi_{G_a,p}(\mathbb{R}^n).$$

Condition (i), in accordance with Theorem 1.21, implies that $\Psi_{G,p}(\mathbb{R}^n)$ is dense in the Sobolev spaces $H^s(\mathbb{R}^n)$ and $H^{s-m}(\mathbb{R}^n)$. Therefore, in order to prove the Theorem we need to estimate the $A(x,D)f(x)$, $f \in \Psi_{G,p}(\mathbb{R}^n)$, in the norms of the corresponding Sobolev spaces. Recall that $f \in H^s(\mathbb{R}^n)$ if $(1+|\xi|^2)^{s/2}F[f](\xi) \in L_2(\mathbb{R}^n)$, or equivalently, $(I-\Delta)^{s/2}f(x) \in L_2(\mathbb{R}^n)$. The norm of $f \in H^s(\mathbb{R}^n)$ is (see Parseval's equality (1.34))

$$\|f|H^s\| = \|(1+|\xi|^2)^{s/2}F[f]|L_2\| = (2\pi)^n \|F^{-1}\left[(1+|\xi|^2)^{s/2}F[f]\right]|L_2\|.$$

Let $f \in \Psi_{G,p}(\mathbb{R}^n)$ with $\operatorname{supp} f = G_0 \Subset G$. Then, taking into account (2.63), one has

$$\begin{aligned}\|A(x,D)f|H^{s-m}\| &= (2\pi)^n \|F^{-1}\left[(1+|\eta|^2)^{\frac{s-m}{2}} F[A(x,D)f](\eta)\right]|L_2\| \\ &= \left\| \int_{\mathbb{R}^n} (1+|\eta|^2)^{\frac{s-m}{2}} \left(\int_G \hat{a}(\eta-\xi,\xi)F[f](\xi)d\xi \right) e^{-i\eta x} d\eta \right\|_{L_2},\end{aligned} \tag{2.70}$$

where $\hat{a}(\zeta,\xi)$ is defined in (2.64). Further, changing order of integration and using substitution $\eta - \xi = \zeta$, one can reduce (2.70) to

$$\begin{aligned}\|A(x,D)f|H^{s-m}\| &= (2\pi)^n \left\| F^{-1}\left[\int_{\mathbb{R}^n} (1+|\xi+\zeta|^2)^{\frac{s-m}{2}} \hat{a}(\zeta,\xi) e^{-i\zeta x} d\zeta \right] F[f](\xi) \right\|_{L_2} \\ &= \|q(x,\xi)F[f](\xi)\|_{L_2},\end{aligned} \tag{2.71}$$

where

$$q(x,\xi) = \int\limits_{K_a} (1+|\xi+\zeta|^2)^{\frac{s-m}{2}} \hat{a}(\zeta,\xi) e^{-i\zeta x} d\zeta.$$

Here K_a is a compact set due to condition (ii). Further, it follows from condition (2.69) and equation (2.64) that

$$|\hat{a}(\zeta,\xi)| \le \frac{1}{(2\pi)^n} \int_{\mathbb{R}^n} |a(x,\xi)| dx \le \frac{\|k(x)\|_{L_1}}{(2\pi)^n} (1+|\xi|^2)^{\frac{m}{2}}. \qquad (2.72)$$

Moreover, for all $\xi, \zeta \in \mathbb{R}^n$, the inequality $1+|\xi+\eta|^2 \le 2(1+|\xi|^2)(1+|\zeta|^2)$ holds. Taking into account this inequality and estimate (2.72), one has

$$|q(x,\xi)| \le \frac{(\sqrt{2})^{s-m}\|k(x)\|_{L_1}}{(2\pi)^n} (1+|\xi|^2)^{\frac{s}{2}} \int\limits_{K_a} (1+|\zeta|^2)^{\frac{s-m}{2}} d\zeta \le C(1+|\xi|^2)^{\frac{s}{2}},$$

where C is a positive constant. Thanks to this estimate for $q(x,\xi)$, it follows from (2.71) that

$$\|A(x,D)f|H^{s-m}\| \le C\|(1+|\xi|^2)^{\frac{s}{2}} F[f](\xi)\|_{L_2} = C\|f|H^s\|.$$

Theorem 2.15. *Let $1 < p,q < \infty$, $-\infty < s < +\infty$. Let the set $G \subset \mathbb{R}^n$ and the symbol $a(x,\xi)$ of a pseudodifferential operator $A(x,D)$ satisfy the following conditions:*

(a) $m_n(R^n \backslash G) = 0$;
(b) $a(x,\xi)$ has a compact support K_a in the variable x for all $\xi \in G$;
(c) for all $x \in K_a$ and $\xi \in G$ there exist numbers $s,l \in \mathbb{R}$ and a constant $C_\alpha > 0$, such that

$$|\xi|^{|\alpha|} |D_\xi^\alpha a(x,\xi)| \le C_\alpha (1+|\xi|)^{s-\ell}, C_\alpha > 0, |\alpha| \le [\frac{n}{2}]+1.$$

Then

(i) there exists a unique continuous extension $\hat{A}(x,D) : B^s_{pq}(\mathbb{R}^n) \to B^\ell_{pq}(\mathbb{R}^n)$ of the operator $A(x,D)$;
(ii) there exists a unique continuous closed restriction $\hat{A}_c(x,-D) : B^{-\ell}_{p'q'}(\mathbb{R}^n) \to B^{-s}_{p'q'}(\mathbb{R}^n)$ of the operator $A^w(x,-D)$;
(iii) the equality $\hat{A}^+_c(x,-D) = \hat{A}(D)$, where $\hat{A}^+_c(-D)$ is the conjugate to $\hat{A}_c(x,-D)$, is valid.

2.6 ΨDOSS in spaces of periodic functions and periodic distributions

In this section we briefly consider pseudo-differential operators with singular symbols introduced above in the spaces of periodic functions and distributions.

Let $\mathbb{Z}^n$ be the n-dimensional integral lattice and $\mathbb{T}^n$ be an n-dimensional torus, namely $\mathbb{T}^n \equiv \{x \in \mathbb{R}^n : |x_j| \leq \pi, j = 1, \ldots, n\}$ is an n-dimensional cube whose opposite sides are identified. Let $\varphi \in L_1(\mathbb{T}^n)$ and its formal Fourier series is

$$\varphi(x) \sim \sum_{k \in \mathbb{Z}^n} \varphi_k e^{ikx}, \quad x \in \mathbb{T}^n, \tag{2.73}$$

where φ_k, $k \in \mathbb{Z}^n$, are Fourier coefficients of φ, i.e.

$$\varphi_k = (2\pi)^{-n} \int_{\mathbb{T}^n} \varphi(x) e^{-ikx} dx.$$

Since $\varphi \in L_1(\mathbb{T}^n)$ its Fourier coefficients are finite: $|\varphi_k| \leq (2\pi)^{-n} \|\varphi\|_{L_1}$. However, the Fourier series on the right-hand side of (2.73), in general, may not converge to $\varphi(x)$ in the norm of $L_1(\mathbb{T}^n)$. In fact, Kolmogorov [Kol26] constructed an example of a function $f \in L_1[-\pi, \pi]$ whose Fourier series is divergent everywhere on $[-\pi, \pi]$. Even for continuous function its Fourier series may not converge (see Additional Remarks to this chapter). In multidimensional case the situation is much more complicated. Now, convergence of the Fourier series depends on how partial sums are formed. We refer the reader to survey papers [Ste58, AP89, Wei12] where the convergence of Fourier series in different norms (uniform, L_p, etc.) for various forms of partial sums (spherical, cubic, rectangular, etc.) are discussed.

In order to avoid such difficulties we consider only functions $\varphi \in L_2(\mathbb{T}^n)$. In this case the Fourier series on the right side of (2.73) converges to $\varphi(x)$ in the norm of $L_2(\mathbb{T}^n)$, no matter how the corresponding partial sum is formed. The inner product of $f, g \in L_2(\mathbb{T}^n)$ is denoted by (f,g). For $f, g \in L_2(\mathbb{T}^n)$, Parseval's equality reads

$$(f,g) = (2\pi)^n \sum_{k \in \mathbb{Z}^n} f_k \overline{g}_k,$$

which immediately implies the following relation for the norm:

$$\|f|L_2(\mathbb{T}^n)\|^2 = (2\pi)^n \sum_{k \in \mathbb{Z}^n} |f_k|^2.$$

The Sobolev space with the smoothness order s and denoted by $W_2^s \equiv W_2^s(\mathbb{T}^n)$ is defined as the set of functions $f \in L_2(\mathbb{T}^n)$, such that

$$\|f|W_2^s\|^2 = \sum_{k \in \mathbb{Z}^n} (1 + |k|^2)^s |f_k|^2 < \infty.$$

Now we develop periodic analogs of the spaces $\Psi_{G,2}(\mathbb{R}^n)$ and $\Psi'_{-G,2}(\mathbb{R}^n)$. Let $\mathfrak{M}_2$ be the set of entire functions of the form

$$a(\xi) = \sum_{|\alpha|=0}^{\infty} a_\alpha \xi^{2\alpha}, \quad \xi \in \mathbb{R}^n,$$

where $a_\alpha \geq 0$ and $\lim_{\ell \to \infty} \sqrt[\ell]{\sum_{|\alpha|=\ell} a_\alpha} = 0$. In other words, any function $a \in \mathfrak{M}_2$ admits an analytic (entire) continuation to $\mathbb{C}^n$.

Definition 2.4. We introduce the space

$$W_{\mathbb{T}^n}^{+\infty} = \{\varphi \in L_2(\mathbb{T}^n) : \|\varphi\|_a^2 = \sum_{k \in Z^n} a(k)|\varphi_k|^2 < \infty, \forall\, a \in \mathfrak{M}_2\},$$

where φ_k, $k \in \mathbb{Z}^n$, are Fourier coefficients of φ, and $a(k)$ is the value of $a(\xi)$ at the lattice point $k = (k_1, \ldots, k_n) \in \mathbb{Z}^n$. A sequence $\varphi_m(x) \in W_{\mathbb{T}^n}^{+\infty}$ is said to converge to φ in $W_{\mathbb{T}^n}^{+\infty}$ if for every $a \in \mathfrak{M}_2$ the sequence $\|\varphi_m - \varphi\|_a \to 0$.

Theorem 2.16. *A function φ belongs to $W_{\mathbb{T}^n}^{+\infty}$ if and only if there exists integer $N(\varphi)$ such that $\varphi_k = 0$ if $|k| > N$.*

Proof. The sufficiency is obvious. Let us show the necessity. We will prove it by contradiction. Let $\varphi \in L_2(\mathbb{T}^n)$. Assume that for every natural N there exists $k \in Z^n$ such that $|k| > N$ and $\varphi_k \neq 0$. We choose two sequences R_l and σ_l which satisfy the following conditions:

i) $R_l \to \infty, l \to \infty$;
ii) $\sum_{l=1}^{\infty} \sigma_l \tau_l = \infty$, where $\tau_l = \sum_{k \in Z_l^n} |\varphi_k|^2$, $Z_l^n = \{k \in Z^n : R_l \leq |k| \leq R_{l+1}\}$.

Let for $k \in \mathbb{Z}_l^n$ the inequality

$$a_l(k) = \sum_{|\alpha|=m_{l-1}}^{m_l} \frac{k^{2\alpha}}{R_l^{2|\alpha|}} \geq \sigma_l. \tag{2.74}$$

holds. This inequality can always be achieved by an appropriate choice of numbers m_l. Consider the function

$$a_0(\xi) = \sum_{l=1}^{\infty} a_l(\xi) = \sum_{l=1}^{\infty} \sum_{|\alpha|=m_{l-1}}^{m_l} \frac{\xi^{2\alpha}}{R_l^{2|\alpha|}}.$$

In fact, $a_0(\xi)$ is an entire function. Indeed, due to condition i) we have

$$\lim_{k \to \infty} \lim_{|\alpha| \to \infty} \left(\frac{1}{R_l^{2|\alpha|}} \right)^{\frac{1}{2|\alpha|}} = 0,$$

which shows that the radius of convergence of the power series expansion of a_0 is infinite. Now taking into account assumptions i), ii), and inequality (2.74), we obtain

$$\|\varphi\|_{a_0}^2 = \sum_{|\alpha|=0}^{\infty} a_0(k)|\varphi_k|^2 = \sum_{l=1}^{\infty} \sum_{k \in Z_l^n} a_l(k)|\varphi_k|^2 \geq \sum_{l=1}^{\infty} \sigma_l \tau_l = +\infty,$$

i.e., φ does not belong to $W_{\mathbb{T}^n}^{+\infty}$.

Definition 2.5. $W_{\mathbb{T}^n}^{-\infty}$ is the space of linear continuous functionals defined on $W_{\mathbb{T}^n}^{+\infty}$ and endowed with the weak convergence. The value of $f \in W_{\mathbb{T}^n}^{-\infty}$ on the element $\varphi \in W_{\mathbb{T}^n}^{\infty}$ will be written in the form $< f, \varphi >$.

Let $f \in W_{T^n}^{-\infty}$ and the series $\sum f_k e^{ikx}$ be its formal Fourier series. Here

$$f_k = (2\pi)^{-n} < f(x), e^{-ikx} >, \quad k \in \mathbb{Z}^n.$$

One can show that this series converges to f in the weak sense and the equality

$$< f, \varphi >= (2\pi)^n \sum_{k \in \mathbb{Z}^n} f_{-k} \varphi_k, \quad \varphi \in W_{\mathbb{T}^n}^{+\infty},$$

holds.

Theorem 2.17. *Let $s \in \mathbb{R}$. Then the embedding $W_{\mathbb{T}^n}^{+\infty} \hookrightarrow W_2^s(\mathbb{T}^n)$ is continuous and dense.*

Proof. Since the embeddings

$$W_2^s(\mathbb{T}^n) \hookrightarrow L_2(\mathbb{T}^n) \hookrightarrow W_2^{-s}(\mathbb{T}^n), \quad s > 0,$$

are continuous and dense, it is sufficient to show that $W_{\mathbb{T}^n}^{+\infty}$ is continuously and densely embedded to $L_2(\mathbb{T}^n)$. It follows from Theorem 2.16 that the convergence of a sequence φ_m to φ_0 in $W_{\mathbb{T}^n}^{+\infty}$ is equivalent to the following:

(i) there exists a natural number N such that $\varphi_k^{(m)} = 0$ for all $k : |k| > N$, and $m = 1, 2, \ldots$;
(ii) $\|\varphi_m - \varphi_0\|_{L_2} \to 0$, $m \to \infty$.

This immediately implies the continuity of the embedding in the theorem. Further, suppose $\varphi \in W_2^s(\mathbb{T}^n)$. We take as the approximating sequence

$$\varphi_N(x) = \sum_{|k| \le N} \varphi_k e^{ikx} \in W_{\mathbb{T}^n}^{\infty}.$$

Then

$$\|\varphi - \varphi_N\|_{L_2}^2 = (2\pi)^n \sum_{|k| \le N} |\varphi_k|^2 (1 + |k|^2)^s \to 0, \; N \to \infty.$$

Definition 2.6. Let $A(k) : \mathbb{Z}^n \to \mathbb{C}$ be a discrete function defined on the integer lattice $\mathbb{Z}^n$. We define a pseudo-differential operator with the symbol $A(k)$ by

$$A(D)f(x) = \sum_{k \in Z^n} A(k) f_k e^{ikx}, \quad x \in \mathbb{T}^n. \tag{2.75}$$

Theorem 2.18. *The space $W_{\mathbb{T}^n}^{+\infty}$ is invariant with respect to the operator $A(D)$. Moreover, the mapping $A(D) : W_{\mathbb{T}^n}^{+\infty} \to W_{\mathbb{T}^n}^{+\infty}$ is continuous.*

Proof. Let $f \in W_{\mathbb{T}^n}^{+\infty}$, i.e. there is a natural number N_f, so that

$$f(x) = \sum_{|k| \le N_f} f_k e^{ikx}.$$

Then

$$A(D)f(x) = \sum_{|k|\le N_f} A(k) f_k e^{ikx} \in W_{\mathbb{T}^n}^{+\infty}.$$

Moreover, one has the estimate

$$\|A(D)f\|_a^2 = \sum_{k\in\mathbb{Z}^n} a(k)|[A(D)f]_k|^2 = \sum_{|k|\le N_f} a(k)|A(k)|^2|f_k|^2 \le C\|f\|_a^2,$$

where $C = \max_{|k|\le N_f} |A(k)|$. This inequality immediately implies the continuity of the mapping $A(D) : W_{\mathbb{T}^n}^{+\infty} \to W_{\mathbb{T}^n}^{+\infty}$.

Further, we determine a weak extension of $A(D)$ to the dual space $W_{\mathbb{T}^n}^{-\infty}$ by the formula

$$< A^w(D)f, \varphi > = < f, A(-D)\varphi >, \ f \in W_{\mathbb{T}^n}^{-\infty}, \quad \varphi \in W_{\mathbb{T}^n}^{+\infty}.$$

Theorem 2.18 implies the following corollary.

Corollary 2.19 *$W_{\mathbb{T}^n}^{-\infty}$ is invariant with respect to the operator $A^w(D)$. Moreover, the mapping*

$$A^w(D) : W_{\mathbb{T}^n}^{-\infty} \to W_{\mathbb{T}^n}^{-\infty}$$

is continuous.

Theorem 2.20. *Let $A(D) : W_{\mathbb{T}^n}^{+\infty} \to W_{\mathbb{T}^n}^{+\infty}$ be a pseudo-differential operator defined in (2.75). This operator has a unique closed extension $\hat{A}(D) : W_2^s(\mathbb{T}^n) \to W_2^l(\mathbb{T}^n)$ if and only if the condition $|A(k)| \le C(1+|k|^2)^{\frac{s-l}{2}}$, $C > 0$, $k \in \mathbb{Z}^n$, is fulfilled.*

Proof. *Sufficiency.* Let $f \in W_{\mathbb{T}^n}^{+\infty}$. Then

$$\begin{aligned}\|A(D)f\|_l^2 &= (2\pi)^n \sum_{k\in\mathbb{Z}^n} |A(k)|^2|f_k|^2(1+|k|^2)^l \\ &\le (2\pi)^n \sum_{k\in Z^n} C^2|f_k|^2(1+|k|^2)^s = C^2\|f\|_s^2. \end{aligned} \tag{2.76}$$

If $f \in W_2^s(\mathbb{T}^n)$, then according to Theorem 2.17 there exists a sequence $f_m \in W_{\mathbb{T}^n}^{+\infty}$ converging to f in the norm of $W_2^s(\mathbb{T}^n)$. We put $g_k = A(D)f_k$. Due to Theorem 2.18 $g_k \in W_{T^n}^{+\infty}$. It follows from (2.76) that $\|g_k - g_m\|_l \le C\|f_k - f_m\|_s$. Since f_m is convergent in $W_2^s(\mathbb{T}^n)$, it follows that g_m is fundamental in $W_2^l(\mathbb{T}^n)$. Hence, there exists a function $g \in W_2^s(\mathbb{T}^n)$ such that $A(D)f_k \to g$ in the norm of $W_2^s(\mathbb{T}^n)$. We set $\hat{A}f = g$. It is easy to see that this definition does not depend on the choice of f_m. Closing the estimate (2.76) we obtain that $\hat{A}(D)$ is bounded.

Necessity. Assume that there is a sequence $k_N, N = 1,2,\ldots$, such that $|k_N| \to \infty$, and

$$|A(k_N)| > N(1+|k_N|^2)^{\frac{s-l}{2}}. \tag{2.77}$$

Consider the sequence of functions

$$f_N(x) = f_{k_N} e^{ik_N x}, x \in \mathbb{T}^n,$$

with $f_{k_N} = (1+|k_N|^2)^{-s/2}$. It is clear that $\|f_N\|_s = 1$ for all N, i.e., $f_N \in W_2^s$. Moreover, due to inequality (2.77), one has the estimate

$$\|A(D)f_N\|_l^2 = (2\pi)^n |A(k_N)|^2 |f_{k_N}|^2 (1+|k_N|^2)^l > N^2.$$

This estimate contradicts to the continuity of $A(D)$.

2.7 Pseudo-differential operators with complex arguments

In this section we briefly discuss pseudodifferential operators with symbols $a(\zeta)$ depending on complex variables $\zeta = (\zeta_1, \ldots, \zeta_n) \in \mathbb{C}^n$. If one follows the procedure

$$A(D)u(z) = F^{-1}\Big[a(\zeta)F[u(\zeta)]\Big](z) \tag{2.78}$$

for construction of classes of pseudo-differential operators, then one needs to know what is the Fourier transform F and its inverse F^{-1} for functions depending on complex variables. There are two classical alternatives for the complex Fourier transform: the Fourier-Laplace transform and the Borel transform. We have seen the Fourier-Laplace transform, when we formulated the Paley-Wiener-Schwartz theorem, Theorem 1.12. Below we introduce the Fourier-Laplace transform in general case and establish a connection with the Borel transform. For simplicity we show this connection in the case $n = 1$. Then we introduce a complex version of the Fourier transform F, which actually is an extension of the Borel transform to the class of analytic functionals. The operator F is very convenient for construction of pseudo-differential operators using the procedure (2.78). In Chapter 9 we will go further and will study pseudo-differential operators and equations with singular symbols depending on n complex variables.

Let $\mathscr{O}(\mathbb{C}^n)$ be the space of entire analytic functions endowed with the topology of uniform convergence. Taking a sequence of compacts $K_n \subset \mathbb{C}^n$, such that $K_n \subset K_{n+1}$, and $\mathbb{C}^n = \cup_{n=1}^{\infty} K_n$, and introducing Banach spaces $\mathscr{A}_n$ of functions f analytic in a neighborhood of K_n with the norm $\|f|\mathscr{A}_n\| = \max_{K_n}|f(z)|$, one can see that $\mathscr{O}(\mathbb{C}^n)$ can be represented as the inductive limit of $\mathscr{A}_n$. Therefore, due to Proposition 1.21 $\mathscr{O}(\mathbb{C}^n)$ is a locally convex topological vector space of Fréchet type. The conjugate space of $\mathscr{O}(\mathbb{C}^n)$ with the weak topology is called *analytic functionals* and denoted by $\mathscr{O}'(\mathbb{C}^n)$. The value of $\mu \in \mathscr{O}'(\mathbb{C}^n)$ on $f \in \mathscr{O}(\mathbb{C}^n)$ we denote by $\mu(f)$ or $<\mu, f>$. For each analytic functional μ there exists a compact set K_μ determining μ (see, [Hor90]). Also, linear combinations of functions of the form $\exp(z,\zeta)$, where $\zeta \in \mathbb{C}^n$ and $(z,\zeta) = z_1\zeta_1 + \cdots + z_n\zeta_n$, form a dense subset of $\mathscr{O}(\mathbb{C}^n)$ [Hor90]. Let $Exp(\mathbb{C}^n)$ be the space of entire functions of finite exponential type, that is

$$Exp(\mathbb{C}^n) \equiv \{f \in \mathscr{O}(\mathbb{C}^n) : |f| \leq Ce^{r_1|z_1| + \ldots + r_n|z_n|}\},$$

with some constant $C > 0$ and $r = (r_1, \dots, r_n)$, $r_1 > 0, \dots, r_n > 0$. Taking a sequence $r^{(k)} = (r_1^k, \dots, r_n^{(k)}), k = 1, 2, \dots,$ such that $r_j^{k+1} > r_j^{(k)}$, $j = 1, \dots, n$, and $r^k \to \infty$, as $k \to \infty$, one has a sequence of Banach spaces

$$\mathscr{X}_k \equiv \{f \in \mathscr{O}(\mathbb{C}^n) : |f| \leq C_k e^{r_1^{(k)}|z_1| + \dots + r_n^{(k)}|z_n|}\},$$

with the norm $\|f|\mathscr{X}_k\| = \sup_{z \in \mathbb{C}^n}(e^{-r_1^{(k)}|z_1| - \dots - r_n^{(k)}|z_n|}|f(z)|)$. It is easy to see that $\mathscr{X}_k \subset \mathscr{X}_{k+1}$ is a continuous inclusion for each k, and $Exp(\mathbb{C}^n)$ is the inductive limit of $\mathscr{X}_k$. Hence, $Exp(\mathbb{C}^n)$ is a locally convex topological vector space of Fréchet type. The conjugate of $Exp(\mathbb{C}^n)$ denoted by $Exp'(\mathbb{C}^n)$ with the weak topology is called a space of *exponential functionals.*

Definition 2.7. The Fourier-Laplace transform of an analytic functional μ is

$$\tilde{\mu}(\zeta) = \mu(\exp(z, \zeta)) = < \mu(z), \exp(z, \zeta) >, \quad \zeta \in \mathbb{C}^n.$$

The Fourier-Laplace transform of $\mu \in \mathscr{O}'(\mathbb{C}^n)$ is well defined due to above-mentioned denseness of linear combinations of $\exp(z, \zeta)$ in $\mathscr{O}(\mathbb{C}^n)$. Moreover, one can readily see that $\tilde{\mu}(\zeta)$ is an entire analytic function (see, e.g., [Hor90]).

Further, assume $n = 1$. By definition, the Borel transform $B[f]$ of an entire function $f \in Exp(\mathbb{C})$ with the power series representation

$$f(z) = \sum_{k=0}^{\infty} a_k z^k \tag{2.79}$$

is

$$B[f](\zeta) = \sum_{k=0}^{\infty} \frac{a_k k!}{\zeta^{k+1}}. \tag{2.80}$$

The series in (2.80) converges absolutely if $|\zeta| > r$, where r is the exponential type of f. Indeed, for coefficients a_k of f, an entire function of exponential type r, one has the estimate [Leo76]

$$|a_k| \leq \frac{e^k (r + \varepsilon)^k}{k^k}, \tag{2.81}$$

where ε is an arbitrary positive number. Using Stirling's inequality $k! \leq e^{1-k}\sqrt{k}k^k$, it follows from (2.81) that

$$\frac{k!|a_k|}{|\zeta|^{k+1}} \leq \frac{e}{|\zeta|}\sqrt{k}\left(\frac{r+\varepsilon}{|\zeta|}\right)^k,$$

for all k sufficiently large. Hence, for the absolute convergence of (2.80), one needs

$$\lim_{k \to \infty} \left(\frac{k!|a_k|}{|\zeta|^{k+1}}\right)^{1/k} = \frac{r+\varepsilon}{|\zeta|} < 1,$$

which yields $|\zeta| > r$.

The inverse Borel transform is given by (see, e.g., [Hor90, Leo76])

$$f(z) = \frac{1}{2\pi i} \int_\gamma B[f](\zeta) \exp(z\zeta) d\zeta, \quad z \in \mathbb{C}, \tag{2.82}$$

where the closed contour γ lies in the region $|\zeta| > r$. For a function g analytic in the domain $|\zeta| > r$ the expression

$$L[g](z) = \frac{1}{2\pi i} \int_\gamma g(\zeta) \exp(z\zeta) d\zeta, \quad z \in \mathbb{C},$$

defines a linear functional. The latter can be extended to an analytic functional with defining compact $K = \{\zeta : |\zeta| \leq r\}$ [Hor90], that is $L[g](z) =< g, \exp(z, \zeta) >$. Therefore, one can rewrite formula (2.82) for the inverse Borel transform in the form

$$f(z) =< B[f](\zeta), \exp(z\zeta) >, \quad |\zeta| > r,$$

In other words, the inverse Borel transform is the Fourier-Laplace transform.

Proposition 2.2. *Let $f \in \mathscr{X}_k$. Then for its Borel transform $B[f](\zeta)$ the estimate*

$$|B[f](\zeta)| \leq \frac{Ce^{r^{(k+1)}|\zeta|}}{|\zeta| - r} \|f|\mathscr{X}_{k+1}\|, \quad |\zeta| > r, \tag{2.83}$$

holds.

Proof. First we obtain an estimate for the coefficients a_j of the power series representation of f. Using the relation [GR09]

$$a_j = \frac{1}{2\pi i} \int_{|\zeta|=r} \frac{f(\zeta) d\zeta}{\zeta^{j+1}}, \quad j = 0, 1, \ldots,$$

where $r^{(k)} < r < r^{(k+1)}$, one has

$$|a_j| = \frac{1}{2\pi} \left| \int_{|\zeta|=r} \frac{f(\zeta) e^{-z\zeta} e^{z\zeta} d\zeta}{\zeta^{j+1}} \right| \leq C\|f|X_{k+1}\| \frac{r^j}{j!} e^{r|z|}$$
$$\leq C\|f|X_{k+1}\| \frac{r^j}{j!} e^{r^{(k+1)}|z|} \quad j = 0, 1, \ldots.$$

Taking this into account, for $|\zeta| > r$,

$$|B[f](\zeta)| \leq \sum_{j=0}^{\infty} \frac{|a_j| j!}{|\zeta|^{j+1}} \leq \frac{C}{|\zeta|} \|f|X_{k+1}\| e^{r^{(k+1)}|z|} \sum_{j=0}^{\infty} \left(\frac{r}{|\zeta|}\right)^j,$$

obtaining (2.83).

Example 2.3. An entire function f of the form

$$f(z) = c_1 e^{a_1 z} + \cdots + c_m e^{a_m z} \tag{2.84}$$

is called a *quasi-polynomial.* We want to find the Borel transform of f. Since the Borel transform is a linear operator, it suffices to calculate $B[e^{az}](\zeta)$. Using the Taylor expansion of the exponential function e^{az}, one can easily obtain that $B[e^{az}](\zeta) = (\zeta - a)^{-1}$. Hence, the Borel transform of the quasi-polynomial f in (2.84) is

$$B[f](\zeta) = \sum_{j=1}^{m} \frac{c_j}{\zeta - a_j}, \quad |\zeta| > \max\{a_1, \ldots, a_m\}.$$

Let $a \in \mathcal{O}(\mathbb{C}^n)$ with a power series representation $a(\zeta) = \sum_{|\alpha|=0}^{\infty} c_\alpha \zeta^\alpha$. Define a differential operator of infinite order $a(D)$, where $D \equiv \left(\frac{\partial}{\partial z_1}, \ldots, \frac{\partial}{\partial z_n}\right)$, in the form

$$a(D) = \sum_{|\alpha|=0}^{\infty} c_\alpha D^\alpha. \tag{2.85}$$

Proposition 2.3. *Let $a \in \mathcal{O}(\mathbb{C}^n)$. Then the mapping*

$$a(D) : Exp(\mathbb{C}^n) \to Exp(\mathbb{C}^n) \tag{2.86}$$

is continuous.

Proof. Let $f \in Exp(\mathbb{C}^n)$. Applying the operator $a(D)$ in (2.85) to f one has $a(D) \in \mathcal{O}(\mathbb{C}^n)$. This immediately follows from the fact that any analytic function f satisfies the Cauchy-Riemann equation $\overline{\partial}_j f = 0$, $j = 1, \ldots, n$. Applying the Cauchy-Riemann operator $\overline{\partial}_j$ to $a(D)f$, one has

$$\overline{\partial}_j [a(D) f(z)] = \sum_{|\alpha|=0}^{\infty} c_\alpha D^\alpha \overline{\partial}_j f(z) = 0, \ j = 1, \ldots, n,$$

which yields $a(D)f \in \mathcal{O}(\mathbb{C}^n)$.

Further, let $f \in Exp(\mathbb{C}^n)$. Then $f \in \mathscr{X}_k \subset \mathscr{X}_{k+1}$ for some $k \in \mathbb{N}$. We show that

$$\|a(D)f|\mathscr{X}_{k+1}\| \leq C\|f|\mathscr{X}_{k+1}\|, \tag{2.87}$$

which implies that $a(D)f \in Exp(\mathbb{C}^n)$. We have

$$\begin{aligned} a(D)f(z) &= \sum_{|\alpha|=0}^{\infty} c_\alpha D^\alpha f(z) = \sum_{|\alpha|=0}^{\infty} D^\alpha f(z) \Big[\frac{1}{(2\pi i)^n} \int_{\Pi(0,r^{(k)})} \frac{a(\zeta) d\zeta}{\zeta^{\alpha+1}}\Big] \\ &= \sum_{|\alpha|=0}^{\infty} \frac{D^\alpha f(z)}{\alpha!} \Big[\frac{\alpha!}{(2\pi i)^n} \int_{\Pi(0,r^{(k)})} \frac{a(\zeta) d\zeta}{\zeta^{\alpha+1}}\Big] \\ &= \frac{1}{(2\pi i)^n} \int_{\Pi(0,r^{(k)})} \Big(\sum_{|\alpha|=0}^{\infty} \frac{D^\alpha f(z)}{\alpha!} \frac{\alpha!}{\zeta^{\alpha+1}}\Big) a(\zeta) d\zeta. \end{aligned} \tag{2.88}$$

Here $\Pi(0, r^{(k)})$ is the polydisc with the center 0 and polyradius $r^{(k)}$. The expression

$$\sum_{|\alpha|=0}^{\infty} \frac{D^\alpha f(z)}{\alpha!} \frac{\alpha!}{\zeta^{\alpha+1}}$$

is the Borel transform of the entire function

$$h(w) = \sum_{|\alpha|=0}^{\infty} \frac{D^\alpha f(z)}{\alpha!} w^\alpha, \quad w \in \mathbb{C}^n,$$

which is related to f through $h(u) = f(u+z)$. Therefore, (2.88) takes the form

$$a(D)f(z) = \frac{1}{(2\pi i)^n} \int_{\Pi(0,r^{(k)})} B[f(z+\cdot)](\zeta)a(\zeta)d\zeta.$$

Due to Proposition 2.2, it follows from the latter that

$$|a(D)f(z)| \leq C \max_K |a| \|f|\mathscr{X}_{k+1}\| \frac{e^{r^{(k+1)}|z+u|}}{\prod_{j=1}^n (r_j^{(k+1)} - r_j^{(k)})} \leq C' \|f|\mathscr{X}_{k+1}\| e^{r^{(k+1)}|z|},$$

implying estimate (2.87).

Finally, to show the continuity of the mapping (2.86) we assume that a sequence $f_m \in Exp(\mathbb{C}^n)$ converges to zero in the topology of $Exp(\mathbb{C}^n)$. That is, $f_m \in \mathscr{X}_{k+1}$ for some k for all $m \in \mathbb{N}$, and $f_m \to 0$ in the norm of $\mathscr{X}_k$. Due to estimate (2.87), one has

$$\|a(D)f_m|\mathscr{X}_{k+1}\| \leq C\|f_m|\mathscr{X}_{k+1}\| \to 0,$$

as $m \to \infty$, hence $a(D)f_m \to 0$ in the topology of $Exp(\mathbb{C}^n)$.

Proposition 2.4. *Let $a \in \mathscr{O}(\mathbb{C}^n)$. Then the mapping $a(-D) : Exp'(\mathbb{C}^n) \to Exp'(\mathbb{C}^n)$ is continuous.*

Proof. The proof follows from Proposition 2.3 by duality, since

$$< a(-D)f, \varphi > = < f, a(D)\varphi >$$

for arbitrary $f \in Exp'(\mathbb{C}^n)$ and $\varphi \in Exp(\mathbb{C}^n)$.

The following propositions can be proved similar to Propositions 2.3 and 2.4.

Proposition 2.5. *Let $a \in Exp(\mathbb{C}^n)$. Then the mapping $a(D) : \mathscr{O}(\mathbb{C}^n) \to \mathscr{O}(\mathbb{C}^n)$ is continuous.*

Proposition 2.6. *Let $a \in Exp(\mathbb{C}^n)$. The mapping $a(-D) : \mathscr{O}'(\mathbb{C}^n) \to \mathscr{O}'(\mathbb{C}^n)$ is also continuous.*

Definition 2.8. Let $f \in \mathcal{O}(\mathbb{C}^n)$. Then the exponential functional

$$\hat{f}(\zeta) = (2\pi)^n f(-D)\delta(\zeta) \tag{2.89}$$

is called a Fourier transform of the analytic function f. Similarly, if $f \in Exp(\mathbb{C}^n)$, then the analytic functional $\hat{f}(\zeta)$ defined by the same formula in (2.89) is called a Fourier transform of the entire function of a finite exponential type f. In either case we use both notations $\hat{f}(\zeta)$ and $F[f](\zeta)$ for the Fourier transform of f.

Proposition 2.7. *The following statements hold:*

1. *The mappings $F : \mathcal{O}(\mathbb{C}^n) \to Exp'(\mathbb{C}^n)$ and $F : Exp(\mathbb{C}^n) \to \mathcal{O}'(\mathbb{C}^n)$ are continuous;*
2. *If $f \in Exp(\mathbb{C}^n)$, then $F[f] = (2\pi)^n B[f]$, where $B[f]$ is the Borel transform of f;*
3. *The inverse Fourier transform F^{-1} satisfies the relation*

$$F^{-1}[\hat{f}](z) = f(z) = (2\pi)^{-n} < \hat{f}, exp(z,\zeta) >,$$

where $\hat{f} = F[f]$.

Proof. Part 1 immediately follows from Propositions 2.4 and 2.6. To prove Part 2, suppose f is an entire function of a finite exponential type, that is satisfies the condition $|f(z)| \leq Ce^{r_1|z_1|+\cdots+r_n|z_n|}$, where $r_1 > 0, \ldots, r_n > 0$. We show that in this case the equality $< F[f], \varphi >= (2\pi)^n < B[f], \varphi >$ holds for an arbitrary function $\varphi \in \mathcal{O}(\mathbb{C}^n)$. Indeed, let $a_\alpha = \frac{D^\alpha f(0)}{\alpha!}$. Using the Taylor expansion of f, one has

$$\begin{aligned} < F[f](\zeta), \varphi(\zeta) > &= (2\pi)^n < \sum_{|\alpha|=0}^{\infty} a_\alpha (-D)^\alpha \delta(\zeta), \varphi(\zeta) > \\ &= (2\pi)^n \sum_{|\alpha|=0}^{\infty} a_\alpha < \delta(\zeta), D^\alpha \varphi(\zeta) > \\ &= (2\pi)^n \sum_{|\alpha|=0}^{\infty} a_\alpha D^\alpha \varphi(0). \end{aligned} \tag{2.90}$$

Due to the Cauchy formula,

$$D^\alpha \varphi(0) = (2\pi i)^{-n} \alpha! \int_\gamma \frac{\varphi(\zeta) d\zeta}{\zeta^{\alpha+1}},$$

where $\gamma = \gamma_1 \otimes \cdots \otimes \gamma_n$, the direct product of closed contours $\gamma_j, j = 1, \ldots, n$, containing the discs $|\zeta_j| < r_j, j = 1, \ldots, n$, respectively, and $\alpha + 1 = (\alpha_1 + 1, \ldots, \alpha_n + 1)$. Using this fact in (2.90), and taking into account the absolute and uniform convergence of the series $\sum_\alpha a_\alpha \alpha! \zeta^{-\alpha-1}$ outside of the polydisc $\Pi(0,r) = \otimes_{j=1}^n \{|\zeta_j| < r_j\}$ with the center at the origin and polyradius $r = (r_1, \ldots, r_n)$, one obtains

$$< F[f](\zeta), \varphi(\zeta) >= \frac{1}{(i)^n} \int_\gamma \Big(\sum_{|\alpha|=0}^{\infty} \frac{a_\alpha \alpha!}{\zeta^{\alpha+1}} \Big) \varphi(\zeta) d\zeta = (2\pi)^n < B[f](\zeta), \varphi(\zeta) >,$$

as desired.

To show Part 3, one has

$$\frac{1}{(2\pi)^n} < \hat{f}, \exp(z,\zeta) > = < f(-D_\zeta)\delta(\zeta), \exp(z,\zeta) >$$
$$= < \delta(\zeta), f(D_\zeta)\exp(z,\zeta) > = < \delta(\zeta), f(z)\exp(z,\zeta) >$$
$$= f(z) < \delta(\zeta), \exp(z,\zeta) > = f(z).$$

Remark 2.5. Proposition 2.7 Part 2 shows that the complex Fourier transform F introduced in Definition 2.8 is an extension of the Borel transform to the space of exponential functionals $Exp'(\mathbb{C}^n)$. Part 3 of this proposition says that the inverse of the complex Fourier transform is exactly the Fourier-Laplace transform, that is

$$F^{-1}[g(\zeta)](z) = \frac{1}{(2\pi)^n} < g(\zeta), e^{(\zeta,z)} > . \tag{2.91}$$

Formula (2.91) can be used to prove shift, similarity, and delay properties of complex Fourier transform. Indeed, for $a \in \mathbb{C}^n$, one has

$$F^{-1}[e^{-(a,\zeta)}F[f]](z) = \frac{1}{(2\pi)^n} < e^{-(a,\zeta)}F[f], e^{(z,\zeta)} > = \frac{1}{(2\pi)^n} < F[f], e^{(\zeta,z-a)} >$$
$$= f(z-a).$$

This implies the shift formula: $F[f(z-a)](\zeta) = e^{(-a,\zeta)}F[f](\zeta)$. Similarly, one can easily verify that

$$F[f(bz)](\zeta) = \frac{1}{b^n}F[f](\frac{\zeta}{b}), \quad b \in \mathbb{C}, b \neq 0,$$

and

$$F[e^{(a,z)}f(z)](\zeta) = F[f](\zeta - a), \quad a \in \mathbb{C}^n. \tag{2.92}$$

The following properties of the complex Fourier transform also follow from the definition of F and Proposition 2.7:

1. $F[af+bg](\zeta) = aF[f](\zeta) + bF[g](\zeta)$, for any constants $a, b \in \mathbb{C}$;
2. $F[D_z^\alpha f](\zeta) = \zeta^\alpha F[f](\zeta)$;
3. $F[\zeta^\alpha f](\zeta) = (-D_\zeta)^\alpha F[f](\zeta)$;

Further properties of the complex Fourier transform in some Banach spaces of analytic and exponential functions and functionals will be discussed in Chapter 9.

Example 2.4. 1. Let $f(z) \equiv 1 \in \mathscr{O}(\mathbb{C}^n)$. Then $F[1](\zeta) = (2\pi)^n\delta(\zeta)$.
2. Let $f(z) = e^{-z}, z \in \mathbb{C}$. Then

$$F[e^{-z}](\zeta) = 2\pi \exp(D)\delta(\zeta) = 2\pi \sum_{n=0}^{\infty} \frac{D^n\delta(\zeta)}{n!}.$$

This is an analytic (or exponential) functional acting on analytic (or exponential) functions. The formula (2.92) shows that this functional can be written in the form $F[e^{-z}](\zeta) = 2\pi\delta_0(\zeta+1) = 2\pi\delta_{-1}(\zeta)$, or

$$< F[e^{-z}](\zeta), \varphi(\zeta) >= 2\pi\varphi(\zeta-1)$$

for arbitrary entire (or exponential) function $\varphi(\zeta)$.

3. Now let $f(z) = \exp(z^2)$. This function does not belong to $Exp(\mathbb{C})$, but is an entire function. Therefore, we find its Fourier transform as of an element of $\mathscr{O}(\mathbb{C})$. Due to Proposition 2.7 the Fourier transform of f is an analytic functional, that is $F[e^{z^2}](\zeta) \in Exp'(\mathbb{C})$. We use the following representation for the Dirac delta function [Dub96]:

$$\delta_0(\zeta) = \frac{1}{2\sqrt{a\pi}} e^{-aD_\zeta^2}[e^{-\frac{\zeta^2}{4}}]. \tag{2.93}$$

Applying the operator $2\pi e^{D_\zeta^2}$ to both sides of (2.93) with $a = 1$, we have

$$F[e^{z^2}](\zeta) = \sqrt{\pi} e^{-\frac{\zeta^2}{4}}, \quad \zeta \in \mathbb{C},$$

obtaining the complex analog of relation (1.13).

2.8 Functional calculus with singular symbols

Previous sections discussed pseudo-differential operators with singular symbols in various cases. In this section we develop the abstract theory of pseudo-differential operators with singular symbols. In the abstract case we assume that singularities occur in the spectrum of a generic operator A defined on a Banach space. Let X be a reflexive Banach space with a norm $\|v\|$, $v \in X$. Let A be a closed linear operator with a domain $\mathscr{D}(A)$ dense in X and with a spectrum $\sigma(A) \subset \mathbb{C}$. We assume that $\sigma(A)$ is not empty.

We will develop an operator calculus $f(A)$ for analytic functions $f(\lambda)$ in an open domain $G \subset \mathbb{C}$. If G contains $\sigma(A)$, then we define

$$f(A) = \int_\nu \mathscr{R}(\zeta, A) f(\zeta) d\zeta, \tag{2.94}$$

where ν is a contour in G containing $\sigma(A)$, and $\mathscr{R}(\zeta, \mathscr{A})$, $\zeta \in \mathbb{C} \setminus \sigma(\mathscr{A})$, is the resolvent operator of A.

In the case when f has singular points in the spectrum $\sigma(A)$ of the operator A the construction in equation (2.94) is invalid. Below we show how $f(A)$ can be constructed in presence of singularities of f in the spectrum $\sigma(A)$. Assume that G is an open set in $\mathbb{C}$ not necessarily containing $\sigma(A)$. Further, let $0 < r \leq +\infty$ and $\nu < r$. Denote by $\mathrm{Exp}_{A,\nu}(\mathrm{X})$ the set of elements $u \in \cap_{k\geq 1}\mathscr{D}(A^k)$ satisfying the inequalities

$\|A^k u\| \le C\nu^k$ for all $k = 1,2,\ldots$, with a constant $C > 0$ not depending on k. An element $u \in \mathrm{Exp}_{A,\nu}(X)$ is said to be a *vector of exponential type* ν. A sequence of elements u_n, $n = 1,2,\ldots$, is said to converge to an element $u_0 \in \mathrm{Exp}_{A,\nu}(\mathrm{X})$ if:

1) All the vectors u_n are of exponential type $\nu < r$, and
2) $\|u_n - u_0\| \to 0$, $n \to \infty$.

Obviously, $\mathrm{Exp}_{A,\nu_1}(\mathrm{X}) \subset \mathrm{Exp}_{A,\nu_2}(\mathrm{X})$, if $\nu_1 < \nu_2$. Let $\mathrm{Exp}_{A,r}(\mathrm{X})$ be the inductive limit of spaces $\mathrm{Exp}_{A,\nu}(\mathrm{X})$ when $\nu \to r$. For basic notions of topological vector spaces including inductive and projective limits we refer the reader to [R64]. Set $A_\lambda = A - \lambda I$, where $\lambda \in G$, and denote $\mathrm{Exp}_{A,r,\lambda}(\mathrm{X}) = \{u_\lambda \in X : u_\lambda \in \mathrm{Exp}_{A_\lambda,r}(\mathrm{X})\}$, with the induced topology. Finally, for arbitrary $G \subset \sigma(A)$, denote by $\mathrm{Exp}_{A,G}(\mathrm{X})$ the space whose elements are the locally finite sums of elements in $\mathrm{Exp}_{A,r,\lambda}(\mathrm{X})$, $\lambda \in G$, $r < dist(\lambda, \partial G)$, with the corresponding topology. Namely, any $u \in \mathrm{Exp}_{A,G}(\mathrm{X})$, by definition, has a representation $u = \sum_\lambda u_\lambda$ with a finite sum. It is clear that $\mathrm{Exp}_{A,G}(\mathrm{X})$ is a subspace of the space of vectors of exponential type if $r < +\infty$, and coincides with it if $r = +\infty$. $\mathrm{Exp}_{A,G}(\mathrm{X})$ is an abstract analog of the space $\Psi_{G,p}(\mathbb{R})$ introduced in Chapter 1, where $A = -i\frac{d}{dx}$, $G \subseteq \mathbb{R}$, $\mathrm{X} = L_p(\mathbb{R})$, $1 < p < \infty$. In the case $A = -i\frac{d}{dx}$, $\mathrm{X} = L_2(\mathbb{R})$, the corresponding space is $H^\infty(G)$ [Dub82].

Further, let $f(\lambda)$ be an analytic function on G, represented as a finite sum. Then for $u \in \mathrm{Exp}_{A,G}(\mathrm{X})$ with the representation $u = \sum_{\lambda \in G} u_\lambda$, $u_\lambda \in \mathrm{Exp}_{A,r,\lambda}(\mathrm{X})$, the operator $f(A)$ is defined by the formula

$$f(A)u = \sum_{\lambda \in G} f_\lambda(A)u_\lambda, \text{ where } f_\lambda(A)u_\lambda = \sum_{n=0}^{\infty} \frac{f^{(n)}(\lambda)}{n!}(A - \lambda I)^n u_\lambda. \tag{2.95}$$

In other words, each f_λ represents f locally in a neighborhood of $\lambda \in G$, and for u_λ the operator $f_\lambda(A)$ is well defined.

Additionally assume that there exists a one-parameter family of bounded invertible operators

$$U_\lambda : X \to X \tag{2.96}$$

such that $AU_\lambda - U_\lambda A = \lambda U_\lambda$, or the same

$$A - \lambda I = U_\lambda A U_\lambda^{-1}, \quad \lambda \in \sigma(A). \tag{2.97}$$

Obviously, U_λ maps $Exp_{A,r,\lambda}(X)$ onto $Exp_{A,r}(X)$ and it is a bijective mapping.

Example 2.5. For example, if $X = L_2 \equiv L_2(\mathbb{R})$ and $A = -i\frac{d}{dx} : L_2 \to L_2$ with domain $\mathscr{D}(A) = \{v \in L_2 : Av \in L_2\}$, then the operators $U_\lambda : v(x) \to e^{i\lambda x}v(x)$ satisfy

$$\begin{aligned} AU_\lambda v(x) &= -i\frac{d}{dx}(e^{i\lambda x}v(x)) = \lambda e^{i\lambda x}v(x) - ie^{i\lambda x}\frac{dv}{dx} \\ &= \lambda U_\lambda v(x) + U_\lambda A v(x), \end{aligned}$$

obtaining (2.97). The relationship $A - \lambda I = U_\lambda A U_\lambda^{-1}$ follows from the latter multiplying from the right by the inverse operator U_λ^{-1}, which is the multiplication operator by $e^{-i\lambda x}$.

This example tells us that condition (2.97) is essentially a shift of the spectrum of operator A to λ. It follows from equation (2.97) that $(A - \lambda I)^n = U_\lambda A^n U_\lambda^{-1}$, for all $n = 1, 2, \ldots$, yielding

$$f(A)u = \sum_{\lambda \in G} \sum_{n=0}^{\infty} \frac{f^{(n)}(\lambda)}{n!} U_\lambda A^n U_\lambda^{-1} u_\lambda = \sum_{\lambda \in G} U_\lambda f(A) U_\lambda^{-1} u_\lambda.$$

Let X^* denote the dual of X, and $A^* : X^* \to X^*$ be the operator adjoint to A. Further, denote by $\mathrm{Exp}'_{A^*,G^*}(\mathrm{X}^*)$ the space of linear continuous functionals defined on $\mathrm{Exp}_{A,G}(\mathrm{X})$, with respect to weak convergence. Specifically, a sequence $u_m^* \in \mathrm{Exp}'_{A^*,G^*}(\mathrm{X}^*)$ converges to an element $u^* \in \mathrm{Exp}'_{A^*,G^*}(\mathrm{X}^*)$ if for all $v \in \mathrm{Exp}_{A,G}(\mathrm{X})$ the convergence $< u_m^* - u, v > \to 0$ holds as $m \to \infty$. For an analytic function f^w defined on $G^* = \{z \in \mathbb{C} : \bar{z} \in G\}$, we define *a weak extension* of $f(A)$ as follows:

$$< f^w(A^*)u^*, u > = < u^*, f(A)u >, \quad \forall u \in \mathrm{Exp}_{A,G}(\mathrm{X}),$$

where $u^* \in \mathrm{Exp}'_{A^*,G^*}(\mathrm{X}^*)$.

Lemma 2.3. *Let X be a reflexive Banach space and A be a closed operator defined on $\mathscr{D}(A) \subset X$. Let f be an analytic function defined on an open connected set $G \subset \mathbb{C}$. Then the following mappings are well defined and continuous:*

1. $f(A) : \mathrm{Exp}_{A,G}(\mathrm{X}) \to \mathrm{Exp}_{A,G}(\mathrm{X})$,
2. $f^w(A^*) : \mathrm{Exp}'_{A^*,G^*}(\mathrm{X}^*) \to \mathrm{Exp}'_{A^*,G^*}(\mathrm{X}^*)$.

Proof. We will prove that $f(A)$ maps $\mathrm{Exp}_{A,G}(\mathrm{X})$ into itself. Let $u \in \mathrm{Exp}_{A,G}(\mathrm{X})$ has a representation $u = \sum_\lambda u_\lambda$, $u_\lambda \in \mathrm{Exp}_{A_\lambda,\nu}(\mathrm{X})$. Then for $f(A)u$ defined in (2.95), one has the following estimate

$$\|A_\lambda^k f_\lambda(A) u_\lambda\| \leq \sum_{n=0}^{\infty} \frac{|f^n(\lambda)|}{n!} \|(A - \lambda I)^n A_\lambda^k u_\lambda\| \leq C \nu^k \|u_\lambda\|, \tag{2.98}$$

with some $\nu < r$. It follows that $f_\lambda(A)u_\lambda \in \mathrm{Exp}_{A_\lambda,\nu}(\mathrm{X})$ with the same ν. Hence, $f(A)u$ has a representation $\sum_\lambda w_\lambda$, where $w_\lambda = f_\lambda(A)u_\lambda \in \mathrm{Exp}_{A_\lambda,\nu}(\mathrm{X})$, and therefore $f(A)u \in \mathrm{Exp}_{A,G}(\mathrm{X})$. The estimate (2.98) also implies continuity of the mapping $f(A)$ in the topology of $\mathrm{Exp}_{A,G}(\mathrm{X})$.

Now assume that a sequence $u_n^* \in \mathrm{Exp}'_{A^*,G^*}(\mathrm{X}^*)$ converges to 0 in the weak topology of $\mathrm{Exp}'_{A^*,G^*}(\mathrm{X}^*)$. Then for arbitrary $u \in \mathrm{Exp}_{A,G}(\mathrm{X})$ we have

$$< f^w(A^*)u_n^*, u > = < u_n^*, f(A)u > = < u_n^*, v >,$$

where $v = f(A)u \in \mathrm{Exp}_{A,G}(\mathrm{X})$ due to the first part of the proof. Hence, $f^w(A^*)u_n^* \to 0$, as $n \to \infty$, in the weak topology of $\mathrm{Exp}'_{A^*,G^*}(\mathrm{X}^*)$.

Remark 2.6. Note that if $\sigma(A)$ is discrete then the space $\mathrm{Exp}_{A,G}(\mathrm{X})$ consists of the root lineals of eigenvectors corresponding to the part of $\sigma(A)$ with nonempty intersection with G. If the spectrum $\sigma(A)$ is empty, then an additional investigation is required for solution spaces to be nontrivial [Uma88].

Definition 2.9. Let $\mathscr{X}$ be a topological vector space and two operators A and B linear operators mapping $\mathscr{X}$ into $\mathscr{X}$. Let X_1 and X_2 be closed subspaces of $\mathscr{X}$. Suppose, there is a linear bounded invertible operator U defined on $\mathscr{X}$, with bounded inverse U^{-1}, and $U : X_1 \to X_2$, such that $A = UBU^{-1}$. Then operators A and B are called *similar* and U is called a *transformation operator* for A and B. If U is unitary, then A and B are called unitary equivalent.

Example 2.6. 1. The operator U_λ defined in equations (2.96) and (2.97) is a transformation operator for similar operators $A - \lambda I$ and A. Here $\mathscr{X} = Exp_{A,G}(X)$, $X_1 = Exp_{A_\lambda,v}(X)$, and $X_2 = Exp_{A,v}(X)$.
2. Let $P(D), D = -i(\partial/\partial x_1, \ldots, \partial/\partial x_n)$ be a differential operator and $p(\xi)$ its symbol. Then for all $f \in \mathscr{G}(\mathbb{R}^n)$ one has $P(D)f = F^{-1}[p(\xi)F[f]]$, where F is the Fourier transform. Hence, the differential operator $P(D)$ and the multiplication operator by $p(\xi)$ are unitary equivalent with the transformation operator F.
3. Let $\mathscr{X} = C^1[0,\infty)$ with the topology of uniform convergence on any compact interval in $[0,\infty)$. Consider operators $A_j = -\frac{d^2}{dx^2} + q_j(x)$, $j = 1,2$, where $q_j \in C[0,\infty)$. Further, let $X_j = \{f \in \mathscr{X} : f'(0) = \alpha_j f(0)\}$, $\alpha_j \in \mathbb{C}$, $j = 1,2$, and $\alpha_1 \neq \alpha_2$. Then operators A_1 and A_2 are similar with the transformation operator

$$Uf(x) = f(x) + \int_0^x K(x,y)f(y)dy, \quad x \in [0,\infty),$$

where the kernel $K(x,y)$ satisfies the partial differential equation

$$K_{xx} - K_{yy} = (q_2(y) - q_2(x))K,$$

and conditions:

$$K(x,x) = (\alpha_2 - \alpha_1) + \frac{1}{2}\int_0^x [q_1(s) - q_2(s)]ds,$$

$$\frac{\partial K(x,0)}{\partial y} - \alpha_1 K(x,0) = 0.$$

For the proof the reader is referred to [LS70], where many other examples, similar operators, and their applications to harmonic analysis can be found.

It follows from Definition 2.9 that if A and B are similar with the transformation operator U, then operators A^n and B^n are similar for any $n \in \mathbb{N}$ with the same transformation operator U, namely

$$A^n = UB^nU^{-1}, \quad n = 2,3,\ldots. \tag{2.99}$$

This implies the following proposition

Proposition 2.8. *Let operators A and B are similar. Then*
1. *the spaces* $Exp_{A,G}(X)$ *and* $Exp_{B,G}(X)$ *are isomorphic;*
2. *for any* f *analytic in* G *both mappings* $f(A): Exp_{A,G}(X) \to Exp_{A,G}(X)$ *and* $f(B): Exp_{B,G}(X) \to Exp_{B,G}(X)$ *are continuous simultaneously.*

Proof. Let $u \in Exp_{A,\nu}(X)$. Then $w = U^{-1}u \in Exp_{B,\nu}(X)$. Indeed, using (2.99) we have

$$\|B^n w\| = \|U^{-1}A^n U w\| = \|U^{-1}A^n u\| \le C\|A^n u\| \le C_1 \nu^n \|u\| \le C_2 \nu^n \|w\|,$$

which implies both statements of the proposal.

Further, let $\Pi_{\nu,\lambda}$ be a projection operator of the space X to a subspace $Exp_{A,\nu,\lambda}(X)$, $\lambda \in G$, $\nu < dist(\lambda, \partial G)$. We want to build a projector-valued measure E with the help of $\Pi_{\nu,\lambda}$. It is easy to see that $\sigma(\Pi_{\nu,\lambda}) \subseteq D_\nu(\lambda)$, where $D_\nu(\lambda) \subset G$ is the disc with the radius ν and center λ. Let Ω be a union of finite number nonintersecting discs $D_{r_j}(\lambda_j)$, $j = 1, \ldots, M$. We set $\tilde{E}(\Omega) = E_1 + \ldots + E_M$, where $E_j = \Pi_{r_j,\lambda_j}$. It is evident that $\tilde{E}(\Omega)$ is a projector and if $\Omega_1 \cap \Omega_2 = \emptyset$, then

$$\tilde{E}(\Omega_1) \circ \tilde{E}(\Omega_2) = \Theta, \text{ and } \tilde{E}(\Omega_1 \cup \Omega_2) = \tilde{E}(\Omega_1) + \tilde{E}(\Omega_2). \tag{2.100}$$

Let Σ be the smallest σ-algebra containing all possible finite unions of discs $D_r(\lambda)$. Due to conditions (2.100) there exists an extension of $\tilde{E}$ to Σ, which is denoted by E.

Lemma 2.4. *Let* $E(\sigma(A) \setminus G) = \Theta$ *and* $D_\nu(0) \cap \sigma(A) \cap G \subseteq \bigcup_{j=1}^{\infty} D_{r_j}(\lambda_j)$, *where* $\lambda_j \in G$, $r_j < dist(\lambda_j, G)$, $j \ge 1$. *Then the limit*

$$\lim_{M\to\infty} \|\Pi_{\nu,0} - (E_1 + \ldots + E_M)\| = 0$$

holds.

Proof. We set

$$V_M = \bigcup_{j=M+1}^{\infty} D_{r_j}(\lambda_j).$$

Taking into account

$$\Pi_\nu = -(2\pi i)^{-1} \int_{\Gamma_{\nu+\varepsilon}} R_\lambda(A) d\lambda,$$

where $R_\lambda(A)$ is the resolvent of A and $\Gamma_{\nu+\varepsilon} = \partial D_{\nu+\varepsilon}(0)$ is the circle of the radius $\nu + \varepsilon$, and the condition of the lemma, we have

$$\lim_{M\to\infty} \|\Pi_{\nu,0} - \sum_{j=1}^{M} E_j\| = \lim_{M\to\infty} \|E(V_M)\| = \|E((\sigma(A) \setminus G) \cap K_\nu(0))\| = 0.$$

In the theorem below $Exp_A(X)$ is the space of vectors of exponential type, corresponding to the operator A, and $G = \mathbb{C}$.

Theorem 2.21. *For density of the space $Exp_{A,G}(X)$ in $Exp_A(X)$ it is sufficient (if X is a Hilbert space, then it is necessary) that $E(\sigma(A) \setminus G) = \Theta$.*

Proof. Sufficiency. Let $x \in Exp_A(X)$, i.e., there exists $\nu > 0$, such that $x \in Exp_{A,\nu,0}(X)$. We set $x_M = (E_1 + \ldots + E_M)x$. It is clear that $x_M \in Exp_{A,G}(X)$. Moreover, it follows from Lemma 2.4 that $x_M \to x$ as $M \to \infty$ in the norm of X.

Necessity. Suppose that X is a Hilbert space and $E(\sigma(A) \setminus G) = E^\circ > \Theta$. One can find a number ν such that $0 \neq x \in (E^\circ \circ \Pi_{\nu,0})X$. For arbitrary element $y \in Exp_{A,G}(X)$ we have

$$\|x - y|X\|^2 = (x-y, x-y) = \|x|X\|^2 - 2Im(x,y) + \|y|X\|^2 \geq \|x|X\|^2,$$

since due to the selection of x the equality $(x,y) = 0$ holds.

Denote by $\mathscr{A}$ the class of operators for which $Exp_A(X)$ is dense in X. For instance, it is known [39,41] that if A is a generator of a bounded strongly continuous group of operators, then $Exp_A(X)$ is dense in X. Theorem 2.21 implies the following statement.

Theorem 2.22. *Let $A \in \mathscr{A}$. For density of $Exp_{A,G}(X)$ in X it is sufficient (if X is a Hilbert space, then it is necessary also) that $E(\sigma(A) \setminus G) = \Theta$.*

2.9 Additional notes

1. Pseudo-differential operators in its modern form first was introduced by Kohn and Nirenberg [KN65] and Hörmander [Hor65]. The algebra of pseudo-differential operators constructed in these works, which now became classic, contained differential operators, singular integral operators of Calderon and Zygmund [CZ58], and parametrices of elliptic operators. A symbol $a(x,\xi)$ of such a pseudo-differential operator was assumed to be infinite differentiable on co-tangent bundle of Ω, that is $a(x,\xi) \in C^\infty(\Omega \times \mathbb{R}^n)$. The class of symbols of pseudo-differential operators of order m was denoted by $S^m(\Omega)$. Hörmander [Hor68] also introduced the class of symbols $S^m_{\rho,\delta}(\Omega)$ containing parametrices of hypo-elliptic operators. Many researchers contributed to the classic theory of pseudo-differential operators; see, for instance, books [Hor83, Tay81, Tre80, Shu78, Ego67, RSc82] and the references therein. Pseudo-differential operators with non-smooth or singular symbols were studied by Cordes and Williams [CW77], Plamenevskii [Pla86], and Dubinskii [Dub82], using different approaches. Cordess and Williams used abstract algebra methods for construction of pseudo-differential operators with non-regular symbols. Plamenevskii considered symbols with pole type singularities and used the Mellin transform to build corresponding pseudo-differential operators. Dubinskii used differential operators of infinite order for construction of pseudo-differential operators. The corresponding symbols are defined on a domain $G \subset \mathbb{R}^n$ and locally analytic functions, with arbitrary type of singularities on the boundary of G. For properties of such defined pseudo-differential operators and their applications, we refer the reader to books [Dub91, VH94]. Our approach developed in this chapter combines the ideas of differential operators of infinite order with the original form of pseudo-differential operators, and considered in the papers [Uma97, Uma98, GLU00]. In Chapter 9 of this book we will also present a version of complex pseudo-differential operators with analytic and meromorphic symbols studied in the papers [Uma91-1, Uma91-2, Uma92, Uma14]. There we will apply this theory to general boundary value problems for systems of complex pseudo-differential operators.

2. *Hypersingular Integrals.* A class of pseudo-differential operators, qualified as ΨDOSS, is given by hypersingular integrals [Sam80, Rub96]. An example of such operators is the operator

$$D_0^\alpha f(x) = -\frac{1}{d(\alpha,l)}\int_{\mathbb{R}^n}\frac{\Delta_y^l f(x)}{|y|^{n+\alpha}}dy, \quad 0<\alpha<l,$$

introduced in Example 2.2 with the symbol $-|\xi|^\alpha$ (see Proposition 3.4 in Chapter 3). This operator is a model case of the class of hypersingular integrals. The general form of hypersingular integrals is

$$A_\alpha f(x) = \int_{\mathbb{R}^n}\frac{\omega(\frac{y}{|y|})\Delta_y^l f(x)}{|y|^{n+\alpha}}dy, \quad \alpha>0,$$

where $\omega(x)$ is a kernel function defined on the unit sphere and satisfying some conditions. The symbol of A_α has the representation through the surface integral [Sam80]:

$$\sigma_{A_\alpha}(\xi) = \frac{\Gamma(\frac{n+\alpha}{2})}{(2\pi)^{\frac{n-1}{2}}\Gamma(\frac{1+\alpha}{2})}\int_{\Sigma_{n-1}}\omega(\sigma)|(\xi,\sigma)|^\alpha \mathrm{sign}(\xi,\sigma)d\sigma,$$

where Σ_{n-1} is the unit sphere in $\mathbb{R}^n$ with the center at the origin, and $(x,\sigma)=x_1\sigma_1+\cdots+x_n\sigma_n$. Hypersingular integrals with variable $\alpha(x)$ are studied by Almeida and Samko in their recent paper [AS09]. These hypersingular integrals are examples of variable fractional order differential operators. In Chapter 3 we will introduce time-variable fractional order derivatives. Variable order fractional derivatives based on the Liouville-Riemann fractional derivative were studied by Lorenzo and Hurtley in [LH02]. Variable order derivatives based on the Caputo-Djrbashian fractional derivative, their memory properties, and the Cauchy problem for associated differential equations are studied in [US09] by Umarov and Steinberg. In Chapter 3 we also introduce a pseudo-differential operator of the form

$$Af(x) = \int_0^2 \mathbb{D}_0^\alpha f(x)d\rho(\alpha), \quad x\in\mathbb{R}^n,$$

where ρ is a finite measure defined on the interval $[0,2]$. This operator, called a distributed order differential operator, has the symbol

$$\Psi(\xi) = -\int_0^2 |\xi|^\alpha d\rho(\alpha), \quad \xi\in\mathbb{R}^n.$$

3. Another interesting class of pseudo-differential operators, also containing D_0^α, consists of infinitesimal generators of strongly continuous semigroups, associated with, so-called, Feller processes, discussed, for instance, in Applebaum [App09], Jacob [Jac01], Jacob and Schilling [JSc02], and Hoh [Hoh00]. These operators have the following generic form [App09, JSc02]

$$A\varphi(x) = c_0(x)\varphi(x) + \sum_{j=1}^n b_j(x)\frac{\partial\varphi(x)}{\partial x_j} - \sum_{j,k=1}^n a_{jk}\frac{\partial^2\varphi(x)}{\partial x_j\partial x_k}$$
$$+\int_{\mathbb{R}^n\setminus\{0\}}\Big[\varphi(x+y)-\varphi(x)-\chi_{|y|\le 1}(y)(\nabla\varphi(x),y)\Big]d\nu(y), \qquad (2.101)$$

where functions c_0 and b_j, $j=1,\ldots,n$, are continuous from $\mathbb{R}^n$ to $\mathbb{R}$, such that $c_0(x)\le 0$ for all $x\in\mathbb{R}^n$; mappings $a_{ij}:\mathbb{R}^n\to\mathbb{R}$, $i,j=1,\ldots,n$, such that each $(a_{ij}(x))$ is a positive symmetric matrix for each $x\in\mathbb{R}^n$ and the map $x\to(y,a(x)y)$ is upper semicontinuous for each $y\in\mathbb{R}^n$; ν is a Lévy measure, that is a measure satisfying the condition $\int_{\mathbb{R}^n}\min(1,|x|^2)d\nu(x)<\infty$; and $\chi_{|y|\le 1}(y)$ is the indicator function of the unit ball $B_1(0)\subset\mathbb{R}^n_y$. Symbols of these operators are not differentiable, therefore, according to our terminology, these operators can be classified as ΨDOSS. The symbol corresponding to the operator A in (2.101) has the form (see, e.g., [JSc02])

$$p(x,\xi) = c_0(x) + (b(x),\xi) - (\mathscr{A}(x)\xi,\xi) + \int_{\mathbb{R}^n\setminus\{0\}} \left[e^{i(\xi,y)} - 1 - i\chi_{|y|\leq 1}(y)(\xi,y)\right] d\nu(y).$$

Symbols of this form, by construction, are negative definite. By definition, a continuous function $\phi(\xi)$ defined on $\mathbb{R}^n$ is called negative definite, if for any $\xi_1,\dots,\xi_m \in \mathbb{R}^n$ and $z_1,\dots,z_m \in \mathbb{C}$, the inequality

$$\sum_{j,k=1}^{m} [\phi(\xi_j) + \overline{\phi(\xi_k)} - \phi(\xi_j - \xi_k)] z_j \bar{z}_k \geq 0$$

holds. It turns out that the class of negative definite symbols is in one-to-one correspondence with the class of pseudo-differential operators which are infinitesimal generators of strongly continuous semigroups. We refer the reader to [App09, Jac01] for further properties of operators of the form (2.101) and their relation with stochastic processes. We will return to these operators in Chapter 7 in connection with fractional Fokker-Planck-Kolmogorov equations. Taira [Tai91] considered boundary value problems to describe Markovian diffusion processes in a bounded domain.

4. The Fourier transform plays a vital role in the theory of pseudo-differential operators of real variables. A pseudo-differential operator $A(D)$ with the symbol $A(\xi)$ in Definition 2.2 can be written in the form $A(D)f = F^{-1}[A(\xi)F[f]]$, that is both direct and inverse Fourier transforms are involved. Therefore, introduction of a complex Fourier transform with properties similar to the real Fourier transform, and for which the inverse Fourier transform is explicitly determined, would allow to develop complex theory of pseudo-differential operators. In 1984 Dubinskii [Dub84] introduced a complex Fourier transform of a complex function in the form

$$F[f](\zeta) = f(-D)\delta(\zeta). \tag{2.102}$$

This Fourier transform inherits many properties of the real Fourier transform and was used for the construction of PsDO with analytic symbols [Dub84, Dub90, Dub96]. The complex Fourier transform F defined in Section 2.7 is equivalent to (2.102) and differs from it only by the constant factor $(2\pi)^n$. Due to this constant factor the complex Fourier transform becomes consistent with its real version (1.8), thus generalizing it. In Chapter 9 we apply the complex Fourier transform F to develop complex ΨDOSS with meromorphic symbols and study systems of pseudo-differential equations.

5. In Section 2.8 we introduced the Frechet type topological vector space $Exp_{A,G}(X)$ and its dual $Exp'_{A,G}(X)$, where G is an open subset of the complex plain $\mathbb{C}$. These spaces represent an abstract modification of the space $\Psi_{G,p}(\mathbb{R}^n)$ and ψ-distributions $\Psi'_{G,p}(\mathbb{R}^n)$. In the particular case $G = \mathbb{C}$ the spaces $Exp_{A,\mathbb{C}}(X)$ and $Exp'_{A,\mathbb{C}}(X)$ were used in [Rad82] as solution spaces of the Cauchy problem for abstract differential-operator equations. We note that in some particular cases of the operator A the space $Exp_{A,G}(X)$, and hence its dual can effectively be described. Below we consider some examples of such operators. Let A be a linear closed operator defined in a Hilbert space H, that is $\mathfrak{D}(A) \subset H$. A is called normal, if $A^*A = A^*A$. For any normal operator there exists an operator-valued measure E_λ, such that

$$A = \int_{\sigma(A)} \lambda dE_\lambda,$$

where $\sigma(A)$ is the spectrum of A. The widest class of operators, for which a spectral decomposition formula holds true, is the class of spectral operators studied by Dunford and Schwartz [DS88]. Spectral operators defined in a Banach space X, in general, have the form $A = S + N$ (see [DS88]), where S is the spectral part, and N is the quasi-nilpotent part of A. If A is a spectral operator, then for any function f analytic in a neighborhood of the spectrum of the spectral part S, the representation

$$f(A)u = \sum_{k=0}^{\infty} \frac{N^k}{k!} \int\limits_{\sigma(A)} f^{(k)}(s)E(ds)u$$

holds for all $u \in X$ in the domain of $f(A)$. In particular, if $f(s) = s^n$, then

$$A^n u = \sum_{k=0}^{n} \binom{n}{k} N^k \int\limits_{\sigma(A)} s^{n-k} E(ds)u.$$

We show that $u \in Exp_{A,\nu}(X)$ if $u = E(|s| \leq \nu - \varepsilon)x$, where $x \in X$, for some $\varepsilon > 0$. Indeed, this follows from the following estimate

$$\|A^n u\| \leq \sum_{k=0}^{n} \binom{n}{k} \|N\|^k \left\| \int\limits_{|s| \leq \nu - \varepsilon} |s|^{n-k} E(ds)x \right\| \leq \left(\|N\| + \nu - \varepsilon\right)^n \|x\| \leq \nu^n \|x\|,$$

since the norm of a quasi-nilpotent operator is arbitrarily small ([DS88]), $\|N\| < \delta$, where $0 < \delta < \varepsilon$.

Chapter 3
Fractional calculus and fractional order operators

3.1 Introduction

Fractional order differential equations are an efficient tool to model various processes arising in science and engineering. Fractional models adequately reflect subtle internal properties, such as memory or hereditary properties, of complex processes that the classical integer order models neglect. In this chapter we will discuss the theoretical background of fractional modeling, that is the fractional calculus, including recent developments - distributed and variable fractional order differential operators.

Fractional order derivatives interpolate integer order derivatives to real (not necessarily fractional) or complex order derivatives. There are different types of fractional derivatives not always equivalent. The first attempt to develop the fractional calculus systematically was taken by Liouville (1832) and Riemann (1847) in the first half of the nineteenth century, even though discussions on non-integer order derivatives had been started long ago.[1] In the 1870s Letnikov and Grünwald independently used an approach for the definition of the fractional order derivative and integral different from that of Riemann and Liouville. The Cauchy problem for fractional order differential equations with the Riemann-Liouville derivative is not well posed (Section 3.3), that is the Cauchy problem in this case is unphysical. In the 1960s Caputo and Djrbashian introduced independently, so-called, a regularization of the Riemann-Liouville fractional derivative, which was later named a fractional derivative in the sense of Caputo-Djrbashian (Section 3.5). The usefulness of the Caputo-Djrbashian derivative is that the Cauchy problem for fractional order differential equations with the Caputo-Djrbashian derivative is well posed. Thus the definition used by Caputo and Djrbashian returned the "physicality" of the

[1] The first existing documented record on fractional derivatives goes back to year 1695. Leibniz in his letter (dated September 30, 1695) to L'Hôpital wrote on the derivative of order 1/2 of the function $f(t) = t$.

S. Umarov, *Introduction to Fractional and Pseudo-Differential Equations with Singular Symbols*, Developments in Mathematics 41,
DOI 10.1007/978-3-319-20771-1_3

Cauchy problem for fractional order differential equations. For the detailed history of fractional calculus, we refer the reader to books [OS74, SKM87].

Starting from the 1960s an intensive growth of the fractional modeling has been observed. A number of new approaches have been developed and extensive applications in various fields have been found. Sections 3.11, 3.12 present two novel concepts appeared relatively recently, namely distributed and variable order fractional differential operators. Initial and boundary value problems with distributed and variable order differential operators and their applications will be discussed in Chapters 6–8. Distributed order differential equations model ultraslow diffusion [CGSG03, MS06], fractional kinetics with accelerating super-diffusion and decelerating sub-diffusion, macromolecule movement in cell membrane [SJ97, AUS06], etc. They model stochastic processes with mixed diffusion regimes; see Chapter 7. Variable order differential equations are used in modeling of processes arising in viscoelastic materials [LH02], in modeling of relaxation processes and reaction kinetics of proteins [GN95], in the study of rheological properties of fluids [KM67], etc.

What is a fractional order derivative and what is the key idea behind the definition of it? To answer this question let us first review the usual integer order derivative and integral. Let $D = \frac{d}{dt}$ be the differentiation operator and J be the integration operator, that is $Jf(t) = \int_0^t f(\tau)d\tau$. Then the fundamental theorem of calculus states that for a continuous function f

$$DJf(t) = \frac{d}{dt}\int_0^t f(\tau)d\tau = f(t).$$

In the operators language one can write the latter in the form $DJ = I$, where I is the identity operator, which means that the operator D is a left inverse to the operator J. One can easily check that D is not a right inverse to J, since, according to the same fundamental theorem of calculus, for any differentiable function f the equality $JDf(t) = f(t) - f(0)$ holds. The similar relations are true by induction for operators D^n and J^n, where $D^n = \frac{d^n}{dt^n}$, "n-th derivative," and J^n is the n-fold integration operator. Namely,

$$D^n J^n f(t) = f(t), \tag{3.1}$$

and

$$J^n D^n f(t) = f(t) - \sum_{k=0}^{n-1} \frac{D^k f(0)}{k!} t^k. \tag{3.2}$$

Thus D^n is the left inverse to J^n, and is the right inverse to J^n in the class of functions satisfying additional conditions: $f^{(k)}(0) = 0, k = 0,\dots,n-1$. These relations between "differentiation" and "integration" operators valid for $n = 1,2,\dots,$ form the basis for the definitions of fractional derivatives in the sense of Riemann-Liouville and Caputo-Djrbashian, as soon as the fractional order integration operator is defined.

3.2 Fractional order integration operator

In this section we introduce the fractional order integration operator of order $\alpha > 0$ ($\Re(\alpha) > 0$). One can verify (by changing order of integration) that the n-fold integration operator

$$J^n f(t) = \underbrace{\int_0^t \cdots \int_0^{\tau_2}}_{n-\text{times}} f(\tau_1) d\tau_1 \cdots d\tau_n$$

can be represented as

$$J^n f(t) = \frac{1}{(n-1)!} \int_0^t (t-\tau)^{n-1} f(\tau) d\tau.$$

Taking into account the relationship $\Gamma(n) = (n-1)!$, where

$$\Gamma(s) = \int_0^\infty e^{-t} t^{s-1} dt, \quad \Re(s) > 0,$$

is the Euler's gamma-function (see Section 1.4, Chapter 1), one can define a fractional order integral for any α, $\Re(\alpha) > 0$, by

$$J^\alpha f(t) = \frac{1}{\Gamma(\alpha)} \int_0^t (t-\tau)^{\alpha-1} f(\tau) d\tau. \tag{3.3}$$

Incidentally, we recall also Euler's beta-function defined for all $\alpha, \beta \in \mathbb{C}$, with $\Re(\alpha) > 0, \Re(\beta) > 0$, as $B(\alpha,\beta) = \int_0^1 s^{\alpha-1}(1-s)^{\beta-1} ds$, which is connected with the gamma-function through the following relationship (Section 1.4)

$$B(\alpha,\beta) = \frac{\Gamma(\alpha)\Gamma(\beta)}{\Gamma(\alpha+\beta)}. \tag{3.4}$$

Before embarking on the world of fractional calculus let us make two important notes. The first note is about integration end-points, which are also called terminal points. So far we have used only the origin for the lower terminal point. In the general case the fractional integration operator can be defined with arbitrary lower terminal point $a \in [-\infty, \infty)$. Indicating terminal points in the notation, the fractional integration operator can be written in the form

$$_aJ_t^\alpha f(t) = \frac{1}{\Gamma(\alpha)} \int_a^t (t-\tau)^{\alpha-1} f(\tau) d\tau, \tag{3.5}$$

where f is a continuous function in the interval (a,b). In what follows, if the lower terminal point is finite, then we preferably work with the interval $(a,b) = (0,\infty)$ and denote $_0J_t^\alpha = J^\alpha$, unless otherwise specified. This is convenient for the study of initial value problems, as well.

Second, the family $\{J^\alpha, \alpha > 0\}$ (as well as the family $\{_aJ_t^\alpha, \alpha > 0\}$ for an arbitrary $-\infty \le a < \infty$) possesses the semigroup property: $J^\alpha J^\beta = J^{\alpha+\beta}$. Indeed, for any $\alpha > 0$ and $\beta > 0$ changing order of integration, we have

$$J^\alpha J^\beta f(t) = \frac{1}{\Gamma(\alpha)} \int_0^t (t-\tau)^{\alpha-1} \Big(\frac{1}{\Gamma(\beta)} \int_0^\tau (\tau - u)^{\beta-1} f(u) du \Big) d\tau$$
$$= \frac{1}{\Gamma(\alpha)\Gamma(\beta)} \int_0^t f(u) \Big(\int_u^t (t-\tau)^{\alpha-1} (\tau-u)^{\beta-1} d\tau \Big) du.$$

The internal integral after substitution $\tau = t - (t-u)s$ takes the form

$$(t-u)^{\alpha+\beta-1} \int_0^1 s^{\alpha-1}(1-s)^{\beta-1} ds = B(\alpha,\beta)(t-u)^{\alpha+\beta-1}.$$

Therefore, due to (3.4) we obtain the equality $J^\alpha J^\beta f(t) = J^{\alpha+\beta} f(t)$.

Example 3.1. Let $\alpha > 0$ and $\gamma > -1$. Then

$$J^\alpha t^\gamma = \frac{\Gamma(1+\gamma)}{\Gamma(1+\gamma+\alpha)} t^{\gamma+\alpha}, \quad t > 0. \tag{3.6}$$

To show this, one needs to use the substitution $\tau = ts$ in the integral $J^\alpha[t^\gamma]$, and property (3.4). Then,

$$J^\alpha t^\gamma = \frac{1}{\Gamma(\alpha)} \int_0^t (t-\tau)^{\alpha-1} \tau^\gamma d\tau = \frac{t^{\alpha+\gamma}}{\Gamma(\alpha)} \int_0^1 (1-s)^{\alpha-1} s^\gamma ds$$
$$= \frac{t^{\alpha+\gamma}}{\Gamma(\alpha)} B(\alpha, \gamma+1) = \frac{\Gamma(1+\gamma)}{\Gamma(1+\gamma+\alpha)} t^{\gamma+\alpha}.$$

In the particular case $\gamma = 0$, we have

$$J^\alpha 1 = \frac{t^\alpha}{\Gamma(1+\alpha)}.$$

Example 3.1 shows that if $-1 < \gamma < 0$ and $\alpha + \gamma < 0$, then $\lim_{t\to 0} J^\alpha[t^\gamma] = \infty$. However, if $\gamma \ge 0$, then for any $\alpha > 0$, the limit $\lim_{t\to 0} J^\alpha[t^\gamma] = 0$. The latter is valid for any function continuous up to zero. In fact, the following statement holds.

Proposition 3.1. *Let $T > 0$ be an arbitrary number and $f \in C[0,T]$. Then*

$$\lim_{t\to 0+} J^\alpha f(t) = 0, \quad 0 < t < T, \tag{3.7}$$

for any $\alpha > 0$.

Proof. We can assume that $T \le 1$. Making use of the substitution $\tau = ts$ in the integral

$$J^\alpha[f](t) = \frac{1}{\Gamma(\alpha)} \int_0^t (t-\tau)^{\alpha-1} f(\tau) d\tau,$$

one obtains

$$J^{\alpha}[f](t) = \frac{t^{\alpha}}{\Gamma(\alpha)} \int_0^1 (1-s)^{\alpha-1} f(ts)ds.$$

Due to continuity of f on the interval $[0,1]$, we have $|f(st)| \leq M < \infty$, $0 \leq s \leq 1$. Therefore, the estimate

$$|J^{\alpha} f(t)| \leq = \frac{Mt^{\alpha}}{\Gamma(\alpha)} \int_0^1 (1-s)^{\alpha-1} ds = \frac{Mt^{\alpha}}{\alpha\Gamma(\alpha)}$$

holds, yielding (3.7) when $t \to 0+$.

The fractional integral ${}_aJ_t^{\alpha}$ is called *left-sided*. The fractional integral, called *right-sided*, and defined as

$$ {}_tJ_b^{\alpha} f(t) = \frac{1}{\Gamma(\alpha)} \int_t^b (b-\tau)^{\alpha-1} f(\tau)d\tau, \tag{3.8}$$

is also frequently used. In fact, the operators ${}_aJ_t^{\alpha}$ and ${}_tJ_b^{\alpha}$ are mutually adjoint operators. Namely, for arbitrary functions $u, v \in L_2(a,b)$, the relation [SKM87]

$$({}_aJ_t^{\alpha} u(t), v(t)) = (u(t), {}_tJ_b^{\alpha} v(t))$$

holds.

The proposition below states the continuity of the operators ${}_aJ_t^{\alpha}$ and ${}_tJ_b^{\alpha}$ in L_p and Hölder spaces.

Proposition 3.2. *([SKM87]) Fractional order integration operators ${}_aJ_t^{\alpha}$ and ${}_tJ_b^{\alpha}$ are continuous mappings:*

$$ {}_aJ_t^{\alpha} : L_p(a,b) \to L_p(a,b), \quad {}_tJ_b^{\alpha} : L_p(a,b) \to L_p(a,b),$$
$$ {}_aJ_t^{\alpha} : C^{\lambda}[a,b] \to C^{\lambda+\alpha}[a,b], \quad {}_tJ_b^{\alpha} : C^{\lambda}[a,b] \to C^{\lambda+\alpha}[a,b],$$

where $p \geq 1$, and $0 < \lambda < 1$, $\lambda + \alpha \neq 1$. Moreover, the following estimates hold:

$$\|{}_aJ_t^{\alpha} f(t) | L_p(a,b)\| \leq \frac{(b-a)^{\alpha}}{\alpha\Gamma(\alpha)} \|f | L_p(a,b)\|,$$
$$\|{}_tJ_b^{\alpha} f(t) | L_p(a,b)\| \leq \frac{(b-a)^{\alpha}}{\alpha\Gamma(\alpha)} \|f | L_p(a,b)\|.$$

In particular, the mappings ${}_aJ_t^{\alpha} : C[a,b] \to C^{\alpha}[a,b]$ and ${}_tJ_b^{\alpha} : C[a,b] \to C^{\alpha}[a,b]$ are continuous.

3.3 The Riemann-Liouville fractional derivative

Riemann and Liouville defined the fractional derivative as the left inverse to the fractional integration operator of order $\alpha > 0$.

Suppose that $0 < \alpha < 1$ and consider the operator $DJ^{1-\alpha}$. We claim that this operator is the left-inverse to J^α. In fact, using the above-mentioned semigroup property, one has $DJ^{1-\alpha}J^\alpha = DJ = I$. Hence, for $0 < \alpha < 1$ the definition of the fractional derivative of order α in the sense of Riemann-Liouville would be

$$D^\alpha = DJ^{1-\alpha}. \tag{3.9}$$

This operator, as we checked above, satisfies $D^\alpha J^\alpha = I$, extending the relation $D^n J^n = I$ to any real number $\alpha \in (0,1)$. Obviously, this form of fractional derivative extends to $\alpha = 1$ giving $D^1 = D$. However, in the explicit form written below one has to assume that $0 < \alpha < 1$, accepting conventionally $D^1 = D$.

Definition 3.1. The fractional derivative of order α, $0 < \alpha < 1$, of a function f defined on $[0,\infty)$ in the Riemann-Liouville sense, is

$$D^\alpha f(t) = \frac{1}{\Gamma(1-\alpha)} \frac{d}{dt} \int_0^t \frac{f(\tau)d\tau}{(t-\tau)^\alpha}, \quad t > 0, \tag{3.10}$$

provided the right-hand side exists.

In a similar manner for α satisfying $m-1 < \alpha < m$, $m = 2,3,\ldots$, one can easily verify that the operator

$$D^\alpha = \frac{d^m}{dt^m} J^{m-\alpha} \tag{3.11}$$

is the left-inverse to J^α, that is $D^\alpha J^\alpha = I$. Since $0 < m - \alpha < 1$, due to (3.9) $\frac{d}{dt} J^{m-\alpha} = D^{\alpha-m+1}$. Therefore, one can write (3.11) in the form

$$D^\alpha = \frac{d^{m-1}}{dt^{m-1}} D^{\alpha-m+1}, \quad m-1 < \alpha < m.$$

Thus, one has the following definition of the fractional derivative in the sense of Riemann-Liouville for an arbitrary $\alpha \geq 0$.

Definition 3.2. Let $\alpha \geq 0$. The Riemann-Liouville derivative of order α of a function f is

$$D^\alpha f(t) = \begin{cases} \frac{1}{\Gamma(1-\alpha)} \frac{d}{dt} \int_0^t \frac{f(\tau)d\tau}{(t-\tau)^\alpha}, & \text{if } 0 < \alpha < 1 \text{ (associated with } m = 1), \\ \frac{d^{m-1}}{dt^{m-1}} D^{\alpha-m+1} f(t), & \text{if } m-1 < \alpha < m,\ m = 2,3,\ldots, \\ \frac{d^m f(t)}{dt^m}, & \alpha = m = 0,1,\ldots, \end{cases} \tag{3.12}$$

provided the right-hand side exists in each case. Sometimes, to avoid a confusion, we denote the Riemann-Liouville fractional derivative of order α by D_+^α.

Remark 3.1. 1. The explicit form of $D^\alpha f(t)$ for $\alpha \in (m-1,m)$, as it follows from equations (3.10) and (3.11), is

$$D^\alpha f(t) = \frac{1}{\Gamma(m-\alpha)} \frac{d^m}{dt^m} \int_0^t \frac{f(\tau)d\tau}{(t-\tau)^{\alpha-m+1}}, \quad t>0. \tag{3.13}$$

Here one can put $\alpha = m-1$. In this case, as it is readily seen, $D^\alpha f(t) = \frac{d^m}{dt^m} Jf(t) = \frac{d^{m-1}}{dt^{m-1}} f(t)$. However, in the explicit form (3.13) one cannot replace α with its upper bound m, since in this case a strong singularity appears in the integral. On the other hand, the operator form of the fractional derivative in equation (3.13) is

$$D^\alpha = \frac{d^m}{dt^m} J^{m-\alpha}, \tag{3.14}$$

where α satisfies the condition $m-1<\alpha<m$. Here one can formally put $\alpha = m$ obtaining $D^\alpha = D^m = \frac{d^m}{dt^m}$, which is consistent with Definition 3.2. In our further considerations we sometimes write $m-1 \le \alpha < m$, assuming informal setting $\alpha = m-1$ in the explicit form (3.13), or $m-1 < \alpha \le m$, assuming formal setting $\alpha = m$ in the operator form (3.14) of the fractional derivative of order α.

2. The fractional derivative operator defined in (3.14) is called *left-sided*, since the corresponding fractional integral $J^{m-\alpha}$ is left-sided. A right-sided Rieman-Liouville fractional order derivative is defined in a similar manner: ${}_tD_T^\alpha = (-D)^m \, {}_tJ_T^{m-\alpha}$ (see formula (3.16) below).
3. The fractional derivative D^α for non-integer α, unlike the integer order differentiation operator, is not a local operator. It depends on the whole interval $(0,t)$. If the interval where the fractional derivative is defined is (a,b), then the corresponding left-sided Riemann-Liouville fractional derivative is denoted by ${}_aD_t^\alpha$, and defined as

$$ {}_aD_t^\alpha f(t) = \frac{1}{\Gamma(m-\alpha)} \frac{d^m}{dt^m} \int_a^t \frac{f(\tau)d\tau}{(t-\tau)^{\alpha-m+1}}, \quad t \in (a,b). \tag{3.15}$$

Similarly, the right-sided Riemann-Liouville fractional derivative is defined by

$$ {}_tD_b^\alpha f(t) = \frac{(-1)^m}{\Gamma(m-\alpha)} \frac{d^m}{dt^m} \int_t^b \frac{f(\tau)d\tau}{(t-\tau)^{\alpha-m+1}}, \quad t \in (a,b). \tag{3.16}$$

Example 3.2. Let $f(t) = t^\gamma$, where $\gamma > -1$. Then for $\alpha > 0$ one has

$$D^\alpha f(t) = \frac{\Gamma(1+\gamma)}{\Gamma(1+\gamma-\alpha)} t^{\gamma-\alpha}, \quad t>0. \tag{3.17}$$

Suppose $m-1<\alpha \le m$. Then formula (3.17) follows immediately, if one uses (3.6) in the relation $D^\alpha[t^\gamma](t) = D^m J^{m-\alpha}[t^\gamma](t)$. In (3.17) one needs to use the analytic extension of Euler's gamma function to the left complex plane, if $1+\gamma-\alpha<0$, which has simple poles at points $-1,-2,\ldots$. Therefore, $D^\alpha[t^\gamma] = 0$, if $\alpha = 1+\gamma+k$, $k = 0,1,\ldots$.

Consider two important particular cases in equation (3.17):

(1) $\gamma = 1$. In this case

$$D^{\alpha} 1 = \frac{1}{\Gamma(1-\alpha)} t^{-\alpha}. \tag{3.18}$$

This shows that the fractional derivative of a constant is not zero! (unlike the usual derivative). However, recall that the function $\Gamma(s)$ has the analytic extension to the complex plane except points $\{0,-1,-2,\dots\}$, which are simple poles. So,

$$\Gamma(s) = \infty \text{ if } s = 0,-1,-2,\dots. \tag{3.19}$$

Therefore, it follows from (3.18) that $D^{\alpha} 1 = 0$ if $\alpha = 1,2,\dots$.

(2) $\gamma = \alpha - 1$. In this case

$$D^{\alpha} t^{\alpha-1} = 0. \tag{3.20}$$

This shows that the kernel of the operator D^{α} (with the domain in $L_1(0,1)$) is not trivial!

Example 3.3. Consider the fractional order homogeneous differential equation

$$D^{\alpha} u(t) = 0, \quad t > 0, \tag{3.21}$$

where $0 < \alpha < 1$. Introduce the set $\mathscr{K}(\alpha) = \{h(t) : h(t) = Ct^{\alpha-1}, C \in \mathbb{R}\}$. It follows from (3.20) that any function $h \in \mathscr{K}(\alpha)$ satisfies equation (3.21). In fact, the kernel of operator D^{α} consists of all functions $h \in \mathscr{K}(\alpha)$ and not other functions, i.e., $Ker(D^{\alpha}) = \mathscr{K}(\alpha)$. Assuming $Ker(D^{\alpha}) = \mathscr{K}(\alpha)$, one can see that the initial value problem

$$D^{\alpha} u(t) = 0, \quad t > 0, \tag{3.22}$$

$$u(0) = a, \tag{3.23}$$

where $a \neq 0$, has no solution. Therefore, the Cauchy problem (3.22)–(3.23) is ill-posed. On the other hand, if the initial condition (3.23) is replaced by

$$(J^{1-\alpha} u)(0) = a, \tag{3.24}$$

then due to Example 3.1 for a function $h(t) = Ct^{\alpha-1} \in \mathscr{K}(\alpha)$ one has $J^{1-\alpha} h(t) = C\Gamma(\alpha)$. (Notice that h is not continuous at $t = 0$, therefore Proposition 3.1 is not applicable here.) Therefore, the choice $C = \frac{a}{\Gamma(\alpha)}$ provides a unique solution to the Cauchy type problem (3.22)–(3.24). We note that the initial value problem (3.22), (3.24) is not a Cauchy problem, since the initial condition $(J^{1-\alpha} u)(0) = a$ is not a Cauchy condition. This crucial observation is valid for the homogeneous fractional order differential equation

$$D^{\alpha} u(t) = 0, \quad t > 0,$$

for any $\alpha > 0$, $m-1 < \alpha < m$, with the initial conditions

$$(D^k J^{m-\alpha} u)(0) = a_k, k = 0, \ldots, m-1.$$

In this case $\mathscr{K}(\alpha) = \{h(t) : h(t) = \Sigma_{k=0}^{m-1} C_K t^{\alpha-m+k}\}$.

Example 3.4. Let $h \in \mathscr{K}(\alpha)$, $\alpha > 0$. Then (3.20) immediately implies that for $\beta \in (0,\alpha)$, such that $\alpha - \beta$ is not integer:

1. $0 = D^\beta D^\alpha h(t) \neq D^\alpha D^\beta h(t) \neq 0$;
2. $0 = D^\beta D^\alpha h(t) \neq D^{\alpha+\beta} h(t) \neq 0$.

A natural question arising in connection with the Riemann-Liouville derivative is for what class of functions f one can ensure the existence of the fractional derivative $D^\alpha f(t)$ of order α. Below we reproduce one well-known and useful statement (see Samko et al. [SKM87], p. 239) answering this question.

Proposition 3.3. *([SKM87]) Let b be any positive number and $f \in C^\lambda[0,b]$, $0 < \lambda \leq 1$. Then for any $\alpha < \lambda$ the fractional derivative $D^\alpha f(t)$ exists and can be represented in the form*

$$D^\alpha f(t) = \frac{f(0)}{\Gamma(1-\alpha)t^\alpha} + \psi(t), \tag{3.25}$$

where $\psi \in C^{\lambda-\alpha}[0,b]$, and $\psi(0) = 0$. Moreover, the estimate $\|\psi|C^{\lambda-\alpha}\| \leq C\|f|C^\lambda\|$ holds with some constant $C > 0$.

Representation (3.25) carries an important information related to Riemann-Liouville fractional derivatives. Namely, if f is continuous at $t = 0$ and its Riemann-Liouville derivative of order α exists, then this derivative has the singularity of order α at $t = 0$. Therefore, the fact that the Cauchy problem (3.22)–(3.23) is not well posed is not surprising.

Proposition 3.4. *Let $\alpha > 0$. Then the Laplace transform of $J^\alpha f(t)$ is*

$$L[J^\alpha f](s) = s^{-\alpha} L[f](s), \quad s > 0. \tag{3.26}$$

Proof. By definition,

$$L[J^\alpha f](s) = \frac{1}{\Gamma(\alpha)} \int_0^\infty e^{-st} \int_0^t (t-\tau)^{\alpha-1} f(\tau) d\tau dt, \quad s > 0. \tag{3.27}$$

Changing the order of integration (Fubuni is allowed) the right-hand side of (3.27) can be written as

$$\frac{1}{\Gamma(\alpha)} \int_0^\infty f(\tau) \int_\tau^\infty (t-\tau)^{\alpha-1} e^{-st} dt d\tau, \quad s > 0. \tag{3.28}$$

The substitution $t - \tau = u/s$ in the internal integral reduces it into

$$\int_\tau^\infty (t-\tau)^{\alpha-1} e^{-st} dt = \Gamma(\alpha) \frac{e^{-s\tau}}{s^\alpha}, \quad s > 0.$$

The latter and equations (3.27), (3.28) imply (3.26).

Proposition 3.5. *Let $m-1<\alpha\le m$, $m=1,2,\dots$. Then the Laplace transform of $D^\alpha f(t)$ is*

$$L[D^\alpha f](s)=s^\alpha L[f](s)-\sum_{k=0}^{m-1}(D^kJ^{m-\alpha}f)(0)s^{m-1-k}. \tag{3.29}$$

Proof. Making use of the equation $D^\alpha=D^mJ^{m-\alpha}$ and Proposition 3.4, one has

$$L[D^\alpha f](s)=L[D^mJ^{m-\alpha}f](s)=s^mL[J^{m-\alpha}f](s)-\sum_{k=0}^{m-1}(D^kJ^{m-\alpha}f)(0)s^{m-1-k}$$

$$=s^\alpha L[f](s)-\sum_{k=0}^{m-1}(D^kJ^{m-\alpha}f)(0)s^{m-1-k}.$$

Let $0<\alpha\le 1$. Then formula (1.22) with $m=1$ reduces to

$$L[D^\alpha f](s)=s^\alpha L[f](s)-(J^{1-\alpha}f)(0). \tag{3.30}$$

Example 3.5. In Example 3.3 we introduced the set $\mathscr{K}(\alpha)$ of functions $h(t)=Ct^{\alpha-1}, t>0$, satisfying the fractional order differential equation $D^\alpha u(t)=0$, $0<\alpha<1$. Now we show that $\mathscr{K}(\alpha)$ exhausts all possible solutions of this equation. Suppose $h(t)$ satisfies the equation $D^\alpha h(t)=0, t>0$. Then in accordance with formula (3.30) one has

$$L[D^\alpha h](s)=s^\alpha L[h](s)-(J^{1-\alpha}h)(0)=0,\ s>0.$$

Solving this for $L[h](s)$,

$$L[h](s)=\frac{b}{s^\alpha},\ s>0,$$

where $b=(J^{1-\alpha}h)(0)$ is a constant. Hence, see Example 1.3, $h(t)=\frac{b}{\Gamma(\alpha)}t^{\alpha-1}\in\mathscr{K}(\alpha)$. This yields $Ker(D^\alpha)\equiv\mathscr{K}(\alpha)$.

Example 3.6. Consider Abel's integral equation

$$J^\alpha u(t)=h(t),\quad t>0,\ 0<\alpha<1, \tag{3.31}$$

where $h\in C^{\alpha+\lambda}[0,b]$, $b>0$, with λ satisfying $0<\lambda<1-\alpha$, and $h(0)=0$. This equation has a unique solution in $C^\lambda[0,b]$. Moreover, the solution has the representation

$$u(t)=D^\alpha h(t). \tag{3.32}$$

To see this we take the Laplace transform of both sides of (3.31), obtaining $s^{-\alpha}L[u](s)=L[h](s)$. This can be written as

$$L[u](s)=s^\alpha L[h](s)=s^\alpha L[h](s)-(J^{1-\alpha}h)(0). \tag{3.33}$$

since $J^{1-\alpha}h(0)=0$ due to Example 3.1 and continuity of h at $t=0$. Equation (3.33) immediately implies $u(t)=D^\alpha h(t)$, in accordance with (3.30). The fact that $u(t)\in$

$C^\lambda[0,b]$ follows from Proposition 3.3. The case $\alpha > 1$ can be reduced to (3.31) by differentiating.

If h is not continuous up to $t = 0$, then we cannot rely on the fact that $J^{1-\alpha}h(0) = 0$. In the general case the following statement holds.

Proposition 3.6. *([Djr66]) Let $h \in L_1(0,b)$. Then equation* (3.31) *has a unique solution $u(t) \in L_1(0,1)$, which is represented in the form* (3.32)*, if and only if*

1. *$D^{m-1}J^{m-\alpha}h(t)$ is absolute continuous on $[0,b]$;*
2. $J^{m-\alpha}h(0) = \cdots = (D^{m-1}J^{m-\alpha}h)(0) = 0$.

Example 3.7. Consider the fractional differential equation

$$D^\alpha u(t) = h(t), \quad t > 0, \tag{3.34}$$

subject to the initial conditions

$$J^{m-\alpha}u(0) = a_0, \ \ldots, \ D^{m-1}J^{m-\alpha}u(0) = a_{m-1}, \tag{3.35}$$

where $m - 1 < \alpha < m$, and h is a continuous function. This problem has a unique solution

$$u(t) = J^\alpha[h](t) + \sum_{k=0}^{m-1} \frac{a_k}{\Gamma(\alpha - m + k + 1)} t^{\alpha-m+k}, \quad t > 0. \tag{3.36}$$

Indeed, applying the Laplace transform to equation (3.34), one has

$$s^\alpha L[u](s) - \sum_{k=0}^{m-1} D^k(J^{m-\alpha}u)(0)s^{m-k-1} = L[h](s), \quad s > 0,$$

or solving it for $L[u](s)$ and taking into account conditions (3.35),

$$L[u](s) = \frac{L[h](s)}{s^\alpha} + \sum_{k=0}^{m-1} \frac{a_k}{s^{\alpha-m+k+1}}, \quad s > 0.$$

Now (3.36) follows from the latter due to Proposition 3.4 and (1.18).

3.4 Mittag-Leffler function and its Laplace transform

The Mittag-Leffler function $E_\alpha(z)$, $z \in \mathbb{C}$, is an entire function defined as a power series

$$E_\alpha(z) = \sum_{n=0}^{\infty} \frac{z^n}{\Gamma(\alpha n + 1)},$$

where $\Gamma(\cdot)$ is Euler's gamma function. Obviously, $E_\alpha(z) = \exp(z)$ if $\alpha = 1$. For real $z = x$ the Mittag-Leffler function $E_\alpha(x)$ increases exponentially ($\sim \exp(x^{1/\alpha})$), when $x \to \infty$, and tends to 0 behaving like $E_\alpha(x) \sim 1/|x|$, when $x \to -\infty$. Therefore, the Laplace transform of the function $e(t) = E_\alpha(-t)$ defined on the half-axis $\mathbb{R}_+$, is well defined. In our considerations the function $e(\lambda t^\alpha) = E_\alpha(-\lambda t^\alpha)$, i.e.

$$E_\alpha(-\lambda t^\alpha) = 1 - \frac{\lambda t^\alpha}{\Gamma(\alpha+1)} + \frac{\lambda^2 t^{2\alpha}}{\Gamma(2\alpha+1)} + \dots \tag{3.37}$$

for $\lambda \in \mathbb{C}$, will emerge frequently. Therefore we find its Laplace transform.

Proposition 3.7. *For the Laplace transform of* $E_\alpha(-\lambda t^\alpha)$, $t \geq 0$, *where* $\lambda \in \mathbb{C}$, *the following formula holds:*

$$L[E_\alpha(-\lambda t^\alpha)](s) = \frac{s^{\alpha-1}}{\lambda + s^\alpha}, \quad s > |\lambda|^{1/\alpha}. \tag{3.38}$$

Proof. Using the formula (see (1.18))

$$\int_0^\infty e^{-st} t^{\alpha n} dt = L[t^{\alpha n}](s) = \frac{\Gamma(\alpha n + 1)}{s^{\alpha n+1}}, \quad s > 0,$$

one has

$$\begin{aligned} L[E_\alpha(-\lambda t^\alpha)](s) &= \sum_{n=0}^\infty \frac{(-1)^n \lambda^n}{\Gamma(\alpha n + 1)} \int_0^\infty e^{-st} t^{\alpha n} dt \\ &= \frac{1}{s} \sum_{n=0}^\infty \left(-\frac{\lambda}{s^\alpha}\right)^n = \frac{s^{\alpha-1}}{\lambda + s^\alpha}. \end{aligned} \tag{3.39}$$

The geometric series in (3.39) converges if $|\lambda / s^\alpha| < 1$, or the same, if $s > |\lambda|^{1/\alpha}$.

Let $0 < \alpha < 1$. Differentiating $E_\alpha(-\lambda t^\alpha)$ in (3.37) in the variable t, one has

$$\frac{d}{dt} E_\alpha(-\lambda t^\alpha) = -\frac{\lambda t^{\alpha-1}}{\Gamma(\alpha)} + \frac{\lambda^2 t^{2\alpha-1}}{\Gamma(2\alpha)} + \dots,$$

which implies $|E_\alpha(-\lambda t^\alpha)| = O(\frac{1}{t^{1-\alpha}})$, $t \to 0$. Similarly, if $m-1 < \alpha < m$, then m times differentiating $E_\alpha(-\lambda t^\alpha)$, one obtains that $\left|\frac{d^m}{dt^m} E_\alpha(-\lambda t^\alpha)\right| = O(\frac{1}{t^{m-\alpha}})$, when $t \to 0$. Thus we have proved the following proposition.

Proposition 3.8. *Let* $m-1 < \alpha < m$. *Then for each fixed* $\lambda \in \mathbb{C}$ *the asymptotic behavior*

$$\left|\frac{d^m}{dt^m} E_\alpha(-\lambda t^\alpha)\right| = O\left(\frac{1}{t^{m-\alpha}}\right), \quad t \to 0, \tag{3.40}$$

holds.

An integral representation of the function $E_\alpha(-t^\alpha)$ was provided by R. Gorenflo and F. Mainardi in [GMM02]. The proposition below reproduces this representation for $E_\alpha(-\lambda t^\alpha)$ with an arbitrary positive real λ. Introduce the function

$$K_\alpha(r) = \frac{1}{\pi}\frac{r^{\alpha-1}\sin(\alpha\pi)}{r^{2\alpha}+2r^\alpha\cos(\alpha\pi)+1}, \quad r>0. \tag{3.41}$$

Proposition 3.9. *Let λ be a positive real number. Then for $E_\alpha(-\lambda t^\alpha)$ the following representations hold:*

1. if $0<\alpha<1$, then

$$E_\alpha(-\lambda t^\alpha) = \int_0^\infty e^{-\lambda^{2/\alpha} rt} K_\alpha(r)dr. \tag{3.42}$$

2. if $1<\alpha<2$, then

$$E_\alpha(-\lambda t^\alpha) = \int_0^\infty e^{-rt\lambda^{2/\alpha}} K_\alpha(r)dr + \frac{2}{\alpha} e^{t\lambda^{1/\alpha}\cos\frac{\pi}{\alpha}} \cos\left(t\lambda^{1/\alpha}\sin\frac{\pi}{\alpha}\right) \tag{3.43}$$

Proof. It follows from Proposition 3.38 that

$$E_\alpha(-\lambda t^\alpha) = L^{-1}\left[\frac{s^{\alpha-1}}{s^\alpha+\lambda}\right](t) = \frac{1}{2\pi i}\int_{\sigma-i\infty}^{\sigma+i\infty}\frac{s^{\alpha-1}e^{st}}{\lambda+s^\alpha}ds,$$

where $\sigma > \lambda^{1/\alpha}$. Since α is not an integer, the function $H(s) = \frac{s^\alpha e^{st}}{s^\alpha+1}$ under the integral has a branching point at $s=0$. We take values in the main branch. Moreover, $H(s)$ has poles at points $s_k = \lambda^{1/\alpha}e^{i\pi/\alpha(2k+1)}$, $k\in\mathbb{Z}$. The poles which are in the main branch satisfy the condition $-\pi<\arg(s_k)<\pi$. Further, deforming integration path to the Hankel path H_ε, which starts at $-\infty$ along the lower side of the cut negative real axis of the complex s-plane, along the circle of radius ε centered at zero, and then goes to $-\infty$ along the upper side of negative axis, one can reduce the above integral to

$$E_\alpha(-\lambda t^\alpha) = \frac{1}{2\pi i}\int\limits_{s=\tau e^{\pm i\pi}}\frac{s^{\alpha-1}e^{st}}{\lambda+s^\alpha}ds + \sum_{s_k\in\mathscr{P}} e^{s_k t} Res_{|s=s_k}\left[\frac{s^{\alpha-1}}{\lambda+s^\alpha}\right],$$

where $\mathscr{P}$ is the set of relevant poles of the function $H(s)$, and $Res_{|s=a}$ means the residue at the pole $s=a$. Obviously, the function $H(s)$ has simple poles and $Res_{|s=s_k}\left[\frac{s^{\alpha-1}}{\lambda+s^\alpha}\right] = 1/\alpha$. Hence, $E_\alpha(-\lambda t^\alpha) = M_\alpha(t)+N_\alpha(t)$, where

$$M_\alpha(t) = \frac{1}{2\pi i}\int\limits_{s=\tau e^{\pm i\pi}}\frac{s^{\alpha-1}e^{st}}{\lambda+s^\alpha}ds \quad\text{and}\quad N_\alpha(t) = \frac{1}{\alpha}\sum_{s_k\in\mathscr{P}} e^{s_k t}.$$

In computing the integral along two sides of the negative axis, one can see that the real parts cancel out and the imaginary parts are added, so

$$M_\alpha(t) = \frac{1}{\pi}\int\limits_0^\infty e^{-\lambda^{1/\alpha}\tau t}\Im\left(\frac{s^{\alpha-1}}{\lambda+s^\alpha}_{|s=\tau e^{i\pi}}\right)dr, \tag{3.44}$$

where $\Im(z)$ means the imaginary part of z. It is easy to verify that

$$\Im\left(\frac{s^{\alpha-1}}{\lambda+s^{\alpha}}\Big|_{s=\tau e^{i\pi}}\right)=\frac{\lambda\tau^{\alpha-1}\sin(\alpha\pi)}{\lambda^2+2\lambda\tau^{\alpha}\cos(\alpha\pi)+\tau^{2\alpha}}.$$

Substituting this into (3.44) and changing $\tau=r\lambda^{1/\alpha}$, one obtains

$$M_{\alpha}(t)=\int_0^{\infty}e^{-\lambda^{2/\alpha}rt}K_{\alpha}(r)dr,$$

where $K_{\alpha}(r)$ is defined in (3.41). Now notice that if $0<\alpha<1$, there are no poles of $H(s)$ satisfying $-\pi<\arg(s_k)<\pi$. Hence $N_{\alpha}(t)=0$, and in this case we have $E_{\alpha}(-\lambda t^{\alpha})=M_{\alpha}(t)$, obtaining (3.42).

If $1<\alpha<2$, then there are two poles $\lambda^{1/\alpha}e^{\pm i\pi/\alpha}$ in the main branch. This implies

$$N_{\alpha}(t)=\frac{2}{\alpha}e^{t\lambda^{1/\alpha}\cos\frac{\pi}{\alpha}}\cos(t\lambda^{1/\alpha}\sin\frac{\pi}{\alpha}).$$

Hence, in this case $E_{\alpha}(-\lambda t^{\alpha})=M_{\alpha}(t)+N_{\alpha}(t)$, obtaining (3.43).

Remark 3.2. The function $E_{\alpha}(-t^{\alpha})$ has the following asymptotes near zero and infinity [GM97]:

$$E_{\alpha}(-t^{\alpha})\sim 1-\frac{t^{\alpha}}{\Gamma(\alpha+1)}+\frac{t^{2\alpha}}{\Gamma(2\alpha+1)}+\ldots,\quad t\to+0,$$
$$E_{\alpha}(-t^{\alpha})\sim\frac{t^{-\alpha}}{\Gamma(1-\alpha)}-\frac{t^{-2\alpha}}{\Gamma(1-2\alpha)}+\ldots,\quad t\to\infty,\quad(0<\alpha<2).$$

3.5 The Caputo-Djrbashian fractional derivative

The fractional derivative in the Caputo sense, introduced by Caputo [Cap67] in 1967, contrary to Riemann-Liouville derivative, is more convenient for the study of the Cauchy problem. This is important in the study of fractional models of various physical problems. On the other hand, the Caputo derivative is restrictive to compare to the Riemann-Liouville one, since to define the derivative of order $\alpha\in(m-1,m)$ in the sense of Caputo, one requires the existence of the derivative of order m.

Definition 3.3. By definition, the Caputo fractional derivative of order $\alpha\geq 0$ is

$$(D_*^{\alpha})f(t)=\begin{cases}f(t), & \alpha=0,\\ \frac{1}{\Gamma(1-\alpha)}\int_0^t\frac{f'(\tau)d\tau}{(t-\tau)^{\alpha}}, & 0<\alpha<1\text{ (associated with }m=1),\\ D_*^{\alpha-m+1}\frac{d^{m-1}}{dt^{m-1}}f(t), & m-1\leq\alpha<m,\ m=2,3,\ldots\end{cases}\qquad(3.45)$$

provided the right-hand side exists in each case.

The third line of this definition requires some explanation. If $m-1 \leq \alpha < m$, then $0 \leq \alpha - m + 1 < 1$. Hence, one uses the first or second line to define the fractional derivative $D_*^{\alpha-m+1}g(t)$ of the function $g(t) = f^{(m-1)}(t)$.

Let $0 < \alpha < 1$. Then it follows from Definition 3.3 (second line) that

$$D_*^\alpha = J^{1-\alpha} D. \tag{3.46}$$

This is an operator form of D_*^α. Equation (3.46) implies that

$$J^\alpha D_*^\alpha f(t) = J^\alpha J^{1-\alpha} D f(t) = JDf(t) = f(t) - f(0), \tag{3.47}$$

which means that D_*^α is the right-inverse to J^α up to the additive term $-f(0)$. Recall that the fractional derivative D^α in the sense of Riemann-Liouville was the (exact) left-inverse to the operator J^α. In general, if $m-1 < \alpha < m$, $m = 1, 2, \ldots$, then the operator form of D_*^α becomes

$$D_*^\alpha = J^{m-\alpha} \frac{d^m}{dt^m}, \tag{3.48}$$

and relation (3.47) takes the form

$$J^\alpha D_*^\alpha f(t) = J^\alpha J^{m-\alpha} D^m f(t) = J^m D^m f(t) = f(t) - \sum_{k=0}^{m-1} \frac{D^k f(0)}{k!} t^k, \tag{3.49}$$

showing that D_*^α is the right-inverse to J^α up to the additive polynomial

$$-\sum_{k=0}^{m-1} \frac{D^k f(0)}{k!} t^k.$$

Now again assume that $0 < \alpha < 1$. Then the definition of $D_*^\alpha f(t)$ uses the first derivative $f'(t)$. Therefore, if one works with Definition 3.3, then one must ensure the existence of the first derivative of f in some sense. However, the relationship

$$D_*^\alpha f(t) = D^\alpha [f(t) - f(0)]. \tag{3.50}$$

between D^α and D_*^α, obtained by applying D^α to both sides of (3.47), can be used to overcome the restriction connected with the presence of the first derivative. Obviously, the expression on the right-hand side of (3.50) exists representing an extension of D_*^α to the class of functions f for which the fractional derivative in the sense of Riemann-Liouville exists, and $f(0)$ is finite. If additionally, f is differentiable (or absolutely continuous), then $D_*^\alpha f(t)$ in (3.50) and in Definition 3.3 coincide. Moreover, equation (3.50) shows that D_*^α and D^α coincide in the class of functions satisfying $f(0) = 0$.

In the general case the Caputo fractional derivative is related to the Riemann-Liouville one through the following formula

$$D_*^\alpha f(t) = D^\alpha \left(f(t) - \sum_{k=0}^{m-1} \frac{t^k f^{(k)}(0)}{k!} \right), \tag{3.51}$$

which shows that these two derivatives coincide if $f^{(n)}(0) = 0, n = 0, \ldots, m-1$. To prove (3.51) let us denote $g(t) = D_*^\alpha f(t)$. Applying the fractional integration operator J^α, one has (see (3.49))

$$J^\alpha g(t) = J^\alpha J^{m-\alpha} D^m f(t) = J^m D^m f(t) = f(t) - \sum_{k=0}^{m-1} \frac{f^{(k)}(0)}{k!} t^k.$$

Now since D^α is the left-inverse to J^α, one obtains

$$D^\alpha J^\alpha g(t) = g(t) = D^\alpha \left(f(t) - \sum_{k=0}^{m-1} \frac{t^k f^{(k)}(0)}{k!} \right),$$

thus (3.51).

The relationship (3.51) was noted by M. Djrbashian in his 1966 monograph [Djr66] with the aim to prove existence of the Riemann-Liouville derivative of order α. The fractional derivative D_*^α justly called the Caputo-Djrbashian fractional derivative. From now on we will also call D_*^α the *Caputo-Djrbashian fractional derivative.*

Proposition 3.10. *(Djrbashian [Djr66]) Let $D^{m-1} f(t)$ be absolutely continuous on $[0,b]$ for any $b > 0$. Then $D^\alpha f(t)$ exists a.e. in $(0,b)$ for any $\alpha \in (0,m]$. And if $m-1 < \alpha \le m$, then*

$$D_*^\alpha f(t) = D^\alpha f(t) - \sum_{k=0}^{m-1} \frac{f^{(k)}(0)}{\Gamma(1+k-\alpha)} t^{k-\alpha}, \tag{3.52}$$

Proof. Formula (3.52) follows from (3.51) immediately if one takes into account

$$D^\alpha [t^k] = \frac{k!}{\Gamma(1+k-\alpha)} t^{k-\alpha};$$

see Example 3.2.

Remark 3.3. 1. Notice that the functions $t^{k-\alpha}$, $k = 0, \ldots, m-1$, in (3.52) are singular at $t = 0$. The Caputo-Djrbashian derivative "removes" all the possible singularities of the RL derivative arising intrinsically as an action of D^α. Therefore, $D_*^\alpha f(t)$ is also called a *regularization* of the RL derivative $D^\alpha f(t)$.
2. If $m-1 < \alpha \le m$, $f(t)$ is defined on the interval $[a,b]$, and $D^{m-1} f(t)$ is absolutely continuous on $[a,b]$, then the relationship between the Riemann-Liouville and Caputo-Djrbashian fractional derivatives takes the form

$${}_aD_t^\alpha f(t) = \sum_{k=0}^{m-1} \frac{f^{(k)}(a)(t-a)^{k-\alpha}}{\Gamma(1+k-\alpha)} + {}_aD_{*t}^\alpha f(t). \tag{3.53}$$

Here ${}_aD_t^\alpha$ is defined in (3.15), and

$$ {}_aD^\alpha_{*\,t}f(t) = \frac{1}{\Gamma(m-\alpha)} \int_a^t \frac{f^{(m)}(\tau)}{(t-\tau)^{\alpha-m+1}} d\tau, \quad t \in (a,b). \tag{3.54}$$

3. The derivative ${}_aD^\alpha_{*\,t}$ in (3.54) is the left-sided Caputo-Djrbashian fractional derivative. The right-sided Caputo-Djrbashian fractional derivative is defined as

$$ {}_tD^\alpha_{*\,b}f(t) = \frac{(-1)^m}{\Gamma(m-\alpha)} \int_t^b \frac{f^{(m)}(\tau)}{(t-\tau)^{\alpha-m+1}} d\tau, \quad t \in (a,b).$$

Proposition 3.11. *Let $m-1 < \alpha \leq m$. The Laplace transform of the Caputo-Djrbashian derivative of a function $f \in C^m[0,\infty)$ is*

$$ L[D_*^\alpha f](s) = s^\alpha L[f](s) - \sum_{k=0}^{m-1} f^{(k)}(0) s^{\alpha-1-k}, \quad s > 0. \tag{3.55}$$

Proof. Since $D_*^\alpha f = J^{m-\alpha} D^m f$ we have

$$\begin{aligned} L[D_*^\alpha f](s) &= L[J^{m-\alpha} D^m f](s) = s^{-(m-\alpha)} L[D^m f](s) \\ &= s^{-(m-\alpha)} \left(s^m L[f](s) - \sum_{k=0}^{m-1} f^{(k)}(0) s^{m-1-k} \right) \\ &= s^\alpha L[f](s) - \sum_{k=0}^{m-1} f^{(k)}(0) s^{\alpha-1-k}, \quad s > 0. \end{aligned}$$

Example 3.8. 1. It follows from the definition of the Caputo-Djrbashian fractional derivative that $D_*^\alpha 1 = 0, \forall \alpha > 0$;
2. Let $\alpha > 0$ and $\gamma > 0$. Then

$$ D_*^\alpha [t^\gamma] = \frac{\Gamma(\gamma+1)}{\Gamma(\gamma+1-\alpha)} t^{\gamma-\alpha}.$$

Indeed, suppose $m-1 < \alpha \leq m$. Then, using the formula

$$ D^m [t^\gamma] = \frac{\Gamma(\gamma+1)}{\Gamma(\gamma-m+1)} t^{\gamma-m},$$

and relationship (3.6), one obtains

$$ D_*^\alpha [t^\gamma] = J^{m-\alpha} D^m [t^\gamma] = \frac{\Gamma(\gamma+1)}{\gamma+1-m} J^{m-\alpha} [t^{\gamma-m}] = \frac{\Gamma(\gamma+1)}{\Gamma(\gamma+1-\alpha)} t^{\gamma-\alpha}.$$

Example 3.9. Let $m-1 < \alpha \leq m$. Consider the following fractional order differential equation

$$ D_*^\alpha u(t) = h(t), \; t > 0,$$

where $h(t) \in C[0,T]$. We solve this equation under the assumption that $u(t)$ satisfies the initial conditions

$$u^{(k)}(0) = a_k, \quad k = 0, \ldots, m-1.$$

Applying the Laplace transform and taking into account formula (3.55), one has

$$L[u](s) = s^{-\alpha} L[h](s) + \sum_{k=0}^{m-1} u^{(k)}(0) s^{-k-1}.$$

Inverting the latter, one obtains a unique solution

$$u(t) = J^{\alpha} h(t) + \sum_{k=0}^{m-1} \frac{a_k}{k!} t^k.$$

Example 3.10. Consider the following Cauchy problem:

$$D_*^{\alpha} u(t) + \lambda u(t) = 0, \quad t > 0,$$

$$u^k(0) = a_k, \quad k = 0, \ldots, m-1,$$

where λ is a complex number. Applying the Laplace transform, one has

$$(s^{\alpha} + \lambda) L[u](s) = \sum_{k=0}^{m-1} a_k s^{\alpha-k-1}$$

Hence, due to Proposition 3.7,

$$u(t) = \sum_{k=0}^{m-1} a_k L^{-1} \left[\frac{1}{s^k} \frac{s^{\alpha-1}}{s^{\alpha} + \lambda} \right] = \sum_{k=0}^{m-1} a_k J^k E_{\alpha}(-\lambda t^{\alpha}). \tag{3.56}$$

where $E_{\alpha}(z)$ is the Mittag-Leffler function.

3.6 The Liouville-Weyl fractional derivative

Consider the pseudo-differential operator $A_{\alpha}(D)$ for $\alpha > 0$ with the symbol $\sigma_{A_{\alpha}}(\xi) = (i\xi)^{-\alpha}$, i.e.

$$A_{\alpha}(D) f(x) = \frac{1}{2\pi} \int_{-\infty}^{\infty} \frac{F[f](\xi) e^{-ix\xi} d\xi}{(i\xi)^{\alpha}}. \tag{3.57}$$

Obviously, A_{α} is a ΨDOSS whose symbol has a strong singularity at $\xi = 0$ if $\alpha \geq 1$. In Section 2.3 we saw that this operator is well defined on the space $\Psi_{G,p}(\mathbb{R})$, or on its dual $\Psi'_{-G,p'}(\mathbb{R})$. Moreover, since the symbol of A is isolated and is of finite order, this operator can be extended to a Lizorkin type spaces introduced in Section 1.12 with the base spaces of Besov or Lizorkin-Triebel. The number of orthogonality conditions depends on α.

Now we show that the operator $A_\alpha(D)$ in (3.57) is related to the fractional integration operator of order α. Let $0 < \alpha < 1$. Due to Corollary 2.11 (ii) one can write the symbol in the form

$$\frac{1}{(i\xi)^\alpha} = \sigma_{A_\alpha}(\xi) = (A_\alpha(D)e^{-ix\xi})_{|x=0}, \quad \xi \neq 0. \tag{3.58}$$

In equation (3.58) one can assume that the operator $A_\alpha(D)$ acts on the space $\Psi_{G,p}(\mathbb{R})$, or on $\Psi'_{-G,p'}(\mathbb{R})$. On the other hand, by virtue of formula (1.21), one has

$$\begin{aligned}\frac{1}{(i\xi)^\alpha} &= \frac{1}{\Gamma(\alpha)}\frac{\Gamma(\alpha)}{(i\xi)^\alpha} = \frac{1}{\Gamma(\alpha)}L[y^{\alpha-1}](i\xi) \\ &= \frac{1}{\Gamma(\alpha)}\int_0^\infty y^{\alpha-1}e^{iy\xi}dy = \frac{1}{\Gamma(\alpha)}\int_{-\infty}^0 (-y)^{\alpha-1}e^{-iy\xi}dy \\ &= \lim_{x\to 0}\frac{1}{\Gamma(\alpha)}\int_{-\infty}^x (x-y)^{\alpha-1}e^{-iy\xi}dy. \end{aligned}\tag{3.59}$$

Equation (3.59) shows that $(i\xi)^{-\alpha}$ is also the symbol of the fractional integral with the lower terminal point $-\infty$:

$$_{-\infty}J^\alpha f(x) = \frac{1}{\Gamma(\alpha)}\int_{-\infty}^x (x-y)^{\alpha-1}f(y)dy. \tag{3.60}$$

Indeed, one can rewrite (3.59) in the form

$$\frac{1}{(i\xi)^\alpha} = ({}_{-\infty}J^\alpha e^{-ix\xi})_{|x=0}, \quad \xi \neq 0.$$

Hence, the two operators $A_\alpha(D)$ and $_{-\infty}J^\alpha$ have the same symbol, and therefore coincide, if one considers them on the spaces $\Psi_{G,p}(\mathbb{R})$ or $\Psi'_{-G,p'}(\mathbb{R})$, $p \geq 1$, or Lizorkin type spaces.

The dual to $A_\alpha(D)$ operator $A^*_\alpha(D)$ has the symbol $(-i\xi)^{-\alpha}$. Analogously, one can verify that the dual fractional integral operator can be written in the form

$$_xJ^\alpha_\infty f(x) = \frac{1}{\Gamma(\alpha)}\int_x^\infty (y-x)^{\alpha-1}f(y)dy. \tag{3.61}$$

Example 3.11. 1. Let $a > 0$. Then for an arbitrary $\alpha > 0$ one has $_{-\infty}J^\alpha[e^{ax}] = a^{-\alpha}e^{ax}$.
2. Similarly, for $a > 0$ one has $_xJ^\alpha_\infty[e^{-ax}] = a^{-\alpha}e^{-ax}$.

Further, similar to the Riemann-Liouville fractional derivatives, one can define fractional derivatives exploiting the fractional integrals $_{-\infty}J^\alpha$ and $_xJ^\alpha_\infty$ with terminal points $\pm\infty$. Let $m-1 < \alpha \leq m$, $m = 1,2,\ldots$. Introduce the operators

$$_{-\infty}D^{\alpha} = D^{m}\, _{-\infty}J^{m-\alpha},$$

and

$$_{x}D^{\alpha}_{\infty} = (-D)^{m}\, _{x}J^{m-\alpha}_{\infty},$$

which are called *Liouville-Weyl forward (LWf) and backward (LWb) fractional derivatives,* respectively. Explicit forms of LWf and LWb fractional derivatives are

$$_{-\infty}D^{\alpha} f(x) = \frac{1}{\Gamma(m-\alpha)} \frac{d^m}{dx^m} \int_{-\infty}^{x} \frac{f(y)}{(x-y)^{\alpha-m+1}} dy, \tag{3.62}$$

and

$$_{x}D^{\alpha}_{\infty} f(x) = \frac{(-1)^m}{\Gamma(m-\alpha)} \frac{d^m}{dx^m} \int_{x}^{\infty} \frac{f(y)}{(y-x)^{\alpha-m+1}} dy. \tag{3.63}$$

Notice that the LWf and LWb fractional derivatives are related to the usual integer order derivatives through $_{-\infty}D^{m} = D^{m}$ and $_{x}D^{m}_{\infty} = (-D)^{m}$. It follows from the explicit representations (3.62) and (3.63) that for the existence of Liouville-Weyl fractional derivatives the function f must satisfy some differentiability and decay conditions. Namely, it is not hard to see that

(a) LWf derivative exists on $(-\infty, a]$ if $f \in C^{\lambda}(-\infty, a]$ has the asymptotic behavior $|f(x)| \sim |x|^{-m+\alpha-\varepsilon}$, $x \to -\infty$, and
(b) LWb derivative exists on $[b, \infty)$ if $f \in C^{\lambda}[b, \infty)$ has the asymptotic behavior $|f(x)| \sim |x|^{-m+\alpha-\varepsilon}$, $x \to +\infty$,

where C^{λ}, $\lambda > \alpha$, is the Hölder space, a and b finite numbers, and ε is an arbitrary positive number. Moreover, Liouville-Weyl fractional derivatives can be interpreted as ΨDOSS. Namely, for functions $f \in \Psi_{G,p}(\mathbb{R}^n)$, $0 \notin G$, we can write

$$_{-\infty}D^{\alpha} f(x) = D^{m} A_{m-\alpha}(D) f(x),$$

as well as for ψ-distributions F,

$$_{x}D^{\alpha}_{\infty} F(x) = (-D)^{m} A^{*}_{m-\alpha}(D) F(x).$$

Using the properties of the Fourier transform and symbols of Liouville-Weyl integrals one can easily derive the symbols of LWf and LWb fractional derivatives. Namely,

$$F[_{-\infty}D^{\alpha} f](\xi) = F[D^{m}\, _{-\infty}J^{m-\alpha} f](\xi) = (i\xi)^{m} \frac{1}{(i\xi)^{m-\alpha}} F[f](\xi)$$
$$= (i\xi)^{\alpha} F[f](\xi),$$

and

$$F[_{x}D^{\alpha}_{\infty} f](\xi) = F[(-1)^{m} D^{m}\, _{x}J^{m-\alpha}_{\infty} f](\xi) = (-1)^{m} (i\xi)^{m} \frac{1}{(-i\xi)^{m-\alpha}} F[f](\xi)$$
$$= (-i\xi)^{\alpha} F[f](\xi).$$

Proposition 3.12. *The symbols of operators* $_{-\infty}D^{\alpha}$ *and* $_{x}D^{\alpha}_{\infty}$ *are*

$$\sigma_{_{-\infty}D^{\alpha}}(\xi) = (i\xi^{\alpha}) \text{ and } \sigma_{_{x}D^{\alpha}_{\infty}}(\xi) = (-i\xi)^{\alpha}. \tag{3.64}$$

Example 3.12. Show that for an arbitrary $a > 0$ the following formulas are valid:

1. $_{-\infty}D^{\alpha}e^{ax} = a^{\alpha}e^{ax}$;
2. $_{x}D^{\alpha}_{\infty}e^{-ax} = a^{\alpha}e^{-ax}$.

3.7 Riesz potential. Connection with ΨDOSS

Let $0 < \alpha < 1$. The integral

$$\mathscr{R}_{\alpha}f(x) = C_{\alpha}\int_{\mathbb{R}} \frac{f(y)dy}{|x-y|^{1-\alpha}} \tag{3.65}$$

defined for $f \in L_1(\mathbb{R})$ is called the *Riesz potential.* Here the constant C_{α} is normalized in such a way that $\sigma_{\mathscr{R}_{\alpha}}(\xi) = |\xi|^{-\alpha}$. Since $\mathscr{R}_{\alpha}f(x) = (\mathscr{K} * f)(x)$, that is a convolution operator with $\mathscr{K} = \frac{C_{\alpha}}{|x|^{1-\alpha}}$, Proposition 1.20 implies that

$$\sigma_{\mathscr{R}_{\alpha}}(\xi) = F[\mathscr{K}](\xi) = C_{\alpha}F\left[\frac{1}{|x|^{1-\alpha}}\right](\xi) = |\xi|^{-\alpha}, \tag{3.66}$$

where we set $C_{\alpha} = \frac{1}{b_{\alpha}}$. Therefore, in order to determine C_{α} one needs to know b_{α}, which appears in the Fourier transform of $|x|^{-(1-\alpha)}$, see Example 1.6, 8. However, instead of computing b_{α}, one can find C_{α} applying the symbolic calculus for ΨDOSS introduced in Chapter 2.

Indeed, we show that there is a constant d_{α} such that

$$d_{\alpha}\{\sigma_{A_{\alpha}}(\xi) + \sigma_{A^*_{\alpha}}(\xi)\} = \sigma_{\mathscr{R}_{\alpha}}(\xi).$$

First, notice that $\phi(\xi) = \sigma_{A_{\alpha}}(\xi) + \sigma_{A^*_{\alpha}}(\xi) = 2\Re\{\sigma_{A_{\alpha}}(\xi)\} = 2\Re\{\frac{1}{(i\xi)^{\alpha}}\}$, where $\Re$ stands for the real part. Second, the function $\phi(\xi)$ is symmetric, i.e., $\phi(-\xi) = \phi(\xi)$. Therefore, it suffices to find $\Re\{\frac{1}{(i\xi)^{\alpha}}\}$ only for $\xi > 0$. We have

$$\Re\{\frac{1}{(i\xi)^{\alpha}}\} = \Re\{e^{-i\frac{\pi}{2}\alpha}\frac{1}{\xi^{\alpha}}\} = \frac{1}{|\xi|^{\alpha}}\cos\frac{\pi\alpha}{2}, \quad \xi > 0.$$

This immediately implies that

$$d_{\alpha}\{\sigma_{A_{\alpha}}(\xi) + \sigma_{A^*_{\alpha}}(\xi)\} = 2\frac{1}{|\xi|^{\alpha}}\cos\frac{\pi\alpha}{2},$$

which yields

$$d_\alpha = \frac{1}{2\cos\frac{\pi\alpha}{2}}.$$

Proposition 3.13. *Let* $\alpha \in (0,1)$. *Then for the one-dimensional Riesz potential* $\mathscr{R}_\alpha$ *the following relations hold:*

1. $\sigma_{\mathscr{R}_\alpha}(\xi) = \frac{1}{2\cos\frac{\pi\alpha}{2}}\{\sigma_{A_\alpha}(\xi) + \sigma_{A^*_\alpha}(\xi)\} = |\xi|^{-\alpha}$;
2. $\mathscr{R}_\alpha = \frac{1}{2\cos\frac{\pi\alpha}{2}}\{A_\alpha(D) + A^*_\alpha(D)\}$;
3. $\mathscr{R}_\alpha = \frac{1}{2\cos\frac{\pi\alpha}{2}}\{{}_{-\infty}J^\alpha + {}_xJ^\alpha_\infty\}$.

Since

$$\begin{aligned}\mathscr{R}_\alpha f(x) &= \{\frac{{}_{-\infty}J^\alpha + {}_xJ^\alpha_\infty}{2\cos\frac{\pi\alpha}{2}}\}f(x)\\ &= \frac{1}{2\Gamma(\alpha)\cos\frac{\pi\alpha}{2}}\Big(\int_{-\infty}^x \frac{f(y)dy}{(x-y)^{1-\alpha}} + \int_x^\infty \frac{f(y)dy}{(y-x)^{1-\alpha}}\Big)\\ &= \frac{1}{2\Gamma(\alpha)\cos\frac{\pi\alpha}{2}}\int_{-\infty}^\infty \frac{f(y)dy}{|x-y|^{1-\alpha}},\end{aligned}$$

obviously,

$$C_\alpha = \frac{1}{2\Gamma(\alpha)\cos\frac{\pi\alpha}{2}}.$$

Hence, the constant b_α in Example 1.6, 8 is

$$b_\alpha = 2\Gamma(\alpha)\cos\frac{\pi\alpha}{2}.$$

Now we extend the definition of the Riesz potential for arbitrary α setting

$$\mathscr{R}_\alpha = \frac{1}{2\cos\frac{\pi\alpha}{2}}\{{}_{-\infty}J^\alpha + {}_xJ^\alpha_\infty\}, 0 < \alpha \neq 2k+1, k = 0,1,\dots. \tag{3.67}$$

The operator $\mathscr{R}_\alpha$ with the symbol $\sigma_{R_\alpha}(\xi) = \frac{1}{|\xi|^\alpha}$, where $\alpha > 0$ and $\alpha \neq 2k-1, k = 1,2,\dots$, has the strong singularity at $\xi = 0$, if $\alpha \geq 1$. Hence, $\mathscr{R}_\alpha$ is a ΨDOSS if we consider it on the space of ψ-distributions $\Psi'_{-G,p'}(\mathbb{R}^n)$. Therefore, all the properties related to ΨDOSS are valid for $\mathscr{R}_\alpha$. The operator $\mathscr{R}_\alpha$ is well defined on the Lizorkin spaces $\Phi(\mathbb{R}^n)$, as well. Additionally, we note that $\mathscr{R}_\alpha$ is the inverse operator to the fractional power α of the operator $D^2 = -\frac{d^2}{dx^2}$, i.e., $(D^2)^{\alpha/2}$, whose symbol is $|\xi|^\alpha$. The Riesz-Feller derivatives, which we are going to introduce, tell us how to construct such fractional powers.

By definition, *the Riesz-Feller derivative of order* α (analogously to (3.67)) is

$$\mathscr{D}^\alpha_0 f(x) = \frac{-1}{2\cos\frac{(\alpha\pi)}{2}}({}_{-\infty}D^\alpha f(x) + {}_xD^\alpha_\infty f(x)), \quad 0 < \alpha \neq 2k+1, k = 0,1,\dots, \tag{3.68}$$

where ${}_{-\infty}D^{\alpha}$ and ${}_{x}D^{\alpha}_{\infty}$ are LWf and LWb fractional derivatives of order α defined in (3.62) and (3.63), respectively.

Proposition 3.14. *The symbol of the Riesz-Feller derivative is*

$$\sigma_{\mathscr{D}_0^{\alpha}}(\xi) = -|\xi|^{\alpha}.$$

Proof. Taking into account (3.64) one has

$$\begin{aligned}
\sigma_{\mathscr{D}_0^{\alpha}}(\xi) &= -\frac{1}{2\cos\frac{\pi\alpha}{2}}\{\sigma_{{}_{-\infty}D^{\alpha}}(\xi) + \sigma_{{}_{x}D^{\alpha}_{\infty}}(\xi)\} \\
&= -\frac{1}{2\cos\frac{\pi\alpha}{2}}\{(i\xi)^{\alpha} + (-i\xi)^{\alpha}\} = -\frac{1}{2\cos\frac{\pi\alpha}{2}} 2\Re\{(i\xi)^{\alpha}\} \\
&= -|\xi|^{\alpha}\frac{1}{\cos\frac{\pi\alpha}{2}}\cos\frac{\pi\alpha}{2} = -|\xi|^{\alpha}.
\end{aligned}$$

Introduce the shift operator $\tau_h f(x) = f(x+h)$, where $0 < |h| < 1$. Then the ℓ-th order central finite difference operator Δ_h^{ℓ} can be defined in the form

$$\Delta_h^{\ell} f(x) = (\tau_{\frac{h}{2}} - \tau_{\frac{h}{2}})^{\ell} f(x). \tag{3.69}$$

In particular, for $\ell = 0, 1$, and 2, the operators $\Delta_h^0 f(x)$, $\Delta_h^1 f(x)$, and $\Delta_h^2 f(x)$ take the form

$$\begin{aligned}
\Delta_h^0 f(x) &= If(x) = f(x); \\
\Delta_h^1 f(x) &= f(x+h/2) - f(x-h/2); \\
\Delta_h^2 f(x) &= f(x+h) - 2f(x) + f(x-h).
\end{aligned}$$

Proposition 3.15. *Let $0 < \alpha < 2$. Then the Riesz-Feller derivative of order α of a function f in the Hölder space $C^{\lambda}(\mathbb{R}) \cap L_{\infty}$, $\lambda > \alpha$, has the representation*

$$\mathscr{D}_0^{\alpha} f(x) = \omega_{\alpha}\int_0^{\infty}\frac{\Delta_h^2 f(x)}{h^{1+\alpha}}dh, \tag{3.70}$$

where $\omega_{\alpha} = (1/\pi)\Gamma(\alpha+1)\sin\frac{\pi\alpha}{2}$.

Proof. We show (3.70) for $0 < \alpha < 1$. The general case follows from Theorem 3.4 (in the particular case $n = 1$). By definition,

$$\begin{aligned}
\mathscr{D}_0^{\alpha} f(x) &= \frac{-1}{2\cos\frac{\alpha\pi}{2}}({}_{-\infty}D^{\alpha} + {}_{x}D^{\alpha}_{\infty})f(x) \\
&= \frac{-D}{2\cos\frac{\alpha\pi}{2}}({}_{-\infty}J^{1-\alpha} + {}_{x}J^{1-\alpha}_{\infty})f(x) \\
&= \frac{-D}{2\Gamma(1-\alpha)\cos\frac{\alpha\pi}{2}}\left(\int_{-\infty}^{x}\frac{f(y)dy}{(x-y)^{\alpha}} - \int_{x}^{\infty}\frac{f(y)dy}{(y-x)^{\alpha}}\right).
\end{aligned} \tag{3.71}$$

The substitutions $x-y=u$ and $y-x=v$ in two integrals on the right-hand side of (3.71) give

$$\mathscr{D}_0^{\alpha} f(x)=\frac{-D}{2\Gamma(1-\alpha)\cos\frac{\alpha\pi}{2}}\left(\int_0^{\infty}\frac{f(x-u)du}{u^{\alpha}}-\int_0^{\infty}\frac{f(x+u)du}{(u)^{\alpha}}\right).$$

By virtue of the equality $u^{-\alpha}=\alpha\int_u^{\infty}\frac{dh}{h^{\alpha+1}}$ and changing the order of integration, we have

$$\frac{-\alpha}{2\Gamma(1-\alpha)\cos\frac{\alpha\pi}{2}}\left(\int_0^{\infty} f'(x-u)\int_u^{\infty}\frac{1}{h^{\alpha+1}}dhdu-\int_0^{\infty} f'(x+h)\int_u^{\infty}\frac{1}{h^{\alpha+1}}dhdu\right)$$
$$=\frac{\alpha}{2\Gamma(1-\alpha)\cos\frac{\alpha\pi}{2}}\int_0^{\infty}\frac{f(x-h)-2f(x)+f(x+h)}{h^{1+\alpha}}dh.$$

Now using the known property

$$\Gamma(1-\alpha)\Gamma(1+\alpha)=\frac{\pi\alpha}{\sin\pi\alpha}$$

of Euler's gamma function, we obtain

$$\frac{\alpha}{2\Gamma(1-\alpha)\cos\frac{\alpha\pi}{2}}=(1/\pi)\Gamma(\alpha+1)\sin\frac{\pi\alpha}{2},$$

and, consequently, the representation (3.70).

Theorem 3.1. *For the Fourier transforms of $\mathscr{R}_0^{\alpha} f(x)$ and $\mathscr{D}_0^{\alpha} f(x)$ for $\alpha\geq 0$ the following formulas are valid:*

1. $F[\mathscr{R}_0^{\alpha} f](\xi)=|\xi|^{-\alpha}\hat{f}(\xi);$
2. $F[\mathscr{D}_0^{\alpha} f](\xi)=-|\xi|^{\alpha}\hat{f}(\xi).$

Part 2 of this theorem implies that the α values in the operator $\mathscr{D}_0^{\alpha} f(x)$ can naturally be extended to $\alpha=2$ as $\mathscr{D}_0^2=\frac{d^2}{dx^2}$.

3.8 Multidimensional Riesz potentials and their inverses

The natural generalizations of the one-dimensional Riesz potential $\mathscr{R}_0^{\alpha}$ and Riesz-Feller derivative $\mathscr{D}_0^{\alpha}$ to the n-dimensional case, respectively, are the n-dimensional Riesz potential (with the normalizing constant $C_{n,\alpha}$)

$$\mathbb{R}_0^{\alpha} f(x)=C_{n,\alpha}\int_{\mathbb{R}^n}\frac{f(y)dy}{|x-y|^{n-\alpha}},\ \alpha>0,\ \alpha\neq n,n+2,\ldots, \tag{3.72}$$

and the n-dimensional hyper-singular integral (the inverse Riesz potential) defined as (see (2.48))

$$\mathbb{D}_0^\alpha f(x) = \frac{1}{d(\alpha,l)} \int_{\mathbb{R}^n} \frac{\Delta_y^l f(x)}{|y|^{n+\alpha}} dy, \; 0 < \alpha \neq 2m, \; m \in \mathbb{N}. \tag{3.73}$$

The normalizing constant in (3.72) is

$$C_{n,\alpha} = \frac{\Gamma(\frac{n-\alpha}{2})}{2^\alpha \pi^{n/2} \Gamma(\frac{\alpha}{2})}.$$

In equation (3.73) l is integer, $l > \alpha$, Δ_y^l is the centered finite difference of the order l in the y direction, and $d(\alpha,l)$ is constant defined by (2.49). For our further considerations it suffices to consider the case $l = 2$ and $0 < \alpha < 2$. In this case the operator $\mathbb{D}_0^\alpha$ takes the form (see Section 2.4)

$$\mathbb{D}_0^\alpha f(x) = B_{n,\alpha} \int_{\mathbb{R}^n} \frac{f(x-y) - 2f(x) + f(x+y)}{|y|^{n+\alpha}} dy, \quad 0 < \alpha < 2, \tag{3.74}$$

where

$$B_{n,\alpha} = \frac{\alpha \Gamma(\frac{\alpha}{2}) \Gamma(\frac{n+\alpha}{2}) sin \frac{\alpha\pi}{2}}{2^{2-\alpha} \pi^{1+n/2}}. \tag{3.75}$$

Below we will derive (3.75) (see Theorem 3.4). It is seen from equation (3.75) that the value $\alpha = 2$ is degenerate.

Theorem 3.2. *Let $\alpha > 0$ and $\alpha \neq n, n+2, \ldots$. Then for the symbol $\sigma_{\mathbb{R}_0^\alpha}(\xi)$ of the n-dimensional Riesz potential the formula*

$$\sigma_{\mathbb{R}_0^\alpha}(\xi) = |\xi|^{-\alpha}, \quad \xi \in \mathbb{R}^n, \tag{3.76}$$

holds.

Proof. The Riesz potential can be expressed as a convolution. Namely,

$$\mathbb{R}_0^\alpha f(x) = \left(\frac{C_{n,\alpha}}{|x|^{n-\alpha}} * f \right)(x).$$

Applying the Fourier transform, one has

$$F[\mathbb{R}_0^\alpha f](\xi) = \sigma_{\mathbb{R}_0^\alpha}(\xi) \cdot F[f](\xi),$$

where

$$\sigma_{\mathbb{R}_0^\alpha}(\xi) = C_{n,\alpha} F\left[\frac{1}{|x|^{n-\alpha}}\right](\xi).$$

In Proposition 1.20 of Chapter 1 setting $\sigma = n - \alpha$, one has

$$F\left[\frac{1}{|x|^{n-\alpha}}\right](\xi) = b_{\alpha,n} |\xi|^{-\alpha},$$

where

$$b_{\alpha,n} = \frac{2^\alpha \pi^{n/2}\Gamma(\frac{\alpha}{2})}{\Gamma(\frac{n-\alpha}{2})}.$$

Obviously, $C_{n,\alpha} b_{\alpha,n} \equiv 1$, and we obtain (3.76).

Theorem 3.3. *Let $1 < p,q < \infty$, $s \in \mathbb{R}$. Then the following mappings are continuous:*

1. *if $0 < \alpha < \frac{n(p-1)}{p}$, then*

$$\mathbb{R}_0^\alpha : \overset{\circ}{B}{}^s_{pq}(\mathbb{R}^n) \to B^{s+\alpha}_{pq}(\mathbb{R}^n); \tag{3.77}$$

2. *if $\frac{n(p-1)}{p} + 2(m-1) < \alpha < \frac{n(p-1)}{p} + 2m - 1$, $m \in \mathbb{N}$, then*

$$\mathbb{R}_0^\alpha : \overset{\circ}{B}{}^s_{pq,2m-1}(\mathbb{R}^n) \to B^{s+\alpha}_{pq}(\mathbb{R}^n); \tag{3.78}$$

3. *if $\frac{n(p-1)}{p} + 2m - 1 \le \alpha < \frac{n(p-1)}{p} + 2m$, $m \in \mathbb{N}$, then*

$$\mathbb{R}_0^\alpha : \overset{\circ}{B}{}^s_{pq,2m}(\mathbb{R}^n) \to B^{s+\alpha}_{pq}(\mathbb{R}^n). \tag{3.79}$$

Proof. One can easily verify that the symbol of the operator $\mathbb{R}_0^\alpha$ considered as a ΨDOSS defined on $\Psi_{G,p}$, where $G = \mathbb{R}^n \setminus \{0\}$, satisfies the condition (2.36) with $l = s + \alpha$. Due to Theorem 3.2 the symbol of $\mathbb{R}_0^\alpha$ has a singularity of order α at $\xi = 0$. Let $\alpha \in [0, n(p-1)p^{-1})$ and $\varphi \in B^s_{pq}(\mathbb{R}^n)$ with $\operatorname{supp}\varphi \subset Q$.. Further, let $\{\phi_j\}_{j=0}^\infty$ with $\{F^{-1}\phi_j\}_{j=0}^\infty \in \Phi$, define the norm of the Besov space $B^s_{pq}(\mathbb{R}^n)$. Then we have

$$\begin{aligned} \|\mathbb{R}_0^\alpha \varphi(x)|B^{s+\alpha}_{pq}\|^q &\equiv \Big(\int_{\mathbb{R}^n} |F^{-1}\phi_0 \frac{1}{|\xi|^\alpha} F\varphi|^p dx\Big)^{\frac{p}{q}} \\ &+ \sum_{j=1}^\infty 2^{(s+\alpha)jq} \|F^{-1}\phi_j \frac{1}{|\xi|^\alpha} F\varphi|L_p\|^q. \end{aligned} \tag{3.80}$$

Since $\alpha p' < n$, the first term on the right of (3.80), in accordance with Theorem 1.24 of Chapter 1, can be estimated as

$$\int_{\mathbb{R}^n} |F^{-1}\phi_0 \frac{1}{|\xi|^\alpha} F\varphi|^p dx \le C\|\varphi|B^s_{pq}\|^p,$$

where C is a positive constant depending on Q, α, and p. Further, since $\operatorname{supp}\phi_j \subseteq \{2^j \le |\xi| \le 2^{j+1}\}$, $j = 1, \ldots,$ the second term in (3.80) can be estimated as

$$\begin{aligned} \sum_{j=1}^\infty 2^{(s+\alpha)jq} \|F^{-1}\phi_j \frac{1}{|\xi|^\alpha} F\varphi|L_p\|^q &\le C \sum_{j=1}^\infty 2^{sjq} \|F^{-1}\phi_j F\varphi|L_p\|^q \\ &= C\|\varphi(x)|B^s_{pq}\|^q. \end{aligned} \tag{3.81}$$

Estimates (3.80)–(3.81) imply $\|\mathbb{R}_0^\alpha \varphi(x)|B_{pq}^{s+\alpha}\| \le \|\varphi(x)|B_{pq}^s\|$, and thus, the continuity of the mapping in (3.77).

Now suppose that $\alpha \in (2(m-1)+n(p-1)p^{-1}, 2m-1+n(p-1)p^{-1})$ and $\varphi \in \overset{\circ}{B}{}^{s}_{pq,2m-1}$, or $\alpha \in (2m-1+n(p-1)p^{-1}, 2m+n(p-1)p^{-1})$ and $\varphi \in \overset{\circ}{B}{}^{s}_{pq,2m}$. Then due to Theorem 1.23 there exists a function $v \in B_{pq}^{s+\bar{m}}$ such that $F\varphi = |\xi|^{\bar{m}} Fv$. Here $\bar{m} = 2m-1$ in the first case, and $\bar{m} = 2m$ in the second case. In both cases this fact yields the following estimate for the first term of (3.80)

$$C\Big(\int_{\operatorname{supp}\phi_0} |\xi|^{(\bar{m}-\alpha)p'} d\xi\Big)^{p-1} \|v|B_{pq}^{s+\bar{m}}\|^p \le C\|\varphi|B_{pq}^s\|^p,$$

where C is a positive constant depending on Q, α, p, and m. One can estimate the second term on the right side of (3.80) as in the previous case, and hence, obtain the continuity of mappings in (3.78) and (3.79). The proof is complete.

Theorem 3.4. *Let $0 < \alpha < 2$. Then for the symbol $\sigma_{\mathbb{D}_0^\alpha}(\xi)$ of the n-dimensional Riesz-Feller derivative the formula*

$$\sigma_{\mathbb{D}_0^\alpha}(\xi) = -|\xi|^\alpha, \quad \xi \in \mathbb{R}^n, \tag{3.82}$$

holds.

Proof. Due to Corollary 2.11, Part ii) for the symbol of $\mathbb{D}_0^\alpha$, we have

$$\sigma_{\mathbb{D}_0^\alpha}(\xi) = \mathbb{D}_0^\alpha e^{-ix\xi}|_{x=0} = B_{n,\alpha} \int_{\mathbb{R}^n} \frac{\Delta_y^2 e^{-ix\xi}}{|y|^{n+\alpha}} dy|_{x=0},$$

where $B_{n,\alpha}$ is defined in (3.75). Therefore, in order to prove the proposition we need to show that

$$\int_{\mathbb{R}^n} \frac{\Delta_y^2 e^{-ix\xi}}{|y|^{n+\alpha}} dy|_{x=0} = -\frac{|\xi|^\alpha}{B_{n,\alpha}}. \tag{3.83}$$

For the left-hand side we have

$$\begin{aligned}
\int_{\mathbb{R}^n} \frac{\Delta_y^2 e^{-ix\xi}}{|y|^{n+\alpha}} dy|_{x=0} &= \int_{\mathbb{R}^n} \frac{e^{-i(x+y)\xi} - 2 + e^{-i(x-y)\xi}}{|y|^{n+\alpha}} dy|_{x=0} \\
&= \int_{\mathbb{R}^n} \frac{e^{-iy\xi} - 2 + e^{iy\xi}}{|y|^{n+\alpha}} dy \\
&= F[\frac{1}{|y|^{n+\alpha}}](-\xi) - 2F[\frac{1}{|y|^{n+\alpha}}](0) + F[\frac{1}{|y|^{n+\alpha}}](\xi) \\
&= \Delta_\xi^2 F[\frac{1}{|y|^{n+\alpha}}](0).
\end{aligned}$$

In Proposition 1.20 taking $\sigma = n+\alpha$, we have

$$F\left[\frac{1}{|x|^{n+\alpha}}\right](\xi) = \frac{2^{-\alpha}\pi^{n/2}\Gamma(-\frac{\alpha}{2})}{\Gamma(\frac{n+\alpha}{2})}|\xi|^{\alpha},$$

where $\Gamma(-\frac{\alpha}{2})$ is the value at $z=-\alpha/2\in(-1,0)$ of the analytic continuation of the Euler gamma-function $\Gamma(z)$ to the interval $(-1,0)$. It follows from this equality that

$$\Delta_{\xi}^{2} F\left[\frac{1}{|y|^{n+\alpha}}\right](0) = \frac{2^{1-\alpha}\pi^{n/2}\Gamma(-\frac{\alpha}{2})}{\Gamma(\frac{n+\alpha}{2})}|\xi|^{\alpha}. \tag{3.84}$$

Finally, using the relationship (see [AS64], formula 6.1.17)

$$-\frac{\alpha}{2}\Gamma\left(-\frac{\alpha}{2}\right)\Gamma\left(\frac{\alpha}{2}\right) = \frac{\pi}{\sin\frac{\pi\alpha}{2}}$$

in equation (3.84), we obtain (3.83).

Theorem 3.5. *Let* $1<p,q<\infty$, $s\in\mathbb{R}$, *and* $0<\alpha<2$. *Then the mapping*

$$\mathbb{D}_0^{\alpha} : B^{s}_{p,q}(\mathbb{R}^n) \to B^{s-\alpha}_{pq}(\mathbb{R}^n)$$

is continuous.

Proof. Due to Theorem 3.4 operator $\mathbb{D}_0^{\alpha}$ can be considered as a ΨDOSS defined on $\Psi_{G,p}(\mathbb{R}^n)$ with $G=\mathbb{R}^n\setminus\{0\}$. Now the proof follows from Theorem 2.8, since the symbol of $\mathbb{D}_0^{\alpha}$ satisfies all the conditions of this theorem.

Remark 3.4. 1. Riesz potentials form a semigroup. Namely, if $\Re(\alpha)>0$, $\Re(\beta)>0$, and $\Re(\alpha+\beta)>0$, then $\mathbb{R}_0^{\alpha}\mathbb{R}_0^{\beta} f = \mathbb{R}_0^{\alpha+\beta} f$ for $f\in\mathcal{G}$. It is not hard to see that the mapping $\mathbb{R}_0^{\alpha}:\mathcal{G}\to\mathcal{G}$ is continuous. By duality, the mapping $\mathbb{R}_0^{\alpha}:\mathcal{G}'\to\mathcal{G}'$ is continuous, and thus the semigroup property extends to $\mathcal{G}'$, as well.
2. The operator $\mathbb{D}_0^{\alpha}$ in (3.74) is defined for $0<\alpha<2$. As was noted above in this definition the value $\alpha=2$ is degenerate. However, the symbol of this operator is meaningful for $\alpha=2$, as well:

$$\lim_{\alpha\to 2}\sigma_{\mathbb{D}_0^{\alpha}}(\xi) = -|\xi|^2.$$

We know that this function is the symbol of the Laplace operator. Therefore, it is natural to extend $\mathbb{D}_0^{\alpha}$ to $\alpha=2$, setting $\mathbb{D}_0^{2}=\Delta$.

3.9 Fractional powers of positive definite operators

The relationship between the symbols of the operator $\mathbb{D}_0^{\alpha}$ and the (negative) Laplace operator $-\Delta$, i.e.

$$\sigma_{\mathbb{D}_0^\alpha}(\xi) = -|\xi|^\alpha = -(|\xi|^2)^{\alpha/2} = -[\sigma_{-\Delta}(\xi)]^{\alpha/2},$$

obtained in the previous section, at least at the formal level, gives an idea to represent $\mathbb{D}_0^\alpha$ as a fractional power of $-\Delta$, namely $D_0^\alpha = -(-\Delta)^{\alpha/2}$. Therefore, in this section we briefly discuss fractional powers of positive definite operator. By definition, a closed linear operator A with a domain $\mathscr{D}(A)$, dense in a Banach space X, is called *positive definite,* if its spectrum does not contain the negative axis $(-\infty,0]$, and the estimate $\|(A-\lambda I)^{-1}\| \le C(1+|\lambda|)^{-1}$ holds for $\lambda \in (-\infty,0]$.

First, as is known from the spectral theory of linear operators [DS88], that for an arbitrary positive self-adjoint operator A defined in a Hilbert space H and with a spectrum $\Sigma(A)$ and a spectral decomposition E_λ, one can define the fractional powers of A :

$$A^\alpha = \int\limits_{\Sigma(A)} \lambda^\alpha dE_\lambda,$$

whose domain is

$$D(A^\alpha) = \{f \in H : \int_{\Sigma(A)} \lambda^{2\alpha}(dE_\lambda f, f) < \infty\}.$$

For instance, the Laplace operator $A = -\Delta$ in $L_2(R^n)$ is a positive self-adjoint operator, whose spectrum is $\Sigma(-\Delta) = (0,\infty)$, so one can define its fractional powers in the form

$$(-\Delta)^{\alpha/2} = \int_0^\infty \lambda^{\alpha/2} dE_\lambda, \quad \alpha > 0,$$

where E_λ is the spectral decomposition corresponding to $-\Delta$. Though this approach is nice from the theoretical point of view, however it is not always practical. To use this approach one needs to know how to find and work with the spectral decomposition of A.

The following statement will be used in our further considerations.

Proposition 3.16. *Let $\mu > 0$ and $0 < \alpha < m$. Then*

$$\int\limits_0^\infty \frac{s^{\alpha-1}ds}{(\mu+s)^m} = \mu^{\alpha-m}\frac{\Gamma(\alpha)\Gamma(m-\alpha)}{\Gamma(m)}. \tag{3.85}$$

Proof. The substitution $s = \frac{\mu u}{1+u}$ in the integral yields

$$\int\limits_0^\infty \frac{s^{\alpha-1}ds}{(\mu+s)^m} = \mu^{\alpha-m}\int\limits_0^\infty u^{\alpha-1}(1-u)^{m-\alpha-1}du = \mu^{\alpha-m}B(\alpha, m-\alpha),$$

where $B(\cdot,\cdot)$ is Euler's beta function. Now using formula (3.4), one obtains (3.85).

Let A be a positive definite operator and $0 < \alpha < 1$. Then for $f \in \mathscr{D}(A)$ fractional powers of A is defined by

$$A^{\alpha} f = \frac{1}{\Gamma(\alpha)\Gamma(1-\alpha)} \int_0^{\infty} \lambda^{\alpha-1}(A+\lambda I)^{-1} A f d\lambda. \tag{3.86}$$

This definition was first given by Balakrishnan in 1960 [Bal60]. To feel this definition better let us assume that $A = t$, that is $Af = tf$, where t is a positive number, then the expression on the right side of (3.86) becomes $t^{\alpha} f \, (= A^{\alpha} f)$. Indeed, changing $\lambda / t \to s$, one has

$$\int_0^{\infty} \frac{t\lambda^{\alpha-1}}{t+\lambda} f d\lambda = \left(t^{\alpha} \int_0^{\infty} \frac{s^{\alpha-1} ds}{1+s} \right) f.$$

The integral on the right-hand side equals $\Gamma(\alpha)\Gamma(1-\alpha)$. This follows from (3.85) taking $\mu = 1$ and $m = 1$. Balakrishnan in his paper [Bal60] proved that this idea works for any positive definite operators.

The definition (3.86) can be generalized to the case $0 < \alpha < m$, where $m \in \mathbb{N}$; see [Kom66]. Namely, for arbitrary $f \in \mathscr{D}(A^m)$

$$A^{\alpha} f = \frac{\Gamma(m)}{\Gamma(\alpha)\Gamma(m-\alpha)} \int_0^{\infty} \lambda^{\alpha-1} [(A+\lambda I)^{-1} A]^m f d\lambda.$$

In particular, if $m = 2$, $0 < \alpha < 2$, and $A = -\Delta$, then for $f(x) \in H^2(\mathbb{R}^n)$, where $H^2(\mathbb{R}^n)$ is the Sobolev space, we have

$$-(-\Delta)^{\alpha/2} f(x) = \frac{-1}{\Gamma(\alpha)\Gamma(2-\alpha)} \int_0^{\infty} \lambda^{\frac{\alpha}{2}-1} [(-\Delta+\lambda I)^{-1}(-\Delta)]^2 f(x) d\lambda. \tag{3.87}$$

Applying the Fourier transform to the right side of (3.87), and changing the order of integration, we have

$$\begin{aligned} &\frac{-1}{\Gamma(\alpha)\Gamma(2-\alpha)} \int_0^{\infty} \lambda^{\frac{\alpha}{2}-1} F[[(-\Delta+\lambda I)^{-1}(-\Delta)]^2 f](\xi) d\lambda \\ &= \frac{-1}{\Gamma(\alpha)\Gamma(2-\alpha)} \int_0^{\infty} \lambda^{\frac{\alpha}{2}-1} \frac{|\xi|^4}{(\lambda+|\xi|^2)^2} d\lambda \\ &= -|\xi|^{\alpha}. \end{aligned}$$

In the last stage of this calculation we used (3.85) with $m = 2$ and $\mu = |\xi|^2 > 0$. Thus, the operator $\mathbb{D}_0^{\alpha}$ defined in (3.74) and the fractional power operator $-(-\Delta)^{\alpha/2}$ defined in (3.87) have the same symbol, and therefore these two operators coincide in the space $H^2(\mathbb{R}^n)$.

3.10 Grünwald-Letnikov fractional derivative

In the numerical calculation of fractional derivatives the Grünwald-Letnikov approximation or its modifications are frequently used.

Definition 3.4. By definition, the Grünwald-Letnikov fractional derivative of order $\alpha \geq 0$ is ${}_a\mathscr{D}_t^\alpha$ defined for $t \in (a,b)$ as

$$ {}_a\mathscr{D}_t^\alpha f(t) = \lim_{h\to 0} \frac{(\Delta_h^\alpha f)(t)}{h^\alpha}, \tag{3.88}$$

provided the limit exists, where Δ_h^α means the finite difference of fractional order α with step $h > 0$:

$$(\Delta_h^\alpha f)(t) = \sum_{m=0}^{\lfloor \frac{t-a}{h} \rfloor} (-1)^m \binom{\alpha}{m} f(t-mh). \tag{3.89}$$

Here

$$\binom{\alpha}{m} = \frac{\Gamma(\alpha+1)}{\Gamma(\alpha-m+1)m!}, m = 0,1,\ldots. \tag{3.90}$$

If $\alpha = 0$, then $\Delta_h^0 f(t) = f(t)$, since

$$\binom{0}{m} = \frac{1}{\Gamma(1-m)m!} = \begin{cases} 1 & \text{if } m = 0, \\ 0 & \text{if } m = 1,2,\ldots. \end{cases}$$

due to poles of the gamma function $\Gamma(z)$ at points $z = 0,-1,-2,\ldots$. Hence, ${}_a\mathscr{D}_t^0 = I$, where I is, as usual, the identity operator. If $\alpha = 1$, in a similar way we get $\Delta_h^1 f(t) = f(t) - f(t-h)$, and therefore, in this case the Grünwald-Letnikov derivative coincides with the first order derivative:

$$ {}_a\mathscr{D}_t^1 f(t) = \lim_{h\to 0} \frac{f(t)-f(t-h)}{h} = f'(t).$$

Analogously, it can easily be verified that ${}_a\mathscr{D}_t^n f(t) = f^{(n)}(t)$ for every integer $\alpha = n$:

$$ {}_a\mathscr{D}_t^n f(t) = \lim_{h\to 0} \frac{1}{h^n} \sum_{m=0}^{n} (-1)^m \binom{n}{m} f(t-mh), \tag{3.91}$$

where

$$\binom{n}{m} = \frac{n!}{m!(n-m)!} = \frac{\Gamma(n+1)}{m!\Gamma(n-m+1)}.$$

What concerns non-integer α, the definition of ${}_a\mathscr{D}_t^\alpha$ in (3.88) and (3.89) is essentially obtained from (3.91) replacing n by α. Hence, the binomial coefficients take the form (3.90), and apart from the integer case, the binomial coefficients never vanish. However, if $m > (t-a)/h$, then $t - mh < a$, that is $t - mh$ will be out of interval (a,b). Therefore, in the case of non-integer α, the upper limit of the sum in the Grünwald-Letnikov derivative is $N = \lfloor (t-a)/h \rfloor$, and $N \to \infty$ when $h \to 0$. If the interval (a,b) coincides with the real axis $(-\infty,\infty)$, then the upper limit in the summation becomes ∞. Thus in this case the Grünwald-Letnikov derivatives are defined through infinite series.

We note that ${}_a\mathscr{D}_t^\alpha$ defined in (3.88) and (3.89) is called a *forward Grünwald-Letnikov derivative*, in accordance with the forward finite difference in its definition. If one replaces the step size h to $-h, h > 0$, in the definition of ${}_a\mathscr{D}_t^\alpha$, then the obtained expression is called a *backward Grünwald-Letnikov derivative* and is denoted by ${}_t\mathscr{D}_b^\alpha$. Hence, by definition, the backward Grünwald-Letnikov derivative is

$$ {}_t\mathscr{D}_b^\alpha f(t) = (-1)^\alpha \Gamma(\alpha+1) \lim_{h\to 0} \left[\frac{1}{h^\alpha} \sum_{m=0}^{\lfloor \frac{b-t}{h} \rfloor} \frac{(-1)^m}{m!\Gamma(\alpha-m+1)} f(t+mh) \right]. \quad (3.92) $$

Now we present two important assertions related to Grünwald-Letnikov derivatives without proof. The first one is due to Letnikov proved in his original work [Let68] published in 1868.

Proposition 3.17. *(Letnikov) Let $n-1 < \alpha < n$ and $f \in C^n[a,b]$. Then*

$$ {}_a\mathscr{D}_t^\alpha f(t) = \sum_{k=0}^{n-1} \frac{f^{(k)}(a)(x-a)^{k-\alpha}}{\Gamma(k-\alpha+1)} + \frac{1}{\Gamma(n-\alpha+1)} \int_a^t \frac{f^{(n)}(\tau)d\tau}{(t-\tau)^{\alpha-n}}. \quad (3.93) $$

Comparing (3.93) with (3.53) one can see that the Grünwald-Letnikov derivative ${}_a\mathscr{D}_t^\alpha$ and the Riemann-Liuoville derivative ${}_aD_t^\alpha$ coincide in the class of functions satisfying the conditions of Proposition 3.17. Therefore, both Riemann-Liuoville and Caputo-Djrbashian derivatives can be numerically evaluated using the Grünwald-Letnikov approximation.

The second result is on the order of accuracy of the Grünwald-Letnikov approximation. The reader is referred to paper [Gor97] for details.

Proposition 3.18. *Let $f \in C^n[a,b]$. Then for all $\alpha \in (0,n]$,*

$$ {}_a\mathscr{D}_t^\alpha f(t) = \frac{(\Delta_h^\alpha f)(t)}{h^\alpha} + O(h), \quad h \to 0, $$

where $(\Delta_h^\alpha f)(t)$ is defined in (3.89).

Remark 3.5. 1. Similar results hold for the backward GL derivatives and approximations

2. If $(a,b) = (-\infty,\infty)$, then the forward and backward Grünwald-Letnikov derivatives of order $\alpha \in (m-1,m)$ take the forms

$$ {}_{-\infty}\mathscr{D}_t^\alpha f(t) = \lim_{h\to 0} \frac{1}{h^\alpha} \sum_{m=0}^{\infty} (-1)^m \binom{\alpha}{m} f(t-mh), \quad (3.94) $$

$$ {}_t\mathscr{D}_\infty^\alpha f(t) = \lim_{h\to 0} \frac{1}{h^\alpha} \sum_{m=0}^{\infty} (-1)^{m+\alpha} \binom{\alpha}{m} f(t+mh), \tag{3.95} $$

respectively, and coincide with the corresponding forward and backward Liuoville-Weyl derivatives of order α in the class of suitable functions (see [SKM87]).

3.11 Generalized fractional order operators. Distributed order operators

A distributed fractional order differential operator generalizes fractional order derivatives.

Definition 3.5. Let $T > 0$ be an arbitrary number and function $f \in C^m[0,T]$. Let μ be a bounded measure defined on the interval $[0,\nu]$, where $\nu \in (m-1,m]$, $m \in \mathbb{N}$, and such that the function $\alpha \to D_*^\alpha f(t)$, where D_*^α is the Caputo-Djrbashian derivative of order α, is μ-measurable for all $t \in [0,T]$. The operator D_μ defined by

$$ D_\mu f(t) = \int_0^\nu D_*^\alpha f(t) d\mu(\alpha), \quad 0 < t \leq T, \tag{3.96} $$

is called a distributed fractional order differential operator with mixing measure μ.

Since the integral in (3.96) is carried out with respect to a measure, some explanation is needed. First, for any function $f \in C^m[0,T]$ the function $\varphi(\alpha,t) = D_*^\alpha f(t)$ is an analytic function of α on the complex domain $\Re(\alpha) < m$ (see [OS74], page 49). Therefore, in general, μ can be a measure defined on the complex plane with the support on $\Re(\alpha) \leq m$. However, we will consider only measures defined on the interval $(0,m]$, to have a generalization of fractional order differential equations.

Second, since $\varphi(\alpha,t)$ is an analytic function of α for each fixed t the integral (3.96) is well defined for any Borel measure $\mu \in \mathscr{B}[0,m]$. In particular, if μ is a linear combination of the Dirac delta functions concentrated on integer points $j_k \in (0,m]$, then D_μ defines a differential operator.

Third, the specification of (3.96) is that the integration is carried out in the variable α, the order of differentiation. Therefore, it is named a distributed order differential operator. In what follows we use the abbreviation DODO for distributed fractional order differential operators, and DODE for distributed fractional order differential equations. Models with DODO arise naturally in various fields, for instance, in the kinetic theory [CGSG03] when the exact scaling is lacking or when diffusion is too slow (ultra-slow diffusion), in the theory of elasticity [LH02] for description of rheological properties of composite materials. Caputo [Cap67] was first who introduced DODO to model waves in viscoelastic media.

Fourth, Definition 3.5 is based on the Caputo-Djrbashian fractional derivative. Similarly, one can introduce the DODO based on the Riemann-Liouville fractional derivative, i.e.

$$_{RL}D_\mu f(t) = \int_0^\nu D^\alpha f(t) d\mu(\alpha),\ 0 < t \le T, \tag{3.97}$$

There is a connection between these two approaches, which will be established later on; see Proposition 3.23. In Chapter 7 we also introduce DODO based on the Liouville-Weyl fractional derivative.

Consider some examples of DODO.

Example 3.13. 1. Let $\mu = \delta_\beta$, i.e., Dirac's delta with mass on $\beta \in (0, \nu]$. Then one has $D_\mu = D_*^\beta$.

2. Let $\mu = \sum a_j \delta_{\alpha_j}$, where $a_j \in \mathbb{R}$ and $\alpha_j \in (0, \nu]$, $j = 1, \dots, J$. Then

$$D_\mu f(t) = \sum_{j=1}^{J} a_j D_*^{\alpha_j} f(t).$$

3. Let $d\mu(\alpha) = a(t)dt$, where $a \in C[0, \nu]$ is a positive function. Then

$$D_\mu f(t) = \int_0^\nu a(t) D_*^\alpha f(t) dt.$$

In the theory of DODOs the following functions play an important role:

$$\mathscr{K}_{\mu,j}(t) = \int_{j-1}^{j} \frac{t^{j-\alpha-1} d\mu(\alpha)}{\Gamma(j-\alpha)}, \quad t > 0,\ j = 1, \dots, m, \tag{3.98}$$

which will be called kernel functions, and

$$\Phi_u(s) = \int_0^u s^\alpha d\mu(\alpha), \quad \Re(s) > 0,\ u \in (0, \nu]. \tag{3.99}$$

Since $\operatorname{supp}\mu \subset [0, \nu]$, for $\mathscr{K}_m(t)$ one has

$$\mathscr{K}_{\mu,m}(t) = \int_{m-1}^{\nu} \frac{t^{m-\alpha-1} d\mu_j(\alpha)}{\Gamma(m-\alpha)}.$$

The substitution $\alpha - (j-1) = \beta$ reduces the operators K_j to

$$\mathscr{K}_{\mu_j}(t) = \int_0^1 \frac{t^{-\beta} d\mu_j(\beta)}{\Gamma(1-\beta)},$$

where $\mu_j(\beta) = \mu(\beta + (j-1))$. Therefore, it suffices to study the properties of the kernel function

$$\mathscr{K}_\nu(t) = \int_0^1 \frac{t^{-\beta} d\nu(\beta)}{\Gamma(1-\beta)},$$

where ν is a bounded measure defined on $[0, 1]$.

Proposition 3.19. *(Kochubey [Koc08]) Let* $d\nu(\beta) = a(\beta)d\beta, a \in C^3[0, 1]$, *and* $a(1) \neq 0$. *Then*

1. $\mathscr{K}_\nu(t) = \frac{a(1)}{t(\ln t)^2} + O(\frac{1}{t(\ln t)^3})$, $t \to 0$;
2. $\Phi_1(s) = \frac{a(1)}{\ln s} + O(\frac{1}{(\ln s)^2})$, $s \to \infty$,
3. If $a \in C[0,1]$ *and* $a(0) \neq 0$, *then* $\Phi_1(s) \sim \frac{a(0)}{s \ln s}$, $s \to 0$.

Another description of the kernel function $\mathscr{K} \equiv \mathscr{K}_\nu$ is due to Meerschaert and Scheffler [MS06]. Denote by $RV_\infty(\gamma)$ the set of functions regularly varying at infinity with exponent γ, that is eventually positive functions with behavior $g(\lambda t)/g(t) = \lambda^\gamma, t \to \infty$, for any $\lambda > 0$. Similar meaning has the set of functions regularly varying at zero, which is denoted by $RV_0(\gamma)$.

Proposition 3.20. *(MMM) Let* $d\nu(\beta) = a(\beta)d\beta$, *where* $a \in RV_0(\beta - 1)$. *Then there exists* $K^* \in RV_\infty(0)$ *such that* $\mathscr{K}(t) = (\ln t)^{-\beta} K^*(\ln t)$. *Especially,* $\mathscr{K}(t) = M(\ln t)$ *for some* $M \in RV_\infty(-\beta)$ *and* $\mathscr{K} \in RV_\infty(0)$, *so* $\mathscr{K}(t)$ *is slowly varying at infinity. Conversely, if for* $\mathscr{K}(t)$ *we have* $\mathscr{K} = M(\ln t)$ *for some* $M \in RV_\infty(-\beta)$ *and* $\beta > 0$, *then* $a \in RV_0(\beta - 1)$.

Proposition 3.21. *Let* $m \in \mathbb{N}$ *and* $\nu \in (0,m]$. *Then the distributed fractional order operator* D_μ *defined in (3.96) has the representation*

$$D_\mu f(t) = \sum_{j=1}^{m} (\mathscr{K}_j * f^{(j)})(t), \tag{3.100}$$

where $*$ *denotes the convolution operation and the kernel functions* $\mathscr{K}_j \equiv \mathscr{K}_{\mu,j}$, $j = 1,\dots,m$, *are defined by equation (3.98).*

Proof. Dividing the interval $(0,m]$ into subintervals $(j-1,j]$, $j = 1,\dots,m$, one has

$$D_\mu f(t) = \sum_{j=1}^{m} \int_{j-1}^{j} D_*^\alpha f(t) d\mu(\alpha).$$

Since, the Caputo-Djrbashian derivative of order $\alpha \in (j-1,j])$ of a function f is a convolution $K_j^\alpha * f^{(j)}$, $K_j^\alpha(t) = \frac{t^{j-\alpha-1}}{\Gamma(j-\alpha)}$, then

$$D_\mu f(t) = \sum_{j=1}^{m} \int_{j-1}^{j} K_j^\alpha * f^{(j)} d\mu(\alpha) = \sum_{j=1}^{m} (\mathscr{K}_j * f^{(j)})(t),$$

obtaining (3.100).

In the same manner one can prove the following proposition.

Proposition 3.22. *Let* $m \in \mathbb{N}$ *and* $\nu \in (0,m]$. *Then the distributed fractional order operator* ${}_{RL}D_\mu$ *defined in (3.97) has the representation*

$${}_{RL}D_\mu f(t) = \sum_{j=1}^{m} \frac{d^j}{dt^j} (\mathscr{K}_j * f)(t),$$

where $*$ denotes the convolution operation and the kernel functions $\mathscr{K}_j$, $j=1,\dots,m$, are defined by equation (3.98).

The following proposition generalizes relation (3.52) establishing a connection between DODOs based on the Caputo-Djrbashian and Riemann-Liouville derivatives.

Proposition 3.23. *The two DODOs D_μ and ${}_{RL}D_\mu$ are related to each other through the formula*

$$D_\mu f(t) = {}_{RL}D_\mu f(t) - \sum_{k=0}^{m-1} f^{(k)}(0)\mathscr{M}_k(t), \tag{3.101}$$

where

$$\mathscr{M}_k(t) = \int_0^\nu \frac{d\mu(\alpha)}{\Gamma(1+k-\alpha)t^{\alpha-k}}, \quad k=0,\dots,m-1.$$

Proof. The proof of this statement immediately follows from (3.52) by integration with respect to the measure $\mu(\alpha)$.

Proposition 3.24. *For the Laplace transform of the kernel function $\mathscr{K}_j$, $j=1,\dots,m$, the following formulas hold:*

$$\begin{aligned}\mathscr{L}[\mathscr{K}_j](s) &= \int_{j-1}^{j} s^{\alpha-j} d\mu(\alpha) \\ &= \frac{\Phi_j(s)-\Phi_{j-1}(s)}{s^j}, \quad s>0,\ j=1,\dots,m.\end{aligned} \tag{3.102}$$

Proof. The Laplace transform of $\mathscr{K}_j(t)$ is

$$\mathscr{L}[\mathscr{K}_j](s) = \int_0^\infty e^{-st}\left(\int_{j-1}^{j}\frac{t^{j-\alpha-1}d\mu(\alpha)}{\Gamma(j-\alpha)}\right)dt = \int_{j-1}^{j} L\Big[\frac{t^{j-\alpha-1}}{\Gamma(j-\alpha)}\Big]d\mu(\alpha).$$

Now the result follows due to formula (1.18) and the definition of $\Phi_j(s)$ given in equation (3.99).

Proposition 3.25. *Let $f\in C^{(m)}[0,\infty)$. Then for the Laplace transform of $D_\mu f$ the following formula holds:*

$$\mathscr{L}[D_\mu f](s) = \Phi_\nu(s)\mathscr{L}[f](s) - \sum_{k=0}^{m-1} f^{(k)}(0)\frac{\Phi_\nu(s)-\Phi_k(s)}{s^{k+1}}, \quad s>0, \tag{3.103}$$

where $\Phi_u(s)$ is defined in (3.99).

Proof. Due to Proposition 3.21, one has $D_\mu f(t) = \sum_{j=1}^m (\mathscr{K}_j * f^{(j)})(t)$. This implies

$$\mathscr{L}[D_\mu f](s) = \sum_{j=1}^{m} \mathscr{L}[\mathscr{K}_j](s)\mathscr{L}[f^{(j)}](s), \quad s>0.$$

Further, using (3.102) and the first differentiation formula for the Laplace transform (see (1.22)), one obtains

$$\begin{aligned}\mathscr{L}[D_\mu f](s) &= \sum_{j=1}^{m} \frac{\Phi_j(s)-\Phi_{j-1}(s)}{s^j}\left(s^j\mathscr{L}[f](s) - \sum_{k=0}^{j-1} f^{(k)}(0)s^{j-1-k}\right)\\ &= \mathscr{L}[f](s)\sum_{j=1}^{m}\Big(\Phi_j(s)-\Phi_{j-1}(s)\Big)\\ &\quad - \sum_{j=1}^{m}\sum_{k=0}^{j-1}\Big(\Phi_j(s)-\Phi_{j-1}(s)\Big)\frac{f^{(k)}(0)}{s^{k+1}}, \quad s>0. \end{aligned} \tag{3.104}$$

Taking into account $\Phi_0(s) \equiv 0$ and $\Phi_m(s) = \Phi_\nu(s)$ due to the assumption that $\operatorname{supp}\mu \subset (0,\nu]$, where $\nu \in (m-1,m]$, one has $\sum_{j=1}^{m}\Big(\Phi_j(s)-\Phi_{j-1}(s)\Big) = \Phi_\nu(s)$. Therefore, the first term on the right-hand side of (3.104) equals $\Phi_\nu(s)\mathscr{L}[f](s)$. Changing order of summation in the second term,

$$\begin{aligned}\sum_{j=1}^{m}\sum_{k=0}^{j-1}\Big(\Phi_j(s)-\Phi_{j-1}(s)\Big)\frac{f^{(k)}(0)}{s^{k+1}} &= \sum_{k=0}^{m-1}\frac{f^{(k)}(0)}{s^{k+1}}\sum_{j=k+1}^{m}\Big(\Phi_j(s)-\Phi_{j-1}(s)\Big)\\ &= \sum_{k=0}^{m-1}\frac{f^{(k)}(0)}{s^{k+1}}\Big(\Phi_\nu(s)-\Phi_k(s)\Big),\end{aligned}$$

yielding (3.103).

Proposition 3.26. *For the Laplace transform of ${}_{RL}D_\mu f$ the following formula holds:*

$$\mathscr{L}[{}_{RL}D_\mu f](s) = \Phi_\nu(s)\mathscr{L}[f](s) - \sum_{k=0}^{m-1}\mathscr{N}_k s^{m-k-1}, s>0, \tag{3.105}$$

where $\mathscr{N}_k = \int_0^\nu (D^k J^{m-\alpha} f)(0)d\mu(\alpha)$.

Proof. The proof of this statement immediately follows from (3.29) by integrating with respect to the measure $\mu(\alpha)$ on the interval $[0,\nu)$.

3.12 Variable order fractional derivatives and the memory effect

Another generalization of fractional order derivatives are *fractional variable order differential operators.* We will use the abbreviation VODO for fractional variable order differential operators. The study of variable fractional order derivatives and operators started in the middle of the 1990s by N. Jacob et al. [JL93], S. G. Samko et al. [SR93, Sam95], W. Hoh [Hoh00]. A. V. Chechkin et al. [CGS05] used a version of variable order derivatives to describe kinetic diffusion in heterogeneous media. Lorenzo and Hartley [LH02] introduced a wide class of variable fractional order derivatives, which can be used for modeling of processes with various types of memory effects.

Definition 3.6. Let a function $\beta(t)$, $t>0$, satisfies the condition $0<\beta(t)\leq 1$. By definition, a variable order fractional derivative is

$$\mathscr{D}_{\mu,\nu}^{\beta(t)} f(t) = \frac{d}{dt}\int_0^t \mathscr{K}_{\mu,\nu}^{\beta}(t,\tau)f(\tau)d\tau, \tag{3.106}$$

where μ and ν are real parameters, $t>0$, and

$$\mathscr{K}_{\mu,\nu}^{\beta}(t,\tau) = \frac{1}{\Gamma(1-\beta(\mu t+\nu\tau))(t-\tau)^{\beta(\mu t+\nu\tau)}}, \quad 0<\tau<t. \tag{3.107}$$

The function $\beta(t)$ is called an *order function.* If $\beta(t)=1$ for some $t_0>0$, then we agree that the integral on the right-hand side of (3.106) equals $\frac{df(t)}{dt}$ whenever $\mu t+\nu\tau=t_0$. The operator $\mathscr{D}_{\mu,\nu}^{\beta(t)}$ depends on parameters μ and ν. These parameters run in the parallelogram Π shown in Figure 3.1, which we call *Lorenzo-Hartley causality parallelogram* (or LH-parallelogram). Therefore, we call the operator $\mathscr{D}_{\mu,\nu}^{\beta(t)}$ a Riemann-Liouville type (μ,ν)-VODO with the order function β.

Similarly, one can introduce the Caputo-Djrbashian type (μ,ν)-VODO with the order function β.

Definition 3.7. Let a function $\beta(t)$, $t>0$, satisfy the condition $0<\beta(t)\leq 1$. By definition, a Caputo-Djrbashian type (μ,ν)-VODO with the order function β is

$$\mathscr{D}_{*\mu,\nu}^{\beta(t)} f(t) = \int_0^t \mathscr{K}_{\mu,\nu}^{\beta(t)}(t,\tau)\frac{df(\tau)}{d\tau}d\tau. \tag{3.108}$$

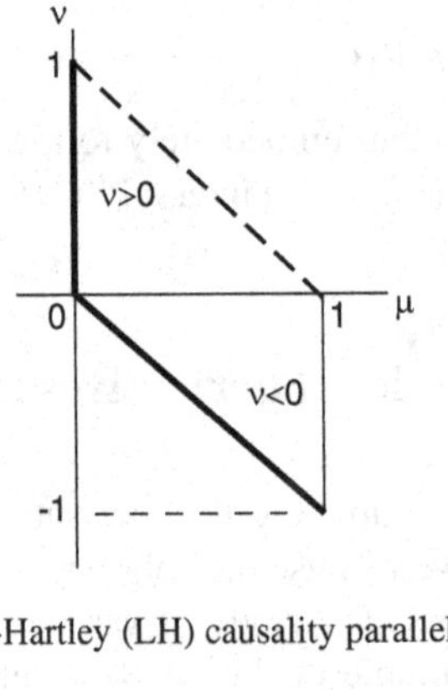

Fig. 3.1 The Lorenzo-Hartley (LH) causality parallelogram Π.

where parameters μ and ν and the kernel function $\mathscr{K}_{\mu,\nu}^{\beta(t)}(t,\tau)$ are defined as in Definition 3.6.

The Caputo-Djrbashian type (μ,ν)-VODOs can be defined for any positive bounded piecewise continuous order functions. Denote by $[\beta(t)]$ the integer part of $\beta(t)$ for each $t>0$, and by $\{\beta(t)\}$ its fractional part. Then one can rewrite $\mathscr{D}^{\beta(t)}_{*\mu,\nu}f(t)$ in equation (3.108) in the form

$$\mathscr{D}^{\beta(t)}_{*\mu,\nu}f(t) = \int_0^t \mathscr{K}^{\{\beta\}}_{\mu,\nu}(t,\tau)f^{[\beta(\mu t+\nu\tau)]+1}(\tau)d\tau. \tag{3.109}$$

Obviously, if $\beta(t)<1$, then $[\beta(\mu t+\nu\tau)]=0$ and $\{\beta(\mu t+\nu t)\}=\beta(\mu t+\nu t)$. Therefore, (3.109) coincides with (3.108). For those values of $t_0>0$ for which $\beta(t_0)=1$, as we did in Definition 3.6, we agree that the integral on the right-hand side of (3.109) equals $\frac{df(t)}{dt}$ whenever $\mu t+\nu\tau=t_0$. The definition (3.109) is valid for any bounded (not necessarily bounded with 1) piecewise continuous function $\beta(t)$ with the additional agreement that if $\beta(t_0)=m\in\mathbb{N}$ for some $t_0>0$, then the right-hand side of (3.109) equals $\frac{d^m f(t)}{dt^m}$ whenever $\mu t+\nu\tau=t_0$. To verify validity of this claim, assume that $m-1<\beta(t)<m$, in some interval (a,b). Then $[\beta(\mu t+\nu\tau)]=m-1$ and $\{\beta(\mu t+\nu\tau)\}=\beta(\mu t+\nu\tau)-m+1$ when $\mu t+\nu\tau\in(a,b)$. The latter makes the integral in (3.109) consistent with the Caputo-Djrbashian form of the fractional derivative.

Remark 3.6. The role of the Lorenzo-Hartley causality parallelogram (see Figure 3.1) $\Pi=\{(\mu,\nu)\in R^2: 0\le\mu\le 1, 0\le\mu+\nu\le 1\}$ in the VODOs (3.106) and (3.108) or, in more general case, in (3.109), is that $\mu t+\nu\tau$ runs in the interval $(0,t)$ when $(\mu,\nu)\in\Pi$. Indeed, the conditions $(\mu,\nu)\in\Pi$ and $\tau\in(0,t)$ yield $\mu t+\nu\tau\in(\mu t,(\mu+\nu)t)\subset(0,t)$. In other words, the condition $(\mu,\nu)\in\Pi$ predetermines the causality, since $0\le\mu t+\nu\tau\le t$ for all $t>0$ and $0\le\tau\le t$.

Remark 3.7. In Chapter 5 we will use VODOs to model complex diffusion processes in heterogeneous media with different diffusion modes in different time intervals (see Definition 3.8). The corresponding mathematical model is the Cauchy problem for a pseudo-differential equation with a singular symbol and a variable fractional order time derivative. Such a model takes into account the memory effects of the past in computing present or future states of the underlying diffusion process. Diffusion processes in heterogeneous media are accompanied by frequent changes of diffusion modes. It is known that a non-Markovian random process possesses a memory of past (see [MK00, Zas02]). For instance, protein movement in cell membrane, as is recorded in [SJ97, Sax01], follows a non-Markovian (anomalous) diffusion process. Descriptions of such processes using random walk models also show the presence of non-Markovian type memory [AUS06, GMM02, LSAT05]. It turns out there is another type of memory noticed first by Lorenzo and Hartley in their paper [LH02] in some particular cases of μ and ν. This kind of memory arises when the diffusion mode changes. Below we study memory effects in the case when the order function is piecewise constant.

The kernel (3.107), and thus, both the operators (3.106) and (3.108) are weakly singular for $(\mu, \nu) \in \Pi$. Further, denote

$$\mathscr{K}(t,\tau,s) = \frac{1}{\Gamma(1-\{\beta(s)\})(t-\tau)^{\{\beta(s)\}}}, \quad t>0, 0<\tau<t, 0<s<t, \quad (3.110)$$

where $0 < \beta(s) \le M$, for some $M < \infty$, is an order function. Let $0 = T_0 < T_1 < \ldots < T_N < T_{N+1} = \infty$ be a partition of the interval $(0,\infty)$ into $N+1$ sub-intervals (T_k, T_{k+1}). Let $\beta(t)$ be a piecewise constant function

$$\beta(t) = \sum_{k=0}^{N} \beta_k \mathscr{I}_k(t), \quad t \in (0,\infty), \quad (3.111)$$

where $\mathscr{I}_k$ is the indicator function of the interval (T_k, T_{k+1}) and $0 < \beta_k \le 1, k = 0,\ldots,N$, are constants. Under these conditions, the function (3.110) becomes

$$K(t,\tau,s) = \sum_{k=0}^{N} \mathscr{I}_k(s) \frac{1}{\Gamma(1-\beta_k)(t-\tau)^{\beta_k}}, \quad t>0, 0<\tau<t, 0<s<t, \quad (3.112)$$

and the kernel of the fractional order operator (3.108) becomes

$$\mathscr{K}_{\mu,\nu}^{\beta(t)}(t,\tau) = K(t,\tau,\mu t+\nu\tau), \quad t>0, \quad 0\le\tau<t. \quad (3.113)$$

with $K(t,\tau,s)$ defined in (3.112).

Theorem 3.6. *Let the order function $\beta(t)$ be a piecewise constant. Then the mapping*

$$\mathscr{D}_{*\mu,\nu}^{\beta(t)} : C^m[0,T] \to C[0,T]$$

is continuous.

Proof. For simplicity we assume $m = 1$. The proof for $m > 1$ does not have an essential difference. Let $f \in C^1[0,T.]$ Then $h(t) = \mathscr{D}_{*\mu,\nu}^{\beta(t)} f(t)$ is continuous. Indeed, exploiting (3.112) and (3.113), we have

$$\begin{aligned}\mathscr{D}_{*\mu,\nu}^{\beta(t)} f(t)| &= \int_0^t K(t,\tau,\mu t+\nu\tau)\frac{df}{dt}dt \\ &= \sum_{k=0}^{N} \frac{1}{\Gamma(1-\beta_k)} \int_0^t \frac{\mathscr{I}_k(\mu t+\nu\tau)\frac{df}{dt}}{(t-\tau)^{\beta_k}} d\tau = \sum_{k=0}^{N} J_k^{\beta_k} Df(t),\end{aligned}$$

where $J_k^{\beta_k}$ is a fractional order integration operator. Since $Df = \frac{df}{dt} \in C[0,T]$, it follows from Proposition 3.2 that $J_k^{\beta_k} Df(t) \in C[0,1]$ for each $k = 0,\ldots,N$. Moreover, since the kernel $K_k(t,\tau) = \mathscr{I}_k(\mu t+\nu\tau)(t-\tau)^{-\beta_k} \in L_1(0,1)$ for each $k = 0,\ldots,N$, it follows that

$$|J_k^{\beta_k} Df(t)| \leq \|K_k|L_1\| \sup_{[0,T]} \left|\frac{df}{dt}\right|, \quad k = 0, \ldots, N,$$

implying the continuity of $\mathscr{D}_{*\mu,\nu}^{\beta(t)}$.

Remark 3.8. With slight modification of the proof one can show that Theorem 3.6 can be extended to any piecewise continuous order functions.

Definition 3.8. Let the order function $\beta(t)$ be defined as in (3.111) and $(\mu,\nu) \in \Pi$. We say that the triplet (β_k,μ,ν) determines a *diffusion mode* in the time interval (T_k, T_{k+1}).

Remark 3.9. If one assumes that the input is a triplet (β_k,μ,ν), then the output is determined by the fact that which values of $\beta(t)$ are used to compute the variable order derivative, or in other words, by the fact that which interval (T_k, T_{k+1}) the point $s = \mu t + \nu\tau$ belongs to. Since $(\mu,\nu) \in \Pi$ implies $\mu t + \nu\tau \in (\mu t, (\mu+\nu)t)$, this means that the operators $\mathscr{D}_{\mu,\nu}^{\beta(t)}$ and $\mathscr{D}_{*\mu,\nu}^{\beta(t)}$ use information taken in the time sub-interval $(\mu t, (\mu+\nu)t)$ if ν is positive and from the sub-interval $((\mu+\nu)t, \mu t)$ if ν is negative. In both cases, the length of this interval is $|\nu|t$.

Now we analyze the memory effects in a special case of a single change of diffusion mode, that is, a diffusion mode given by a triplet $\{\beta_1,\mu,\nu\}$ changes at time T to a diffusion mode corresponding to another triplet $\{\beta_2,\mu,\nu\}$.

Definition 3.9. Let $\{\beta_1,\mu,\nu\}$ and $\{\beta_2,\mu,\nu\}$ be two admissible triplets which determine two distinct diffusion modes. Assume the diffusion mode is changed at time $t = T$ from $\{\beta_1,\mu,\nu\}$-mode to $\{\beta_2,\mu,\nu\}$-mode. Then the process is said to have a 'short-range' (or short) memory, if there is a finite $T^* > T$ such that for all $t > T^*$ the $\{\beta_2,\mu,\nu\}$-mode holds. Otherwise, the process is said to have a 'long-range' (or long) memory.

Remark 3.10. According to Definition 3.9, a diffusion mode has a long memory if the effect of the previous diffusion mode never vanishes, even the diffusion mode is changed, i.e., the particle never forgets its past. In the case of short memory, particle remembers the previous mode until some critical time, and then forgets it fully, recognizing the new mode.

Theorem 3.7. *([US06]) Let $\nu > 0$ and $\mu \neq 0$. Assume the $\{\beta_1,\mu,\nu\}$-diffusion mode is changed at time $t = T$ to the $\{\beta_2,\mu,\nu\}$-diffusion mode (Figure 3.2). Let $T^* = T/\mu$ and $t^* = T/(\mu+\nu)$. Then the process has a short memory. Moreover,*

(i) for all $0 < t < t^*$ the $\{\beta_1,\mu,\nu\}$-diffusion mode holds;
(ii) for all $t > T^*$ the $\{\beta_2,\mu,\nu\}$-diffusion mode holds;
(iii) for all $t^* < t < T^*$ a mix of both $\{\beta_1,\mu,\nu\}$ and $\{\beta_2,\mu,\nu\}$-diffusion modes hold.

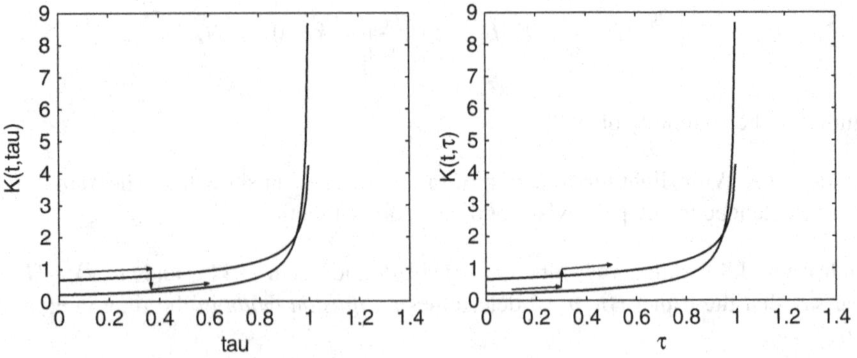

Fig. 3.2 These figures illustrate Theorem 3.7. Moving from the upper curve to the lower curve (left figure), and from lower curve to the upper one (right figure) does not occur at time T, when one diffusion mode changed to another mode. Instead, it occurs at time $T^* > T$ fully accepting a new diffusion mode. The curves in these figures are kernel functions corresponding to two diffusion modes.

Proof. Let $\beta(s) = \beta_1$ for $0 < s < T$ and $\beta(s) = \beta_2$ for $s > T$. Assume $\nu > 0$. Denote $s = \mu t + \nu\tau$. So, the $\{\beta_1, \mu, \nu\}$-diffusion mode holds if $\mu t + \nu\tau < T$. Let $0 < t < t^* = T/(\mu+\nu)$. Then for every $\tau \in (0,t)$ we have $\mu t + \nu\tau < (\mu+\nu)t < T$. This means that the order operator $\beta(s)$ in $\mathscr{D}^{\beta(t)}_{*\{\mu,\nu\}}$ takes the value β_1 giving (i). If $t > T/\mu$, then for all $\tau > 0$, $\mu t + \nu\tau > T$. Hence, $\beta(s) = \beta_2$, obtaining (ii). Now assume $T/(\mu+\nu) < t < T/\mu$. Denote $\tau_0 = (T - \mu t)/\nu$. Obviously $\tau_0 > 0$. It follows from $(\mu+\nu)t > T$ dividing by ν that $t > T/\nu - t\mu/\nu = \tau_0$, i.e., $0 < \tau_0 < t$. It is easy to check that if $0 < \tau < \tau_0$ then $\mu t + \nu\tau \in (\mu t, T) \subset (0,T)$, giving $\beta(s) = \beta_1$, while if $\tau_0 < \tau < t$ then $\mu t + \nu\tau \in (T, (\mu+\nu)t) \subset (T,\infty)$, giving $\beta(s) = \beta_2$. Hence, in this case the mix of both $\{\beta_1, \mu, \nu\}$ and $\{\beta_2, \mu, \nu\}$-diffusion modes is present.

Theorem 3.8. *([US06]) Let $\nu < 0$ and $\mu + \nu \neq 0$. Assume the $\{\beta_1, \mu, \nu\}$-diffusion mode is changed at time $t = T$ to the $\{\beta_2, \mu, \nu\}$-diffusion mode. Let $t^{*'} = T/\mu$ and $T^{*'} = T/(\mu+\nu)$. Then the process has a short memory. Moreover,*

(i') for all $0 < t < t^{*'}$ the $\{\beta_1, \mu, \nu\}$-diffusion mode holds;
(ii') for all $t > T^{*'}$ the $\{\beta_2, \mu, \nu\}$-diffusion mode holds;
(iii') for all $t^{*'} < t < T^{*'}$ a mix of both $\{\beta_1, \mu, \nu\}$ and $\{\beta_2, \mu, \nu\}$-diffusion modes hold.

Proof. Let $\nu < 0$. Assume again $\beta(s) = \beta_1$ for $0 < s < T$ and $\beta(s) = \beta_2$ for $s > T$. As in the previous theorem, denote $s = \mu t + \nu\tau$. First we notice that if $0 < t < T/\mu$ then $\mu t + \nu\tau < T$, which implies $\beta(s) = \beta_1$, giving (i'). Now let $t > T/(\mu+\nu)$ be any number. Then for $0 < \tau < t$ we have $\mu t + \nu\tau > T$, which yields $\beta(s) = \beta_2$. So, we get (ii'). Now assume $T/\mu < t < T/(\mu+\nu)$. Again denote $\tau_0 = (T - \mu t)/\nu$. Obviously $\tau_0 > 0$. It follows from $(\mu+\nu)t < T$ dividing by $\nu < 0$ that $t > T/\nu - t\mu/\nu = \tau_0$, i.e., $0 < \tau_0 < t$. It is easy to check that if $0 < \tau < \tau_0$ then $\mu t + \nu\tau \in (T, \mu t) \subset (T,\infty)$, giving $\beta(s)=\beta_2$, while if $\tau_0<\tau<t$ then $\mu t+\nu\tau\in((\mu+\nu)t,T)\subset(0,T)$, giving $\beta(s)=\beta_1$. Hence, in this case the mix of both $\{\beta_1,\mu,\nu\}$ and $\{\beta_2,\mu,\nu\}$-diffusion modes is present, obtaining (iii').

Corollary 3.9 *Let $\nu=0$ and $\mu\neq 0$. Assume the $\{\beta_1,\mu,\nu\}$-diffusion mode is changed at time $t=T$ to the $\{\beta_2,\mu,\nu\}$-diffusion mode. Let $T^*=T/\mu$. Then the process has a short memory. Moreover,*

(a) for all $0<t<T^*$ the $\{\beta_1,\mu,\nu\}$-diffusion mode holds;
(b) for all $t>T^*$ the $\{\beta_2,\mu,\nu\}$-diffusion mode holds.

Proof. If $\nu=0$, then we have $\beta(s)=\beta(\mu t)=\beta_1$ for $t<T/\mu$ and $\beta(s)=\beta_2$ for $t>T/\mu$.

Corollary 3.10 *Let $\mu=0$ or $\mu+\nu=0$. Assume the $\{\beta_1,\mu,\nu\}$-diffusion mode is changed at time $t=T$ to the $\{\beta_2,\mu,\nu\}$-diffusion mode. Then the process has the long memory.*

Proof. According to the structure of LH-parallelogram $\mu=0$ implies $\nu>0$. In this case $T^*=\infty$. If $\mu+\nu=0$, then $\nu<0$ and $t^*=\infty$. In both cases we have long memory effect.

Remark 3.11. Notice, that if $\nu=0$ then there is no intervals of mix of modes. Moreover, if $\nu=0,\mu=1$, then $T^*=t^*=T$. In this sense we say that a process has no memory. For all values of $\{\mu,\nu\}$ except the bold lines in the LH-parallelogram. (see Figure 3.1), the operator $D_{*\{\mu,\nu\}}^{\beta(t)}$ has a short memory. Memory is stronger in the region $\nu<0$ and weaker in $\nu>0$. On the dashed line $\mu+\nu=1$ we have $t^*=T<T^*$. The bold lines $\mu=0,\nu\geq 0$ and $\mu+\nu=0$ identify the long range memory.

3.13 Additional notes

1. The starting point of fractional calculus goes back to 1695, when L'Hôpital wrote a letter to Leibnitz asking him about the notation d^nf/dx^n for $f(x)=x$, if $n=1/2$. Leibnitz responded stating that "An apparent paradox, from which one day useful consequences will be drawn." Contributions to factional calculus were made by classics Euler, Laplace, Fourier, Abel, Liouville, Riemann, Grünwald, Letnikov, Hadamard, Heaviside, Weyl, Lévy, Marchaud, Zygmund, M. Riesz, etc. Several books devoted to fractional calculus are written among which we would like to mention the encyclopedic book by Samko, Kilbas, and Marychev [SKM87] published first in Russian in 1987, the earlier books by Djrbashian [Djr66], Oldham and Spanier [OS74], Miller and Ross [MR93], books appeared relatively recently and written by Kiryakova [Kir94], Podlubni [Pod99], Rubin [Rub96], Hilfer [Hil00], Kilbas, Srivastava, and Trijillo [KST06], etc. The survey papers written by Gorenflo and Mainardi [GM97], Metzler and Klafter [MK00], contain a vast material including recent developments and historical facts.

2. Example 3.4 shows that the semigroup property $J^{\alpha}J^{\beta} = J^{\alpha+\beta} = J^{\beta}J^{\alpha}$ valid for fractional integrals and for integer order derivatives, in general, fails in the case of Riemann-Liouville fractional derivatives. However, as is shown in [OS74], if f satisfies the condition $f = J^{\beta}D^{\beta}f$, then the rule $D^{\alpha}D^{\beta}f = D^{\alpha+\beta}f$, called a *composition rule*, holds. Some differentiation rules and properties valid in the classical calculus generalize to the fractional calculus as well. Obviously, all the different versions of fractional derivatives and operators introduced in this chapter, including distributed and variable order derivatives, are linear. The product rule, or in the general case *the Leibniz rule* ($n \in \mathbb{N}$)

$$(fg)^{(n)} = \sum_{k=0}^{n} \binom{n}{k} f^{(n-k)} g^{(k)},$$

in the case of fractional Riemann-Liouville derivatives takes the form [OS74]

$${}_aD_t^{\alpha}[fg](t) = \sum_{k=0}^{\infty} \binom{\alpha}{k} {}_aD_t^{\alpha-k} f(t) g^{(k)}(t),$$

where $\binom{\alpha}{k} = \Gamma(\alpha+1)/(k!\Gamma(\alpha-k+1))$. To generalize the chain rule to fractional derivatives one needs first the chain rule for n-th order derivative, called the Faá di Bruno formula (see [AS64]):

$$[f(g(x))]^{(n)} = n! \sum_{k=1}^{n} f^{(k)} \sum \prod_{j=1}^{n} \frac{1}{P_j!} \left[\frac{g^{(j)}}{j!} \right]^{P_j},$$

where the sum inside is over all combinations of nonnegative integer numbers $P_1, \ldots, P_n$ such that

$$\sum_{j=1}^{n} jP_j = n, \quad \text{and} \quad \sum_{j=1}^{n} P_j = k.$$

Then the fractional generalization of *the chain rule* for the Riemann-Liouville derivative reads [OS74]

$$\begin{aligned} {}_aD_t^{\alpha} f(g(t)) = & \frac{f(g(t))}{\Gamma(1-\alpha)(t-a)^{\alpha}} \\ & + \sum_{n=1}^{\infty} \binom{\alpha}{n} \frac{(t-a)^{n-q}}{\Gamma(n-q+1)} n! \sum_{k=1}^{n} f^{(k)}(g(t)) \sum \prod_{j=1}^{n} \frac{1}{P_j!} \left[\frac{g^{(j)}(t)}{j!} \right]^{P_j} \end{aligned}$$

with the same meaning of P_j, $j = 1, \ldots, n$, as in the Faá di Bruno's formula. For further fractional generalizations of the Leibniz and chain rules, we refer the reader to papers [Osl72, SKM87, FGT12]. In particular, for suitable functions $f(z)$ and $g(z)$ we note the representation of the fractional derivative of order $\alpha > 0$ of $f(z)$ with respect to $g(z)$, that is $D_{g(z)}^{\alpha} f(z) = D_w^{\alpha} F(g^{-1}(w))$ [FGT12]:

$$D_{g(z)}^{\alpha} f(z) = \frac{\Gamma(\alpha+1)}{2\pi i} \int_C \frac{f(t)g'(t)}{\left(g(t)-g(z)\right)^{\alpha+1}} dt,$$

and [FGT12]

$$D_{g(z)}^{\alpha} \left([g(z)]^p f(z) \right) = \frac{e^{-i\pi p}\Gamma(\alpha+1)}{4\pi \sin p\pi} \int_{C_1} \frac{f(t)[g(t)]^p g'(t)}{\left(g(t)-g(z)\right)^{\alpha+1}} dt, \tag{3.114}$$

where contours C and C_1 are special Pochhammer contours.

3. *Proof of formula* (1.7). The formula (3.6) can be used for the proof of the following property of the Gamma-function:

$$\Gamma(\alpha)\Gamma(1-\alpha) = \frac{\pi}{\sin \pi\alpha}, \quad 0 < \alpha < 1. \tag{1.14}$$

First, we note that using the contour integrals method one can easily verify that

$$\int_0^\infty \frac{x^{-\alpha}}{1+x} dx = \frac{\pi}{\sin \pi\alpha}. \tag{3.115}$$

Further, taking $\gamma = -\alpha$ in equation (3.6),

$$J^\alpha t^{-\alpha} = \Gamma(1-\alpha) = \frac{1}{\Gamma(\alpha)} \int_0^t \frac{\tau^{-\alpha} d\tau}{(t-\tau)^{1-\alpha}}.$$

The substitution $\tau = t - t/(x+1)$ in the latter integral implies

$$\Gamma(\alpha)\Gamma(1-\alpha) = \int_0^\infty \frac{x^{-\alpha}}{1+x} dx.$$

Comparing this with equation (3.115) we obtain (1.7).

4. *Marchaud fractional derivative.* One can use the idea of analytic continuation of ${}_{-\infty}J^\alpha f(x)$ defined in (3.60) (or ${}_xJ^\alpha_\infty f(x)$ in (3.61)) to the domain $\Re(\alpha) < 0$, in order to define a fractional order derivative for suitable functions $f(x)$. This idea leads to the following definition of the Marchaud fractional derivative:

Definition 3.10. The fractional order derivative in the sense of Marchaud is defined by

$$(\mathbf{D}^\alpha_+)f(x) = \begin{cases} f(x), & \text{if } \alpha = 0; \\ \frac{\alpha}{\Gamma(1-\alpha)} \int_0^\infty \frac{f(x)-f(x+h)}{h^{1+\alpha}} dh, & \text{if } 0 < \alpha < 1; \\ \frac{d^k}{dx^k} \mathbf{D}^{\alpha-k}_- f(y), & \text{if } k \le \alpha < k+1,\ k = 1,2,\ldots, \end{cases} \tag{3.116}$$

subject to the integral on the right is finite.

Indeed, let $0 < \alpha < 1$. Then, one can write formally

$${}_{-\infty}J^{-\alpha} f(x) = \frac{1}{\Gamma(-\alpha)} \int_0^\infty \frac{f(x-h)}{h^{1+\alpha}} dh.$$

However, the integral on the right diverges due to the strong singularity of the integrand at $h = 0$. Regularizing (and using the equality $\Gamma(1-\alpha) = -\alpha\Gamma(-\alpha)$), one obtains a meaningful integral

$$(\mathbf{D}^\alpha_+)f(x) = \frac{\alpha}{\Gamma(1-\alpha)} \int_0^\infty \frac{f(x) - f(x-h)}{h^{1+\alpha}} dh,$$

convergent for suitable functions. Hence, the Marchaud fractional derivative in Definition 3.10 is "a regularization" of ${}_{-\infty}J^{-\alpha} f(x)$. Similarly, the regularization of ${}_xJ^{-\alpha}_\infty f(x)$, $0 < \alpha < 1$, gives

$$(\mathbf{D}^\alpha_-)f(x) = \frac{\alpha}{\Gamma(1-\alpha)} \int_0^\infty \frac{f(x+h) - f(x)}{h^{1+\alpha}} dh.$$

The derivatives $\mathbf{D}^\alpha_+$ and $\mathbf{D}^\alpha_-$ are called, respectively, a forward and backward Marchaud fractional derivatives. We note that if for a function $f \in L_p(\mathbb{R})$, $1 < p < \infty$, one of the fractional derivatives $(\mathbf{D}^\alpha_+)f(x)$ (Marchaud), ${}_{-\infty}D^\alpha f(x)$ (Liouville-Weyl), and ${}_{-\infty}\mathscr{D}^\alpha_x f(x)$ (Grünwald-Letnikov) exists, then other two also exist and all the three coincide (see [SKM87]):

$$(\mathbf{D}_{+}^{\alpha})f(x) = {}_{-\infty}D^{\alpha}f(x) = {}_{-\infty}\mathscr{D}_{x}^{\alpha}f(x).$$

The same is true for backward versions of Marchaud, Liouville-Weyl, and Grünwald-Letnikov fractional derivatives. Further, it follows from Example 3.12 that if $B_{\alpha}^{\pm}$ is a one-sided Marchaud, Grünwald-Letnikov, or Liouville-Weyl fractional derivative of order α, then

$$B_{\alpha}^{\pm}e^{\pm ax} = a^{\alpha}e^{\pm ax}, \quad a>0. \tag{3.117}$$

This formula will be used in our further considerations.
For functions defined on a finite interval (a,b) the general forms of forward and backward fractional Marchaud derivatives are $(0<\alpha<1)$:

$$(\mathbf{D}_{a+}^{\alpha})f(x) = \frac{f(x)}{\Gamma(1-\alpha)(x-a)^{\alpha}} + \frac{\alpha}{\Gamma(1-\alpha)}\int_a^x \frac{f(x)-f(y)}{(x-y)^{\alpha+1}}dy, \quad x\in(a,b),$$

and

$$(\mathbf{D}_{b-}^{\alpha})f(x) = \frac{f(x)}{\Gamma(1-\alpha)(b-x)^{\alpha}} + \frac{\alpha}{\Gamma(1-\alpha)}\int_x^b \frac{f(y)-f(x)}{(y-x)^{\alpha+1}}dy, \quad x\in(a,b).$$

5. *Generalized Mittag-Leffler function. Mainardi's function.* The Mittag-Leffler function is an entire function of the complex variable $z\in\mathbb{C}$ and depends on two parameters, α and β :

$$E_{\alpha,\beta}(z) = \sum_{n=0}^{\infty}\frac{z^n}{\Gamma(\alpha n+\beta)}, \quad \alpha>0, \beta>0.$$

The function $E_{\alpha}(z)$ introduced in Section 3.4 corresponds to the case $\beta=1$. For various properties of the Mittag-Leffler functions, we refer the reader to [GK14, HMS11]. For the reader's convenience, below we provide some of them, which will be used in subsequent chapters. It is easy to see that in particular cases Mittag-Leffler functions are related to the exponential, cosine, and sine functions. Namely, the equalities $E_1(z)=e^z$, $E_2(-z^2)=\cos z$, and $E_{2,2}(-z^2)=(\sin z)/z$ hold. The function $E_{\alpha,\beta}(z)$ is an entire function of exponential order $1/\alpha$, and the following asymptotic behavior is valid for $0<\alpha<2$, $\beta=1$, as $|z|\to\infty$ through different sectors [GK14]:

$$E_{\alpha}(z) \sim \begin{cases} \frac{1}{\alpha}\exp(z^{1/\alpha}) - \sum_{k=1}^{\infty}\frac{z^{-k}}{\Gamma(1-\alpha k)}, & \text{if } |z|<\frac{\alpha\pi}{2}, \\ \sum_{k=1}^{\infty}\frac{z^{-k}}{\Gamma(1-\alpha k)}, & \text{if } \frac{\alpha\pi}{2}<\arg z<2\pi-\frac{\alpha\pi}{2}, \end{cases} \quad |z|\to\infty.$$

The formula

$$L\left[t^{\beta-1}E_{\alpha,\beta}(-\lambda t^{\alpha})\right](s) = \frac{s^{\alpha-\beta}}{s^{\alpha}+\lambda}, \quad \Re(s)>|\lambda|^{1/\alpha},$$

generalizes (3.38). One can derive from the latter the following useful formula valid for all $k=0,1,\cdots$:

$$L\left[t^{\alpha k+\beta-1}E_{\alpha,\beta}^{(k)}(-\lambda t^{\alpha})\right](s) = \frac{k!s^{\alpha-\beta}}{(s^{\alpha}+\lambda)^{k+1}}, \quad \Re(s)>|\lambda|^{1/\alpha}, \tag{3.118}$$

Further, the function, called M-Wright or Mainardi function, and defined as

$$M_{\beta}(z) = \sum_{n=0}^{\infty}\frac{(-1)^n z^n}{n!\Gamma(-\beta(n+1)+1)}, \quad 0<\beta<1,$$

is useful in description of Lévy distributions. In fact, for $M_{\beta}(t), t\geq 0$, and $0<\beta<1$ the following relations hold [Mai10]:

$$L[M_\beta](s) = E_\beta(-s), \quad L\left[\frac{\beta}{t^{\beta+1}} M_\beta\left(\frac{1}{t^\beta}\right)\right](s) = e^{-s^\beta},$$

$$L\left[\frac{\beta}{t^\beta} M_\beta\left(\frac{1}{t^\beta}\right)\right](s) = s^{\beta-1} e^{-s^\beta}.$$

We note also the following connection with the Gaussian density evolved in time (see (2.13) with $\kappa = 1$)

$$\frac{1}{2\sqrt{t}} M_{1/2}(x/\sqrt{t}) = G_t(x) = \frac{1}{2\sqrt{\pi t}} e^{-\frac{x^2}{4t}}, \quad t > 0,\ x \in \mathbb{R}.$$

6. *Distributed fractional order differential operators.* The idea of a distributed order fractional derivative was first appeared in the paper [Cap69] by Michele Caputo in 1969 in connection with modeling of wave propagation in viscoelastic media. In the papers [Cap95, Cap01] he applied DODE models to other processes arising in filtering, dielectric induction, and diffusion. Distributed order derivatives can be used to model complex processes with a simultaneous effect of different modes, and therefore, become an attractive tool for many researchers. For instance, Chechkin et al. [CGSG03, CSK11] used DODE for modeling of hereditary and ultra-slow diffusion, Podlubny [Pod99] for control and signaling systems, Kazemipour et al. [KAN10] for Klein-Gordon distributed order equation, Andries et al. [AUS06] for cell biology, etc. Mathematical foundations of distributed fractional order derivatives are studied in [BT00] by Bagley and Torvik, [UG05-2] by Umarov and Gorenflo, [MS06] by Meerschaert and Scheffler, and [Koc08] by Kochubei. Section 3.11 also contains new mathematical properties of DODOs. In the papers [SCK04, CSK11] fractional diffusion processes are modeled by two different forms of time-DODEs and two different forms of space-DODEs.

    a. The natural time-DODE form:

    $$\int_0^1 \tau^{\beta-1} w(\beta) D_*^\beta p(t,x) d\beta = K \frac{\partial^2 p(t,x)}{\partial t^2}, \quad t > 0,\ x \in \mathbb{R},$$

    where $w(\beta)$ is a nonnegative dimensionless function satisfying the condition $\int_0^1 w(\beta) d\beta = 1$, $\tau > 0$, and K is the diffusion coefficient;

    b. The modified time-DODE form:

    $$\frac{\partial p(t,x)}{\partial t} = \int_0^1 w(\beta) K(\beta) D_t^{1-\beta} \left[\frac{\partial^2 p(t,x)}{\partial t^2}\right] d\beta, \quad t > 0,\ x \in \mathbb{R},$$

    where $w(\beta)$ has the same properties as in the natural form, and $K(\beta) = \tau^{1-\beta}$;

    c. The natural space-DODE form:

    $$\frac{\partial p(t,x)}{\partial t} = K \int_0^2 \rho(\alpha) \ell^{\alpha-2} \mathbb{D}_0^\alpha \left[\frac{\partial^2 p(t,x)}{\partial t^2}\right] d\alpha, \quad t > 0,\ x \in \mathbb{R},$$

    where ℓ and K are positive constants, and $\rho(\alpha)$ is a nonnegative dimensionless function satisfying the condition $\int_0^2 \rho(\alpha) d\alpha = 1$; and

    d. The modified space-DODE form:

    $$\int_0^2 \rho(\alpha) \ell^{2-\alpha} \mathbb{D}_0^{2-\alpha} p(t,x) d\alpha = -K \frac{\partial^2 p(t,x)}{\partial t^2}, \quad t > 0,\ x \in \mathbb{R},$$

    where ℓ, K, and $\rho(\alpha)$ have the same meaning as in the natural space-DODE form. In all the four cases the initial condition $p(0,x) = \delta_0(x)$ is required.

7. *Variable fractional order differential operators.* Another mathematical apparatus, relatively young and being intensively developed last two decades, is the variable fractional order derivatives. Variable order derivatives can be effectively used for modeling of diffusion processes in a

heterogeneous media, processes with changing regimes in time, etc. A variable fractional order derivative was introduced and studied by S. G. Samko and his collaborators in 1993–95 [SR93, Sam95]. In the papers [LH02] Lorenzo and Hartley introduced several types of fractional variable order derivatives based on the Riemann-Liouville derivative and applied them to engineering problems. A. V. Chechkin at al. [CGS05] used variable fractional order derivatives to describe kinetic diffusion in heterogeneous media. Umarov et al. [US09] studied variable fractional order derivatives based on the Caputo-Djrbashian fractional derivatives. In this paper the memory effects provided by variable order differential operators are studied in detail. For numerical approximations of fractional variable order derivatives, see papers [SCWC11, VC11], and the references therein.

8. *Fractional order integrals.* In Section 3.2 we showed that the families of operators ${}_aJ_t^\alpha$ and ${}_tJ_b^\alpha$, where $\alpha \geq 0$, form semigroups of operators. These semigroups are strongly continuous on Banach spaces $L_p(a,b)$, $p \geq 1$, and $C[a,b]$. See the definition of strongly continuous semigroups in Section 7.3.
In Section 3.7 we established the equality of the ΨDOSS $A_\alpha(D)$ and the fractional Liouville-Weyl integral operator ${}_{-\infty}J_t^\alpha$ in $\Psi_{G,p}(\mathbb{R})$, $0 \notin G$. The equality of these operators remains valid in the scale of Sobolev spaces, as well, with some orthogonality conditions. To feel it better let us consider some examples. Let $1 \leq \alpha < 2$. Then $A_\alpha(D)f(x)$, $f \in H^s(\mathbb{R})$, is meaningful if f is orthogonal to 1, i.e., $< f,1 >= 0$, or, the same, $F[f](0) = 0$. It is easy to see that $A_\alpha(D)f \in H^{s+\alpha}(\mathbb{R})$. On the other hand, ${}_{-\infty}J_t^\alpha f \in H^{s+\alpha}(\mathbb{R})$ also implies $\int_{-\infty}^{\infty} f(x)dx = 0$, that is $F[f](0) = 0$. For example, the function

$$f(x) = \frac{sign(x)}{1+x^2},$$

belongs to $H^0(\mathbb{R}) = L_2(\mathbb{R})$ and satisfies the condition $< f,1 >= 0$. For $\alpha = 1$ we have

$${}_{-\infty}J_t^\alpha f(x) = \begin{cases} -\frac{\pi}{2} + \tan^{-1} x, & \text{if } x > 0 \\ -\frac{\pi}{2} - \tan^{-1} x, & \text{if } x < 0. \end{cases}$$

This function is continuous and has asymptotics $1/x$ as $x \to \pm\infty$. Hence, ${}_{-\infty}J^\alpha f(x) \in H^1(\mathbb{R})$. However, we note that, in general, for the operator ${}_{-\infty}J_t^\alpha$ to be meaningful, it is not necessary that $f \in H^s(\mathbb{R})$ was orthogonal to 1. But, now one cannot guarantee that ${}_{-\infty}J_t^\alpha f \in H^{s+\alpha}(\mathbb{R})$. An example, again for $\alpha = 1$, is the function $g(x) = (1+x^2)^{-1} \in H^0(\mathbb{R})$, but ${}_{-\infty}J^\alpha g(x) = \tan^{-1}(x) \notin H^1(\mathbb{R})$. Obviously the function g does not satisfy the condition $< g,1 >= 0$.

Chapter 4
Boundary value problems for pseudo-differential equations with singular symbols

4.1 Introduction

Let $\Omega \subset \mathbb{R}^n$ be a bounded domain with a smooth boundary or $\Omega = \mathbb{R}^n$. This chapter discusses well-posedness problems of general boundary value problems for pseudo-differential and differential-operator equations of the form

$$L[u] \equiv \frac{\partial^m u}{\partial t^m} + \sum_{k=0}^{m-1} A_k(t) \frac{\partial^k u}{\partial t^k} = f(t,x), \quad t \in (T_1, T_2),\ x \in \Omega, \tag{4.1}$$

$$B_k[u] \equiv \sum_{j=0}^{m-1} b_{kj} \frac{\partial^j u(t_{kj}, x)}{\partial t^j} = \varphi_k(x), \quad x \in \Omega,\ k = 0, \dots, m-1, \tag{4.2}$$

where $f(t,x)$ is defined on $(T_1, T_2) \times \Omega$, $-\infty < T_1 < T_2 \leq \infty$, and $\varphi_k(x)$, $x \in \Omega$, $k = 0, \dots, m-1$, are given functions; $A_k(t)$ and b_{kj}, $k = 0, \dots, m-1$, $j = 0, \dots, m-1$, are operators acting on some spaces (specified below) of functions defined on Ω; and $t_{jk} \in [T_1, T_2]$, $j,k = 0, \dots, m-1$. For example, when $\Omega = \mathbb{R}^n$, the latter operators may act as ΨDOSS defined on the space of distributions $\Psi'_{-G,p'}(\mathbb{R}^n)$ with an appropriate $G \subset R^n$.

Examples discussed in Section 2.2 showed that the solution operators of simplest boundary value problems can be interpreted as ΨDOSS. Moreover, the equation in Examples 2.2.0.4 and 2.2.0.5 (that is, equation (2.24)) is a pseudo-differential equation with a symbol singular in the dual variable.

The classes $OPXS_p(G)$ of ΨDOSS' introduced in Chapter 2) are convenient in the study of boundary value problems of the form (4.1)–(4.2). The role of the set G is to localize singularities of the coefficients, as well as the solution operators. This allows to construct algebras of ΨDOSS' complete in the sense that not only operators $A_k(t)$ and b_{kj}, but also solution operators of boundary value problem (4.1)–(4.2) belong to the same algebra. Moreover, this approach (being a variation of the operator method) works independently of the type of equation (4.1).

S. Umarov, *Introduction to Fractional and Pseudo-Differential Equations with Singular Symbols*, Developments in Mathematics 41,
DOI 10.1007/978-3-319-20771-1_4

In general, due to singularity of symbols of operators $A_k(t)$ and b_{kj}, problem (4.1)–(4.2) is not well posed in the sense of Hadamard in classical function spaces. However, as we have seen in Chapter 2, one can always find a space (possibly too narrow) in which the problem is well posed in the strong sense. By duality, this can be extended to the dual space of distributions, but now the solution is understood in the weak sense (the exact definitions are given in Section 4.2). Depending on the symbol of solution operators sometimes the well-posedness space can be extended up to Sobolev, or Besov and Lizorkin-Triebel spaces. In this chapter we obtain general conditions of well posedness in classical Sobolev, Besov, and Lizorkin-Triebel type spaces.

4.2 General boundary value problems for ΨDOSS: homogeneous case

Consider the following general boundary value problem for a pseudo-differential equation

$$L(t,\frac{\partial}{\partial t},D)u(t,x) = \frac{\partial^m u}{\partial t^m} + \sum_{k=0}^{m-1} A_k(t,D)\frac{\partial^k u}{\partial t^k} = f(t,x), \quad t \in (T_1,T_2),\ x \in \mathbb{R}^n, \tag{4.3}$$

$$B_k(D)[u] = \sum_{j=0}^{m-1} b_{kj}(D)\frac{\partial^j u(t_{kj},x)}{\partial t^j} = \varphi_k(x), \quad x \in \mathbb{R}^n,\ k = 0,\ldots,m-1, \tag{4.4}$$

where $m \geq 1$, $-\infty < T_1 < T_2 \leq \infty$, $t_{kj} \in [T_1,T_2]$, $D = (D_1,\ldots,D_n)$, $D_j = -i\frac{\partial}{\partial x_j}$; operators $A_k(t,D)$, $k = 0,\ldots,m-1$, and $b_{kj}(D)$, $k,j = 0,\ldots,m-1$, are ΨDOSS with respective symbols $A_k(t,\xi)$ and $b_{kj}(\xi)$, $k,j = 0,\ldots,m-1$; the functions $f(t,x)$ and $\varphi_k(x)$, $k = 0,\ldots,m-1$, are given functions in certain spaces that will be specified later. Problem (4.3)–(4.4) cover all the examples (Examples 2.2.0.1–2.2.0.6) discussed in Section 2.2.

Definition 4.1. Let $f \in C^0[T_1,T_2;\Psi_{G,p}(\mathbb{R}^n)]$, and $\varphi_k \in \Psi_{G,p}(\mathbb{R}^n)$, $k = 1,\ldots,m$. A function $u(t,x) \in C^m[T_1,T_2;\Psi_{G,p}(\mathbb{R}^n)]$ is called *a strong solution* to boundary value problem (4.3)–(4.4) if it satisfies both relations (4.3) and (4.4) pointwise.

A ψ-distribution valued function $u(t,x) \in C^m[T_1,T_2;\Psi'_{-G,p'}(\mathbb{R}^n)]$ is said to be *a weak solution* to boundary value problem (4.3)–(4.4) if the following relations hold for every $\phi \in \Psi_{G,p}(\mathbb{R}^n)$ and all $t \in (T_1,T_2)$:

$$< \frac{\partial^m u}{\partial t^m},\phi > + \sum_{k=0}^{m-1} < \frac{\partial^k u}{\partial t^k}, A_k(t,-D)\phi > = < f(t,x),\phi >,$$

$$< B_k(u),\phi > = \sum_{j=0}^{m-1} < \frac{\partial^j u(t_{kj},x)}{\partial t^j}, b_{kj}(-D)\phi) > = < \varphi_k(x),\phi(x) > .$$

If at least two points t_{kj} in the boundary conditions (4.4) are distinct, then (4.3)–(4.4) are nonlocal multi-point boundary value problems. If $t_{kj} \equiv t_0 \in [T_1, T_2]$ for all $k, j = 0, \ldots, m-1$, and operators $b_{kj} = \delta_{kj} I$, where δ_{kj} is the Kronecker's symbol and I is the identity operator, then (4.3)–(4.4) represent the Cauchy problem:

$$L(t, \frac{\partial}{\partial t}, D)u(t,x) = f(t,x), \quad t > t_0, x \in \mathbb{R}^n, \tag{4.5}$$

$$\frac{\partial^k u(t,x)}{\partial t^k}\Big|_{t=t_0} = \varphi_k(x), \quad x \in \mathbb{R}^n,\ k = 0, \ldots, m-1, \tag{4.6}$$

where $L(t, \frac{\partial}{\partial t}, D)$ is the operator defined in (4.3). Cauchy problem (4.5)–(4.6) for an inhomogeneous equation ($f(t,x) \neq 0$) can always be reduced to the Cauchy problem for the corresponding homogeneous equation

$$L(t, \frac{\partial}{\partial t}, D)u(t,x) = 0, \quad t > \tau, \tag{4.7}$$

$$\frac{\partial^k u(\tau,x)}{\partial t^k} = \psi_k(x), \quad k = 0, \ldots, m-1, \tag{4.8}$$

for some $\tau \geq t_0$, and with nonhomogeneous Cauchy conditions with the initial data ψ_k, which depend on functions (or functionals) φ_k and $f(t,x)$. Here the operator $L(t, \frac{\partial}{\partial t}, D)$ is the same as in equation (4.3), however it acts on functions (functionals) defined on the interval (τ, T_2). The reduction procedure was first found by Jean-Marie-Constant Duhamel in the 1830th, and therefore it is called *the Duhamel principle.* We will discuss the Duhamel principle in the general case later on and obtain its generalizations to various classes of boundary value problems, including fractional order pseudo-differential equations. However, here we introduce the notions of the *Duhamel integral* and *Duhamel principle* in a simple case. The classical Duhamel integral (see, e.g., [TS66, BJS64]) is used for representation of a solution of the Cauchy problem for a given inhomogeneous linear partial differential equation with homogeneous initial conditions via the solution of the Cauchy problem for the corresponding homogeneous equation. Consider the Cauchy problem for the second order inhomogeneous differential equation

$$\frac{\partial^2 u}{\partial t^2}(t,x) = A\,u(t,x) + f(t,x), \quad t > 0^1,\ x \in \mathbb{R}^n, \tag{4.9}$$

with homogeneous initial conditions

$$u(0,x) = 0, \quad \frac{\partial u}{\partial t}(0,x) = 0, \tag{4.10}$$

[1] For simplicity here it is assumed that $t_0 = 0$.

where A is a linear differential operator containing the temporal derivatives of order, not higher then 1. Further, let a sufficiently smooth function $v(t,\tau,x)$, $t \geq \tau$, $\tau \geq 0$, $x \in \mathbb{R}^n$, be a solution of the homogeneous equation

$$\frac{\partial^2 v}{\partial t^2}(t,\tau,x) = Av(t,\tau,x),\ t > \tau,$$

satisfying the following conditions:

$$v(t,\tau,x)|_{t=\tau} = 0, \quad \frac{\partial v}{\partial t}(t,\tau,x)|_{t=\tau} = f(\tau,x).$$

Then the solution of Cauchy problem (4.5)–(4.8) is given by means of the Duhamel integral

$$u(t,x) = \int_0^t v(t,\tau,x)d\tau.$$

The formulated statement is called the "Duhamel principle" (see for details, e.g., [BJS64] or [TS66]).

An analogous construction is possible in the case of the Cauchy problem with a homogeneous initial condition for the first order inhomogeneous partial differential equation

$$\frac{\partial u}{\partial t}(t,x) = Bu(t,x) + f(t,x), \quad t > 0,\ x \in \mathbb{R}^n,$$

where B is a linear differential operator containing only spatial derivatives.

Now we prove the Duhamel principle, for simplicity, in the case of second order (in the sense of time-derivatives) pseudo-differential equations with singular symbols. Namely, consider the Cauchy problem

$$\frac{\partial^2 u}{\partial t^2} + A_1(t,x,D_x)\frac{\partial u}{\partial t} + A_0(t,x,D_x)u = h(t,x), \quad t > 0,\ x \in \mathbb{R}^n,$$

$$u(0,x) = \varphi_0(x) \quad \text{and} \quad \frac{\partial u}{\partial t}(0,x) = \varphi_1(x), \quad x \in \mathbb{R}^n.$$

We assume that symbols $a_j(t,x,\xi)$, $j = 0,1$, of the pseudo-differential operators $A_j(t,x,D_x)$, $j = 0,1$, belong to S_G^∞ for some $G \subset \mathbb{R}^n$ for every fixed $t > 0$ and continuous in the variable $t > 0$. In this case the Duhamel principle is formulated as follows.

Proposition 4.1. *Let a twice differentiable in the variable t function $U(t,\tau,x) \in \Psi_{p,G}$, $t > \tau \geq 0$, be a solution of the Cauchy problem for a homogeneous equation*

$$\frac{\partial^2 U}{\partial t^2} + A_1(t,x,D_x)\frac{\partial U}{\partial t} + A_0(t,x,D_x)U = 0, \quad 0 < \tau < t, x \in \mathbb{R}^n, \tag{4.11}$$

$$U(t,\tau,x)|_{t=\tau} = 0, \quad x \in \mathbb{R}^n, \tag{4.12}$$

$$\frac{\partial U}{\partial t}(t,\tau,x)|_{t=\tau} = f(\tau,x), \quad x \in \mathbb{R}^n, \tag{4.13}$$

in the domain $D = \{t > \tau, x \in \mathbb{R}^n\}$*, where* $\tau \geq 0$*. Then the function*

$$u(t,x) = \int_0^t U(t,\tau,x)d\tau \tag{4.14}$$

is a solution of the Cauchy problem

$$\frac{\partial^2 u}{\partial t^2} + A_1(t,x,D_x)\frac{\partial u}{\partial t} + A_0(t,x,D_x)u = h(t,x), \quad t > 0,\ x \in \mathbb{R}^n, \tag{4.15}$$

$$u(0,x) = 0 \quad \text{and} \quad \frac{\partial u}{\partial t}(0,x) = 0, \quad x \in \mathbb{R}^n. \tag{4.16}$$

Proof. Obviously $u(0,x) = 0$. Further, for the first order derivative, one has

$$\begin{aligned}\frac{\partial u}{\partial t}(t,x) &= U(t,t,x) + \int_0^t \frac{\partial}{\partial t}U(t,\tau,x)d\tau \\ &= \int_0^t \frac{\partial}{\partial t}U(t,\tau,x)d\tau,\end{aligned}$$

since $U(t,t,x) = 0$ for any $t > 0$ due to condition (4.12). It follows that $\frac{\partial u}{\partial t}(0,x) = 0$. Therefore, the function $u(t,x)$ in (4.14) satisfies the initial conditions (4.16). Further, for the second derivative, using condition (4.13), one obtains

$$\begin{aligned}\frac{\partial^2}{\partial t^2}u(t,x) &= \frac{\partial}{\partial t}U(t,t,x) + \int_0^t \frac{\partial^2}{\partial t^2}U(t,\tau,x)d\tau \\ &= h(t,x) + \int_0^t \frac{\partial^2}{\partial t^2}U(t,\tau,x)d\tau.\end{aligned}$$

Moreover,

$$\begin{aligned}\frac{\partial^2 u}{\partial t^2} + A_1(t,x,D_x)\frac{\partial u}{\partial t} + A_0(t,x,D_x)u &= h(t,x) + \int_0^t \frac{\partial^2}{\partial t^2}U(t,\tau,x)d\tau \\ &+ A_1(t,x,D_x)\int_0^t \frac{\partial}{\partial t}U(t,\tau,x)d\tau + A_0(t,x,D_x)\int_0^t U(t,\tau,x)d\tau \\ = h(t,x) + \int_0^t \Big[\frac{\partial^2 U}{\partial t^2} &+ A_1(t,x,D_x)\frac{\partial U}{\partial t} + A_0(t,x,D_x)U\Big]d\tau = h(t,x),\end{aligned}$$

since $U(t,\tau,x)$ is a solution to equation (4.11). Hence, $u(t,x)$ in (4.14) satisfies the equation (4.15) as well.

Remark 4.1. The Duhamel principle for abstract differential-operator equations of arbitrary order m will be proved in Section 4.7, and for fractional order differential equations in Sections 5.5 and 6.4.

Unfortunately, the Duhamel principle is not valid for multi-point problems. It is convenient to split the general boundary value problems (4.3)–(4.4) into two partial problems. Namely, the boundary value problem with the homogeneous equation and inhomogeneous boundary conditions

$$L(t,\frac{\partial}{\partial t},D)u(t,x)\frac{\partial^k u}{\partial t^k}=0, \quad t\in(T_1,T_2),\, x\in\mathbb{R}^n, \tag{4.17}$$

$$B_k(D)[u]=\varphi_k(x), \quad x\in\mathbb{R}^n,\, k=0,\dots,m-1, \tag{4.18}$$

and the boundary value problem with the corresponding inhomogeneous equation and homogeneous boundary conditions

$$L(t,\frac{\partial}{\partial t},D)u(t,x)\frac{\partial^k u}{\partial t^k}=f(t,x), \quad t\in(T_1,T_2),\, x\in\mathbb{R}^n, \tag{4.19}$$

$$B_k(D)[u]=0, \quad x\in\mathbb{R}^n,\, k=0,\dots,m-1. \tag{4.20}$$

In this section we will study boundary value problem (4.17)–(4.18). The boundary value problem (4.19)–(4.20) will be studied in Section 4.3. Since the Duhamel principle does not work for general multi-point value problems, we will use the Green's function or fundamental solution approaches to solve boundary value problem (4.19)–(4.20).

Consider in the interval (T_1,T_2) the following ordinary differential equation, called a *characteristic equation* of pseudo-differential equation (4.17):

$$u^{(m)}(t)+\sum_{k=0}^{m-1}A_k(t,\xi)u^{(k)}(t)=0,\ t\in(T_1,T_2), \tag{4.21}$$

which depends on the parameter $\xi\in G$. Assume that the symbols $A_k(t,\xi)$ are continuous in the variable t for each fixed value of the parameter ξ. Then there exist m linearly independent solutions $u_k(t,\xi), k=1,\dots,m$, of equation (4.21) which are defined on the interval (T_1,T_2) and m-times differentiable in this interval, that is $u_k(t,\xi)\in C^{(m)}(T_1,T_2)$ for each fixed value of ξ. The set $\{u_k(t,\xi)\}_{k=1}^m$ is called a fundamental system of solutions of the characteristic equation (4.21). It is clear that a fundamental system of solutions is not defined uniquely. Depending on our purposes we will construct different fundamental systems of solutions.

Let $u_j(t,\xi)$, where $j\in\{0,\dots,m-1\}$, be a solution to differential equation (4.21) satisfying the conditions

$$u_j^{(k)}(t_0,\xi)=\delta_{jk}, \quad k=0,\dots,m-1. \tag{4.22}$$

It is known [Nai67] that $u_j(t,\xi)\in C^m(T_1,T_2)$ exists and unique. Moreover, the set of solutions $u_0(t,\xi),\dots,u_{m-1}(t,\xi)$ are linearly independent, and hence, form *a fundamental system of solutions.*

Theorem 4.1. *Let the symbols $A_k(t,\xi)$, $k=0,\dots,m-1$, and $b_{kj}(\xi)$, $k,j=0,\dots,m-1$, of the operators $A_k(t,D)$ and $b_{k,j}(D)$ in equation (4.17) belong to S_G^∞ for some open set $G\subset\mathbb{R}^n$. Then there is a subset G_0 of G, such that for all $\varphi_k\in\Psi_{G_0,p}(\mathbb{R}^n)$, $1<p<\infty$, there exists a unique strong solution $u(t,x)$ to the* *multi-point value problem* (4.17)–(4.18) *in* $C^{(m)}[(T_1,T_2),\Psi_{G_0,p}(\mathbb{R}^n)]$. *Moreover, the solution has the representation*

$$u(t,x)=\sum_{k=0}^{m-1} S_k(t,D)\varphi_k(x),$$

where solution operators $S_k(t,D), k=0,\dots,m-1$, *have symbols* $s_k(t,\xi)\in S^{\infty}_{G_0}, k=0,\dots,m-1$, *which form a fundamental system of solutions to* (4.21).

Proof. Consider the boundary value problem for ordinary differential equation with a parameter ξ :

$$L(t,\frac{d}{dt},\xi)u(t,\xi)=\frac{d^m u}{dt^m}+\sum_{k=0}^{m-1} A_k(t,\xi)\frac{d^k u}{dt^k}=0, \quad t\in(T_1,T_2), \tag{4.23}$$

$$B_k(\xi)[u]=\sum_{j=0}^{m-1} b_{kj}(\xi)\frac{d^j u(t_{kj},\xi)}{dt^j}=F[\varphi_k](\xi), \quad k=0,\dots,m-1, \tag{4.24}$$

where $F[\varphi_k](\xi)$ is the Fourier transform of $\varphi_k(x)$. We look for a solution of problem (4.23)–(4.24) in the form

$$u^*(t,\xi)=\sum_{k=0}^{m-1} f_k(\xi)u_k(t,\xi), \tag{4.25}$$

where the set $u_k(t,\xi)$, $k=0,\dots,m-1$, is a fundamental system of solutions of (4.21) satisfying Cauchy conditions (4.22). It is clear that $u^*(t,\xi)\in C^m(T_1,T_2)$ and satisfies (4.23). Substituting it into (4.24) we get a system of linear algebraic equations

$$M(\xi)F(\xi)=\hat{\Phi}(\xi). \tag{4.26}$$

Here $M(\xi)$ is the square matrix of order m with entries

$$m_{kl}=\sum_{j=0}^{m-1} B_{kj}(\xi)u_l^{(j)}(t_{kj},\xi), \quad k,l=0,\dots,m-1, \tag{4.27}$$

$F(\xi)=(f_0(\xi),\dots,f_{m-1}(\xi))$ is an unknown vector, and $\hat{\Phi}(\xi)$ is the transpose of the vector $\Big(F[\varphi_0](\xi),\dots,F[\varphi_{m-1}](\xi)\Big)$. Denote by M_0 the set of points $\xi\in G$ at which the determinant of $M(\xi)$ vanishes, that is Det$M(\xi)=0$. If $\xi\notin M_0$, then equation (4.26) has a unique solution

$$F(\xi)=M^{-1}(\xi)\hat{\Phi}(\xi). \tag{4.28}$$

We note that M_0 is the singular set for the symbols of the solution operators. Substituting the representation (4.28) of the vector $F(\xi)=(f_0(\xi),\dots,f_{m-1}(\xi))$ into (4.25) and applying the inverse Fourier transform one obtains the solution of the general multi-point value problem (4.17)–(4.18) in the form

$$u(t,x) = \sum_{k=0}^{m-1} S_k(t,D)\varphi_k(x), \tag{4.29}$$

where the $S_k(t,D)$, $k = 0,\dots,m-1$, are solution pseudo-differential operators of (4.17)–(4.18) with the symbols

$$s_k(t,\xi) = u_k^*(t,\xi) = \Big((M^*)^{-1}(\xi)u(t,\xi)\Big)_k, \quad k = 0,\dots,m-1, \tag{4.30}$$

where the symbol $(\cdot)_k$ means k-th component of a vector $(\cdot)$, $(M^*)^{-1}(\xi)$ is the matrix inverse to the Hermitian conjugate of $M(\xi)$. Indeed, the symbols $s_k(t,\xi) \in C^m(T_1,T_2), k = 0,\dots,m-1$, for all $\xi \in G_0 \subset G \setminus M_0$. This together with Theorem 2.1 implies $u(t,x) \in C^{(m)}[(T_1,T_2),\Psi_{G_0,p}(\mathbb{R}^n)]$. The system $\{s_0(t,\xi),\dots,s_{m-1}(t,\xi)\}$ is linearly independent, otherwise the system $\{u_0(t,\xi),\dots,u_{m-1}(t,\xi)\}$ would not be linearly independent due to the fact that the matrix $M(\xi)$ has nonzero determinant if $\xi \in G_0$. The proof is complete.

Remark 4.2. 1. The vector-function $S(t,\xi) = \Big(s_0(t,\xi),\dots,s_{m-1}(t,\xi)\Big)$, components of which are symbols of solution operators $S_k(t,D)$, $k = 1,\dots,m$, depends on operators $A_k(t,D)$, $k = 0,\dots,m-1$, and $B_{kj}(D)$, $k = 0,\dots,m-1$, $j = 0,\dots,m-1$, given in (4.17), (4.18), respectively. Its behavior may be of different nature: it may have singularities in variables $(\xi_1,\dots,\xi_n)$ in G if $M_0 \cap G \neq \emptyset$, or may increase or decrease when $|\xi| \to \infty$. The well posedness of boundary value problem (4.17)–(4.18) in the classical Sobolev, Besov, and other spaces essentially depends on the behavior of the vector function $S(t,\xi)$; see Section 4.4.
2. It is useful to have a maximal G_0 in this theorem, which is actually $G \setminus M_0$. In accordance with Theorem 1.21, $\Psi_{G_0,p}(\mathbb{R}^n)$ is dense in classical spaces if G_0 is a dense subset of $\mathbb{R}^n$.

Theorem 4.2. *Let $f = 0$ and let the symbols $A_k(t,\xi)$, $k = 0,\dots,m-1$, and $b_{kj}(\xi)$, $k,j = 0,\dots,m-1$, of the operators $A_k(t,D)$ and $b_{k,j}(D)$ in equations* (4.3)–(4.4) *belong to S_G^∞ for some open set $G \subset \mathbb{R}^n$. Then there is a subset G_0 of G, such that for all $\varphi_k \in \Psi'_{-G_0,p'}(\mathbb{R}^n)$, $1 < p < \infty$, there exists a unique weak solution $w(t,x)$ to the multi-point value problem* (4.17)–(4.18) *in $C^{(m)}[(T_1,T_2),\Psi'_{-G_0,p'}(\mathbb{R}^n)]$. Moreover, the solution has a representation*

$$W(t,x) = \sum_{k=0}^{m-1} S_k^w(t,-D)\varphi_k(x), \tag{4.31}$$

where solution operators $S_k^w(t,-D)$, $k = 0,\dots,m-1$, have symbols $s_k(t,\xi)$, $k = 0,\dots,m-1$, which form a fundamental system of solutions to (4.21).

Proof. Let $\varphi_k \in \Psi'_{-G_0,p'}(\mathbb{R}^n)$, $k = 0,\dots,m-1$, and ϕ be an arbitrary function in $\Psi_{G,p}(\mathbb{R}^n)$. We only need to show that $w(t,x)$ defined in (4.31) is a weak solution of boundary value problem (4.17)–(4.18) with weak extensions "$A_k^w(t,-D)$" and "$b_{kj}^w(-D)$" instead of operators "$A_k(t,D)$" and "$b_{kj}(D)$." Let

$$W(t,x) = W_0(t,x) + \dots + W_{m-1}(t,x), \tag{4.32}$$

where $W_j(t,x) = S_j^w(t,-D)\varphi_j(x)$, that is the j-th term in representation (4.31). We have

$$
\begin{aligned}
&< L^w(t,\frac{\partial}{\partial t},-D)W_j(t,x), \phi(x) > \\
&=< \frac{\partial^m W_j}{\partial t^m} + \sum_{k=0}^{m-1} A_k^w(t,-D)\frac{\partial^k W_j}{\partial t^k}, \phi(x) > \\
&=< \frac{\partial^m S_j^w(t,-D)\varphi_j(x)}{\partial t^m} + \sum_{k=0}^{m-1} A_k^w(t,-D)\frac{\partial^k S_j^w(t,-D)\varphi_j(x)}{\partial t^k}, \phi(x) > \\
&=< \varphi_j(x), \left[\frac{\partial^m S_j(t,D)}{\partial t^m} + \sum_{k=0}^{m-1} A_k(t,D)\frac{\partial^k S_j(t,D)}{\partial t^k}\right]\phi(x) > \\
&=< \varphi_j(x), F^{-1}\left[\left(\frac{\partial^m s_j(t,\xi)}{\partial t^m} + \sum_{k=0}^{m-1} A_k(t,\xi)\frac{\partial^k s_j(t,\xi)}{\partial t^k}\right)F[\phi](\xi)\right] > \\
&= 0, \quad \forall t \in (T_1,T_2),
\end{aligned}
$$

since by construction $s_j(t,\xi)$ satisfies equation (4.23). Similarly,

$$
\begin{aligned}
&< B_k^w(-D)[W_j], \phi(x) > =< \sum_{\ell=0}^{m-1} b_{k\ell}^w(-D)\frac{\partial^\ell S_j(t_{k\ell},D)\varphi_j}{\partial t^\ell}, \phi(x) > \\
&=< \varphi_j(x), \sum_{\ell=0}^{m-1} b_{k\ell}(D)\frac{\partial^\ell S_j(t_{k\ell},D)}{\partial t^\ell}\phi(x) > \\
&=< \varphi_j(x), F^{-1}\left[\left(\sum_{\ell=0}^{m-1} b_{k\ell}(\xi)\frac{\partial^\ell s_j(t_{k\ell},\xi)}{\partial t^\ell}\right)F[\phi](\xi)\right] > \\
&=< \varphi_k(x), \delta_{k,j}\phi(x) > =< \delta_{k,j}\varphi_j(x), \phi(x) >, \quad k,j = 0,\ldots,m-1. \qquad (4.33)
\end{aligned}
$$

Now summing up (4.33) by index $j = 0,\ldots,m-1$, and taking into account (4.32), we have $B_k^w(-D)[W] = \varphi_k(x)$ in the sense of $\Psi'_{-G_0,p'}(\mathbb{R}^n)$. Thus, $W(t,x)$ given by in (4.31) satisfies boundary value problem (4.17)–(4.18) in the weak sense.

4.3 General boundary value problems for ΨDOSS: inhomogeneous case

Theorems 4.1 and 4.2 establish well posedness of boundary value problems for homogeneous equations with nonhomogeneous boundary conditions in the spaces $\Psi_{G_0,p}(\mathbb{R}^n)$, and $\Psi'_{-G_0,p'}(\mathbb{R}^n)$, where $G_0 \subseteq G \setminus \overset{\circ}{M}$. Here $\overset{\circ}{M}$ defined as

$$\overset{\circ}{M} = \{\xi \in \mathbb{R}^n : det M(\xi) = 0\}, \qquad (4.34)$$

where $M(\xi)$ is the matrix with entries defined in equation (4.27).

Now consider an inhomogeneous equation with homogeneous boundary conditions

$$L(t,\frac{\partial}{\partial t},D)u = h(t,x), \quad t \in (T_1,T_2), x \in \mathbb{R}^n, \tag{4.35}$$

$$B_k(D)u = 0, \quad x \in \mathbb{R}^n, \; k = 0,\ldots,m-1, \tag{4.36}$$

where operators $L(t,\frac{\partial}{\partial t},D)$ and $B(D)$ are defined in equations (4.3) and (4.4), respectively. As was noted above the Duhamel principle is not applicable in the case of multi-point boundary value problems. Below we show the well posedness of problem (4.35)–(4.36) in the spaces $\Psi_{G_0,p}(\mathbb{R}^n)$ and $\Psi'_{-G_0,p'}(\mathbb{R}^n)$.

Applying the Fourier transform in x in equations (4.35) and (4.36) one obtains the following multi-point boundary value problem for an ordinary differential equation with the parameter $\xi \in \mathbb{R}^n$:

$$L(t,\frac{d}{dt},\xi)v = \hat{h}(t,\xi), \quad t \in (T_1,T_2), \tag{4.37}$$

$$B_k(\xi)v = 0, \quad k = 0,\ldots,m-1, \tag{4.38}$$

where $v(t,\xi) = F[u](t,\xi)$ and $\hat{h}(t,\xi) = F[h](t,\xi)$.

One can find a solution to problem (4.35)–(4.36) using the fundamental solution of the operator

$$L(t,\frac{d}{dt},\xi) = \frac{d^m}{dt^m} + \sum_{k=0}^{m-1} A_k(t,\xi)\frac{d^k}{dt^k}$$

in a suitable space of distributions. Since symbols $A_k(t,\xi)$ by assumption are continuous functions in the variable t for each fixed $\xi \in G \subseteq \mathbb{R}^n$, this operator has a fundamental solution (see, e.g., [Vla79]) $\mathscr{E}(t,s,\xi)$ satisfying the differential equation

$$L(t,\frac{d}{dt},\xi)\mathscr{E}(t,s,\xi) = \delta_s(t), \quad t \in (T_1,T_2), \tag{4.39}$$

for arbitrary $s \in (T_1,T_2)$. From the general theory it follows also that the function $\mathscr{E}(t,s,\xi)$ is m times differentiable on each of the intervals $T_1 < t < s$ and $s < t < T_2$. The function

$$e(t,\xi) = \int_{T_1}^{T_2} \mathscr{E}(t,s,\xi)\hat{h}(s,\xi)ds$$

obviously solves equation (4.37). Let $E(t,x)$ be the inverse Fourier transform of $e(t,\xi)$ in the sense of distributions, that is $E(t,x) = F^{-1}[e](t,x)$. Then, one can easily verify that

$$E(t,x) = \int_{T_1}^{T_2} \mathscr{E}(t,s,D)h(s,x)ds, \tag{4.40}$$

and by construction it satisfies the equation

$$L(t,\frac{\partial}{\partial t},D)E(t,x) = h(t,x).$$

Indeed,

$$L(t,\frac{\partial}{\partial t},D)E(t,x) = \int_{T_1}^{T_2} L(t,\frac{\partial}{\partial t},D)\mathscr{E}(t,s,D)h(s,x)ds$$
$$= \int_{T_1}^{T_2} \delta_t(s)h(s,x)ds = h(t,x).$$

In fact, we would like to have a fundamental solution which would satisfy not only equation (4.39), but also boundary conditions (4.38). Therefore, we fix $\mathscr{E}(t,s,\xi)$ adding an additional term, namely

$$\mathscr{E}_0(t,s,\xi) = \mathscr{E}(t,s,\xi) + W_0(t,s,\xi), \tag{4.41}$$

where $W_0(t,s,\xi)$ is a solution to the boundary value problem

$$L(t,\frac{d}{dt},\xi)w = 0, t \in (T_1,T_2), \quad x \in \mathbb{R}^n, \tag{4.42}$$
$$B_k(\xi)w = \phi_k(s,\xi), \quad x \in G_0,\ k = 0,\dots,m-1, \tag{4.43}$$

with

$$\phi_k(s,\xi) = -B_k(\xi)[\mathscr{E}] = -\sum_{j=0}^{m-1} b_{kj}(\xi)\frac{d^j\mathscr{E}}{dt^j}(t,s,\xi)\Big|_{t=t_{kj}}. \tag{4.44}$$

Since $\xi \notin \overset{\circ}{M}$, boundary value problem (4.42)–(4.43) has a unique solution $W_0(t,s,\xi) \in C^{(m)}(T_1,T_2)$, which represents the second term in (4.41). Moreover, the desired fundamental solution $\mathscr{E}_0(t,s,\xi)$ has the representation

$$\mathscr{E}_0(t,s,\xi) = \mathscr{E}(t,s,\xi) - \sum_{k=0}^{m-1} s_k(t,\xi)\phi_k(s,\xi), \tag{4.45}$$

where $s_k(t,\xi)$, $k = 0,\dots,m-1$, are the symbols of solution operators to problem (4.42)–(4.43). Now one can readily verify that

$$u(t,x) = \int_{T_1}^{T_2} \mathscr{E}_0(t,s,D)h(t,x)ds \tag{4.46}$$

satisfies both equation (4.35) and boundary conditions (4.36). The latter can be represented in the form

$$u(t,x) = E(t,x) + w(t,x,) \tag{4.47}$$

where $E(t,x)$ is defined in (4.40) and

$$w(t,x) = \int_{T_1}^{T_2} W(t,s,D)h(s,x)ds.$$

Here $W(t,s,D)$ is the pseudo-differential operator with the symbol $W_0(t,s,\xi)$.

Another approach to the solution of problem (4.35)–(4.36) is based on the Green's function of the operator $L(t, \frac{d}{dt}, \xi)$ with boundary conditions (4.38) for suitable values of parameter ξ.

By definition, the Green's function of the operator $L(t, \frac{d}{dt}, \xi)$ with boundary conditions (4.38) is a function $G(t,s,\xi)$ defined on $(T_1,T_2)\times(T_1,T_2)\times G$, and satisfying the following conditions:

1. all the derivatives in the variable t up to order $m-2$ are continuous;
2. m times differentiable on intervals $T_1 < t < s$ and $s < t < T_2$;
3. $(m-1)$-st derivative satisfies the jump condition at $t = s$:

$$G_t^{(m-1)}(t,s,\xi)\Big|_{t=s+0} - G_t^{(m-1)}(t,s,\xi)\Big|_{t=s-0} = 1;$$

4. satisfies the equation $L(t, \frac{d}{dt}, \xi)G(t,s,\xi) = 0$ on intervals $T_1 < t < s$ and $s < t < T_2$, and boundary conditions $B_k(D)[G] = 0$.

Let $\{u_0(t,\xi),\ldots,u_{m-1}(t,\xi)\}$ be a fundamental system of solutions of the operator $L(t, \frac{d}{dt}, \xi)$. Then one can look for $G(t,s,\xi)$ in the form

$$G(t,s,\xi) = \begin{cases} G_-(t,s,\xi) & \text{if } T_1 < t < s, \\ G_+(t,s,\xi) & \text{if } s < t < T_2. \end{cases}$$

where

$$G_-(t,s,\xi) = \sum_{j=0}^{m-1} c_j(\xi)u_j(t,\xi), \quad t < s,\ \xi \notin \overset{\circ}{M},$$

$$G_+(t,s,\xi) = \sum_{j=0}^{m-1} d_j(\xi)u_j(t,\xi), \quad t > s,\ \xi \notin \overset{\circ}{M}.$$

Here $c_j(\xi)$, $d_j(\xi)$, $j = 0,\ldots,m-1$, are $2m$ unknown coefficients to be found. Due to the definition of $G(t,s,\xi)$ for these unknowns we have the following $2m$ relations

$$\begin{cases} B_k(D)[G(t,s,\xi)] = 0, \quad k = 0,\ldots,m-1, \\ G_+^j(s+,s,\xi) - G_-^j(s-,s,\xi) = 0 \quad j = 0,\ldots,m-2, \\ G_+^{m-1}(s+,s,\xi) - G_-^{m-1}(s-,s,\xi) = 1. \end{cases}$$

This system of equations has a unique solution if $\xi \notin \overset{\circ}{M}$, thus identifying the Green's function. Once $G(t,s,\xi)$ is found, one can find the solution to problem (4.35)–(4.36) using the formula

$$u(t,x) = \int_{T_1}^{T_2} G(t,s,D)h(s,x)ds.$$

Theorem 4.3. *Let the symbols* $A_k(t,\xi)$, $k = 0,\ldots,m-1$, *and* $b_{kj}(\xi)$, $k,j = 0,\ldots,m-1$, *of the operators* $A_k(t,D)$ *and* $b_{k,j}(D)$ *in boundary value problem* (4.3)–

(4.4) belong to S_G^∞ for some open set $G \subset \mathbb{R}^n$. Then for any $f \in C[(T_1,T_2),\Psi_{G_0,p}(\mathbb{R}^n)]$ and $\varphi_k \in \Psi_{G_0,p}(\mathbb{R}^n)$, $k=0,\dots,m-1$, where $1<p<\infty$ and $G_0 \subseteq G \setminus \overset{\circ}{M}$, there exists a unique strong solution $u(t,x) \in C^{(m)}[(T_1,T_2),\Psi_{G_0,p}(\mathbb{R}^n)]$ of problem (4.3)–(4.4). Moreover, the solution has the representation

$$u(t,x) = \sum_{k=0}^{m-1} S_k(t,D)\varphi_k(x) + \int_{T_1}^{T_2} \mathscr{E}_0(t,s,D)f(s,x)ds, \tag{4.48}$$

where $\mathscr{E}_0(t,s,\xi)$ is the fundamental solution of the operator $L(t,\frac{d}{dt},\xi)$ defined in (4.45) and the solution operators $S_k(t,D)$, $k=1,\dots,m$, have symbols $s_k(t,\xi) \in S_{G_0}^\infty$, $k=1,\dots,m$, which form a fundamental system of solutions to equation (4.21).

Proof. Recall that general boundary value problem (4.3)–(4.4) was split into two problems: problem (4.17)–(4.18) with homogeneous equation and nonhomogeneous boundary conditions, and problem (4.35)–(4.36) with nonhomogeneous equation and homogeneous boundary conditions. Taking into account Theorem 4.1, which summarizes problem (4.17)–(4.18), and equation (4.46), which resumes problem (4.35)–(4.36), one obtains representation (4.48).

By duality it follows from Theorem 4.3 the following theorem.

Theorem 4.4. *Let the symbols $A_k(t,\xi)$, $k=0,\dots,m-1$, and $b_{kj}(\xi)$, $k,j=0,\dots,m-1$, of the operators $A_k(t,D)$ and $b_{k,j}(D)$ in problem (4.3)–(4.4) belong to S_G^∞ for some open set $G \subset \mathbb{R}^n$. Then for all $f \in C[(T_1,T_2),\Psi'_{-G_0,p'}(\mathbb{R}^n)]$, and $\varphi_k \in \Psi'_{-G_0,p'}(\mathbb{R}^n)$, $1<p<\infty$, there exists a unique weak solution $u^w(t,x)$ in the space $C^{(m)}[(T_1,T_2),\Psi'_{-G_0,p'}(\mathbb{R}^n)]$. Moreover, the solution has a representation*

$$w(t,x) = \sum_{k=0}^{m-1} S_k^w(t,-D)\varphi_k(x) + \int_{T_1}^{T_2} \mathscr{E}_0^w(t,s,-D)f(s,x)dx,$$

where symbols $s_k(t,\xi)$, $k=1,\dots,m$, of solution operators, and $\mathscr{E}_0(t,\xi)$ is defined as in Theorem 4.3.

Remark 4.3. In Theorems 4.3 and 4.4 one can replace operators $\mathscr{E}_0(t,s,D)$, $\mathscr{E}_0^w(t,s,-D)$ by $G(t,s,D)$, $G^w(t,s,-D)$ with the symbol being the Green's function $G(t,s,\xi)$, respectively.

4.4 Well posedness of general boundary value problems in Besov and Lizorkin-Triebel spaces

This section discusses well posedness in the sense of Hadamard of boundary value problem (4.3)–(4.4) in the classical function spaces, including Sobolev, Besov, and Lizorkin-Triebel spaces. Let X denote one of these spaces. The approach we want to use to establish well posedness in X is based on the possible closability of solution operators $S_k(t,D), k=1,\dots,m,$ defined on the space $\Psi_{G,p}(\mathbb{R}^n)$ up to bounded operators acting on the scales of X. This strategy requires a verification of two conditions:

1. the denseness of $\Psi_{G,p}(\mathbb{R}^n)$ in X; and
2. the closability of solution operators $S_k(t,D), k=1,\dots,m,$ up to bounded operators acting on the scales of the space X.

The denseness of $\Psi_{G,p}(\mathbb{R}^n)$ in the Besov and Lizorkin-Triebel spaces was studied in Section 1.11 of Chapter 1; see Theorem 1.21. The existence of a closed extension in the Sobolev, Besov, and Lizorkin-Triebel spaces of pseudo-differential operators defined on $\Psi_{G,p}(\mathbb{R}^n)$ was studied in Section 2.3 of Chapter 2; Theorems 2.7–2.9. Thus, we are prepared to investigate conditions for the well posedness of general multi-point boundary value problem (4.3)–(4.4) in the classical function spaces. We start from the definition of well-posed problems in the Besov spaces.

Definition 4.2. Let $\overline{\ell}=(\ell_0,\dots,\ell_m)\in\mathbb{R}^{m+1}$, $\overline{s}=(s_0,s_1,\dots,s_m)\in\mathbb{R}^{m+1}$, and $1<p,q<\infty$. The problem (4.3)–(4.4) is said to be $(\overline{\ell},\overline{s})$ - well posed in the scale of Besov spaces if for every $\varphi_{j-1}(x)\in B^{s_j}_{pq}(\mathbb{R}^n)$, $j=1,\dots,m,$ and for every $h\in C^0[T_1,T_2;B^{s_0}_{pq}(\mathbb{R}^n)]$ there exists a functional $u(t,x)\in C^{(m)}[(T_1,T_2);\Psi'_{-G,p'}(\mathbb{R}^n)]$ for some $G\subseteq\mathbb{R}^n$, such that

(i) $u_t^{(k)}(t,x)\in C^{(m-k)}[(T_1,T_2);B^{l_k}_{pq}(\mathbb{R}^n)]$, $k=0,\dots,m$;
(ii) $u(t,x)$ satisfies the considering problem in the weak sense;
(iii) the estimate

$$\sup_{t\in(T_1,T_2)}\sum_{k=0}^{m}\|u_t^{(k)}|B^{l_k}_{pq}\|\le C\sum_{j=1}^{m}\|\varphi_{j-1}|B^{s_j}_{pq}\|+\sup_{t\in(T_1,T_2)}\|h(t,x)|B^{s_0}_{pq}\|, \tag{4.49}$$

holds with $C>0$ independent on $\varphi_j(x)$, $j=0,\dots,m-1$, and $h(t,x)$. Recall that here $\|\cdot|B^{\ell}_{pq}\|$ is a norm of the Besov space $B^{\ell}_{pq}(\mathbb{R}^n)$.

Similarly one can define a $(\overline{\ell},\overline{s})$ - well-posed problem in the scale of Lizorkin-Triebel spaces $F^s_{pq}(\mathbb{R}^n)$. For shortness, we will say that boundary value problem (4.3)–(4.4) is $(\overline{\ell},\overline{s};B)$-well posed and $(\overline{\ell},\overline{s};F)$-well posed if it is $(\overline{\ell},\overline{s})$-well posed in the scale of Besov and Lizorkin-Triebel spaces, respectively.

In formulations of the well-posedness theorems below $s_j(t,\xi)$, $j=0,\dots,m-1$, are symbols of the solution operators constructed in Theorem 4.1.

Theorem 4.5. *Let $\varphi_j(x) \in B^{s_j}_{pq}(\mathbb{R}^n)$, $j = 0,\dots,m-1$, and $f \in C^0[T_1,T_2;B^{s_0}_{pq}(\mathbb{R}^n)]$, where $1 < p,q < \infty$. Let the symbols $s_j(t,\xi)$, $j = 0,\dots,m-1$, and $\mathscr{E}(t,\xi)$ for all $k = 0,\dots,m$, and $|\alpha| \le [\frac{n}{2}]+1$ satisfy the estimates*

$$|\xi|^{|\alpha|}|D^\alpha_\xi \frac{\partial^k s_j(t,\xi)}{\partial t^k}| \le C_\alpha(1+|\xi|)^{s_j-\ell_k}, t \in [T_1,T_2],\ \xi \in R^n,\ j = 0,\dots,m-1, \tag{4.50}$$

$$|\xi|^{|\alpha|}|D^\alpha_\xi \frac{\partial^k \mathscr{E}_0(t,\xi)}{\partial t^k}| \le C_\alpha(1+|\xi|)^{s_0-\ell_k}, t \in [T_1,T_2],\ \xi \in R^n, \tag{4.51}$$

where C_α is a positive constant. Then boundary value problem (4.3)–(4.4) is $(\bar{\ell},\bar{s};B)$-well posed.

Proof. Let $\varphi_j \in B^{s_j}_{pq}(\mathbb{R}^n)$, $j = 0,\dots,m-1$. Consider $\varphi_0,\dots,\varphi_{m-1}$ as ψ-distributions in $\Psi'_{-G,p'}(\mathbb{R}^n)$, with some G dense in $\mathbb{R}^n$. Then, in accordance with Theorem 4.4, the problem has a unique solution $u(t,x) \in C^{(m)}[(T_1,T_2),\Psi'_{-G_0,p'}(\mathbb{R}^n)]$. Moreover, $u(t,x)$ has the representation

$$u(t,x) = \sum_{k=0}^{m-1} S^w_k(t,-D)\varphi_k(x) + \int_{T_1}^{T_2} \mathscr{E}^w_0(t-s,-D)h(s,x)dx,$$

Further, it follows from Theorem 2.8 that there are closed restrictions $\hat{S}_{jc}(t,-D)$, $j = 0,\dots,m-1$, of operators $S^w_j(t,-D)$ mapping continuously the space $B^{-\ell_k}_{p'q'}$ to the space $B^{-s_j}_{p'q'}$, and a closed restriction $\hat{\mathscr{E}}_{0c}(t-s,-D)$ of the operator $\mathscr{E}^w_0(t-s,-D)$ mapping continuously the space $B^{-\ell_k}_{p'q'}$ to the space $B^{-s_0}_{p'q'}$ for each fixed t and s. Their adjoint operators, $\hat{S}^+_{jc}(t,-D) = \hat{S}_j(t,D) : B^{s_j}_{pq} \to B^{\ell_0}_{pq}$ and $\hat{\mathscr{E}}^+_{0c}(t,-D) = \hat{\mathscr{E}}_0(t,D) : B^{s_0}_{pq} \to B^{\ell_0}_{pq}$ serve as solution operators of the considered problem. That is, for the solution we have the representation

$$u(t,x) = \sum_{k=0}^{m-1} \hat{S}_k(t,D)\varphi_k(x) + \int_{T_1}^{T_2} \hat{\mathscr{E}}_0(t-s,D)h(s,x)ds, \tag{4.52}$$

Indeed, $u(t,x)$ defined in (4.52) satisfies all the three conditions of Definition 4.2. Condition (i) of this definition immediately follows from the conditions 4.50 and 4.51 of the theorem. Condition (ii) is fulfilled due to construction of the solution $u(t,x)$. Condition (iii) follows from conditions 4.50 and 4.51 of the theorem and Theorem 2.8.

Remark 4.4. Theorem 4.5 remains valid for Lizorkin-Triebel spaces, as well.

For $p = 2$ the inverse assertion is fulfilled under weaker assumption.

Theorem 4.6. *Let $p=2$, $1<q<\infty$. The problem (4.3)–(4.4) is $(\overline{\ell},\overline{s};B)$ (or $(\overline{\ell},\overline{s};F)$) -well posed if and only if for any $t\in[T_1,T_2]$ and $k=0,\dots,m$ the estimates*

$$|\frac{\partial^k s_j(t,\xi)}{\partial t^k}|\leq C(1+|\xi|)^{s_j-\ell_k},\quad C>O,\ \xi\in R^n,\ j=0,\dots,m-1.$$

and

$$|\frac{\partial^k \mathscr{E}_0(t,\xi)}{\partial t^k}|\leq C(1+|\xi|)^{s_0-\ell_k},\quad C>O,\ \xi\in R^n,$$

hold.

Proof. We only need to prove the *"only if"* part. Assume that the condition of the Theorem is not fulfilled. For simplicity, assume that the first estimate is not verified. This means that for some component $s_{j_0}(t,\xi)$ of the system $S(t,\xi)$ there exists a neighborhood $U(\xi_*)$ of a point $\xi_*\in\overset{\circ}{M}$, such that for any $N>0$ and $L>0$ the inequality

$$|s_{j_0}(t,\xi)|>N(|\xi-\xi_*|)^{-L},\quad \xi\in U(\xi_*), \tag{4.53}$$

holds. If the condition is not verified at infinity, then for large $|\xi|$ and for some $t_0\in(T_1,T_2)$, $k_0\in\{0,\dots,m\}$, and $j_0\in\{0,\dots,m-1\}$, the inequality

$$|\frac{\partial^{k_0} s_{j_0}(t_0,\xi)}{\partial t^{k_0}}|>N(1+|\xi|)^{s_{j_0}-\ell_{k_0}} \tag{4.54}$$

holds. Inequalities (4.53) or (4.54) imply that

$$\|\frac{\partial^r}{\partial t^r}\hat{S}_{j_0}(t_0,D)\varphi_{j_0}|B^{\ell_r}_{2q}\|>N\|\varphi_{j_0}|B^{s_{j_0}}_{2q}\|, \tag{4.55}$$

where $\varphi_{j_0}\in B^{s_{j_0}}_{2q}$, $suppF\varphi_0\subset U(\xi_*)$ and r is 0 or k_0. Further, setting $\varphi_j=0$ if $j\neq j_0$, it follows from (4.55) that

$$\|\frac{\partial^r}{\partial t^r}u(t_0,x)|B^{\ell_r}_{2q}\|>N\|\varphi_{j_0}|B^{s_{j_0}}_{2q}\|.$$

The latter contradicts to the $(\overline{\ell},\overline{s};B)$-well posedness of boundary value problem (4.3)–(4.4).

4.5 On sufficient conditions for existence of a solution

If conditions of Theorems 4.5 and 4.6 are not fulfilled, then generally speaking, the problem (4.3)–(4.4) is not well posed in the scales of Besov and Lizorkin-Triebel spaces. However, if the structure of the set $\overset{\circ}{M}$ defined in (4.34) is simple, then one can find effective sufficient conditions for $\varphi_j\in B^{s_j}_{pq}(\mathbb{R}^n)$ under which a solution exists. One of such problems is a boundary value problem for uniformly elliptic operator of the order 2ℓ with a boundary operator containing normal or oblique derivatives of higher order. Here for simplicity we assume that $\ell = 1$. Thus, let us consider the problem

$$\frac{\partial^2 u}{\partial t^2} + L(D)u = 0, \quad t > 0, x \in \mathbb{R}^n, \tag{4.56}$$

$$B_\alpha u(t,x)|_{t=0} = \varphi(x), \quad x \in \mathbb{R}^n, \tag{4.57}$$

where $L(D)$ is a second order pseudo-differential operator whose symbol $L(\xi)$ satisfies the two-side estimate $C_1|\xi|^2 \leq -L(\xi) \leq C_2|\xi|^2$, $C_1, C_2 > 0$; $B_\alpha = D_t^\alpha$ or $B_\alpha = D_{x_n}^\alpha$, where $\alpha \geq 0$ is an integer number. For example, let $\alpha = 1$ and $L(D) = \Delta$, the Laplace operator. Then, if the boundary operator is $B_1 = D_t$, then we have the Neumann problem for the Laplace operator. In this case, as is known, the necessary and sufficient condition for the existence of a unique solution is the orthogonality of the boundary function φ to 1. Similarly, if the boundary operator is $B_1 = D_{x_n}$, then we have a boundary problem with the oblique derivative (which is tangent to the boundary in our case). In this case, the solution exists if φ is orthogonal to $1(x_n)$, that is $\int_{\mathbb{R}} \varphi(x)dx_n = 0$, for all $(x_1, \ldots, x_{n-1}) \in \mathbb{R}^{n-1}$.

Fractional generalizations of these boundary conditions will be discussed in detail in Chapter 5.

Remark 4.5. For higher values of α some consistency relationship between equation (4.56) and boundary condition (4.57) may appear. For instance, if $B_\alpha = D_t^\alpha$ and $\alpha \geq 2$, then one has

$$\varphi(x) + D_t^{\alpha-2} L(D)u(t,x)\Big|_{t=0} = 0.$$

The solutions found below (e.g., (4.59)) automatically satisfy this relationship. Therefore, we do not emphasize this condition in formulations of theorems.

Theorem 4.7. *Let $B_\alpha = D_t^\alpha$, $1 < p < \infty$, $1 < q < \infty$ and the following conditions are fulfilled:*

1) if $0 \leq \alpha < n(p-1)p^{-1}$, then $\varphi \in B_{pq}^s$ has a compact support;

2) if $m - 1 + n(p-1)p^{-1} \leq \alpha < m + n(p-1)p^{-1}$, $m \geq 1$ integer, then $\varphi \in \overset{\circ}{B}{}^s_{pq,m}$, where $\overset{\circ}{B}{}^s_{pq,m}$ is a Lizorkin type space defined in Section 1.12.

Then there exists a unique solution $u(t,x)$ of problem (4.56)–(4.57) with $B_\alpha = D_t^\alpha$ in the space $L^\infty(\overline{R}_+; B_{pq,m}^{s+\alpha})$. Moreover, for the solution the estimate

$$\sup_{t>0} \|u(t,x)|B_{pq,m}^{s+\alpha}\| \leq C\|\varphi|B_{pq}^s\|, \tag{4.58}$$

holds, where $C > 0$ depends on the size of $\operatorname{supp} \varphi$.

Remark 4.6. The case 1) is associated with $m = 0$. Hence, in estimate (4.58) one needs to put $B_{pq,m}^{s+\alpha} = B_{pq,0}^{s+\alpha} = B_{pq}^{s+\alpha}$.

Proof. One can easily verify that a bounded solution of problem (4.56)–(4.57) with $B_\alpha = D_t^\alpha$ can be represented in the form

$$u(t,x) = P_\alpha(t,D)\varphi(x), \tag{4.59}$$

where $P_\alpha(t,D)$ is the pseudo-differential operator with the symbol

$$P_\alpha(t,\xi) = \frac{\exp(-t\sqrt{-L(\xi)})}{(-\sqrt{-L(\xi)})^\alpha}, \quad \xi \neq 0. \tag{4.60}$$

The latter has only singular point $\xi = 0$, and its order of singularity equals α. Therefore, $u(t,x)$ is well defined only under some conditions on φ. Depending on α the sufficient conditions for φ are given in items 1) and 2) of the theorem. To verify these conditions one needs to show the validity of estimate (4.58). Let $\alpha \in [0, n(p-1)p^{-1})$ and $\varphi \in B^s_{pq}(\mathbb{R}^n)$. Further, let the collection $\{\phi_j\}_{j=0}^\infty$ with $\{F^{-1}\phi_j\}_{j=0}^\infty \in \Phi$, define the norm of the Besov space $B^s_{pq}(\mathbb{R}^n)$. Using representation (4.59), (4.60) of the solution $u(t,x)$ and the definition of the norm of the Besov space (see, (1.88)), one has

$$\begin{aligned}\|u(t,x)|B^{s+\alpha}_{pq}\|^q \equiv &\left(\int_{\mathbb{R}^n}\left|F^{-1}\left[\phi_0\frac{\exp(-t\sqrt{-L(\xi)})}{(-\sqrt{-L(\xi)})^\alpha}F[\varphi]\right]\right|^p dx\right)^{\frac{p}{q}}\\ &+\sum_{j=1}^\infty 2^{(s+\alpha)jq}\left\|F^{-1}\left[\phi_j\frac{\exp(-t\sqrt{-L(\xi)})}{(-\sqrt{-L(\xi)})^\alpha}F[\varphi]\right]|L_p\right\|^q.\end{aligned} \tag{4.61}$$

We will estimate each term on the right of (4.61). For the first term, since $\alpha p' < n$, in accordance with Theorem 1.24 and Remark 1.10 of Chapter 1, one obtains

$$\int_{\mathbb{R}^n}\left|F^{-1}\left[\phi_0\frac{\exp(-t\sqrt{-L(\xi)})}{(-\sqrt{-L(\xi)})^\alpha}F[\varphi]\right]\right|^p dx \le C\|\varphi|B^s_{pq}\|^p, \tag{4.62}$$

where C is a positive constant depending on the size of the support of φ, α, and p. Further, taking into account the fact $\operatorname{supp}\phi_j \subseteq \{2^j \le |\xi| \le 2^{j+1}\}$, $j = 1,\dots,$ the second term in (4.61) can be estimated by the expression

$$C\sum_{j=1}^\infty 2^{sjq}\|F^{-1}[\phi_j F[\varphi]]|L_p\|^q.$$

The latter together with (4.61) and (4.62) implies estimate (4.58).

Now let us assume $\alpha \in [m-1+n(p-1)p^{-1}, m+n(p-1)p^{-1})$ and $\varphi \in \overset{\circ}{B}{}^s_{pq,m}$. Then in accordance with Theorem 1.23 there exists a function $v \in B^{s+m}_{pq}$ such that $F\varphi = |\xi|^m Fv$. Thanks this fact the first term on the right-hand side of (4.61) can be estimated by

$$C\left(\int_{\operatorname{supp}\phi_0}|\xi|^{(m-\alpha)p'}d\xi\right)^{p-1}\|v|B^{s+m}_{pq}\|^p \le C\|\varphi|B^s_{pq}\|^p,$$

where C is a positive constant depending on the size of the support of φ, α, p, and m. One can estimate the second term on the right side of (4.61) as in the previous case.

The theorem below uses the oblique derivative in the boundary condition in the form $B_\alpha = D^\alpha_{x_n}$. Its proof is similar to the proof of Theorem 4.7. We leave it for the reader as an exercise.

Theorem 4.8. *Let $B_\alpha = D^\alpha_{x_n}$, $1 < p < \infty$, $1 < q < \infty$, and the following conditions are fulfilled:*

1) *if $\alpha = 0$, then $\varphi \in B^{s',s_n}_{pq}$ and has a support in the strip $\{|x_n| \leq N\}$ with a width $2N, N > 0$;*
2) *if $\alpha = 1,2,\ldots$, then $\varphi \in \overset{\circ}{B}{}^{s',s_n}_{pq,\alpha}$, where $\overset{\circ}{B}{}^{s',s_n}_{pq,\alpha}$ is a Lizorkin type space defined in Section 1.12.*

Then there exists a unique solution $u(t,x)$ of the problem (4.56)–(4.57) with $B_\alpha = D^\alpha_{x_n}$ in the space $L^\infty(\overline{R}_+; B^{s',s_n+\alpha}_{pq,\alpha})$. Moreover, the estimate

$$\sup_{t>0} \|u(t,x)|B^{s',s_n+\alpha}_{pq,\alpha}\| \leq C\|\varphi|B^{s',s_n}_{pq}\|, \; C > 0,$$

holds.

4.6 Examples and applications

In this section we demonstrate a few examples and discuss some applications of ΨDOSS. We start with a brief analysis of Problems (examples) 2.2.0.1–2.2.0.5 discussed in Section 2.2 with a nonhomogeneous term. Then we will discus applications of established theorems to the analysis of boundary values of harmonic functions
(hyperfunctions) and the uniqueness problem of a solution of polyharmonic equation.

4.6.1 Examples

1. *Problem 2.2.0.1:* This is the Cauchy problem

$$\frac{\partial^2 u(t,x)}{\partial t^2} + \lambda^2 D^2 u(t,x) = h(t,x), \quad t > 0, x \in \mathbb{R},$$

$$u(0,x) = \varphi(x), \quad \frac{\partial u(0,x)}{\partial t} = \psi(x), \quad x \in \mathbb{R},$$

where $\lambda = \mu + i\nu$ and $D = \partial/i\partial x$. Here the Duhamel principle is applicable. Therefore, we first consider the case $h(t,x) = 0$. In this case the symbols of solution operators are $s_0(\lambda,t,\xi) = \cos(\xi\lambda t)$ and $s_1(\lambda,t,\xi) = \frac{\sin(\xi\lambda t)}{\lambda\xi}$. These symbols satisfy the conditions of Theorem 4.5 only if λ is real, which corresponds to the Cauchy problem for the wave equation. Moreover, in accordance with the Duhamel principle we need to solve the following Cauchy problem for arbitrary $\tau > 0$:

$$\frac{\partial^2 V(t,\tau,x)}{\partial t^2} + \lambda^2 D^2 V(t,\tau,x) = 0, \quad t > \tau, x \in \mathbb{R},$$

$$V(\tau,\tau,x) = 0, \quad \frac{\partial V(\tau,\tau,x)}{\partial t} = h(\tau,x), \quad x \in \mathbb{R},$$

Exploiting the solution operators with symbols $s_0(\lambda,t-\tau,\xi)$ and $s_1(\lambda,t-\tau,\xi)$, we have

$$V(t,\tau,x) = \left[\frac{\sin\lambda(t-\tau)D}{\lambda D}\right] h(\tau,x).$$

Hence, if λ is real, then the problem is well posed in the Besov spaces. Namely, for any $\varphi \in B^s_{p,q}(\mathbb{R}^n)$, $\psi \in B^{s+1}_{p,q}$ and $h \in C[t \geq 0; B^s_{p,q}(\mathbb{R}^n)]$ there exists a unique solution $u(t,x)$ in the space

$$C^2[t > 0; B^s_{p,q}(\mathbb{R}^n)] \cap C[t \geq 0; B^s_{p,q}(\mathbb{R}^n)] \cap C^1[t \geq 0; B^{s+1}_{p,q}(\mathbb{R}^n)].$$

If λ has nonzero imaginary part, and in particular, purely imaginary, then the problem cannot be well posed in Besov or Lizorkin-Triebel spaces, because of an exponential growth at infinity of the symbols of solution operators.

2. *Problem 2.2.0.2:* This is the Dirichlet problem

$$\frac{\partial^2 u(t,x)}{\partial t^2} + \lambda^2 D^2 u(t,x) = h(t,x),$$

$$u(0,x) = \varphi(x), \quad u(1,x) = \psi(x),$$

where again $\lambda = \mu + i\nu$ and $D = \partial / i\partial x$. The Duhamel principle is not applicable for this problem. Again, we first assume that $h(t,x) = 0$. Then the symbols of solution operators are $s_0(\lambda,t,\xi) = \frac{\sin\,(1-t)\lambda\xi}{\sin\,\lambda\xi}$ and $s_1(\lambda,t,\xi) = \frac{\sin\,t\lambda\xi}{\sin\lambda\xi}$. These symbols belong to the class $C^\infty S_p(G)$, where $G = \mathbb{R} \setminus \{\xi_k = \frac{\pi k}{\lambda}, k = \pm 1, \pm 2, \ldots\}$ for a fixed $\lambda \in \mathbb{C}$ and $t \geq 0$. Obviously, conditions of Theorems 4.5 and 4.6 are not verified if λ is real (the Dirichlet problem for the wave equation). If $\lambda \in \mathbb{C} \setminus \mathbb{R}$, then conditions of Theorem 4.6 is verified. Moreover, one can readily verify that the Green function $G(t,\tau,\xi)$ for the operator

$$L(\frac{d}{dt},\xi) = \frac{d^2}{dt^2} + \lambda^2\xi^2, \quad v(0) = v(1) = 0,$$

has the form

$$G(t,\tau,\xi) = \begin{cases} \frac{\sin\lambda\xi t \sin(\tau-1)\lambda\xi}{\lambda\xi\sin\lambda\xi}, & \text{if } t < \tau, \\ \frac{\sin(t-1)\lambda\xi\sin\lambda\xi}{\lambda\xi\sin\lambda\xi}, & \text{if } t > \tau. \end{cases}$$

Therefore, for any $\varphi,\,\psi \in H^s(\mathbb{R})$ and $h \in C[[0,1];B^s_{p,q}(\mathbb{R}^n)]$ there exists a unique solution of the problem in the space $C^2[(0,1);H^s(\mathbb{R})] \cap C[[0,1];H^s(\mathbb{R})]$. If λ is real, then the problem cannot be well posed in Besov or Lizorkin-Triebel spaces, because of pole type singularities of the symbols of solution operators.

3. *Problem 2.2.0.3:* This is the Cauchy problem

$$\frac{\partial u(t,x)}{\partial t} = \kappa \Delta u(t,x) + h(t,x), \quad t>0,\; x\in\mathbb{R}^n,$$
$$u(0,x) = f(x), \quad x \in \mathbb{R}^n.$$

We apply the Duhamel principle. The symbol of the solution operator of this problem when $h(t,x)=0$ is $s(t,\xi) = \exp(-t\kappa|\xi|^2)$ and belongs to $C^\infty S_p(\mathbb{R}^n)$ for $t \geq 0$. The conditions of Theorems 4.5 and 4.6 are verified if $\kappa > 0$ (forward heat equation). The function $V(t,\tau,x)$, which solves the Cauchy problem

$$\frac{\partial V(t,\tau,x)}{\partial t} = \kappa \Delta V(t,\tau,x), \quad t>\tau,\; x\in\mathbb{R}^n,$$
$$V(\tau,\tau,x) = h(\tau,x), \quad x \in \mathbb{R}^n.$$

has the form
$$V(t,\tau,x) = e^{\kappa(t-\tau)\Delta} h(\tau,x).$$

Hence, in the case $\kappa > 0$ for any functions $\varphi \in B^s_{p,q}(\mathbb{R}^n)$ and $h \in C[t\geq 0;B^s_{p,q}(\mathbb{R}^n)]$, $1<p,q<\infty$, $s\in\mathbb{R}$, there is a unique solution

$$u(t,x) \in C^2[(0,1),B^s_{p,q}(\mathbb{R}^n)] \cap C[[0,1],B^s_{p,q}(\mathbb{R}^n)].$$

In fact, it is known from the classical theory that $u(t,x)$ is infinite differentiable in $\mathbb{R}^{n+1}_+ \equiv \{(t,x): t>0,\, x\in\mathbb{R}^n\}$. If $\kappa < 0$ (backward heat equation), then evidently the conditions of Theorems 4.5 and 4.6 are not verified. Due to Theorem 4.1, in this case a solution exists for $\varphi \in \Psi_{G,p}(\mathbb{R}^n)$, but the solution operator cannot be closed up to Besov and Lizorkin-Triebel spaces.

4. *Problem 2.2.0.4:* We consider this problem in a particular case, namely the Cauchy problem

$$\frac{\partial u(t,x)}{\partial t} = \mathbb{D}^\alpha_0 u(t,x) + h(t,x), \quad t>0,\; x\in\mathbb{R}^n, \tag{4.63}$$
$$u(0,x) = \varphi(x), \quad x \in \mathbb{R}^n, \tag{4.64}$$

Equation (4.63) is a fractional order differential equation. Fractional order differential equations are studied in detail in the next chapter. Here, we consider the case $\alpha = 1$. The symbol of the solution operator of the Cauchy problem (4.63)–(4.64) when $h(t,x)=0$ is $s(t,\xi) = \exp(-t|\xi|)$. This symbol is not differentiable at the origin. Thus, we have $s(t,\xi) \in CS_p(\mathbb{R}^n)$ for every fixed $t\geq 0$. We note that the integrand $V(t,\tau,x)$ in Duhamel's principle has the form

$$V(t,\tau,x) = e^{-(t-\tau)\sqrt{-\Delta}} h(\tau,x).$$

Like the previous example the symbol of the solution operator also satisfies the conditions of Theorems 4.5 and 4.6, and hence, is well posed in the Besov and Lizorkin-Triebel spaces. However, there is a crucial difference between solutions of these two problems. The inverse Fourier transform of the symbol $s(t,\xi) = \exp(-t|\xi|)$ due to formula (1.12) is $Ct(t^2+|x|^2)^{-1}$. Thanks to this fact the fundamental solution $\mathscr{E}(t,x)$ has a power law decay when $|x| \to \infty$, while the fundamental solution of the previous problem has an exponential decay at infinity. This is true in the general case of $0 < \alpha < 2$, as well.

5. Consider the following boundary value problem with a nonlocal integral boundary condition:

$$\frac{\partial u(t,x)}{\partial t} = \Delta u(t,x) + h(t,x), \quad 0 < t < 1,\ x \in \mathbb{R}^n, \tag{4.65}$$

$$\int_0^1 u(t,x)dt = f(x), \quad x \in \mathbb{R}^n. \tag{4.66}$$

The Duhamel principle is not applicable for this problem. It is not hard to verify that the symbol of the solution operator of this problem in the homogeneous case ($h(t,x) = 0$) is

$$s(t,\xi) = \frac{|\xi|^2 e^{-t|\xi|^2}}{1-e^{-|\xi|^2}},$$

and the Green function for the operator

$$L(\frac{d}{dt},\xi) = \frac{d}{dt} + |\xi|^2, \quad \int_0^1 v(t)dt = 0,$$

has the form

$$G(t,\tau,\xi) = \begin{cases} \frac{e^{-t|\xi|^2} - e^{-(1-\tau+t)|\xi|^2}}{e^{-|\xi|^2}-1}, & \text{if } t < \tau, \\ \frac{e^{-t|\xi|^2} - e^{-(t-\tau)|\xi|^2}}{e^{-|\xi|^2}-1}, & \text{if } t > \tau. \end{cases}$$

From the forms of $s(t,\xi)$ and $G(t,\tau,\xi)$ it follows that the problem in (4.65), (4.66) is well posed in the sense of Hadamard in the Besov and Lizorkin-Triebel spaces.

4.6.2 *The Cauchy problem for the Schrödinger equation of a relativistically free particle*

This is Problem 2.2.0.5 in Section 2.2. The state function of a relativistically free particle, as is shown in [BD64], satisfies the following Cauchy problem for a Schrödinger equation

$$
\begin{aligned}
i\frac{\partial u(t,x)}{\partial t} &= \frac{\omega}{\hbar}\sqrt{I-\omega^2\Delta}\; u(t,x), \quad t>0,\, x\in\mathbb{R}^n,\\
u(0,x) &= \varphi(x), \quad x\in\mathbb{R}^n,
\end{aligned}
$$

where I is the identity operator; $\omega = \frac{c}{m\hbar}$, c is the speed of light, m is the mass of the particle, $\hbar$ is the Planck's constant. It is easy to see that in this case the symbol of the solution operator is $s(t,\xi) = \exp(-i\omega t\hbar^{-1}\sqrt{1+\omega^2|\xi|^2})$ and the solution $u(t,x)$ belongs to $C[\mathbb{R}_+;B^s_{pq}(\mathbb{R}^n)]\cap C^1[\mathbb{R}_+;B^{s-1}_{pq}(\mathbb{R}^n)]$, provided $\varphi\in B^s_{pq}(\mathbb{R}^n)$. The well posedness of this problem in $H^{\pm\infty}(\pm G)$ was studied in [Sam83], in the Sobolev spaces $H^s(\mathbb{R}^n)$ in [Uma98].

4.6.3 On uniqueness of a solution of the polyharmonic equation

Edenhofer [Ede75] proved the uniqueness theorem for the m-polyharmonic equation with zero levels on m given concentric hyperspheres with the center at the origin. Applying Theorems 4.1 and 4.2, one can prove similar result in the case of half-space $\mathbb{R}^{n+1}_+\equiv\{(t,x): t>0, x\in\mathbb{R}^n\}$ with zero levels on m given hyperplanes parallel to $t=0$. Namely, consider

$$
\Big(\frac{\partial^2}{\partial t^2}+\Delta_x\Big)^m u(t,x) = 0, \quad t>0,\, x\in\mathbb{R}^n, \tag{4.67}
$$

$$
u(t_j,x) = \varphi_j(x), \quad x\in\mathbb{R}^n,\ j=1,\dots,m, \tag{4.68}
$$

$$
|u(t,x)| = O(1), \quad t\to\infty, \tag{4.69}
$$

where $\Delta_x = \frac{\partial^2}{\partial x_1^2}+\ldots+\frac{\partial^2}{\partial x_n^2}$ is the Laplace operator, and $0\le t_1<\ldots<t_m<\infty$. The characteristic equation corresponding to (4.67) is

$$
(\frac{d^2}{dt^2}-|\xi|^2)v(t,\xi) = 0.
$$

Its m linearly independent solutions, due to constraint (4.69), are $t^k e^{-t|\xi|}, k = 0,\dots,m-1$. Therefore, in order to find the symbols $s_j(t,\xi)$, $j=1,\dots,m$, of solution operators one needs to solve the system of linear algebraic equations

$$
\begin{bmatrix}
1 & t_1 & t_1^2 & \dots & t_1^{m-1}\\
1 & t_2 & t_2^2 & \dots & t_2^{m-1}\\
\dots & \dots & \dots & \dots & \dots\\
1 & t_j & t_j^2 & \dots & t_j^{m-1}\\
\dots & \dots & \dots & \dots & \dots\\
1 & t_m & t_m^2 & \dots & t_m^{m-1}
\end{bmatrix}
\begin{bmatrix}
c_0(\xi)\\ c_1(\xi)\\ \dots\\ c_j(\xi)\\ \dots\\ c_{m-1}(\xi)
\end{bmatrix}
=
\begin{bmatrix}
0\\ 0\\ \dots\\ e^{t_j|\xi|}\\ \dots\\ 0
\end{bmatrix}.
$$

One can write this system in the form $VC(\xi) = B_j e^{t_j|\xi|}$, where V is the Vandermonde matrix, and $B_j \in \mathbb{R}^m$ is the vector, whose the only nonzero component is $b_j = 1$. Since, $Det(V) = \prod_{1\leq k<j\leq m}(t_k - t_j) \neq 0$, this system has a unique solution

$$C_j(\xi) = v_{kj}^{-1} e^{t_j|\xi|}, \quad k = 0,\dots,m-1,$$

where v_{kj}^{-1}, $k = 0,\dots,m-1$, $j = 1,\dots,m$, are entries of the inverse matrix V^{-1}. It follows from this fact that the symbols $s_j(t,x)$, $j = 1,\dots,m$, have representations

$$s_j(t,\xi) = \Big(\sum_{k=0}^{m-1} v_{kj}^{-1} t^k\Big) e^{-(t-t_j)|\xi|}, \quad j = 1,\dots,m.$$

Obviously, the symbol $s_j(t,x)$ has a singularity at infinity if $0 < t < t_j$, and has no singularities if $t \geq t_j$. Hence, corresponding operators $S_j(t,D)$, $j = 1,\dots,m$, are ΨDOSS defined on $\Psi_{G,p}(\mathbb{R}^n)$ and $\Psi'_{-G,p'}(\mathbb{R}^n)$ for any $G \subseteq \mathbb{R}^n$ and $1 < p < \infty$. Now it follows from Theorem 4.1 (Theorem 4.2) that for $\varphi_j \in \Psi_{G,p}(\mathbb{R}^n)$ ($\varphi \in \Psi'_{-G,p'}(\mathbb{R}^n)$) there exists a unique solution $u(t,x) \in C^2[\mathbb{R}_+;\Psi_{G,p}(\mathbb{R}^n)]$ $(u(t,x) \in C^2[\mathbb{R}_+;\Psi'_{-G,p'}(\mathbb{R}^n)])$ of boundary value problem (4.67)–(4.69). In particular, if $\varphi_j(x) \equiv 0$ (zero levels), then $u(t,x) \equiv 0$ in $\mathbb{R}^{n+1}_+$. We note that the solution operator $S_j(t,D)$ is closable up to the class of functions $\varphi \in \mathscr{G}'$, such that

$$|F\varphi_j(\xi)| \leq Ce^{-(t_j+\varepsilon)|\xi|}, \quad j = 1,\dots,m.$$

where $C > 0$, $\varepsilon > 0$ are some constants.

4.6.4 *On derivatives of harmonic functions with a given trace (hyperfunction)*

The space of hyperfunctions, introduced by Mikio Sato [Sat59, Sat60] in 1959, contains distributions, ultra-distributions, and analytic functionals as subclasses. A hyperfunction on an open set $\Omega \subset \mathbb{R}^n$ is defined as a boundary values of a pair of holomorphic functions (F_+, F_-) defined on "upper" and "lower" tubular neighborhoods $\mathfrak{D}_\pm \subset \mathbb{C}^n$ of Ω. In the theory of hyperfunctions it is well known [Sat59, SKK73] that hyperfunctions can be represented as boundary values of harmonic functions. In particular, for any hyperfunction $h(x)$ there exists a defining harmonic function $u(t,x)$ defined on $(0,T) \times \mathbb{R}^n$, such that $u(t,x) \to h(x)$, as $t \to 0+$, in a suitable inductive topology.

Here we consider the problem of existence of a function $u(t,x)$, harmonic in R^{n+1}_+, whose derivative of order $\alpha \in \mathbb{N}_0$ tends to a given trace $\varphi \in \Psi'_{2,-G}(\mathbb{R}^n)$ in a certain topology. We note that in two cases, namely if $0 \notin G$ or $\alpha = 0$, it follows from Theorem 4.4 (in the case of $p = q = 2$) that for any $\varphi \in \Psi'_{-G,2}(\mathbb{R}^n)$ there exists a harmonic function $u(t,x)$, such that $D_t^\alpha u(t,x) \to \varphi(x)$ as $t \to 0+$ in $\Psi'_{-G,2}(\mathbb{R}^n)$. Below we assume that $0 \in G$ and $\alpha \in \mathbb{N}$, the cases which are not covered by Theorem 4.4. The case of arbitrary real $\alpha > 0$ will be discussed in Chapter 5.

Definition 4.3. Denote by $H_\beta(G)$, $\beta > -\frac{n}{2} + \alpha$, the class of functions $\varphi \in \Psi_{2,G}(\mathbb{R}^n)$, such that for any $\varepsilon \in (0, \varepsilon_1)$, $0 < \varepsilon_1 < 1$, the estimate

$$|\xi|^{-\beta} |F[\varphi](\xi)| \leq C < \infty, \quad |\xi| < \varepsilon,$$

holds in the ε-neighborhood of the origin. Evidently, if $\beta_1 > \beta_2 > \alpha - n/2$, then $H_{\beta_1}(G) \subset H_{\beta_2}(G)$. We introduce the space $Z_\alpha^+(G)$ as an inductive limit of $H_\beta(G)$ as $\beta \to (\alpha - n/2)$ from the right, i.e.

$$Z_\alpha^+(G) = ind \lim_{\beta \searrow (-n/2+\alpha)} H_\beta(G).$$

Let $H_\beta^*(-G)$ be the space, conjugate to $H_\beta(G)$. Then, it follows from Proposition 1.23 that the topological dual to $Z_\alpha^+(G)$ is the projective limit

$$Z_\alpha^-(-G) = pr \lim_{\beta \searrow (-n/2+\alpha)} H_\beta^*(-G).$$

Since $Z_\alpha^+(G) \hookrightarrow \Psi_{G,p}(\mathbb{R}^n)$, then the topological inclusion $\Psi'_{-G,2}(\mathbb{R}^n) \hookrightarrow Z_\alpha^-(-G)$ is valid.

Lemma 4.1. *The pseudo-differential operator $P(\alpha;t,D)$ with the symbol*

$$p(\alpha;t,\xi) = (\exp(-t|\xi| - i\pi\alpha))/|\xi|^\alpha \tag{4.70}$$

for any fixed $t > 0$, is continuous as the mapping

$$P(\alpha;t,D) : Z_\alpha^+(G) \to \Psi_{G,2}(\mathbb{R}^n).$$

Proof. Let $\varphi \in Z_\alpha^+(G)$, i.e., there is the $\beta_0 > -\frac{n}{2} + \alpha$ such that $|\xi|^{-\beta_0}|F[\varphi](\xi)| \leq C$ for $|\xi| < \varepsilon$, where $\varepsilon > 0$. It follows that the Fourier transform $F[P(\alpha;t,D)\varphi(x)](\xi)$ has the asymptotic behavior $O(|\xi|^{\beta_0 - \alpha})$ as $|\xi| \to 0$. Since $\alpha - \beta_0 < \frac{n}{2}$, then we have $F[P(\alpha;t,D)\varphi] \in L_2(\mathbb{R}^n_\xi)$. Hence, due to Parseval's equality, $P(\alpha;t,D)\varphi \in L_2(\mathbb{R}^n_x)$. Moreover, it is obvious that

$$\operatorname{supp} F[P(\alpha;t,D)\varphi] \subseteq \operatorname{supp} F[\varphi] \Subset G. \tag{4.71}$$

Thus, $P(\alpha;t,D)\varphi \in \Psi_{G,2}(\mathbb{R}^n)$. Further, let the sequence $\varphi_\ell \in Z_\alpha^+(G)$, $\ell = 1,2,\ldots$, converge to zero in the topology of $Z_\alpha^+(G)$. This means that there are a general compact $K_0 \subset G$, a number $\beta_0 > n/2 - \alpha$, and a number $\varepsilon_0 \in (0,1)$, such that

1. $\operatorname{supp} \varphi_\ell \subseteq K_0$ for all $\ell = 1,2,\ldots$,
2. $|\xi|^{-\beta_0}|F\varphi_\ell(\xi)| \leq C$ provided $|\xi| < \varepsilon < \varepsilon_0$, and
3. $\|\varphi_\ell\|_{L_2} \to 0$ as $\ell \to \infty$.

It follows from (4.71) that $\operatorname{supp} F[P(\alpha;t,D)\varphi_\ell] \subseteq K_0$ for all $\ell \geq 1$. Further, since $2(\alpha - \beta_0) < n$, one has the estimate

$$\|P(\alpha;t,D)\varphi_\ell|L_2\|^2 = \frac{1}{(2\pi)^n}\|\frac{e^{(-t|\xi|-i\pi\alpha)}}{|\xi|^\alpha}F[\varphi_\ell]|L_2\|^2 \\ \leq C\int_{|\xi|<\varepsilon}\frac{d\xi}{|\xi|^{2(\alpha-\beta_0)}} + K_\varepsilon\|\varphi_\ell|L_2\|^2, \tag{4.72}$$

where $C > 0$, $K_\varepsilon > 0$ are positive constants. Now suppose that δ is an arbitrary preassigned positive number. Choosing ε small enough so that the first term on the right-hand side of (4.72) is less than $\delta/2$, and ℓ large enough so that the second term is less than $\delta/2$, we obtain that $P(\alpha;t,D)\varphi_\ell \to 0$ as $\ell \to \infty$, in the topology of $\Psi_{G,2}(\mathbb{R}^n)$.

By duality we immediately obtain the following statement.

Lemma 4.2. *A pseudo-differential operator $P(\alpha;t,D)$ with the symbol in (4.70) is continuous for every fixed t as the mapping*

$$P(\alpha;t,D) : \Psi'_{-G,2}(\mathbb{R}^n) \to Z^-_\alpha(-G).$$

Lemma 4.3. *Let $\varphi(x) \in Z^+_\alpha(G)$. Then there exists a harmonic function $u(t,x)$ such that*

$$\lim_{t\to+0} D^\alpha_t u(t,x) = \varphi(x)$$

in the topology of $\Psi_{G,2}(\mathbb{R}^n)$.

Proof. Suppose that the harmonic function $u(t,x)$ solves the boundary value problem

$$(D^2_t + \Delta)u(t,x) = 0,\, t \in (0,T),\, x \in \mathbb{R}^n, \\ D^\alpha_t u(0,x) = \varphi(x), \quad |u(T,x)| < \infty,\, x \in \mathbb{R}^n,$$

where $D_t = \partial/\partial t$ and Δ is the Laplace's operator. One can then easily verify (see also the proof of Theorem 4.7) that $u(t,x) = P(\alpha;t,D)\varphi$. Due to Lemma 4.1 $u(t,x) \in \Psi_{G,2}(\mathbb{R}^n)$ for every fixed $t \in (0,T)$. Moreover, its derivative $D^\alpha_t u(t,x) = P(0;t,D)\varphi(x)$ represents the Poisson's integral for $\varphi(x)$. It follows from this fact that $\lim_{t\to+0} D^\alpha_t u(t,x) = \varphi(x)$ in the topology of $\Psi_{G,2}(\mathbb{R}^n)$.

Theorem 4.9. *Let $h(x) \in \Psi'_{-G,2}(\mathbb{R}^n)$. Then there exists a harmonic function $u(t,x)$ such that $\lim_{t\to+0} D^\alpha_t u(t,x) = h(x)$ in the topology of $Z^-_\alpha(-G)$.*

Proof. By duality, $u(t,x) = P(\alpha;t,-D)h(x)$ is harmonic in the weak sense in $(0,T)\times\mathbb{R}^n$. In accordance with the Weyl's lemma [Shu78] it is an ordinary harmonic function. Moreover, it follows from Lemma 4.2 that it is as an element of the space $Z^-_\alpha(-G)$ for any fixed t. Let v be an arbitrary function in $Z^+_\alpha(G)$. One has

$$< D_t^{\alpha} u(t,x) - h(x), v(x) > = < h(x), [D_t^{\alpha} P(\alpha;t,D) - I] v(x) > .$$

Now Lemma 4.3 implies that $D_t^{\alpha} u(t,x) \to h(x)$ as $t \to +0$.

4.6.5 Boundary values of harmonic functions

We have seen above that ψ-distributions $\Psi'_{-G,2}(\mathbb{R}^n)$ are boundary values of harmonic functions in a special topology. Continuing the discussion on boundary values of harmonic functions, we note that Gorbachuk [G84] found the necessary and sufficient conditions for boundary values of harmonic functions to belong to the Sobolev spaces $W_2^s(\mathbb{R}^n)$ for arbitrary $s \in \mathbb{R}$. Below we will study boundary values of harmonic functions in spaces $B_{2q}^s(\mathbb{R}^n)$ (or $F_{2q}^s(\mathbb{R}^n)$) for arbitrary $s \in \mathbb{R}$ and $1 < q < \infty$, generalizing the results of [G84].

Theorem 4.10. *A harmonic on $(0,T) \times \mathbb{R}^n$ function, $u(t,x)$ has boundary values as $t \to +0$ belonging to the space $B_{2q}^s(\mathbb{R}^n)$, $s \geq 0$, if and only if the estimate*

$$\sup_{0<t<\varepsilon} \|u(t,x)|B_{2q}^s\| \leq C < \infty \tag{4.73}$$

holds for some $\varepsilon > 0$.

Proof. The necessity of condition (4.73) is the particular case of Theorem 4.7, corresponding to the case of $\alpha = 0$ and $p = 2$. Suppose condition (4.73) is fulfilled. The sufficiency of this condition follows from the following estimate

$$\begin{aligned} C \geq \|u(t,x)|B_{2q}^s\|^q &= \sum_{j=o}^{\infty} 2^{sjq} \|F^{-1}\left[\varphi_j e^{-t|\xi|} F[u(0,x)]\right] |L_2\|^q \\ &\geq C_1 \sum_{j=0}^{\infty} 2^{sjq} \|F^{-1}\left[\varphi_j F[u(0,x)]\right] |L_2\|^q \\ &= C_1 \|u(0,x)|B_{2q}^s\|^q. \end{aligned} \tag{4.74}$$

Here we used the inequality $e^{-t|\xi|} \leq 1$ on the support of each φ and the Parseval's equality.

Theorem 4.11. *A harmonic in $(0,T) \times \mathbb{R}^n$ function $u(t,x)$ has boundary values as $t \to +0$ belonging to $B_{2q}^{-s}(\mathbb{R}^n)$, $s > 0$, if and only if the estimate*

$$\int_0^{\varepsilon} t^{qs-1} \|u(t,x)|B_{2q}^0\|^q dt \leq C < \infty \tag{4.75}$$

holds for some $\varepsilon > 0$.

Proof. Necessity. Suppose for the harmonic function $u(t,x)$ its limit $u(0,x)$ as $t \to 0+$ in the norm of $B^{-s}_{2q}(\mathbb{R}^n)$ exists and belongs to $B^{-s}_{2q}(\mathbb{R}^n)$. Then, as we have seen above,

$$u(t,x) = P(0;t,D)u(0,x),$$

where the operator $P(0;t,D)$ has the symbol $P(0;t,\xi) = e^{-t|\xi|}$. Let a collection $\{\varphi_j(\xi)\}_{j=0}^{\infty}$ define the norm of the Besov space $B^{-s}_{2q}(\mathbb{R}^n)$ in accordance with (1.88). Recall that $\operatorname{supp}\varphi_j \subset \{2^{j-1} \le |\xi| \le 2^{j+1}\}$. Further, for $\xi \in \operatorname{supp}\varphi_j$ the inequality

$$e^{-t2^{j-1}} \le e^{-t|\xi|} \le e^{-t2^{j+1}} \tag{4.76}$$

holds for the symbol $P(0;t,\xi)$. Therefore, using the right inequality in (4.76) and the Parseval's equality, one has

$$\begin{aligned}\int_0^{\varepsilon} t^{qs-1}\|u(t,x)|B^0_{2q}\|^q dt &= \int_0^{\varepsilon} t^{qs-1}\sum_{j=0}^{\infty}\left\|F^{-1}\left[\varphi_j e^{-t|\xi|}F[u(0,x)]\right]|L_2\right\|^q \\ &\le C\sum_{j=0}^{\infty}\left(\int_0^{\varepsilon} t^{qs-1}e^{-t2^{j+1}q}dt\right)\|F^{-1}[\varphi_j F[u(0,x)]]|L_2\|^q. \end{aligned}\tag{4.77}$$

Now consider the function

$$\mu(\lambda) = \int_0^{\varepsilon} t^{qs-1}e^{-\lambda t}dt, \quad \lambda \ge \frac{1}{2}. \tag{4.78}$$

The substitution $\lambda t = \tau$ in this integral yields the estimate

$$\mu(\lambda) = \lambda^{-qs}\int_0^{\lambda\varepsilon} \tau^{qs-1}e^{-\tau}d\tau \le \Gamma(qs)\lambda^{-qs}, \tag{4.79}$$

where $\Gamma(\cdot)$ is the Euler's gamma function. Using the latter estimate with $\lambda = q2^{j+1}$ in inequality (4.77), one obtains

$$\begin{aligned}\int_0^{\varepsilon} t^{qs-1}\|u(t,x)|B^0_{2q}\|^q dt &\le C\sum_{j=0}^{\infty} 2^{-sjq}\|F^{-1}[\varphi_j F[u(0,x)]]|L_2\|^q \\ &= C\|u(0,x)|B^{-s}_{2q}\|^q < \infty, \end{aligned}$$

proving (4.75).

Sufficiency. Assume that (4.75) holds for some $\varepsilon > 0$. We need to show that $u(0,x) \in B^{-s}_{2q}(\mathbb{R}^n)$. Using the left inequality in (4.76) and the Parseval's equality, we have

$$\begin{aligned}\int_0^{\varepsilon} t^{qs-1}\|u(t,x)|B^0_{2q}\|^q dt &= \int_0^{\varepsilon} t^{qs-1}\sum_{j=0}^{\infty}\left\|F^{-1}\left[\varphi_j e^{-t|\xi|}F[u(0,x)]\right]|L_2\right\|^q \\ &\ge C\sum_{j=0}^{\infty}\left(\int_0^{\varepsilon} t^{qs-1}e^{-t2^{j-1}q}dt\right)\left\|F^{-1}[\varphi_j F[u(0,x)]]|L_2\right\|^q. \end{aligned}\tag{4.80}$$

Further, replacing λ by its lowest value $1/2$ in the upper integration endpoint in the integral in (4.79), we have $\mu(\lambda) \geq C_\varepsilon \lambda^{-qs}$. Now using this estimate with $\lambda = q2^{j-1}$ in inequality (4.80), we obtain

$$\infty > \int_0^\varepsilon t^{qs-1} \|u(t,x)|B^0_{2q}\|^q dt \geq C_\varepsilon \sum_{j=0}^{\infty} 2^{-sjq} \left\| F^{-1}\left[\varphi_j F[u(0,x)]\right] |L_2\right\|^q$$
$$= C_\varepsilon \|u(0,x)|B^{-s}_{2q}\|^q,$$

which shows that $u(0,x) \in B^{-s}_{2q}(\mathbb{R}^n)$.

4.7 Duhamel principle for differential-operator equations

Recall that in Section 2.8 of Chapter 2 we introduced the space of exponential elements $\mathrm{Exp}_{A_\lambda,G}(\mathrm{X})$ and its dual $\mathrm{Exp}'_{A^*,G^*}(\mathrm{X}^*)$, where X is a reflexive Banach space and A is a closed operator with a dense domain $\mathscr{D}(A) \subset X$, and defined operators of the form $f(A)$ with symbols f analytic in a domain G. Using this construction one can study abstract boundary value problems for differential operator equations of the form

$$L(t,\frac{d}{dt},A)u(t) = u^{(m)}(t) + \sum_{k=0}^{m-1} f_k(t,A)u^{(k)}(t) = h(t), \quad t \in (T_1,T_2), \tag{4.81}$$

$$B_k(A)[u] = \sum_{j=0}^{m-1} b_{kj}(A)u^{(j)}(t_{kj}) = y_k, \quad k = 1,\ldots,m, \tag{4.82}$$

where $h(t)$ and y_k, $k = 1,\ldots,m$, are given elements. Boundary value problem (4.81)–(4.82) generalizes problem (4.3)–(4.4) for ΨDOSSs considered above to the case of abstract differential-operator equations. The next section (Section 4.8) presents generalizations of the results obtained for ΨDOSSs to the abstract case of boundary value problem (4.81)–(4.82). In this section we will discuss the Duhamel principle for abstract Cauchy problem for differential-operator equations.

Consider the Cauchy problem

$$u^{(m)}(t) + \sum_{k=0}^{m-1} f_k(A)u^{(k)}(t) = h(t), \quad t > 0, \tag{4.83}$$

$$u^{(k)}(0) = \varphi_k, \quad k = 0,\ldots,m-1. \tag{4.84}$$

The Duhamel principle establishes a connection between the solutions of the Cauchy problem for nonhomogeneous equation (4.83) with the homogeneous initial conditions

$$u^{(k)}(0) = 0, \quad k = 0,\ldots,m-1, \tag{4.85}$$

and the Cauchy problem for the corresponding homogeneous equation

$$\frac{\partial^m U}{\partial t^m}(t,\tau)+\sum_{k=0}^{m-1} f_k(A)\frac{\partial^k U}{\partial t^k}(t,\tau)=0, \quad t>\tau, \tag{4.86}$$

$$\frac{\partial^k U}{\partial t^k}(t,\tau)|_{t=\tau+0}=0, \quad k=0,\dots,m-2, \tag{4.87}$$

$$\frac{\partial^{m-1} U}{\partial t^{m-1}}(t,\tau)|_{t=\tau+0}=h(\tau). \tag{4.88}$$

Note that if $h(t)$ is a continuous $\mathrm{Exp}_{A_\lambda,G}(\mathrm{X})$-valued ($\mathrm{Exp}'_{A^*,G^*}(\mathrm{X}^*)$-valued) function then the solution of (4.86)–(4.88) is an m times differentiable $\mathrm{Exp}_{A_\lambda,G}(\mathrm{X})$-valued ($\mathrm{Exp}'_{A^*,G^*}(\mathrm{X}^*)$-valued) function (see [Uma98]). Taking this fact into account, in the following theorem we assume that the vector-functions $h(t)$ and $U(t,\tau)$ are $\mathrm{Exp}_{A_\lambda,G}(\mathrm{X})$-, or $\mathrm{Exp}'_{A^*,G^*}(\mathrm{X}^*)$-valued, $h(t)$ is continuous, $U(t,\tau)$ is m times differentiable with respect to the variable t, and the derivatives $\frac{\partial^j U(t,\tau)}{\partial t^j}$, $0\le j\le k-1$, are jointly continuous in the topology of $\mathrm{Exp}_{A_\lambda,G}(\mathrm{X})$, or of $\mathrm{Exp}'_{A^*,G^*}(\mathrm{X}^*)$, respectively.

Lemma 4.4. *Suppose $v(t,\tau)$ is a X-valued function defined for all $t\ge\tau\ge 0$, the derivatives $\frac{\partial^j v(t,\tau)}{\partial t^j}$, $0\le j\le k-1$, are jointly continuous in the X-norm, and $\frac{\partial^k v(t,\tau)}{\partial t^k}\in L_1(0,t;X)$ for all $t>0$. Let $u(t)=\int_0^t v(t,\tau)d\tau$. Then*

$$\frac{d^k}{dt^k}u(t)=\sum_{j=0}^{k-1}\frac{d^j}{dt^j}\Big[\frac{\partial^{k-1-j}}{\partial t^{k-1-j}}v(t,\tau)_{|_{\tau=t}}\Big]+\int_0^t\frac{\partial^k}{\partial t^k}v(t,\tau)d\tau. \tag{4.89}$$

Proof. For a fixed $t>0$ and small h one can easily verify that

$$\begin{aligned}\frac{u(t+h)-u(t)}{h}&=\frac{1}{h}\Big(\int_0^{t+h}v(t+h,\tau)d\tau-\int_0^t v(t,\tau)d\tau\Big)\\&=\frac{1}{h}\int_t^{t+h}v(t+h,\tau)d\tau+\int_0^t\frac{v(t+h,\tau)-v(t,\tau)}{h}d\tau.\end{aligned} \tag{4.90}$$

Due to the continuity and differentiability conditions of the lemma, in the X-norm we have

$$\begin{aligned}\|\frac{1}{h}\int_t^{t+h}v(t+h,\tau)d\tau-v(t,t)\|_X&=\|\frac{1}{h}\int_t^{t+h}[v(t+h,\tau)d-v(t,t)]d\tau\|_X\\&\le\sup_{t<\tau<t+h}\|v(t+h,\tau)-v(t,t)\|_X=o(h),\quad h\to 0,\end{aligned} \tag{4.91}$$

$$\|\int_0^t\frac{v(t+h,\tau)-v(t,\tau)}{h}d\tau-\int_0^t\frac{\partial v(t,\tau)}{\partial t}d\tau\|_X=o(h),\quad h\to 0. \tag{4.92}$$

Now, letting $h\to 0$, estimates (4.91), (4.92) and equation (4.90) imply the formula

$$\frac{d}{dt}u(t)=v(t,t)+\int_0^t\frac{\partial}{\partial t}v(t,\tau)d\tau. \tag{4.93}$$

Formula (4.89) follows from (4.93) by repeated differentiation.

In the general case of abstract differential-operator equations the Duhamel principle is formulated as follows.

Theorem 4.12. *Let* $U(t,\tau)$ *be a solution of the Cauchy problem* (4.86)–(4.88). *Then a solution of the Cauchy problem* (4.83), (4.85) *is represented via Duhamel's integral*

$$u(t) = \int_0^t U(t,\tau)d\tau. \tag{4.94}$$

Proof. Let $u(t)$ be as defined by (4.94). Obviously $u(0) = 0$. Further, for the first order derivative of $u(t)$, using (4.89) in the case $k = 1$, one has

$$\frac{du}{dt}(t) = U(t,t) + \int_0^t \frac{\partial U}{\partial t}(t,\tau)d\tau,$$

By virtue of (4.87) ($k = 0$) the latter implies that $\frac{du}{dt}(0) = 0$. Further, differentiating k times,

$$\frac{d^k u}{dt^k}(t) = \frac{\partial^{k-1} U}{\partial t^{k-1}}(t,t) + \int_0^t \frac{\partial^k U}{\partial t^k}(t,\tau)d\tau,$$

which due to condition (4.87) implies that $\frac{d^k u}{dt^k}(0) = 0$, $k = 2,\dots,m-1$. Therefore, the function $u(t)$ in (4.94) satisfies initial conditions (4.85). Moreover, substituting (4.94) to (4.83), and taking into account (4.88), we have

$$\begin{aligned}
&u^{(m)}(t) + \sum_{k=0}^{m-1} f_k(A)u^{(k)}(t)\\
&= \frac{d^m}{dt^m}\int_0^t U(t,\tau)d\tau + \sum_{k=0}^{m-1} f_k(A)\frac{d^k}{dt^k}\int_0^t U(t,\tau)d\tau\\
&= \frac{\partial^{m-1} U}{\partial t^{m-1}}(t,t) + \int_0^t \frac{\partial^m U}{\partial t^m}(t,\tau)d\tau + \sum_{k=0}^{m-1} f_k(A)\int_0^t \frac{\partial^k U}{\partial t^k}(t,\tau)d\tau\\
&= h(t) + \int_0^t \left[\frac{\partial^m U}{\partial t^m}(t,\tau) + \sum_{k=0}^{m-1} f_k(A)\frac{\partial^k U}{\partial t^k}(t,\tau)\right]d\tau = h(t).
\end{aligned}$$

Hence, $u(t)$ in (4.94) satisfies equation (4.83) as well.

Remark 4.7. In Chapters 5 and 6 we will discuss fractional generalizations of the Duhamel principle for a wide class of fractional and distributed order differential equations.

4.8 Well posedness of general boundary value problems for differential-operator equations

In this section we prove an abstract analog of Theorems 4.3 and 4.5. On the base of these results we consider broad class of boundary value problems (see § 6).

Consider the following nonlocal boundary value problem for differential operator equations

$$u^{(m)}(t)+\sum_{k=0}^{m-1} a_k(t,A)u^{(k)}(t)=0, \quad t\in(T_1,T_2), \tag{4.95}$$

$$\sum_{j=0}^{m-1} b_{kj}(A)u^{(j)}(t_{kj})=y_k, \quad k=1,\dots,m, \tag{4.96}$$

where the $a_k(t,A)$, $k=0,\dots,m-1$, and the $b_{kj}(A)$, $k,j+1=1,\dots,m$, are operators defined in the sense of (2.95), Section 2.8, by functions $a_k(t,\lambda)$ and $b_{kj}(\lambda)$, analytic in G; $t_{kj}\in[T_1,T_2]$. Assume that the operator A commutes with $\frac{d}{dt}$.

Let $\{U_0(t,\lambda),\dots,U_{m-1}(t,\lambda)\}$ be a fundamental system of solutions to the characteristic equation

$$u^{(m)}(t)+\sum_{k=0}^{m-1} a_k(t,\lambda)u^{(k)}(t)=0, \quad \lambda\in\mathbb{C},$$

corresponding to equation (4.95), and satisfying the Cauchy conditions

$$u_j^{(k)}(0)=\delta_{j,k}, \quad j,k=0,\dots,m-1.$$

Further, introduce the set

$$\overset{\circ}{M}=\{\lambda\in\mathbb{C}:\det M(\lambda)=0\}\subset\mathbb{C}, \tag{4.97}$$

where $M(\lambda)$ is the $m\times m$ matrix with entries

$$m_{kl}=\sum_{j=0}^{m-1} b_{kj}(\lambda)u_l^{(j)}(t_{kj},\lambda), \quad k,l=0,\dots,m-1.$$

We also introduce a vector-function $U^*(t,\lambda)=(u_0^*(t,\lambda),\dots,u_{m-1}^*(t,\lambda))$ defined as

$$U^*(t,\lambda)=(M^*(\lambda))^{-1}U(t,x), \tag{4.98}$$

where M^* is the Hermitian adjoint of the matrix $M(\lambda)$, and $U(t,\lambda)$ is the vector-function with components $u_0(t,\lambda),\dots,u_{m-1}(t,\lambda)$.

Theorem 4.13. *Let $y_k\in Exp_{A,G}(X)$ and $\overset{\circ}{M}\bigcap G=\emptyset$, where $\overset{\circ}{M}$ is defined in (4.97). Then there exists a unique solution $u(t)$ of the problem (4.95)–(4.96) in the space $C^{(m)}[(T_1,T_2);Exp_{A,G}(X)]$, and for the solution the representation*

$$u(t) = \sum_{k=1}^{m} u_k^*(t,A) y_k, \tag{4.99}$$

holds. Here $u_k^*(t,A)$, $k = 0,\ldots,m-1$, *are operators with the corresponding symbols* $u_k(t,\lambda)$, *k-th component of the vector-function* $U^*(t,\lambda)$ *in* (4.98).

Proof. We set

$$w_j(t) = u_j^*(t,A)y_j = \sum_{\lambda \in G} u_{j\lambda}^*(t,A) y_{\lambda j}, \quad j \in \{1,\ldots,n\},$$

where $y_{\lambda j} \in Exp_{A,\nu,\lambda}(X)$, $\nu < R(\lambda)$. By substituting $w_j(t)$ to (4.95) and taking into account the equality

$$w_j^{(k)}(t) = \sum_{\lambda \in G} u_{j\lambda}^{*(k)}(t,A) y_{\lambda j},$$

we have

$$w_j^{(m)}(t) + \sum_{k=0}^{m-1} a_k(t,A) w_j^{(k)}(t) = \sum_{\lambda \in G} \{ \sum_{n=0}^{\infty} \frac{1}{n!} D_\lambda^n \Big[u_j^{*(m)}(t,\lambda)$$

$$+ \sum_{k=0}^{m-1} a_k(t,\lambda) u_j^{*(k)}(t,\lambda) \Big] (A - \lambda I)^n \} y_{\lambda j} \equiv 0, \tag{4.100}$$

because the expression in the square brackets in (4.100) vanishes. Similarly, one can verify that the solution $u(t)$ defined in (4.99) satisfies boundary conditions (4.96), as well.

Remark 4.8. If one changes the operators $a_k(t,A)$ and $b_{kj}(A)$ in (4.95)–(4.96) to their weak extensions, then the similar assertion is valid in the dual space. Namely, for any $y_k \in Exp'_{A^*,G^*}(X^*)$, $k = 1,\ldots,m$, under the condition $\overset{\circ}{M} \cap G = \emptyset$, there exists a unique weak solution in the space $C^{(m)}[(T_1,T_2); Exp'_{A^*,G^*}(X^*)]$. In this case in representation (4.99) the operators $u_k^*(t,A)$ also change to their weak extensions.

In the next theorem we assume that for A the space $Exp_A(X)$ is dense in X and $\ell_k \geq 0$, $k = 0,\ldots,m$, $s_j \geq 0, j = 1,\ldots,m$.

Theorem 4.14. *Let the set* $\overset{\circ}{M}$ *does not contain unremovable singularities of the vector-function* $U^*(t,\lambda)$. *Moreover, let for* $|\lambda| \geq L$, $\lambda \in \sigma(A)$, $L > 0$, *the estimate*

$$|\frac{\partial^k}{\partial t^k} u_j^*(t,\lambda)| \leq C|\lambda|^{s_j - \ell_k}, \quad j = 1,\ldots,m, \; k = 0,\ldots,m \; t \in [T_1,T_2], \;\; C > 0,$$

holds. Then for any $y_j \in D(A^{s_j})$, $j = 1,\ldots,m$, *there exists a unique solution* $u(t) \in C^{(m)}[(T_1,T_2); D(A^{\ell_0})]$, $u^{(k)}(t) \in C^{(m-k)}[(T_1,T_2); D(A^{\ell_k})]$, $k = 1,\ldots,m$, *of the problem* (4.95)–(4.96), *for which the estimate*

$$\max_{t\in[T_1,T_2]}[\sum_{k=0}^{m}\sum_{q=0}^{\ell_k}(\|A^q u^{(k)}(t)|X\|)] \le C\sum_{j=1}^{m}\sum_{q=0}^{s_j}\|A^q u_j|X\|, \quad C>0, \tag{4.101}$$

holds.

Proof. We sketch the proof, since it is similar to the proof of Theorem 4.5. We first assume that $y_j \in Exp_{A,G}(X)$. Then, due to Theorem 4.13 there exists a unique solution $u(t)$ in the form (4.99) through the solution operators $u_k^*(t,A)$, $k=0,\dots,m-1$. Further, if $u_k^*(t,\lambda)$ satisfies the condition of the theorem, then due to denseness of $Exp_{A,G}(X)$ in X, there exists a unique closure $\hat{u}_k^*$ to the space $D(A^{s_k})$ of the operator $u_k^*(t,A)$, and consequently, estimate (4.101) holds. In conclusion we note that the construction of the closure $\hat{U}^*(t,A)$ is standard.

Remark 4.9. 1. If $\ell<0$ and $s_j<0$ for certain $j\in\{1,\dots,m\}$, then one can show that the estimate

$$\sup_{x\neq 0}\frac{|<\hat{u}_{jc}^{*+}(t,A)y_j,x>|}{\sum_{q=0}^{-\ell}\|A^q x|X\|} \le C_j \sup_{x\neq 0}\frac{|<y_j,x>|}{\sum_{q=0}^{-s_j}\|A^q x|X\|},$$

holds. Here $\hat{u}_{jc}^*(t,A)$ is the closed restriction of $u_j^{*w}(t,A)$ which is the weak extension of $u_j^*(t,A)$. Taking this into account we can conclude that Theorem 4.14 remains valid in this case also, however with appropriate understanding of estimate (4.101).

2. If A is defined in a Hilbert space and self-adjoint, then one can show that the condition of the theorem is also necessary for well posedness of the problem (4.95)–(4.96). This statement extends for arbitrary spectral operators of the scalar type, in particular, for normal operators.

As an application of Theorems 4.13 and 4.14 let us consider two examples: the general boundary value problem in the space of periodic functions and a differential-operator equation with a self-adjoint elliptic operator.

Example 4.1. 1. The first application is to the theory of periodic boundary value problems. In other words, we set $X=L_2(\mathbb{T}^n)$, where $\mathbb{T}^n$ is the n-dimensional torus, and $A=(\frac{\partial}{i\partial x_1},\dots,\frac{\partial}{i\partial x_n})$. Consider the following general boundary value problem for a homogeneous pseudo-differential equation on $\mathbb{T}^n$:

$$\frac{\partial^m u}{\partial t^m}+\sum_{k=0}^{m-1}A_k(t,D)\frac{\partial^k u}{\partial t^k}=0, \quad t\in(T_1,T_2), x\in\mathbb{T}^n \tag{4.102}$$

$$B_k(u)=\sum_{j=0}^{m-1}b_{kj}(D)\frac{\partial^j u(t_{kj},x)}{\partial t^j}=\varphi_k(x), \quad k=0,\dots,m-1, \tag{4.103}$$

where $t_{kj}\in[T_1,T_2)$, $m\ge 1$; $D=(D_1,\dots,D_n)$, $D_j=-i\frac{\partial}{\partial x_j}$; the operators $A_k(t,D)$ and $b_{kj}(D)$ are defined in Section 2.6, and $\varphi_k(x)$ are given periodic functions in certain spaces indicated below. Recall that the spaces $W_{\mathbb{T}^n}^{\pm\infty}$ of periodic test functions and functionals, as well as Sobolev spaces $W^s(\mathbb{T}^n)$ were introduced in Section 2.6.

Theorem 4.15. *a. Let $\varphi_k \in W_{\mathbb{T}^n}^{\infty}, k = 1,\dots,m$. Then there exists a unique function $u(t,x) \in C^m[T_1,T_2;W_{\mathbb{T}^n}^{\infty}]$ that satisfies problem* (4.102)–(4.103) *pointwise. Moreover, for the solution $u(t,x)$ the following representation holds:*

$$u(t,x) = \sum_{k=0}^{m-1} S_k(t,D)\varphi_k(x),$$

where $S_k(t,D)$ is a pseudo-differential operator with the symbol $s_k(t,m), m \in \mathbb{Z}^n$, defined in (4.30).

b. Suppose there exist a unique weak solution $u(t,x) \in C^m[(T_1,T_2);W_{\mathbb{T}^n}^{-\infty}]$ of problem (4.102)–(4.103)*, such that for all $\psi \in W_{\mathbb{T}^n}^{\infty}$ and all $t \in (T_1,T_2)$:*

$$< \frac{\partial^m u}{\partial t^m}, \psi > + \sum_{k=0}^{m-1} < \frac{\partial^k u}{\partial t^k}, A_k(t,-D)\psi >= 0,$$

$$< B_k(u), \psi >= \sum_{j=0}^{m-1} < \frac{\partial^j u(t_{kj},x)}{\partial t^j}, b_{kj}(-D)\psi) >=< \varphi_k(x), \psi(x) > .$$

c. Let $\overline{\ell} = (\ell_0,\dots,\ell_{m-1})$, $\overline{s} = (s_0,s_1,\dots,s_{m-1}) \in \mathbb{R}^m$. Let symbols $s_k(t,\lambda), k = 0,\dots,m-1$, satisfy the conditions

$$|s_k(t,m)| \le C(1+|m|^2)^{\frac{s_k-l_k}{2}}, m \in \mathbb{Z}^n,$$

Then problem (4.102)–(4.103) *is $(\overline{\ell},\overline{s})$-well posed in the scale of Sobolev spaces $W_2^s(\mathbb{T}^n)$.*

2. As the second application consider the following example. Let $\Omega \subset R^n$ be a bounded domain with a smooth boundary $S = \partial\Omega$. Let $A \equiv A(x,D)$ be an elliptic self-adjoint operator of order $2m$ of the form

$$A(x,D) = \sum_{|\alpha|\le 2m} a_\alpha(x)D^\alpha,$$

where $a_\alpha(x)$ are smooth functions on $\overline{\Omega}$. The domain $D(A) = W_2^{2m}(\Omega) \cap \overset{\circ}{W}{}_2^m(\Omega)$. Consider the boundary value problem

$$\frac{\partial^k u(t,x)}{\partial t^k} + A(x,D)u(t,x) = 0, \;\; t \in (0,T), \;\; x \in \Omega, \tag{4.104}$$

$$\left.\frac{\partial^j u(t,x)}{\partial \mathbf{n}^j}\right|_S = 0, \;\; j = 0,\dots,m-1, \;\; t \in (0,T), \tag{4.105}$$

$$u(0,x) - \mu u(1,x) = \varphi_0(x), \;\; x \in \Omega, \tag{4.106}$$

$$\frac{\partial u(0,x)}{\partial t} = \varphi_1(x), \;\; x \in \Omega, \tag{4.107}$$

where T is a positive number, $T > 1$; $\mathbf{n}$ is the normal to S, μ is a complex parameter, $k = 1$ or $k = 2$. If $k = 1$, then the condition (4.107) has to be removed. We introduce the space

$$W^s = W_2^{2sm}(\Omega) \cap \left\{ u \in W_2^{(2s-1)m}(\Omega) : \frac{\partial^j}{\partial \mathbf{n}^j} \circ A^k u \Big|_S = 0, \right.$$

$$\left. k = 0, \ldots, s-1, \; j = 0, \ldots, m-1 \right\},$$

with the norm induced from $W_2^{2sm}(\Omega)$. In order to apply Theorem 4.14 to this problem one should reduce it to the differential-operator form. We assume that $A = A(x,D)$ and $X = \{\varphi \in D(A) : \frac{\partial^j \varphi(x)}{\partial \mathbf{n}^j}\Big|_S = 0, \; j = 0, \ldots, m-1\}$. Then one can easily calculate symbols of solution operators. Namely,

(i) if $k = 1$, then

$$s(t,\lambda) = e^{-t\lambda}(1 - \mu e^{-\lambda})^{-1}, \quad \lambda \in \sigma(A) \subseteq [0,\infty);$$

(ii) if $k = 2$, then

$$s_1(t,\lambda) = \cos t\sqrt{\lambda}(1 - \mu \cos\sqrt{\lambda})^{-1},$$

$$s_2(t,\lambda) = \frac{\sin t\sqrt{\lambda} + \mu \sin\sqrt{\lambda}(1-t)}{\mu\sqrt{\lambda}(1 - \mu\cos\sqrt{\lambda})}.$$

Let $k = 1$. It is obvious that if $\mu \in [1,\infty)$, then there exists the $\lambda_\circ \in R_+^1$ (since $\sigma(A) \subset R_+^1$, it is sufficient to consider only $\lambda \in R_+^1$), namely the $\lambda_\circ = \ln\mu$, which is an unremovable singular point of $s(t,\lambda)$. Thus, we have the following assertion.

Proposition 4.2. *(The case $k = 1$) Let $\mu \in C^1 \setminus [1,\infty)$. Then for any $\varphi_0 \in W^s$ there exists a unique solution $u(t,x)$ of (4.104)–(4.106) belonging to $C[[0,\infty); W^s]$. Moreover, the estimate*

$$\sup_{t \geq 0} \|u(t)|W_2^s(\Omega)\| \leq C\|\varphi_0|W_2^s(\Omega)\|, \quad C > 0,$$

holds.

For $k = 2$, if $\mu \in (-\infty,-1] \cup [1,\infty)$, then $s_j(t,\lambda)$, $j = 1,2$, have no unremovable singularities. Moreover, it is easy to verify that the symbols satisfy the following estimates: $|s_1| \leq C$, $|s_{1t}| \leq C(1+\lambda)^{1/2}$, $|s_2| \leq C(1+\lambda)^{-1/2}$, $|s_2| \leq C$. Therefore, it follows from Theorem (4.14) the following statement.

Proposition 4.3. *(The case $k = 2$) Let $\mu \in C^1 \setminus [(-\infty,-1] \cup [1,\infty)]$. Then for any $\varphi_0 \in W^s$ and $\varphi_1 \in W^{s-1}$ there exists a unique solution $u(t,x) \in C[(0,T); W^s] \cap C^1[(0,T); W^{s-1}]$ of (4.104)–(4.107) and the estimate*

$$\sup_{t\in(0,T)}\left(\|u(t,x)|W_2^s(\Omega)\|+\|u_t(t,x)|W_2^{(s-1)m}(\Omega)\|\right)$$
$$\leq C\left(\|\varphi_0|W_2^s(\Omega)\|+\|\varphi_1|W_2^{(s-1)m}(\Omega)\|\right)$$

holds.

4.9 Additional notes

1. *General well-posedness conditions.* The well-posedness condition for general boundary value problems for differential and pseudo-differential equations was a focus of many researchers. There is a rich literature on this topic; see, e.g., [Pet96, Hor83, ADN69, Tre80, Går98]. From the general results we mention Petrovsky's "A-condition" (1945–48) [Pet96] for the well posedness of the Cauchy problem for higher order hyperbolic equations and for $2b$-parabolic equations. The Shapiro-Lopatinskiĭ condition (1953) [Sha53, Lop53] provides the well-posedness condition of general boundary value problems for elliptic equations. Assuming the equations and boundary conditions have coefficients not depending on x, these conditions are given in terms of roots of characteristic equations and, in essence, eliminate strong singularities arising in solution formulas. The boundary value problems in (4.3)–(4.4), in general, represent multi-point nonlocal boundary value problems for general pseudo-differential equations (with ΨDOSS coefficients), the type of which is not specified. Therefore, Theorem 4.5 generalizes both cases.

   Indeed, if $b_{kj}(\xi)=\delta_{jk}$, where δ_{jk} is the Kronecker symbol, and $t_{kj}=t_0$, then conditions (4.4) become the Cauchy conditions. In this case the condition of the well-posedness theorem (Theorem 4.5) essentially represents Petrovsky's "A-condition." Indeed, the latter declares the equation is hyperbolic, if $\Re(\lambda_j(\xi))\leq C$, where $\lambda_j(\xi)$ are roots of the characteristic equation. For instance, for the Laplace equation $\Re(\lambda_j(\xi))=\pm|\xi|,\ j=1,2$, which does not satisfy the "A-condition," and hence is not hyperbolic. Similarly, for the heat equation $\Re(\lambda(\xi))=|\xi|^2$. However, for the wave equation $\Re(\lambda_j(\xi))=0,\ j=1,2$, and hence it is hyperbolic. Now if one applies Theorem 4.5 to the Cauchy problem for an equation hyperbolic in the sense of Petrovsky, then

   1) the set $\overset{\circ}{M}$ consist of only 0, so $G_0=\mathbb{R}^n\setminus\overset{\circ}{M}$ is dense in $\mathbb{R}^n$, thereby getting well posedness in $\Psi_{G_0,p}(\mathbb{R}^n)$; and
   2) the symbols of solution operators have the form $s_j(t,\xi)=h_j(\xi)e^{t\lambda_j(\xi)}$ with functions $h_j(\xi)$ of polynomial growth, and hence the conditions in (4.50) are verified with some ℓ_k and s_j, thereby obtaining boundedness of solution operators in Sobolev spaces.

   These imply the well posedness of the Cauchy problem for hyperbolic equations in appropriate Sobolev spaces. Schwartz [Sch51] showed that the "A-condition" is necessary and sufficient for the well posedness of the Cauchy problem for equations with coefficients depending on t smoothly in the space of tempered distributions.
   What concerns elliptic (local) boundary value problems the condition of Theorem (4.5) essentially represents the Shapiro-Lopatinskiĭ condition. We note that any elliptic equation necessarily is of even order, that is $m=2p$. Moreover, the characteristic equation of an elliptic equation has p roots with positive imaginary parts, and p roots with negative imaginary parts. This implies that the number of boundary conditions is p. Under these assumptions, if one applies Theorem 4.5, then

   a) $\overset{\circ}{M}=\emptyset$, so $G_0=\mathbb{R}^n$, thereby getting well posedness in $\Psi_{\mathbb{R}^n,p}(\mathbb{R}^n)$; and

b) the symbols of solution operators have exponential decay at infinity, and hence the conditions in (4.50) are verified with some ℓ_k and s_j, thereby obtaining boundedness of solution operators in Sobolev spaces.

These imply the well posedness of boundary value problems for elliptic equations, satisfying the Shapiro-Lopatinskiĭ condition, in appropriate Sobolev spaces.

2. *General nonlocal boundary value problems.* As is was noted above, the boundary value problems in (4.3)–(4.4) are general nonlocal boundary value problems for ΨDOSS equations of any type, and therefore, in general, are not well posed in the sense of Hadamard. Singularities arising in solution formulas, also called a "small denominators problem," can be treated with the help of Diophantine equations/approximations [Pta84]. The construction and estimation of Green's function for Vallee-Poussin boundary value problem, that is with multi-point conditions $u(t_j) = a_j,\ t_j \in [a,b]$, for ordinary, or partial differential equations, are studied in the papers [Pok68, Tsk94] and for convolution type operators in [Nap82, N12]. Uniqueness classes for multi-point nonlocal boundary value problems for differential equations are studied in [Bor69, Bor71]. In works [Pta84, FJG08, Pul99, Zh14] various nonlocal boundary value problems, including multi-point and integral ones, and their applications, are studied. Theorem 4.5 is not valid if $p = \infty$. The main barrier here is non-denseness of $\Psi_{G,p}(\mathbb{R}^n)$ in Besov and Lizorkin-Triebel type spaces, if $p = \infty$. Saydamatov [Say06, Say07] modified the method developed in this chapter to the case $p = \infty$ and obtained existence results. Nazarova [Naz97] studied multi-point boundary value problems generated by a singular Bessel type operators.
3. *Uniqueness of polyharmonic function with given zero levels.* The question of uniqueness of polyharmonic function of order m, vanishing at m pairwise distinct hypersurfaces $S_j, j = 1,\dots,m$, is an important question in many applications, including polyharmonic interpolation [HK07], wavelet analysis [BRV05], etc. It is known [HK93] that the uniqueness does not hold for the set of arbitrary smooth hypersurfaces S_j. For instance, in the paper [Ata02] in the 2-D case, the author constructed two curves γ_1 and γ_2, with γ_1 inside γ_2, such that there exists a nonzero biharmonic inside γ_2 function $u(x,y)$, which vanishes on both curves γ_1, γ_2. In the paper [Ede75] of Edenhofer the uniqueness of a polyharmonic function of order m vanishing at m concentric hyperspheres is proved. As is shown in the paper [HK93], the uniqueness holds for m arbitrary (not necessary concentric) hyperspheres as well. Moreover, one of these hyperspheres can be replaced by a smooth hypersurface. The more general result is proved in [Ren08]: Let $\psi_1,\dots,\psi_k$ be nonhyperbolic, sign-changing irreducible polynomials in n variables of degree 2. If the polynomial f vanishes on the pairwise different sets $\{x \in \mathbb{R}^n : \psi_j(x) = 0\}$ for $j = 1,\dots,k$, and $\Delta^k f = 0$, then f is identically zero. The result obtained in Section 4.6.3 represents an analog of the Edenhofer's result. From this point of view the question, whether hyperplanes $\{t = t_k\} \subset \mathbb{R}^{n+1}$ can be replaced by other hyperplanes/hypersurfaces, is a challenging open question.
4. *Duhamel principle.* The role of the classical "Duhamel principle," introduced by Jean-Marie-Constant Duhamel [Du33] in 1833, is well known. The main idea of this famous principle is to reduce the Cauchy problem for a given linear inhomogeneous partial differential equation to the Cauchy problem for the corresponding homogeneous equation, which is more simpler to handle. The classical Duhamel principle is not directly applicable in the case of fractional differential equations. In Chapters 5 and 6 we establish fractional generalizations of the Duhamel principle for wide classes of fractional differential equations and DODEs.

Chapter 5
Initial and boundary value problems for fractional order differential equations

5.1 Introduction

In this chapter we will discuss boundary value problems for fractional order differential and pseudo-differential equations. For methodological clarity we first consider in detail the Cauchy problem for pseudo-differential equations of time-fractional order β, $m-1<\beta<m$, $(m \in \mathbb{N})$

$$D_*^\beta u(t,x) = A(D)u(t,x) + h(t,x), \quad t>0,\ x \in \mathbb{R}^n, \tag{5.1}$$

$$\frac{\partial^k u(0,x)}{\partial t^k} = \varphi_k(x), \quad x \in \mathbb{R}^n,\ k=0,\dots,m-1, \tag{5.2}$$

where $h(t,x)$ and φ_k, $k=0,\dots,m-1$, are given functions in certain spaces described later, $D=(D_1,\dots,D_n)$, $D_j = -i\frac{\partial}{\partial x_j}$, $j=1,\dots,n$, $A(D)$ is a ΨDOSS with a symbol $A(\xi) \in XS_p(G)$ defined in an open domain $G \subset \mathbb{R}^n$, and D_*^β is the fractional derivative of order $\beta>0$ in the sense of Caputo-Djrbashian (see Section 3.5)

$$D_*^\beta f(t) = \frac{1}{\Gamma(m-\beta)} \int_0^t \frac{f^{(m)}(\tau)d\tau}{(t-\tau)^{m-\beta-1}}, \quad t>0. \tag{5.3}$$

Then we will focus on general boundary value problems for distributed order differential equations

$$\int_0^\alpha A(\beta,t,D) D_*^\beta u(t,x)\mu(d\beta) = h(t,x), \quad t>0, x \in \mathbb{R}^n, \tag{5.4}$$

$$\sum_{j=0}^{m-1} \Gamma_{kj}(D) \frac{\partial^k u(t_{kj},x)}{\partial t^k} = \varphi_k(x), \quad k=0,\dots,m-1,\ x \in \mathbb{R}^n, \tag{5.5}$$

where $\alpha \in (m-1,m]$, $A(\beta,t,D)$ is a family of ΨDOSSs with symbols $A(\beta,t,\xi) \in XS_p(G)$, and μ is a finite measure with $\operatorname{supp}\mu = [a,b] \subset (0,\alpha]$, $b>m-1$, and

S. Umarov, *Introduction to Fractional and Pseudo-Differential Equations with Singular Symbols*, Developments in Mathematics 41,
DOI 10.1007/978-3-319-20771-1_5

$\Gamma_{kj}(D)$, $k,j=0,\dots,m-1$, are ΨDOSSs whose symbols $\Gamma_{kj}(\xi)\in XS_p(G), k,j=0,\dots,m-1$, $t_{kj}\in[0,T]$, $0<T<\infty$, and $h(t,x)$ and φ_k, $k=0,\dots,m-1$, are given functions/functionals. Equation (5.4) contains, as a particular case, pseudo-differential equations of fractional order

$$D_*^{\alpha_m}u(t,x)+\sum_{k=1}^{m-1}A_k(D)D_*^{\alpha_k}u(t,x)+A_0(D)u(t,x)=h(t,x),\quad t\in(0,T),x\in\mathbb{R}^n, \tag{5.6}$$

with the highest order $\alpha_m\in(m-1,m]$. In turn, the Cauchy problem (5.1), (5.2) is a particular case of boundary value problem (5.6), (5.5).

For the study of the Cauchy and multi-point boundary value problems we use the properties of pseudo-differential operators with singular symbols developed in Chapter 2 and the properties of fractional derivatives developed in Chapter 3. In the case of Cauchy problem we establish the fractional generalization of the Duhamel principle. The fractional Duhamel principle differs from the classic Duhamel principle. As we have seen in Section 4.7, in the case of integer order differential equations the Duhamel principle moves the source term $h(t,x)$ to the initial condition for the $(m-1)$-th derivative, changing it to $V_t^{(m-1)}(\tau,x)=h(\tau,x)$. In the fractional case the updated boundary condition contains a fractional derivative of the source function. Namely, the updated initial condition appears in the form $V_t^{(m-1)}(\tau,x)=D^{m-\alpha}h(\tau,x)$. This fact will be rigorously proved in Sections 5.5 and 6.4.

We recall that the Duhamel principle is not valid for multi-point problems. Sections 6.2–6.3 discuss general boundary value problems for distributed fractional order differential equations of the form (5.4)–(5.5). Here we derive a representation formula for a solution and study their continuity properties as mappings in appropriate function and distributions spaces.

Boundary value problems for elliptic operators with boundary conditions involving fractional order pseudo-differential operators is a subject of Section 5.7. The results obtained there generalize theorems proved in Section 4.5 of the previous chapter. These results also allow to study limits of fractional derivatives of harmonic functions in certain topologies, leading to a new representations of hyperfunctions as boundary values of fractional derivatives of harmonic functions (cf. with Section 4.6.5). Section 6.7 discusses the Cauchy problem for variable order differential equations with a piecewise constant order function and some of their applications to sub-diffusion processes.

5.2 Some examples of fractional order differential equations

To illustrate the Cauchy problem (5.1), (5.2) consider three examples.

1. **Time-fractional differential equation.** The first example is the time-fractional differential equation

$$D_*^{\alpha}u(t,x)=\Delta u(t,x),\quad t>0,\ x\in\mathbb{R}^n,\ \alpha>0,$$

where Δ is the Laplace operator. The Cauchy problem for this equation represents a fractional model of sub-diffusion processes in the case $0 < \alpha < 1$, and processes intermediate between diffusion and wave propagation in the case $1 < \alpha < 2$. In Section 5.4 we prove the relaxation property of the solution in the case $0 < \alpha < 1$, and the oscillation-relaxation property in the case $1 < \alpha < 2$.

2. **Space-fractional differential equations.** The second example is the space-fractional equation

$$\frac{\partial u(t,x)}{\partial t} = \mathbb{D}_0^{\beta} u(t,x), \quad t > 0,\ x \in \mathbb{R}^n,\ \beta > 0,$$

where $\mathbb{D}_0^{\beta}$ is the operator introduced in Section 3.8 and whose symbol is $-|\xi|^{\beta}$. This equation models jump processes (cf. with Example 2.2.0.4 in Chapter 2) arising in various applied sciences. We recall that the pseudo-differential operator $\mathbb{D}_0^{\beta}$ can be represented as the inverse operator to the fractional Riesz potential and can be written in the form

$$\mathbb{D}_0^{\beta} f(x) = \frac{1}{d_{n,l}(\beta)} \int_{\mathbb{R}^n} \frac{(\Delta_h^l f)(x)}{|h|^{n+\beta}} dh,$$

where Δ_h^l is the centered finite difference of an even order $l > \beta$ with the vector-step $h \in \mathbb{R}^n$ and $d_{n,l}(\beta)$ is the normalizing constant. In the one-dimensional case and under the condition $0 < \beta \leq 2$ this equation describes symmetric Lévy-Feller diffusion processes [Fel52]. Approximating random walk models for Lévy-Feller diffusion processes were presented by Gorenflo and Mainardi in a series of works [GM98-1, GM98-2, GM99].

3. **Space-time fractional differential equations.** The third example is the time- and space-fractional differential equation

$$D_*^{\alpha} u(t,x) = \mathbb{D}_0^{\beta} u(t,x), \quad t > 0,\ x \in \mathbb{R}^n,\ \alpha, \beta > 0.$$

This equation generalizes the equations in the first two examples and models sub-diffusive processes accompanying with jumps. With $\beta = 2$ we obtain the first, with $\alpha = 1$ the second equation. In Section 5.4 we prove the smoothness theorem for a solution of the Cauchy problem for this equation in the Sobolev spaces $H^s(\mathbb{R}^n)$, $s \in \mathbb{R}$, for all values $\alpha \in (0,2]$.

5.3 The Cauchy problem for fractional order pseudo-differential equations

In this section we will discuss the existence and uniqueness of a solution of Cauchy problem (5.1)–(5.2) in the spaces $\Psi_{p,G}(\mathbb{R}^n)$ and $\Psi'_{-G,p'}(\mathbb{R}^n)$, and derive a representation of a solution through the solution operators with symbols through the Mittag-Leffler function. Since the fractional Duhamel principle is valid for Cauchy problem (5.1)–(5.2) (see Section 5.5), it suffices to consider a homogeneous equation, so we will assume that $h(t,x)=0$ in equation (5.1). Thus, consider the Cauchy problem

$$D_*^{\beta} u(t,x) = A(D)u(t,x), \quad t>0,\ x\in\mathbb{R}^n, \tag{5.7}$$

$$\frac{\partial^k u(0,x)}{\partial t^k} = \varphi_k(x), \quad x\in\mathbb{R}^n,\ k=0,\ldots,m-1, \tag{5.8}$$

where $m-1<\beta<m$, and D_*^{β} is the Caputo-Djrbashian fractional derivative, and $A(D)$ is a ΨDOSS with the symbol $A(\xi)\in XS_p(G)$. The class $XS_p(G)$ in this section is either $C^{\infty}S_p(G)$, the class of symbols smooth in G, or class $CS_p(G)$ of symbols continuous in G. Recall that symbols may have arbitrary type of singularities on the boundary of G. Recall also that if $A(\xi)\in C^{\infty}(G)$, then the corresponding operator $A(D)$ is continuous in $\Psi_{G,p}(\mathbb{R}^n)$ (see Section 2.3), and if $A(\xi)\in C(G)$, then $A(D)$ is continuous in $\Psi_G(\mathbb{R}^n)$ (Section 2.4).

It should be noted that, for the dual theory, we always assume that the operator $A(D)$ on the right-hand side of equation (5.7) is replaced with its weak extension $A^w(D)=A(-D)$. Namely,

$$D_*^{\beta} u(t,x) = A(-D)u(t,x), \quad t>0,\ x\in\mathbb{R}^n, \tag{5.9}$$

$$\frac{\partial^k u(0,x)}{\partial t^k} = \varphi_k(x), \quad x\in\mathbb{R}^n,\ k=0,\ldots,m-1, \tag{5.10}$$

First, performing formal manipulations, we get a representation for the solution of Cauchy problem (5.7)–(5.8). We note that the solution operators are again pseudo-differential operators with symbols from the same class $XS_p(G)$. Then we study the properties of their symbols and use them to prove existence and uniqueness theorems. Applying formally the Fourier transform to equations (5.7) and(5.8), we get

$$D_*^{\beta} \hat{u}(t,\xi) = A(\xi)\hat{u}(t,\xi), \quad t>0,\ \xi\in G, \tag{5.11}$$

$$\frac{\partial^k \hat{u}(0,\xi)}{\partial t^k} = \hat{\varphi}_k(\xi), \quad \xi\in G,\ k=0,\ldots,m-1. \tag{5.12}$$

This is an initial value problem for an ordinary differential equation of fractional order β, that depends on the parameter $\xi\in G$. This problem is a particular case of Example 3.10 (in Section 3.5 of Chapter 3) with $\lambda=-A(\xi)$ and $a_k=\varphi_k(\xi), k=0,\ldots,m-1$. Due to formula (3.56), one obtains the representation

$$\hat{u}(t,\xi) = \sum_{k=1}^{m} J^{k-1}E_{\beta}(t^{\alpha}A(\xi))\hat{\varphi}_{k-1}(\xi)$$

for the solution of (5.11)–(5.12). In this formula J^{k-1} is the $(k-1)$-st order integration operator with the lower limit 0, and $E_{\beta}(z)$ is the Mittag-Leffler function; see Section 3.4. Introducing the notation

$$B_k(\beta;t,\xi) = J^{k-1}E_{\beta}(A(\xi)t^{\beta}), \quad k=1,\ldots,m \tag{5.13}$$

and applying the inverse Fourier transform, one obtains the solution of the Cauchy problem (5.7)–(5.8) in the form

$$u(t,x) = \sum_{k=1}^{m} B_k(\beta;t,D)\varphi_{k-1}(x). \tag{5.14}$$

Here the pseudo-differential operator $B_k(\beta;t,D)$, $k = 1,\ldots,m$, has the symbol $B_k(\beta;t,\xi)$ given by (5.13). We call it the k-th solution operator of the Cauchy problem (5.7)–(5.8).

Example 5.1. Let $0 < \beta < 1$. Then the symbol of the solution operator $B(\beta;t,D) \equiv B_1(\beta;t,D)$ is $B(\beta;t,\xi) = E_\beta(-A(\xi)t^\beta)$. It follows, in the particular case of a fundamental solution, corresponding to the initial condition $u(0,x) = \varphi(x) = \delta_0(x)$,

$$u(t,x) = E_\beta(-A(D)t^\beta)\delta_0(x) = \frac{1}{(2\pi)^n}\int_{\mathbb{R}^n} E_\beta(-A(\xi)t^\beta)e^{-ix\xi}d\xi. \tag{5.15}$$

Definition 5.1. *Let $m-1 < \beta < m$, $m \in \mathbb{N}$. A function*

$$u(t,x) \in C^{(m)}(t>0;\Psi_{G,p}(\mathbb{R}^n)) \cap C^{(m-1)}(t \geq 0;\Psi_{G,p}(\mathbb{R}^n))$$

is called a strong solution of the problem (5.7)–(5.8)*, if it satisfies the equation* (5.7) *and the initial conditions* (5.8) *pointwise.*

Definition 5.2. *Let $m-1 < \beta < m$, $m \in \mathbb{N}$. A function*

$$u(t,x) \in C^{(m)}(t>0;\Psi'_{-G,q}(\mathbb{R}^n)) \cap C^{(m-1)}(t \geq 0;\Psi'_{-G,q}(\mathbb{R}^n))$$

is called a weak solution of the problem (5.9)–(5.10)*, if it satisfies the equation* (5.9) *and the conditions* (5.10) *in the following sense: for arbitrary $v \in \Psi_{G,p}(\mathbb{R}^n)$ the equalities*

$$\begin{aligned} < D_*^\beta u(t,x), v(x) > &= < u(t,x), A(D)v(x) >, \quad t>0, \\ lim_{t\to+0} < u^{(k)}(t,x), v(x) > &= < \varphi_k(x), v(x) >, \quad k=0,\ldots,m-1, \end{aligned}$$

hold.

To prove existence and uniqueness theorems we need some auxiliary assertions. First we introduce some notations. Denote by $C^\ell[t \geq 0; C(G)]$ the space of functions $f(t,\xi)$ continuous with respect to ξ at any fixed $t \in [0,\infty)$, and having continuous derivatives up to order ℓ with respect to t on $t \geq 0$ for each fixed ξ. The similar meaning has the space $C^\ell[t>0;C(G)]$. Analogously, we denote by $C^\infty[t>0;C(G)]$ the space of functions infinitely differentiable with respect to the variable t in the interval $t \in (0,\infty)$.

Lemma 5.1. *For $k = 1,\dots,m$ the following assertions are valid:*

(i) $B_k(\beta;t,\xi) \in C^{m+k-2}[t \geq 0; C(G)]$;
(ii) $B_k(\beta;t,\xi) \in C^{\infty}[t > 0; C(G)]$.

Proof. The symbol $B_1(\beta;t,\xi) = E_\beta(-A(\xi)t^\beta)$ is a composition of the Mittag-Leffler function $E_\beta(z)$, which is an entire function, and of the function $\psi(\xi,t) = -A(\xi)t^\beta$. The continuity of $A(\xi)$ implies that for every fixed $t > 0$ the function $B_1(\beta;t,\xi)$ is also continuous in the domain G. Moreover, the function $\phi(t) = t^\beta$ is infinitely often differentiable at any point $t > 0$. Hence, $B_1(\beta;t,\xi) \in C^\infty[t > 0; C(G)]$. For $m-1 < \beta < m$ it is easy to check that $\phi(t) = t^\beta$ has all derivatives up to order $m-1$, which are continuous up to $t = +0$. This implies $B_1(\beta;t,\xi) \in C^{m-1}[t \geq 0; C(G)]$. Further, we have ($J^{k-1}$ acting with respect to the variable t)

$$B_k(\beta;t,\xi) = J^{k-1}E_\beta(A(\xi)t^\beta) = J^{k-1}B_1(\beta;t,\xi).$$

This function is $m+k-2$ times differentiable with respect to t for a fixed ξ due to the fact that the integration operator J^{k-1} increases the order of differentiability of $B_1(\beta;t,\xi)$ by $k-1$. Hence, $B_k(\beta;t,\xi) \in C^{m+k-2}[t \geq 0; C(G)]$.

Lemma 5.2. *Let $m-1 < \beta < m$. Then the following relations hold:*

(i) $\frac{\partial^{k-1}B_k(\beta;t,\xi)}{\partial t^{k-1}} \to 1$ *as* $t \to 0$ *for all* $k = 1,\dots,m$;
(ii) $\frac{\partial^\ell B_k(\beta;t,\xi)}{\partial t^\ell} \to 0$ *as* $t \to 0$ *for all* $\ell = 0,\dots,m-1,\ k = 1,\dots,m,\ l \neq k-1$.

Proof. In accordance with the definition of the solution operators, one has

$$\frac{\partial^{k-1}B_k(\beta;t,\xi)}{\partial t^{k-1}} = E_\alpha(A(\xi)t^\beta), \quad k = 1,\dots,m.$$

This relation and the fact that $E_\beta(0) = 1$ obviously implies (i). Now suppose that $k-1 < \ell \leq m-1$. It is not difficult to verify that the derivative $\frac{\partial^\ell B_k(\beta;t,\xi)}{\partial t^\ell}$ is a linear combination of expressions of the type

$$t^{\beta-j}E_\beta^{(j)}(A(\xi)t^\beta), \quad j \leq \ell. \tag{5.16}$$

Since $m-1 < \beta < m$, then all the functions in (5.16) tend to zero if $t \to +0$. In the case $0 \leq \ell < k-1$, one has

$$\begin{aligned}\frac{\partial^\ell B_k(\beta;t,\xi)}{\partial t^\ell} &= J^{k-\ell-1}E_\beta(A(\xi)t^\beta)\\ &= \frac{1}{(k-\ell-1)!}\int_0^t (t-\tau)^{k-\ell-2}E_\beta(A(\xi)t^\beta)d\tau \to 0 \text{ as } t \to +0.\end{aligned}$$

Theorem 5.1. *Let $m-1 < \beta < m$, $m \in \mathbb{N}$ and $\varphi_j \in \Psi_{G,p}(\mathbb{R}^n)$, $j = 0,\dots,m-1$. Then the Cauchy problem* (5.7)–(5.8) *has a unique strong solution. This solution is given by the representation* (5.14).

Proof. Let $\varphi_j \in \Psi_{G,p}(\mathbb{R}^n)$, $j = 0,\dots,m-1$. By construction each term on the right-hand side of (5.14) satisfies (at least formally) the equation (5.7) and, due to Lemma 5.2, conditions (5.8). It follows from Lemma 5.1 that for the symbol of the k-th solution operator, we have the inclusion

$$B_k(\beta;t,\xi) \in C^{m+k-2}[t \geq 0; C(G)] \cap C^{\infty}[t > 0; C(G)].$$

Theorem 2.1 (or Lemma 2.1 in the case of continuous symbols) yields

$$B_k(\beta;t,D)\varphi_{k-1}(x) \in \Psi_{G,p}(\mathbb{R}^n) \text{ for every fixed } t > 0,$$

$$B_k(\beta;t,D)\varphi_{k-1}(x) \in C^{m+k-2}(t \geq 0) \cap C^m(t > 0) \text{ for every fixed } x \in \mathbb{R}^n.$$

Hence,

$$u(t,x) \in C^m[t > 0; \Psi_{G,p}(\mathbb{R}^n)] \cap C^{m-1}[t \geq 0; \Psi_{G,p}(\mathbb{R}^n),$$

and $u(t,x)$ is a strong solution of the Cauchy problem (5.7)–(5.8). Its uniqueness follows from the representation formula (5.14) and the $\Psi_{G,p}(\mathbb{R}^n)$-continuity of pseudo-differential operators with symbols in $CS_p(G)$.

Theorem 5.2. *Let $m-1 < \beta < m$ and $\varphi_j \in \Psi'_{-G,q}(\mathbb{R}^n)$, $j = 0,\dots,m-1$. Then the Cauchy problem* (5.9)–(5.10) *has a unique weak solution. This solution is given by*

$$u(t,x) = \sum_{k=1}^{m} B_k(\beta;t,-D)\varphi_{k-1}(x), \tag{5.17}$$

where $B_k(\beta;t,-D)$, $k = 1,\dots,m$, is the k-th solution operator with the symbol $B_k(\beta;t,\xi)$.

Proof. Let $\varphi_j \in \Psi'_{-G,q}(\mathbb{R}^n)$, $j=0,\dots,m-1$. Theorem 2.5 implies that each term on the right-hand side of (5.17), namely, $u_k(t,x) = B_k(\beta;t,-D)\varphi_{k-1}(x), k=1,\dots,m$, is a functional in the space $\Psi'_{-G,q}(\mathbb{R}^n)$. Further, to prove the theorem we have to show that $u_k(t,x)$, $k = 1,\dots,m$, satisfies the equation (5.9) and initial conditions (5.10) in the weak sense. Let $v \in \Psi_{G,p}(\mathbb{R}^n)$ be an arbitrary function. We have

$$\begin{aligned} &< D_*^{\beta} u_k(t,x) - A(-D)u_k(t,x), v(x) > \\ &\quad =< [D_*^{\beta} B_k(\beta;t,-D) - A(-D)B_k(\beta;t,-D)]\varphi_{k-1}(x), v(x) > \\ &\quad = < \varphi_{k-1}(x), [D_*^{\beta} B_k(\beta;t,D) - A(D)B_k(\beta;t,D)]v(x) > . \end{aligned}$$

In accordance with the definition of $B_k(\beta;t,D)$, one has $D_*^{\beta} B_k(\beta;t,D) \equiv A(D)B_k$ $(\beta;t,D)$. Hence, $u_k(x,t)$ satisfies the equation (5.9) in the weak sense. Moreover, Lemma 5.2 implies

$$\frac{\partial^{k-1}}{\partial t^{k-1}} B_k(\alpha;0,D) = \delta_{k,\ell} \cdot I, \quad k = 1,\dots,m, \ \ell = 1,\dots,m,$$

where I is the identity operator and $\delta_{k,\ell}$ is Kronecker's symbol. This, in turn, yields initial conditions (5.10) in the weak sense.

Definition 5.3. Denote by $BC(t \geq 0, \Psi_G(\mathbb{R}^n))$ the set of functions $f(t,x)$, such that

1. $f(t,x)$ as a function of x is in the space $\Psi_G(\mathbb{R}^n)$ for every fixed $t > 0$;
2. $f(t,x) \in C[t > 0; \Psi_G(\mathbb{R}^n)]$;
3. $\|f(t,x)\|_2 \leq C$ for all $t \in [0,T]$, where T is an arbitrary (but fixed) positive number and C is a constant not depending on f (C may depend on T).

Similarly, let $BC(t \geq 0, \Psi'_{-G}(\mathbb{R}^n))$ be the set of functions $f(t,x)$ such that

(a) $f(t,x)$ is in $\Psi'_{-G}(\mathbb{R}^n)$ for every fixed $t > 0$;
(b) $f(t,x) \in C[t > 0; \Psi'_{-G}(\mathbb{R}^n)]$;
(c) for a fixed $T < \infty$ the estimate $| < f(t,x), v(x) > | \leq C$ holds for all $t \in [0,T]$ and for all $v \in \Psi_G(\mathbb{R}^n)$ with a constant C not depending on f (C may depend on v and T).

Lemma 5.3. *Let $m-1 < \beta < m$. For all $k = 1, \ldots, m$, and $\xi \in K \Subset G$ there exists a positive constant $C_{\beta,K}$ such that the inequality*

$$\left| D_*^\beta B_k(\beta; t, \xi) \right| \leq C_{\beta,K}, \quad t > 0, \tag{5.18}$$

holds.

Proof. Using the definition $D_*^\beta f = J^{m-\beta} D^m f$, we have

$$\begin{aligned} D_*^\beta B_k(\beta; t, \xi) &= J^{m-\beta} D^m B_k(\beta; t, \xi) \\ &= J^{m-\beta} D^m J^{k-1} E_\beta(A(\xi) t^\beta) = J^{m-\beta} D^{m-k+1} E_\beta(A(\xi) t^\beta) \\ &= \frac{1}{\Gamma(m-\beta)} \int_0^t (t-\tau)^{m-\beta-1} D_\tau^{m-k+1} E_\beta(A(\xi)\tau^\beta) d\tau, \end{aligned}$$

where D_τ is differentiation with respect to the variable τ, and $E_\beta(z)$ is the Mittag-Leffler function. It is easy to see that the most irregular case in the latter integral is $k = 1$, i.e.,

$$D_*^\beta B_1(\beta; t, \xi) = \frac{1}{\Gamma(m-\beta)} \int_0^t (t-\tau)^{m-\beta-1} D_\tau^m E_\beta(A(\xi)\tau^\beta) d\tau. \tag{5.19}$$

Due to Proposition 3.8, for any fixed $\xi \in K \Subset G$, we get the asymptotic behavior (for $A(\xi) \neq 0$)

$$\left| D_t^m E_\beta(A(\xi) t^\beta) \right| = O(\frac{1}{t^{m-\beta}}), \quad t \to 0.$$

Therefore, the integral on the right-hand side of (5.19) is absolute integrable and it does not exceed the expression

$$I_\beta = C_{\alpha,K} \int_0^t \frac{(t-\tau)^{m-1-\beta}}{t^{m-\beta}} d\tau,$$

where $C_{\beta,K}$ is a positive constant dependent on β and the compact K. Using the substitution $\tau = ts$ in this integral, one has

$$I_\beta = C_{\beta,K}\int_0^1 s^{(\beta+1-m)-1}(1-s)^{m-\beta-1}ds$$
$$= C_{\beta,K}B(\beta+1-m, m-\beta) = C_{\beta,K}\Gamma(m-\beta)\Gamma(\beta+1-m),$$

where $B(\cdot,\cdot)$ is Euler's beta function. Hence,

$$\left|D_*^\beta B_1(\beta;t,\xi)\right| \leq C_{\beta,K}\Gamma(\beta+1-m) < \infty,$$

proving (5.18).

Theorem 5.3. *Let $m-1<\beta<m$ and the conditions of Theorem 5.1 be fulfilled. Then the strong solution of the Cauchy problem* (5.7)–(5.8) *given by* (5.14) *possesses the following properties:*

(a) $u(t,x) \in C^\infty[t>0;\Psi_G(\mathbb{R}^n)]$;
(b) $D_*^\beta u(t,x) \in BC[t\geq 0,\Psi_G(\mathbb{R}^n)]$.

Proof. Part (a) is an implication of properties of symbols $B_k(\beta;t,\xi), k=1,\dots,m,$ of the solution operators $B_k(\beta;t,D)$, $k=1,\dots,m$, indicated in Part (ii) of Lemma 5.1. Let us prove Part (b) of the theorem. We need only to show condition 3) of Definition 5.3. Suppose $\operatorname{supp}\varphi_k \subset K_k \Subset G$, $k=0,\dots,m-1$. Using the Parseval equality and Lemma 5.3, we have

$$\|D_*^\beta u(t,x)|L_2\|^2 \leq \sum_{k=1}^m \int_{R^n} |D_*^\beta B_k(\alpha;t,D)\varphi_{k-1}(x)|^2 dx$$
$$= C\sum_{k=1}^m \int_{K_{k-1}} |D_*^\beta B_k(\beta;t,\xi)|^2 |F[\varphi_{k-1}](\xi)|^2 d\xi$$
$$\leq C\sum_{k=1}^m C_{\beta,K_{k-1}}^2 \|\varphi_{k-1}|L_2\|^2 < \infty.$$

Theorem 5.4. *Let $m-1<\beta<m$ and the conditions of Theorem 5.2 be fulfilled. Then the weak solution of the problem* (5.9)–(5.10) *given by* (5.17) *possesses the following properties:*

(i) $u(t,x) \in C^\infty[t>0;\Psi'_{-G}(\mathbb{R}^n)]$;
(ii) $D_*^\beta u(t,x) \in BC[t\geq 0,\Psi'_{-G}(\mathbb{R}^n)]$.

Proof. Let $\varphi_k \in \Psi_G(\mathbb{R}^n)$, $k=0,\dots,m-1$. Again Part (a) is an implication of properties of symbols $B_k(\beta;t,\xi)$, $k=1,\dots,m$, of the solution operators $B_k(\beta;t,-D)$, $k=1,\dots,m$, indicated in Part (ii) of Lemma 5.1. We prove Part (ii). Let ϕ be an arbitrary element of $\Psi_G(\mathbb{R}^n)$, whose Fourier transform has the support $\operatorname{supp}F[\phi]\subseteq K\Subset G$. We have seen in Section 2.4 that $F[\phi]\in H_{com}^{-s}(G)$ for some $s>n/2$. We recall that for $g\in\Psi'_{-G}(\mathbb{R}^n)$ the equality (see equation (1.96))

$$< g(x),\phi(x) > = (2\pi)^{-n} < F[g](\xi),F[\phi](\xi) > \tag{5.20}$$

holds. This equality shows that $F[g] \in H^s_{loc}(G) \subset L_{2,loc}(G)$, that is $F[g]$ is locally square-integrable in G. Moreover, one has the following estimate

$$\begin{aligned}|<g,\phi>| = (2\pi)^{-n}|<F[g](\xi),F[\phi](\xi)>| &\leq C\|F[g]|H^s(K)\|\|F[\phi]|H^{-s}(K)\| \\ &\leq C\|F[g]|L_2(K)\|\|F[\phi]|H^{-s}(K)\|. \end{aligned} \tag{5.21}$$

Taking into account relation (5.20) and estimates (5.21) and (5.18), we have

$$\begin{aligned}|<D^\beta_* u(t,x),\phi(x)>| &\leq \sum_{k=1}^m |<D^\beta_* B_k(\beta;t,-D)\varphi_{k-1}(x),\phi(x)>| \\ &\leq C\sum_{k=1}^m |<F[D^\beta_* B_k(\beta;t,-D)\varphi_{k-1}],F[\phi]>| \\ &\leq C_1\sum_{k=1}^m \|D^\beta_* B_k(\beta;t,\xi)F[\varphi_{k-1}]|L_2(K)\|\|F[\phi]|H^{-s}(K)\| \\ &\leq C_2\sum_{k=1}^m \sup_{\xi\in K}|D^\beta_* B_k(\beta;t,\xi)|\,\|F[\varphi_{k-1}]|L_2(K)\|\,\|F[\phi]|H^{-s}(K)\| \\ &\leq C_2 C_{\beta,K}\sum_{k=1}^m \|F[\varphi_{k-1}]|L_2(K)\|\,\|F[\phi]|H^{-s}(K)\| < \infty. \end{aligned}$$

5.4 Well posedness of the Cauchy problem in Sobolev spaces

In this section we extend the results on the existence and uniqueness obtained in the previous section to Sobolev spaces. We start with establishing a general result.

Consider the symbol $e(\xi) = E_\beta(A(\xi)t^\beta)$, where $t \geq 0$, $E_\beta(\cdot)$ is the Mittag-Leffler function, and $A(\xi)$ is a continuous symbol defined on a domain $G \subseteq \mathbb{R}^n$. Recall (see Remark 3.2 in Section 3.4), that for $0 < \beta < 2$ the Mittag-Leffler function $E_\beta(z)$ has asymptotic behavior $\sim exp(z^{1/\beta})$, $|z| \to \infty$, if $|\arg(z)| \leq \beta\pi/2$, and $E_\beta(z) \sim 1/|z|$, $|z| \to \infty$, if $\beta\pi/2 \leq |\arg(z)| \leq 2\pi - \beta\pi/2$. Therefore, if the symbol $A(\xi)$ is complex-valued, then $e(\xi)$ may have an exponential growth when $|\xi| \to \infty$, even though $A(\xi)$ has a polynomial growth at infinity.

Theorem 5.5. *Let $\varphi_k \in H^{s_k}(\mathbb{R}^n)$, $k = 0,\ldots,m-1$, $s_k \in \mathbb{R}$. Suppose that the estimate*

$$|E_\beta(A(\xi)t^\beta)| \leq C(1+|\xi|)^\ell, \quad 0 \leq t \leq T,\ \xi \in \mathbb{R}^n$$

holds for any $T > 0$ and some $\ell \in \mathbb{R}$. Then there exists a unique solution of the Cauchy problem (5.7)–(5.8) *in the space*

$$C^m[0 < t \leq T; H^{\ell_0}(\mathbb{R}^n)] \cap C^{m-1}[0 \leq t \leq T; H^{\ell_0}(\mathbb{R}^n)],$$

where $\ell_0 = \min\{s_0 - \ell, \dots, s_{m-1} - \ell\}$. *This solution is given by the formula*

$$u(t,x) = \sum_{k=1}^{m} \overline{B}_k(\beta;t,D)\varphi_{k-1}(x), \tag{5.22}$$

where $\overline{B}_k(\beta;t,D)$ *is the closure in the Sobolev space* $H^{s_k}(\mathbb{R}^n)$ *of the k-th solution operator* $B_k(\beta;t,D)$, $k = 0, \dots, m-1$, *with the symbol* $B_k(\beta;t,\xi)$ *defined in (5.13).*

Proof. Let $\varphi_k \in H^{s_k}(\mathbb{R}^n)$, $k = 0, \dots, m-1$. We can choose any domain G whose complement $\mathbb{R}^n \setminus G$ has zero measure. In particular, one can take $G = \mathbb{R}^n$. Then due to Theorem 1.21 the denseness $\overline{\Psi_G(\mathbb{R}^n)} = H^{s_k}(\mathbb{R}^n)$ holds for each $k = 0, \dots, m-1$. Hence, for each φ_k we have an approximating sequence of functions $\varphi_{k,N} \in \Psi_G(\mathbb{R}^n)$, $N = 0, 1, 2, \dots$, such that $\varphi_{k,N} \to \varphi_k$ in the topology of $\Psi_G(\mathbb{R}^n)$. For fixed N, due to Theorem 5.1, there exists a unique solution of the Cauchy problem (5.7)–(5.8) (where the initial data φ_k, $k = 0, 1 \dots, m-1$, are replaced by $\varphi_{k,N}$, $k = 0, 1 \dots, m-1$) represented by the formula

$$u_N(t,x) = \sum_{k=1}^{m} B_k(\beta;t,D)\varphi_{k-1,N}(x).$$

We recall that $B_1(\beta;t,\xi) = E_\beta(A(\xi)t^\beta)$. Since this symbol satisfies the estimate $|B_1(\beta;t,\xi)| \le C(1+|\xi|)^\ell$, $0 \le t \le T$, $\xi \in \mathbb{R}^n$, it follows from Theorem 2.7 (the case $q = 2$) that there exists a unique continuous closure $\overline{B}_1(\beta;t,D)$ of the operator $B_1(\beta;tD)$, such that $\overline{B}_1(\beta;t,D) : H^{s_0}(\mathbb{R}^n) \to H^{s_0-\ell}(\mathbb{R}^n)$ is continuous. Further, it is not difficult to verify that if $E_\beta(A(\xi)t^\beta)$ satisfies the condition of the theorem then its k-th integral with respect to t also satisfies the same condition, namely

$$|B_k(\beta;t,\xi)| \le C(1+|\xi|)^\ell, \quad 0 \le t \le T,\ \xi \in \mathbb{R}^n.$$

Indeed, for $k = 2$ we have

$$\begin{aligned} |B_2(\beta;t,\xi)| = \left|\int_0^t E_\beta(A(\xi)\tau^\beta)d\tau\right| &\le Ct|E_\beta(A(\xi)t^\beta)| \\ &\le C_1(1+|\xi|)^\ell, \quad 0 \le t \le T, \xi \in \mathbb{R}^n. \end{aligned}$$

Therefore, there exists a unique continuous closure $\overline{B}_2(\beta;t,D) : H^{s_1}(\mathbb{R}^n) \to H^{s_1-\ell}(\mathbb{R}^n)$ of the operator $B_2(\beta;tD)$. By induction, for each $k = 3, \dots, m-1$, there is a unique continuous closure $\overline{B}_k(\beta;t,D) : H^{s_{k-1}}(\mathbb{R}^n) \to H^{s_{k-1}-\ell}(\mathbb{R}^n)$ of the operator $B_k(\beta;tD)$. Thus for the solution $u(t,x)$ we have representation (5.22). The k-th term in this representation is an element of $H^{s_{k-1}-\ell}(\mathbb{R}^n)$ for each fixed $t \in [0,T]$. Therefore, $u(t,x) \in H^{\ell_0}(\mathbb{R}^n)$, where $\ell_0 = \min\{s_0 - \ell, \dots, s_{m-1} - \ell\}$.

Now we apply Theorem 5.5 to the particular case $A(D) = \mathbb{D}_0^\alpha$, where $0 < \alpha < 2$ and the operator $\mathbb{D}_0^\alpha$ is defined in Section 3.8. For $\alpha = 2$ we assume that $A(D)$ is the Laplace operator, that is $\mathbb{D}_0^2 = \Delta$. With this convention one can assume that

$0 < \alpha \leq 2$. We also assume $0 < \beta \leq 2$. In other words we consider the Cauchy problem

$$D_*^\beta u(t,x) = \mathbb{D}_0^\alpha u(t,x), \quad t > 0,\ x \in \mathbb{R}^n, \tag{5.23}$$

$$\frac{\partial^k u(0,x)}{\partial t^k} = \varphi_k(x), \quad x \in \mathbb{R}^n,\ k = 0,\dots,m-1, \tag{5.24}$$

where $m = 1$ or $m = 2$. We recall that the operator $\mathbb{D}_0^\alpha$ has the symbol $-|\xi|^2$ (see Theorem 3.4) and acts continuously from $H^s(\mathbb{R}^n)$ to $H^{s-\alpha}(\mathbb{R}^n)$ (Example 2.2).

The Cauchy problem (5.23)–(5.24) model sub-diffusion $(0 < \beta < 1)$ and super-diffusion $(1 < \beta < 2)$ processes with jumps. Therefore, it is convenient to proceed these two cases separately, as well as integer values $\beta = 1$ and $\beta = 2$.

1. **The case** $0 < \beta < 1$. Consider the Cauchy problem for the space-time fractional equation

$$D_*^\beta u(t,x) = \mathbb{D}_0^\alpha u(t,x), \quad t > 0,\ x \in \mathbb{R}^n, \tag{5.25}$$

$$u(0,x) = \varphi(x), \quad x \in \mathbb{R}^n. \tag{5.26}$$

Then we have only one solution operator, namely, $B_1(\beta;t,D)$ whose symbol is $B_1(\beta;t,\xi) = E_\beta\Big(-|\xi|^\alpha t^\beta\Big)$. The Mittag-Leffler function $E_\alpha(-t)$, $t > 0$, has the asymptotic behavior $E_\alpha(-t) = O(1/t)$ when $t \to \infty$ (see Sections 3.4 and 3.13). Using this fact, we obtain

$$|B_1(\beta;t,\xi)| \leq C(1+|\xi|)^{-\alpha}, \quad 0 \leq t \leq T,\ \xi \in \mathbb{R}^n. \tag{5.27}$$

Applying Theorem 5.5 and estimate (5.27) we get the following result:

Theorem 5.6. *Let $0 < \beta < 1$ and $\varphi \in H^s(\mathbb{R}^n)$. Then the Cauchy problem* (5.25)–(5.26) *has a unique solution in the space $C^\infty(t > 0; H^{s+\alpha}(\mathbb{R}^n)) \cap C(t \geq 0; H^s(\mathbb{R}^n))$. This solution is given by the formula*

$$u(t,x) = \overline{B}_1(\beta;t,D)\varphi(x). \tag{5.28}$$

Moreover, there exists a positive constant C, such that for the solution the estimate

$$\|u(t,x)|H^{s+\alpha}\| \leq CT^{-\beta}\|\varphi|H^s\| \tag{5.29}$$

holds for all $t > T$.

Proof. We only need to prove estimate (5.29). Due to Proposition 3.9 in the case $0<\beta<1$ and $\lambda = |\xi|^\alpha > 0$, we have for the symbol $B_1(\beta;t,\xi) = E_\beta(-|\xi|^\alpha t^\beta)$ the representation (see Proposition 3.9)

$$B_1(\beta;t,\xi) = \int_0^\infty e^{-rt|\xi|^{2\alpha/\beta}} K_\beta(r)dr, \tag{5.30}$$

where

$$K_\beta(r) = \frac{1}{\pi} \frac{r^{\beta-1}\sin(\beta\pi)}{r^{2\beta} + 2r^\beta \cos(\beta\pi) + 1}.$$

For $t \geq T$ the representation (5.30) can be rewritten in the form

$$B_1(\beta;t,\xi) = \int_0^\infty e^{-rT|\xi|^{\frac{2\alpha}{\beta}}} e^{-r(t-T)|\xi|^{\frac{2\alpha}{\beta}}} K_\beta(r)dr,$$

which gives the estimate[1]

$$|B_1(\beta;t,\xi)| \leq \int_0^\infty e^{-rT|\xi|^{\frac{2\alpha}{\beta}}} K_\beta(r)dr, \quad t \geq T. \tag{5.31}$$

Notice that the right-hand side of (5.31) represents the Laplace transform of $K_\beta(r)$ evaluated at $T|\xi|^{2\alpha/\beta}$:

$$|B_1(\beta;t,\xi)| \leq L[K_\beta]\left(T|\xi|^{2\alpha/\beta}\right), \quad t \geq T.$$

Moreover, $K_\beta(r) \sim r^{\beta-1}$, as $r \to 0$. Therefore, due to Watson's lemma (Proposition 1.10), we obtain for large values of $T|\xi|^{\frac{2}{\alpha}}$ the following asymptotic relation for large T :

$$L[K_\beta]\left(T|\xi|^{2\alpha/\beta}\right) = O\left(\frac{1}{T^\beta|\xi|^{2\alpha}}\right), \quad T >> 1.$$

This implies

$$|B_1(\beta;t,\xi)| \leq \frac{C_\beta}{T^\beta}(1+|\xi|^2)^{-\alpha}, \quad t > T, \xi \in \mathbb{R}^n.$$

Using the latter, we obtain

$$\begin{aligned} \|u(t,x)|H^{s+\alpha}\|^2 &= \int_{\mathbb{R}^n} |B_1(\beta;t,\xi)|^2|\hat{\varphi}|^2(1+|\xi|^2)^{s+\alpha}d\xi \\ &\leq \frac{C_\beta^2}{T^{2\beta}}\|\varphi|H^s\|^2, \quad t \geq T, \end{aligned}$$

proving (5.29).

Corollary 5.1. *Under the conditions of Theorem 5.6 for the solution of the Cauchy problem* (5.25)–(5.26) *the following asymptotic relation holds:*

$$\lim_{t\to\infty} \|u(t,x)H^{s+\alpha}\| = 0. \tag{5.32}$$

Remark 5.1. If the initial function $\varphi(x) \geq 0$ and $\int_{\mathbb{R}^n} \varphi(x)dx = 1$, then the solution $u(t,x) > 0$ for all $t > 0$. This fact follows from (5.28) and Bochner's theorem, since

[1] $K_\beta(r) > 0$ for all $r > 0$ if $0 < \beta < 1$.

$F[\varphi](\xi)$ is positive definite and $K_\beta(r) > 0$, $r > 0$. Therefore, property (5.32) of the solution of the Cauchy problem (5.25)–(5.26) expresses its relaxation property. We note also that if $s > \frac{n}{2} - \alpha$, then it follows from the Sobolev embedding theorem that the convergence to zero is uniform for all $x \in \mathbb{R}^n$.

2. **The case** $1 < \beta < 2$. Consider the Cauchy problem for the space-time fractional differential equation

$$D_*^\alpha u(t,x) = \mathbb{D}_0^\alpha u(t,x), \quad t > 0,\ x \in \mathbb{R}^n, \tag{5.33}$$

$$u(0,x) = \varphi(x),\ u_t(0,x) = \psi(x), \quad x \in \mathbb{R}^n. \tag{5.34}$$

In this case we have two solution operators, $B_1(\beta;t,D)$ and $B_2(\beta;t,D)$ with the symbols $B_1(\beta;t,\xi) = E_\beta(-|\xi|^\alpha t^\beta)$ and $B_2(\beta;t,\xi) = JE_\beta(-|\xi|^\alpha t^\beta)$, respectively. The symbol $B_2(\beta;t,\xi)$ has the same asymptotics (5.27) for $|\xi| \to \infty$ as $B_1(\beta;t,\xi)$.

Theorem 5.7. *Let* $1 < \beta < 2$, $\varphi \in H^s(\mathbb{R}^n)$, *and* $\psi \in H^s(\mathbb{R}^n)$. *Then the Cauchy problem* (5.33)–(5.34) *has a unique solution*

$$u(t,x) \in C^\infty[t > 0; H^{s+\alpha}(\mathbb{R}^n)] \cap C^1[t \geq 0; H^s(\mathbb{R}^n)].$$

This solution is given by the formula

$$u(t,x) = \overline{B}_1(\beta;t,D)\varphi(x) + \overline{B}_2(\beta;t,D)\psi(x). \tag{5.35}$$

Moreover, there exists a positive constant C, such that for the solution the estimate

$$\|u(t,x)|H^{s+\alpha}\| \leq \frac{C}{T^{\beta-1}}\left(\frac{\|\varphi|H^s\|}{T} + \|\psi|H^s\|\right), \tag{5.36}$$

holds for all $t > T$.

Proof. Again we need only to prove estimate (5.36). Other conclusions of the theorem follow from Theorem 5.5 and estimate (5.27). Due to Proposition 3.9 for the symbol $B_1(\beta;t,\xi) = E_\beta(-|\xi|^\alpha t^\beta)$ with $1 < \beta < 2$ and $\lambda = |\xi|^\alpha > 0$, the representation

$$B_1(\beta;t,\xi) = \int_0^\infty e^{-rt|\xi|^{\frac{2\alpha}{\beta}}} K_\beta(r)dr + \frac{2}{\beta} e^{t|\xi|^{\frac{2\alpha}{\beta}} \cos\frac{\pi}{\beta}} \cos\left(t|\xi|^{\frac{2\alpha}{\beta}} \sin\frac{\pi}{\beta}\right) \tag{5.37}$$

holds. One can verify by integration of (5.37) that for the symbol $B_2(\beta;t,\xi) = JE_\beta(-|\xi|^\alpha t^\beta)$ the following representation

$$\begin{aligned} B_2(\beta;t,\xi) &= |\xi|^{-2\alpha/\beta} \int_0^\infty e^{-rt|\xi|^{\frac{2\alpha}{\beta}}} K_{\beta,1}(r)dr \\ &+ \frac{2|\xi|^{2\alpha/\beta}}{\beta} e^{t|\xi|^{\frac{2\alpha}{\beta}} \cos\frac{\pi}{\beta}} \cos\left[t|\xi|^{\frac{2\alpha}{\beta}} \sin\left(\frac{\pi}{\beta}\right) - \frac{\pi}{2}\right], \end{aligned} \tag{5.38}$$

holds, where $K_{\beta,1}(r) = -r^{-1}K_\beta(r)$. Obviously $\cos(\pi/\beta) < 0$ for $1 < \beta < 2$, so the second terms in representations (5.37) and (5.38) have exponential decay when $|\xi| \to \infty$. Therefore, asymptotic behaviors of symbols $B_1(\beta;t,\xi)$ and $B_2(\beta;t,\xi)$ are determined by the first terms of these representations. Thus, similar to the previous case, we use Watson's lemma to obtain the asymptotics ($t \geq T$)

$$L[K_\beta]\left(T|\xi|^{2\alpha/\beta}\right) = O\left(\frac{1}{T^\beta|\xi|^{2\alpha}}\right), \; T >> 1,$$

and

$$L[K_{\beta,1}]\left(T|\xi|^{2\alpha/\beta}\right) = O\left(\frac{1}{(T|\xi|^{2\alpha/\beta})^{\beta-1}}\right), \; T >> 1,$$

which imply

$$|B_1(\beta;t,\xi)| \leq \frac{C_1}{T^\beta}(1+|\xi|^2)^{-\alpha}, \quad t > T, \xi \in \mathbb{R}^n.$$

and

$$|B_2(\beta;t,\xi)| \leq \frac{C_2}{T^{\beta-1}}(1+|\xi|^2)^{-\alpha}, \quad t > T, \xi \in \mathbb{R}^n.$$

Taking these estimates into account we have

$$\|B_1(\beta;t,D)\varphi|H^{s+\alpha}\| \leq \frac{C_1}{T^\beta}\|\varphi|H^s\|, \quad \|B_2(\beta;t,D)\psi|H^{s+\alpha}\| \leq \frac{C_2}{T^{\beta-1}}\|\psi|H^s\|,$$

for all $t \geq T$, which imply the estimate (5.36).

Corollary 5.2. *Let the conditions of Theorem 5.7 be verified. Then for the solution of the Cauchy problem* (5.33)–(5.34) *the following asymptotic relation holds:*

$$\|u(t,x)|H^{s+\alpha}\| = o(1), \quad t \to \infty. \tag{5.39}$$

Remark 5.2. The property (5.39) of the solution of the Cauchy problem (5.33)–(5.34) expresses its oscillation-relaxation property. Oscillation of the solution is due to second terms in presentations (5.37) and (5.38) of the symbols of solution operators. We note also that if $s > \frac{n}{2} - \alpha$, then the convergence to zero holds uniformly for all $x \in \mathbb{R}^n$.

Apart from non-integer β, for $\beta = 1$ and $\beta = 2$ we have a pure relaxation and pure oscillation, respectively. We formulate the corresponding results without proofs, which can easily be obtained analogously to the classical cases.

3. **The case** $\beta = 2$. First we consider the Cauchy problem for the space-fractional equation with the time derivative of second order:

$$\frac{\partial^2 u(t,x)}{\partial t^2} = \mathbb{D}_0^\alpha u(t,x), \quad t > 0, \; x \in \mathbb{R}^n, \; \beta > 0, \tag{5.40}$$

$$u(0,x) = \varphi(x), \; u_t(0,x) = \psi(x), \quad x \in \mathbb{R}^n. \tag{5.41}$$

It is easy to see that the solution operators have the symbols

$$B_1(t,\xi) = \cos(t|\xi|^\alpha), \; B_2(t,\xi) = \frac{\sin(t|\xi|^\alpha)}{|\xi|^\alpha},$$

satisfying the estimates

$$|B_1(t,\xi)| \leq 1, \; |B_2(t,\xi)| \leq \frac{1}{1+|\xi|^\alpha}, \quad \xi \in \mathbb{R}^n.$$

Theorem 5.8. *Let $\alpha > 0$, $\varphi \in H^s(\mathbb{R}^n)$, $\psi \in H^{s-\alpha}(\mathbb{R}^n)$. Then the problem (5.40), (5.41) has a unique solution in the space*

$$C[t \geq 0; H^s(\mathbb{R}^n)] \cap C^1[t \geq 0; H^{s-\alpha}(\mathbb{R}^n)].$$

This solution is given by the formula

$$u(t,x) = \overline{B}_1(t,D)\varphi(x) + \overline{B}_2(t,D)\psi(x).$$

4. **The case** $\beta = 1$. The solution operator of the Cauchy problem for the space-fractional equation with the time derivative of the first order

$$\frac{\partial u(t,x)}{\partial t} = \mathbb{D}_0^\alpha u(t,x), \quad t > 0, \, x \in \mathbb{R}^n, \, \alpha > 0, \tag{5.42}$$

$$u(0,x) = \varphi(x), \quad x \in \mathbb{R}^n, \tag{5.43}$$

has the symbol

$$B_1(t,\xi) = e^{-t|\xi|^\alpha},$$

and for every $l > 0$ there exists a positive constant C_l such that the estimate

$$|B_1(t,\xi)| \leq C_l(1+|\xi|)^{-\ell}, \quad \xi \in, \mathbb{R}^n$$

holds. Correspondingly, we arrive at the result:

Theorem 5.9. *Let $\beta > 0$ and $\varphi \in H^s(\mathbb{R}^n)$. Then the problem (5.42), (5.43) has a unique solution in the space*

$$C^\infty[t > 0; \cap_{s \in \mathbb{R}} H^s(\mathbb{R}^n)] \cap C[t \geq 0; H^s(\mathbb{R}^n)].$$

This solution is given by the formula

$$u(t,x) = \overline{B}_1(t,D)\varphi(x).$$

Remark 5.3. The technique used above for the Sobolev spaces $H^s(\mathbb{R}^n)$ remains applicable for the general Sobolev spaces $H_p^s(\mathbb{R}^n)$, Besov spaces $B_{pq}^s(\mathbb{R}^n)$, and Triebel-Lizorkin spaces $F_{pq}^s(\mathbb{R}^n)$ as well, provided $1 < p,q < \infty$.

5.5 Fractional Duhamel principle

In this section we establish a fractional analog of the Duhamel principle, which allows to extend the results obtained in previous sections to inhomogeneous equation. Thus, consider the Cauchy problem for inhomogeneous time-fractional pseudo-differential equations

$$D_*^{\beta} u(t,x) = A(D_x)u(t,x) + h(t,x), \quad t > 0,\ x \in R^n, \tag{5.44}$$

$$\frac{\partial^k u}{\partial t^k}(0,x) = \varphi_k(x), \quad x \in R^n,\ k = 0,\dots,m-1, \tag{5.45}$$

where $\beta \in (m-1,m]$, $m \geq 1$ is an integer; $h(t,x)$ and $\varphi_k(x)$, $k = 0,\dots,m-1$ are given functions in certain spaces defined later; $D_x = (D_1,\dots,D_n)$ $D_j = -i\frac{\partial}{\partial x_j}$, $j = 1,\dots,n$; $A(D_x)$ is a pseudo-differential operator with a symbol $A(\xi)$ defined in an open domain $G \subseteq R^n$.

In this section we also prove the Duhamel principle for fractional order differential equations with the Riemann-Liouville derivative (Subsection 5.5.3). More general case of the Duhamel principle for inhomogeneous distributed order abstract differential-operator equations will be discussed in Section 6.4.

Note that the classical Duhamel principle is not valid for fractional order inhomogeneous differential equations. The fractional generalization of the Duhamel principle established below can be applied directly to inhomogeneous fractional order differential equations reducing them to corresponding homogeneous equations.

Recall the following relationship between the Riemann-Liouville and Caputo-Djrbashian fractional derivatives (cf. (3.53) with $a = 0$):

$$D_+^{\beta} f(t) = D_*^{\beta} f(t) + \sum_{k=0}^{m-1} \frac{f^{(k)}(0)}{\Gamma(k-\beta+1)} t^{k-\beta}. \tag{5.46}$$

Recall also that for the Cauchy problem (5.44), (5.45) in the homogeneous case (i.e., for $f(t,x) \equiv 0$ in equation (5.44)) the following representation formula for a solution was obtained in (5.14):

$$u(t,x) = \sum_{k=1}^{m} J^{k-1} E_{\beta}(t^{\beta} A(D_x))\varphi_{k-1}(x), \tag{5.47}$$

where J^k is the k-th order integral operator, $E_{\beta}(t^{\beta}A(D_x))$ is a pseudo-differential operator with the symbol $E_{\beta}(t^{\beta}A(\xi))$ and $E_{\beta}(z)$ is the Mittag-Leffler function (see Section 3.4).

Lemma 5.4. *For all $\beta \in (m-1,m]$ and $\gamma \geq 0$ the relation*

$$J^{\gamma+\beta} f(t) = J^{\gamma+m} D_+^{m-\beta} f(t) \tag{5.48}$$

holds.

Proof. Obviously, the relationship (5.48) is fulfilled, if $\beta = m$. Let $m-1 < \beta < m$. Then $0 < m-\beta < 1$. It follows from (5.46) that

$$D_+^{m-\beta} f(t) = D_*^{m-\beta} f(t) + \frac{f(0)t^{\beta-m}}{\Gamma(1-m+\beta)}, \quad t > 0. \tag{5.49}$$

Taking into account (5.49) and the definition of the Caputo-Djrbashian fractional derivative $D_*^{m-\beta} = J^{\beta+1-m}D$ (see (3.48)), we have

$$\begin{aligned} J^{\gamma+m}D_+^{m-\beta} f(t) &= J^{\gamma+m}D_*^{m-\beta} f(t) + \frac{f(0)}{\Gamma(1-m+\beta)} J^{\gamma+m} t^{\beta-m} \\ &= J^{\gamma+\beta} J f'(t) + \frac{f(0)}{\Gamma(\beta+\gamma+1)} t^{\beta+\gamma}. \end{aligned}$$

Further, using (3.2) with $n = 1$, we obtain

$$J^{\gamma+m}D_+^{m-\beta} f(t) = J^{\beta+\gamma}\Big[f(t) - f(0)\Big] + \frac{f(0)}{\Gamma(\beta+\gamma+1)} t^{\beta+\gamma}.$$

The last equation immediately implies (5.48), if we take into account the well-known formula $J^\delta 1 = \frac{t^\delta}{\Gamma(\delta+1)}$, $\delta > 0$.

Corollary 5.10 *Assume $f(0) = 0$. Then for all $\beta \in (m-1, m]$ and $\gamma \geq 0$ the relation*

$$J^{\gamma+\beta} f(t) = J^{\gamma+m} D_*^{m-\beta} f(t) \tag{5.50}$$

holds.

Proof. We notice that $m - \beta < 1$. Now the relation (5.50) immediately follows from (5.48) and (5.46).

5.5.1 Fractional Duhamel principle: the case $0 < \beta < 1$

The following heuristic observation is useful in understanding of the fractional Duhamel principle. Assume $0 < \beta < 1$. Consider the Cauchy problem for the non-homogeneous fractional heat equation

$$D_*^\beta u(t,x) = k_\beta \Delta u(t,x) + h(t,x), \tag{5.51}$$

with the initial condition $u(0,x) = 0$. Using the notations introduced in Section 4.7, the solution is represented as the Duhamel integral

$$u(t,x) = \int_0^t V(t,\tau,x)d\tau,$$

where $V(t,\tau,x)$ is a solution of the Cauchy problem for the corresponding homogeneous equation

$$_{\tau}D_*^{\beta}V(t,\tau,x) = k_{\beta}\Delta V(t,\tau,x) = 0, \quad t > \tau, \quad x \in R^n, \tag{5.52}$$

with the initial condition $V(\tau,\tau,x) = H(\tau,x)$. Here $_{\tau}D_*^{\beta}$ is the Caputo-Djrbashian fractional derivative of order β with the initial point τ, and $H(\tau,x)$ is a function related to $h(t,x)$ in a certain way. In order to see this relationship between $H(t,x)$ and $h(t,x)$, suppose that t is small and the initial temperature is zero. Then ignoring the heat flow during the time interval (0,t), that is $k\Delta u \sim 0$, the temperature change is $D_*^{\beta}u(t,x) \sim h(t,x)$, or taking into account $D_*^{\beta} = J^{1-\beta}D$, we have $Du(t,x) \sim D_+^{1-\beta}h(t,x)$. For small t this implies $V(0,0,x) \sim Du(0) \sim D_*^{1-\beta}h(0,x)$. Repeating these heuristic calculations for small time interval $(\tau,\tau+\varepsilon)$ we obtain the relationship $H(\tau,x) \sim D_+^{1-\beta}h(\tau,x)$. Hence, one can expect that the initial condition for $V(t,\tau,x)$ in equation (5.52) has the form

$$V(t,\tau,x)_{|t=\tau} = D_+^{1-\beta}h(\tau,x),$$

where $h(t,x)$ is the function on the right-hand side of equation (5.51). Below we will prove this fact rigorously.

First we formulate a formal fractional generalization of the Duhamel principle and then we discuss applications of this principle in various situations.

Theorem 5.11. *Suppose that $V(t,\tau,x)$, $0 \le \tau \le t, x \in R^n$, is a solution of the Cauchy problem for homogeneous equation*

$$_{\tau}D_*^{\beta}V(t,\tau,x) - A(D_x)V(t,\tau,x) = 0, \quad t > \tau,\ x \in \mathbb{R}^n, \tag{5.53}$$

$$V(\tau,\tau,x) = D_*^{1-\beta}h(\tau,x), \quad x \in \mathbb{R}^n, \tag{5.54}$$

where $h(t,x) \in C^1[t \ge 0; \mathscr{D}(A(D_x))]$, and satisfies the condition $f(0,x) = 0$. Then

$$v(t,x) = \int_0^t V(t,\tau,x)d\tau \tag{5.55}$$

is a solution of the inhomogeneous Cauchy problem

$$D_*^{\beta}v(t,x) - A(D_x)v(t,x) = h(t,x), \tag{5.56}$$

$$v(0,x) = 0. \tag{5.57}$$

Proof. Notice that in accordance with (5.47) a solution of the Cauchy problem (5.53)-(5.54) is represented in the form

$$V(t,\tau,x) = E_{\beta}((t-\tau)^{\beta}A(D_x))D_*^{1-\beta}h(\tau,x). \tag{5.58}$$

Further, we apply the operator J^{β} to both sides of equation (5.56) and use the relation $J^{\beta}D_*^{\beta}v(t,x) = v(t,x) - v(0,x)$, to obtain an integral equation

$$v(t,x) - J^{\beta}A(D_x)v(t,x) = J^{\beta}h(t,x).$$

It follows from the general theory of operator equations that a solution of the last equation can be represented in the form

$$v(t,x) = \sum_{n=0}^{\infty} J^{\beta n+\beta} A^n(D_x)h(t,x). \tag{5.59}$$

Further, (5.50) implies that if $\gamma = \beta n$ and $m = 1$, then for arbitrary function $g(t)$ satisfying the condition $g(0) = 0$, the equality $J^{\beta n+\beta} g(t) = J^{\beta n+1} D_*^{1-\beta} g(t)$ holds. Taking this into account we have

$$\begin{aligned} v(t,x) &= \sum_{n=0}^{\infty} J^{\beta n+1} A^n(D_x) D_*^{1-\beta} h(t,x) \\ &= \int_0^t \sum_{n=0}^{\infty} \frac{(t-\tau)^{\beta n} A^n(D_x)}{\Gamma(\beta n+1)} D_*^{1-\beta} h(\tau,x) d\tau \\ &= \int_0^t E_\beta((t-\tau)^\beta A(D_x)) D_*^{1-\beta} h(\tau,x) d\tau. \end{aligned} \tag{5.60}$$

Due to equation (5.58) the integrand in (5.60) coincides with $V(t,\tau,x)$, and we obtain (5.55).

Remark 5.4. 1. The series in (5.59) converges, for instance, if $h \in C^1[t \geq 0; \Psi_{G,p}(\mathbb{R}^n)]$. In this case there exist positive numbers C and a, not depending on x and t, such that $p_k(A^n(D_x)h) \leq Ca^n p_m(h)$. It follows from (5.59) that

$$\begin{aligned} p_k(v(t,x)) &\leq \sum_{n=0}^{\infty} \left| \frac{1}{\Gamma(\beta n+\beta)} \int_0^t (t-\tau)^{\beta n+\beta-1} p_k(A^n(D_x)h) \right| \\ &\leq C p_m(h) a^{-\beta} \sum_{n=0}^{\infty} \frac{(a^{1/\beta} t)^{n\beta+\beta}}{\Gamma(n\beta+\beta)} = C p_m(h) t^\beta E_{\beta,\beta}(a t^\beta) < \infty, \end{aligned}$$

where $E_{\beta,\beta}(z)$ is the generalized Mittag-Leffler function (see, Section "Additional notes" to Chapter 3). This estimate means that the series on the right of (5.59) converges in the topology of $\Psi_{G,p}(\mathbb{R}^n)$.

2. The condition $h \in C^1[t \geq 0; \Psi_{G,p}(\mathbb{R}^n)]$ for $h(t,x)$ is too strong. In Sections 6.4 and 6.5 we will prove two fractional generalizations of the Duhamel principle for distributed order differential equations, using a different method, weakening conditions for $h(t,x)$. We recall (see Proposition 3.10) that the fractional derivative $D_*^{1-\beta} h(t)$, $0 < \beta < 1$ exists a.e., if $h(t) \in AC[0 \leq t \leq T]$, where T is a positive finite number and $AC[0,T]$ is the class of absolutely continuous functions.
3. The condition $h(0,x) = 0$ in Theorem 3.1 is not essential. If $h(t,x)$ does not satisfy this condition, then in the formulation of the theorem the Cauchy condition (5.54) has to be replaced by

$$V(\tau,\tau,x) = D_+^{1-\beta} h(\tau,x), \quad x \in \mathbb{R}^n,$$

where $D_+^{1-\beta}$ is the operator of fractional differentiation of order $1-\beta$ in the Riemann-Liouville sense. The case $\beta = 1$ recovers the classic Duhamel principle. Theorem 3.1 coincides with the classic Duhamel principle in the set of functions $h(t,x)$ with $h(0,x) = 0$.

Theorem 5.12. *Let $\varphi_0(x) \in \Psi_{G,2}(\mathbb{R}^n)$, $h(t,x) \in C^1[t \geq 0; \Psi_{G,2}(\mathbb{R}^n)]$, and $f(0,x) = 0$. Then the Cauchy problem (5.44)–(5.45) (with $0 < \beta < 1$) has a unique solution*

$$u(t,x) \in C^1[t > 0; \Psi_{G,2}(\mathbb{R}^n)] \cap C[t \geq 0; \Psi_{G,2}(\mathbb{R}^n)].$$

This solution has the representation

$$u(t,x) = E_\alpha(t^\beta A(D_x))\varphi_0(x) + \int_0^t E_\beta((t-\tau)^\beta A(D_x))D_*^{1-\beta}h(\tau,x)d\tau. \tag{5.61}$$

Proof. The representation (5.61) is a simple implication of (5.47) and Theorem 5.11. The first term in (5.61) was studied in Section 5.3 in detail. Denote by $v(t,x)$ the second term in (5.61). For a fixed $t > 0$ making use of the semi-norm of Ψ_N we have

$$\begin{aligned} p_N^2(v(t,x)) &= \|F^{-1}\chi_N F v\|_{L_2}^2 = \|\chi_N F v\|_{L_2}^2 \\ &= \int_{R^n} |\chi_N(\xi)|^2 \cdot |\int_0^t E_\beta((t-\tau)^\beta A(\xi))FD_*^{1-\beta}h(\tau,\xi)d\tau|^2 d\xi. \end{aligned}$$

For $\chi_N(\xi)$ there exists a compact set $K_N \subset G$ such that $supp\, \chi_N(\xi) \subset K_N$. Using the Hölder inequality we get the estimate

$$p_N^2(v(t,x)) \leq \int_{K_N} |\chi_N(\xi)|^2 \cdot \int_0^t |E_\beta((t-\tau)^\beta A(\xi))|^2 d\tau \cdot \int_0^t |FD_*^{1-\beta}h(\tau,\xi)|^2 d\tau d\xi.$$

The function $\int_0^t |E_\beta((t-\tau)^\beta A(\xi))|^2 d\tau$ is bounded on K_N. Consequently, there exists a constant $C_N > 0$, such that

$$\begin{aligned} p_N^2(v(t,x)) &\leq C_N \int_{K_N} |\chi_N(\xi)|^2 \cdot \int_0^t |FD_*^{1-\beta}h(\tau,\xi)|^2 d\tau d\xi \\ &\leq C_N \int_0^t \int_{R^n} |\chi_N(\xi)|^2 \cdot |FD_*^{1-\beta}h(\tau,\xi)|^2 d\xi d\tau \\ &= C_N \int_0^t \|\chi_N(\xi)FD_*^{1-\beta}h(\tau,\xi)\|_{L_2}^2 d\tau = C_N \int_0^t p_N^2(D_*^{1-\beta}h(\tau,x))d\tau. \end{aligned}$$

It follows from the condition $D_*^{1-\beta}h(t,x) \in C[t \geq 0; \Psi_{G,2}(\mathbb{R}^n)]$ that the function $p_N(D_*^{1-\beta}h(\tau,x))$ is continuous with respect to $\tau \in (0;t)$ and for a fixed $t > 0$ and some N_1 the estimate

$$p_N^2(v(t,x)) \leq C_N \cdot t \cdot \sup_{0<\tau<t} p_N^2(D_*^{1-\beta}h(\tau,x)) \leq C_{N_1} \cdot t \cdot \sup_{0<\tau<t} p_{N_1}^2(h(\tau,x))$$

holds. Hence, for every fixed $t \in (0;+\infty)$ the function $v(t,x)$ in (5.55) belongs to the space $\Psi_{G,2}(\mathbb{R}^n)$. The analogous estimate is valid for $\frac{\partial}{\partial t}v(t,x)$. Thus $v(t,x) \in$

$C^1[t>0;\Psi_{G,2}(\mathbb{R}^n)] \cap C[t \geq 0;\Psi_{G,2}(\mathbb{R}^n)]$. Hence, $u(t,x) \in C^1[t>0;\Psi_{G,2}(\mathbb{R}^n)] \cap C[t \geq 0;\Psi_{G,2}(\mathbb{R}^n)]$, as well. The uniqueness of a solution follows from the representation formula for a solution of the homogeneous Cauchy problem.

5.5.2 Fractional Duhamel principle: the case of arbitrary $\beta > 0$

Now we consider the Cauchy problem (5.44)–(5.45) for arbitrary order β satisfying $m-1<\beta \leq m$, $m \in N$. Obviously, in this case $0 \leq m-\beta < 1$.

Theorem 5.13. *Assume $m \geq 1$, $m-1<\beta \leq m$, and $V(t,\tau,x)$ is a solution of the Cauchy problem for the homogeneous equation*

$$ {}_{\tau}D_*^{\beta} V(t,\tau,x) - A(D_x)V(t,\tau,x) = 0, \quad t > \tau, \quad x \in \mathbb{R}^n, \tag{5.62}$$

with the Cauchy conditions

$$\frac{\partial^k V}{\partial t^k}(t,\tau,x)|_{t=\tau} = 0, \quad k = 0,\ldots,m-2, \tag{5.63}$$

$$\frac{\partial^{m-1} V}{\partial t^{m-1}}(t,\tau,x)|_{t=\tau} = D_*^{m-\alpha} h(\tau,x), \tag{5.64}$$

where $h(t,x), t>0, x \in \mathbb{R}^n$, is a given function as in Theorem 5.11. Then

$$v(t,x) = \int_0^t V(t,\tau,x)d\tau \tag{5.65}$$

is a solution of the Cauchy problem for the inhomogeneous equation

$$D_*^{\beta} v(t,x) - A(D_x)v(t,x) = h(t,x), \tag{5.66}$$

with the homogeneous Cauchy conditions

$$\frac{\partial^k v}{\partial t^k}(0,x) = 0, \quad k = 0,\ldots,m-1. \tag{5.67}$$

Proof. It follows from the representation formula (5.47) that

$$V(t,\tau,x) = J^{m-1} E_{\beta}((t-\tau)^{\beta} A(D_x)) D_*^{m-\beta} f(\tau,x) \tag{5.68}$$

solves the Cauchy problem for equation (5.62) with the initial conditions (5.63), (5.64). Further, apply the operator J^{β} to both sides of the equation (5.66) and obtain

$$v(t,x) - \sum_{j=0}^{m-1} \frac{t^j v^j(0,x)}{j!} - J^{\beta} A(D_x)v(t,x) = J^{\beta} h(t,x). \tag{5.69}$$

Taking into account the conditions (5.67), we rewrite equation (5.69) in the form

$$v(t,x) - J^{\beta}A(D_x)v(t,x) = J^{\beta}h(t,x).$$

A solution of this equation is represented as[2]

$$v(t,x) = \sum_{n=0}^{\infty} J^{\beta n+\beta} A^n(D_x)h(t,x).$$

It follows from (5.50) (with $\gamma = \beta n$) that for arbitrary function $g(t)$ satisfying the conditions $g(0) = 0$, one has $J^{\beta n+\beta}g(t) = J^{\beta n+m}(D_*^{m-\beta}g(t))$. Taking this into account, we have

$$\begin{aligned} v(t,x) &= \sum_{n=0}^{\infty} J^{\beta n+1} J^{m-1} A^n(D_x) D_*^{m-\beta} h(t,x) \\ &= \int_0^t J^{m-1} \sum_{n=0}^{\infty} \frac{(t-\tau)^{\beta n} A^n(D_x)}{\Gamma(\beta n+1)} D_*^{m-\beta} h(\tau,x)d\tau \\ &= \int_0^t J^{m-1} E_{\beta}((t-\tau)^{\beta} A(D_x)) D_*^{m-\beta} h(\tau,x)d\tau. \end{aligned} \tag{5.70}$$

Comparing (5.68) and (5.70) we obtain (5.65), and hence, the proof of the theorem.

The condition $h(0,x) = 0$ in the theorem is not essential. If this condition is not verified, then the formulation of the fractional Duhamel principle takes the following form.

Theorem 5.14. *Assume $m \geq 1$, $m-1 < \beta \leq m$, and $V(t,\tau,x)$ is a solution of the Cauchy problem for the homogeneous equation (5.62) with the Cauchy conditions*

$$\frac{\partial^k V}{\partial t^k}(t,\tau,x)|_{t=\tau} = 0, \quad k = 0,\ldots,m-2,$$

$$\frac{\partial^{m-1} V}{\partial t^{m-1}}(t,\tau,x)|_{t=\tau} = D_+^{m-\beta} h(\tau,x),$$

where $h(t,x), t > 0, x \in \mathbb{R}^n$, is a given function. Then $v(t,x)$ defined in (5.65) is a solution of the following Cauchy problem for the inhomogeneous equation

$$D_*^{\beta} v(t,x) - A(D_x)v(t,x) = h(t,x),$$

$$\frac{\partial^k v}{\partial t^k}(0,x) = 0, \quad k = 0,\ldots,m-1.$$

Remark 5.5. Note that if $\beta = m$, then Theorems 5.11 and 5.13 recover the classic Duhamel principle discussed in Section 4.7.

[2] Regarding the convergence of this series see Remark 5.4.

The theorem below generalizes Theorem 5.12 for higher orders β.

Theorem 5.15. *Let* $m-1<\beta\le m$ *and* $\varphi_k(x)\in\Psi_{G,2}(\mathbb{R}^n)$, $k=0,\dots,m-1$, $h(t,x)\in AC[t\ge 0;\Psi_{G,2}(\mathbb{R}^n)]$, $D_*^{m-\beta}h(t,x)\in C[t\ge 0;\Psi_{G,2}(\mathbb{R}^n)]$ *and* $f(0,x)=0$. *Then the Cauchy problem* (5.44)–(5.45) *has a unique solution. This solution is given by the representation*

$$u(t,x)=\sum_{k=1}^{m}J^{k-1}E_\beta(t^\beta A(D_x))\varphi_{k-1}(x)\\ +\int_0^t J^{m-1}E_\beta((t-\tau)^\beta A(D_x))D_*^{m-\beta}h(\tau,x)d\tau. \tag{5.71}$$

Proof. Splitting the Cauchy problem (5.44)–(5.45) into the Cauchy problem for the equation (5.44) with the homogeneous initial conditions and the Cauchy problem for the homogeneous equation corresponding to (5.44) with the initial conditions (5.45), and applying Theorem 5.13 and representation formula (5.47), we obtain (5.71). The fact that

$$\sum_{k=1}^{m}J^{k-1}E_\beta(t^\beta A(D_x))\varphi_{k-1}(x)\in C^{(m)}[t>0;\Psi_{G,2}(\mathbb{R}^n)]\cap C^{(m-1)}[t\ge 0;\Psi_{G,2}(\mathbb{R}^n)]$$

is proved in Section 5.3. Further, since the $m-1$-th derivative with respect to t of the last term in (5.71) belongs to $AC[[0,T];\Psi_{G,2}(\mathbb{R}^n)]$[3], then the estimation obtained in the proof of Theorem 5.12 holds in this case as well.

Remark 5.6. If $h(t,x)$ does not vanish at $t=0$, then in accordance with Theorem 5.14, the representation formula (5.71) takes the form

$$u(t,x)=\sum_{k=1}^{m}J^{k-1}E_\beta(t^\beta A(D_x))\varphi_{k-1}(x)\\ +\int_0^t J^{m-1}E_\beta((t-\tau)^\beta A(D_x))D_+^{m-\beta}f(\tau,x)d\tau.$$

Example 5.2. 1. Let $0<\beta<1$ and $h(t,x)$ be a given suitable function satisfying $f(0,x)=0$. Consider the Cauchy problem for the fractional order heat equation with nonzero external force

$$D_*^\beta u(t,x)=\Delta u(t,x)+h(t,x),\quad t>0, x\in\mathbb{R}^n, \tag{5.72}$$
$$u(0,x)=\varphi_0(x), \tag{5.73}$$

where Δ is the Laplace operator. In accordance with the fractional Duhamel principle the influence of the external force $h(t,x)$ to the output can be counted from the Cauchy problem

[3] T is an arbitrary positive finite number.

$$\begin{aligned} {}_{\tau}D_{*}^{\beta}V(t,\tau,x) &= \Delta V(t,\tau,x), \quad t>\tau, x\in\mathbb{R}^n, \\ V(\tau,\tau,x) &= D_{*}^{1-\beta}h(\tau,x). \end{aligned}$$

The function $V(t,\tau,x) = E_\beta((t-\tau)^\beta\Delta)D_*^{1-\beta}h(\tau,x)$ solves this problem. Hence, the solution of the Cauchy problem (5.72)–(5.73) is given by

$$u(t,x) = E_\beta(t^\beta\Delta)\varphi_0(x) + \int_0^t E_\beta((t-\tau)^\beta\Delta)D_*^{1-\beta}f(\tau,x)d\tau.$$

2. Let $1<\beta<2$, and $F(t,x)$ is a given function. Consider the Cauchy problem

$$\begin{aligned} D_*^\beta u(t,x) &= \Delta u(t,x) + F(t,x), \quad t>0,\ x\in\mathbb{R}^n, \\ u(0,x) &= \varphi_0(x), \qquad u_t(0,x) = \varphi_1(x). \end{aligned}$$

Again in accordance with the fractional Duhamel principle the influence of the external force $F(t,x)$ appears in the form

$$\begin{aligned} {}_{\tau}D_{*}^{\beta}V(t,\tau,x) &= \Delta V(t,\tau,x), \quad t>\tau, x\in\mathbb{R}^n, \\ V(\tau,\tau,x) &= 0, \ \frac{\partial V}{\partial t}(\tau,\tau,x) = D_+^{2-\beta}F(\tau,x). \end{aligned}$$

The unique solution of the latter is $V(t,\tau,x) = JE_\beta((t-\tau)^\beta\Delta)D_+^{2-\beta}F(\tau,x)$. Hence,

$$\begin{aligned} u(t,x) = {} & E_\beta(t^\beta\Delta)\varphi_0(x) + JE_\beta(t^\beta\Delta)\varphi_1(x) \\ & + \int_0^t JE_\beta((t-\tau)^\beta\Delta)D_+^{2-\beta}F(\tau,x)d\tau. \end{aligned}$$

5.5.3 Fractional Duhamel principle: the case of Riemann-Liouville derivative

A fractional generalization of Duhamel's principle is also possible when the fractional order differential equation is given through the Riemann-Liouville fractional derivative. In this section we briefly discuss this important case proving the corresponding theorem in the abstract differential-operator case

$$D^\beta u(t) = Bu(t,x) + h(t),$$

where $0<\beta<1$, and B is a closed operator, independent of t, and with a domain $\mathscr{D}(B)$ dense in a Banach space X. The initial value problem, called the Cauchy type problem, in this case has the form

$$_\tau L[u](t) = h(t), \quad t > 0, \tag{5.74}$$

$$_\tau J^{1-\alpha} u(\tau+) = \varphi \in X. \tag{5.75}$$

where $_\tau L[\cdot] = _\tau D^\beta - B$. The initial condition (5.75) can be rewritten as the weighted Cauchy type initial condition $\lim_{t\to\tau+}(t-\tau)^{1-\alpha}u(t) = \varphi$ (see, e.g., [KST06]).

Theorem 5.16. *Suppose that* $V(t,\tau)$, $t \geq \tau \geq 0$, *is a solution of the Cauchy type problem for the homogeneous equation*

$$_\tau D^\alpha V(t,\tau) + BV(t,\tau) = 0, \quad t > \tau, \tag{5.76}$$

$$_\tau J^{1-\alpha} V(t,\tau)_{|_{t=\tau+}} = h(\tau), \tag{5.77}$$

where $0 < \alpha < 1$ *and* $h(\tau)$, $\tau \geq 0$, *is a continuous vector-function. Then Duhamel's integral*

$$u(t) = \int_0^t V(t,\tau)d\tau \tag{5.78}$$

solves the Cauchy type problem for the inhomogeneous equation

$$D^\alpha u(t) + Bu(t) = h(t), \; t > 0, \tag{5.79}$$

with the homogeneous initial condition $J^{1-\alpha}u(0+) = 0$.

Proof. Let $V(t,\tau)$ satisfy the conditions of the theorem. Then for the Duhamel integral (5.78), by virtue of Lemma 4.4, we have

$$\begin{aligned} D^\alpha u(t) + Bu(t) &= \frac{1}{\Gamma(1-\alpha)} \frac{d}{dt} \int_0^t \frac{\int_0^s V(s,\tau)d\tau}{(t-s)^\alpha} ds + \int_0^t BV(t,\tau)d\tau \\ &= \frac{d}{dt} \int_0^t {}_\tau J^{1-\alpha} V(t,\tau)d\tau + \int_0^t BV(t,\tau)d\tau \\ &= {}_\tau J^{1-\alpha} V(t,\tau)_{|_{\tau=t}} + \int_0^t [{}_\tau D^\alpha V(t,\tau) + BV(t,\tau)]d\tau = h(t). \end{aligned} \tag{5.80}$$

On the other hand, changing the order of integration and using the mean value theorem, we obtain

$$\|J^{1-\alpha}u(t)\| = \|\int_0^t {}_\tau J^{1-\alpha} V(t,\tau)d\tau\| \leq t\|{}_{\tau_*}J^{1-\alpha}V(t,\tau_*)\|, \tag{5.81}$$

where $\tau_* \in (0,t)$, and the operator $_{\tau_*}J^{1-\alpha}$ on the rightmost term of (5.81) acts in the variable t. Condition (5.77) implies that $\lim_{t\to 0+} {}_\tau J^{1-\alpha}V(t,\tau) = h(0)$ in the norm of X. It follows from (5.81) that $\lim_{t\to 0+} J^{1-\alpha}u(t) = 0$ in the norm of X.

Remark 5.7. Theorem 5.16 can be generalized to differential-operator equations of higher order $\alpha > 1$, as well. In Section 6.5 we will generalize the Duhamel principle for higher fractional order distributed order differential-operator equations defined through the Riemann-Liouville fractional derivatives.

5.6 Multi-point value problems for fractional order pseudo-differential equations

Equation (5.1) contains only a single fractional derivative. However, in modeling of real processes equations with several fractional derivatives emerge frequently. In this section we discuss general multi-point value problems for partial differential equation of fractional order of the form

$$D_*^{\alpha_m} u(t,x) + \sum_{k=1}^{m-1} A_k(D) D_*^{\alpha_k} u(t,x) + A_0(D)u(t,x)$$
$$= h(t,x), \quad t \in (0,T),\ x \in \mathbb{R}^n, \tag{5.82}$$
$$\sum_{j=0}^{m-1} \Gamma_{kj}(D) \frac{\partial^k u(t_{kj},x)}{\partial t^k} = \varphi_k(x),\ k = 0,\dots,m-1, \quad x \in \mathbb{R}^n, \tag{5.83}$$

where $A_k(D)$, $k = 0,\dots,m$, and $\Gamma_{kj}(D)$, $k, j = 0,\dots,m-1$, are pseudo-differential operators whose symbols $A_k(\xi)$, $k = 0,\dots,m$, and $\Gamma_{kj}(\xi)$, $k, j = 0,\dots,m-1$, are in the class of symbols $S_p(G)$; $t_{kj} \in [0,T]$, $0 < T < \infty$; and φ_k, $k = 0,\dots,m-1$, are given functions. We assume that the orders of fractional derivatives satisfy the ordering

$$0 < \alpha_1 < 1, \quad 1 < \alpha_2 < 2 \quad ,\dots, \quad m-2 < \alpha_{m-1} < m-1, \quad m-1 < \alpha_m \le m.$$

Boundary value problem (5.82)–(5.83) is a particular case of boundary value problems for distributed order differential equations, studied in Chapter 6.

Denote by $\Delta(s,\xi)$ the "characteristic function" of equation (5.82), namely,

$$\Delta(s,\xi) = s^{\alpha_m} + \sum_{k=1}^{m-1} A_k(\xi) s^{\alpha_k} + A_0(\xi), \quad \xi \in G \setminus \overset{\circ}{M}.$$

Introduce the function

$$c_\delta(t,\xi) = L^{-1}[\frac{s^{\delta-1}}{\Delta(s,\xi)}](t),$$

where $0 < \delta < \alpha_m$, and L^{-1} means the inverse Laplace transform. This function will be used in the construction of symbols of solution operators to the considering problem. Since

$$\frac{s^{\delta-1}}{\Delta(s,\xi)} = o\left(\frac{1}{s}\right), \quad s \to \infty,$$

for each fixed $\xi \in G \setminus \overset{\circ}{M}$, we have $c_\delta(t,\xi) \in C^{(\alpha_m-\delta)}(\mathbb{R}_+)$.

We first consider the Cauchy problem for the equation (5.82) with the initial conditions

$$u^{(k)}(0,x) = \psi_k(x), \quad k = 0,\dots,m-1. \tag{5.84}$$

Applying the Fourier transform with respect to the variable x in equation (5.82) and Cauchy conditions (5.84), we have

$$D_*^{\alpha_m}\hat{u}(t,\xi)+\sum_{k=1}^{m-1}A_k(\xi)D_*^{\alpha_k}\hat{u}(t,\xi)+A_0(\xi)\hat{u}(t,\xi)=0, \qquad t\in(0,T),\ \xi\in G, \tag{5.85}$$

$$\frac{\partial^k\hat{u}(0,\xi)}{\partial t^k}=\hat{\psi}_k(\xi), \quad k=0,\dots,m-1,\ \xi\in G, \tag{5.86}$$

where $\hat{u}(t,\xi)=F[u](t,\xi)$, the Fourier transform of $u(t,x)$. Further, the Laplace transform, due to formula (3.55), reduces equation (5.85) to

$$L_{t\to s}\left[D_*^{\alpha_m}\hat{u}(t,\xi)+\sum_{k=1}^{m-1}A_k(\xi)D_*^{\alpha_k}\hat{u}(t,\xi)+A_0(\xi)\hat{u}(t,\xi)\right](s)$$
$$=s^{\alpha_m}L[\hat{u}](s,\xi)-\sum_{\ell=0}^{m-1}\hat{u}^{(\ell)}(0,\xi)s^{\alpha_m-\ell-1}$$
$$+\sum_{j=1}^{m-1}A_j(\xi)\left[s^{\alpha_j}L[\hat{u}](s,\xi)-\sum_{\ell=0}^{j-1}\hat{u}^{(\ell)}(0,\xi)s^{\alpha_j-\ell-1}\right]+A_0(\xi)L[\hat{u}](s,\xi)=0.$$

It follows from the latter that

$$\left[s^{\alpha_m}+\sum_{k=1}^{m-1}A_k(\xi)s^{\alpha_k}+A_0(\xi)\right]L[\hat{u}](s,\xi)$$
$$=\sum_{\ell=0}^{m-1}\hat{u}^{(\ell)}(0,\xi)s^{\alpha_m-\ell-1}+\sum_{j=1}^{m-1}A_j(\xi)\sum_{\ell=0}^{j-1}\hat{u}^{(\ell)}(0,\xi)s^{\alpha_j-\ell-1},$$

or

$$L[\hat{u}](s,\xi)=\sum_{k=0}^{m-1}\frac{s^{\alpha_m-k-1}+\sum_{j=k+1}^{m-1}A_j(\xi)s^{\alpha_j-k-1}}{\Delta(s,\xi)}\hat{u}^{(k)}(0,\xi).$$

Now, inverting first the Laplace transform and then the Fourier transform, and taking initial conditions (5.86) into account, we have the representation

$$u(t,x)=\sum_{k=0}^{m-1}B_k(t,D)\psi_k(x),$$

for the solution of the Cauchy problem (5.82)–(5.84), where

$$B_k(t,\xi)=c_{\alpha_m-k}(t,\xi)+\sum_{j=k+1}^{m-1}A_j(\xi)c_{\alpha_j-k}(t,\xi). \tag{5.87}$$

We notice that formula (5.87) for the symbols $B_k(t,\xi), k=0,\dots,m-1$, of the solution operators of boundary value problem (5.82)–(5.84), generalizes formula (5.13) for the symbols obtained in the case of one single fractional derivative (see equation (5.7)), that is $\alpha_m=\beta, A_j(\xi)\equiv 0, j=1,\dots,m-1$. Indeed, if $\alpha_m=\beta$, where $m-1<\beta<m$, and $A_j(\xi)=0, j=1,\dots,m-1$, and $A_0(\xi)=-A(\xi)$, then (5.87) implies

$$B_k(t,\xi)=c_\beta(t,\xi)=L^{-1}[\frac{s^{\beta-k-1}}{s^\beta+A_0(\xi)}](t)$$
$$=J^kE_\beta(-A(\xi)t^\beta),\quad k=0,\dots,m-1,$$

recovering (5.13).

Now we construct the,system of solution operators of the general problem (5.82)–(5.83). Applying the Fourier transform to boundary conditions (5.83), we have

$$\sum_{j=0}^{m-1}\Gamma_{kj}(\xi)\frac{\partial^k\hat{u}(t_{kj},\xi)}{\partial t^k}=\hat{\varphi}_k(\xi),\ k=0,\dots,m-1,\ \xi\in G. \tag{5.88}$$

We look for a solution of multi-point value problem (5.85)–(5.88), in the form

$$\hat{u}(t,\xi)=\sum_{k=0}^{m-1}f_k(\xi)B_k(t,\xi) \tag{5.89}$$

with $B_k(t,\xi)$, $k=0,\dots,m-1$, given by (5.87), and unknown coefficients $f_k(\xi)$, $k=0,\dots,m-1$. It is clear that $\hat{u}(t,\xi)$ satisfies (5.85). Substituting it into (5.88) we obtain a system of linear algebraic equations

$$M(\xi)F(\xi)=\hat{\Phi}(\xi), \tag{5.90}$$

where $F(\xi)=(f_0(\xi),\dots,f_{m-1}(\xi))$, $\hat{\Phi}(\xi)=(\hat{\varphi}_0(\xi),\dots,\hat{\varphi}_{m-1}(\xi))$, and $M(\xi)$ is a square matrix of the order m with entries

$$m_{kl}=\sum_{j=0}^{m-1}\Gamma_{kj}(\xi)B_l^{(j)}(t_{kj},\xi),\quad k,l=0,\dots,m-1.$$

Denote by $\overset{\circ}{M}$ the set of all points $\xi\in G$ such that $\operatorname{Det} M(\xi)=0$. If $\xi\notin\overset{\circ}{M}$, then the equation (5.90) has a unique solution

$$F(\xi)=M^{-1}(\xi)\hat{\Phi}(\xi). \tag{5.91}$$

We note that M_0 is the singular set for the symbols of the solution operators. Substituting (5.91) of the vector $F(\xi)=(f_0(\xi),\dots,f_{m-1}(\xi))$ into (5.89) and applying the inverse Fourier transform we get the solution of general multi-point value problem (5.82)–(5.83) as

$$u(t,x) = \sum_{k=0}^{m-1} U_k(t,D)\varphi_k(x), \tag{5.92}$$

where the $U_k(t,D)$, $k = 0,\dots,m-1$ are solution pseudo-differential operators with the symbols

$$U_k(t,\xi) = \Big((M^*)^{-1}(\xi)B(t,\xi)\Big)_k, \quad k = 0,\dots,m-1. \tag{5.93}$$

Here $(M^*)^{-1}(\xi)$ is the matrix inverse to the Hermitian adjoint of $M(\xi)$, and $B(t,\xi)$ is the transpose of the vector row $(B_0(t,\xi),\dots,B_{m-1}(t,\xi))$ with the components given by (5.87).

A behavior of the vector-function $U(t,\xi)$ with components $U_0(t,\xi),\dots,U_{m-1}(t,\xi)$ near singular points and at infinity depends on operators $A_k(D)$, $k=0,\dots,m-1$, and $\Gamma_{kj}(D)$, $k=0,\dots,m-1$, $j=0,\dots,m-1$, in problem (5.82)–(5.83). The well posedness of the multi-point value problem (5.82)–(5.83) in the classical Sobolev, Besov, and other function spaces depends on the behavior of the vector function $U(t,\xi)$. In particular, we have the following result:

Theorem 5.17. *Let $\varphi_k \in \Psi_{G_0,p}(\mathbb{R}^n)$, $k = 0,\dots,m-1$, $G_0 = G \setminus \overset{\circ}{M}$. Then multi-point value problem* (5.82)–(5.83) *has a unique solution in the space $C^m[(0,T); \Psi_{G_0,p}(\mathbb{R}^n)]$. This solution is represented by formula (5.92).*

Similar theorem is valid for the dual problem too. Under the dual problem we mean a problem obtained by replacing the operators $A_k(D)$, $k = 0,\dots,m-1$, and $\Gamma_{kj}(D)$, $k = 0,\dots,m-1$, $j = 0,\dots,m-1$, by their dual operators (see formula (2.45)).

Theorem 5.18. *Let $\varphi_k \in \Psi'_{-G_0,q}(\mathbb{R}^n)$, $k = 0,\dots,m-1$, $G_0 = G \setminus M_0$. Then the dual multi-point value problem has a unique solution in the space $C^m[(0,T);\Psi'_{-G_0,q}(\mathbb{R}^n)]$. This solution is given by the formula*

$$u(t,x) = \sum_{k=0}^{m-1} U_k(t,-D)\varphi_k(x).$$

For the Sobolev spaces H^s we get the following result:

Theorem 5.19. *Let the n-dimensional measure of $\overset{\circ}{M}$ is zero and for all $t \in (T_1,T_2)$ and for the symbols $U_j(t,\xi)$ defined in* (5.93)*, the estimates*

$$\Big|\frac{\partial^k U_j(t,\xi)}{\partial t^k}\Big| \le C(1+|\xi|)^{s_j-\ell_k}, \ C>0, \quad \xi \in \mathbb{R}^n, \ j=0,\dots,m-1$$

hold for all $k=0,\dots,m-1$ with some $s_j \in \mathbb{R}$, $\ell_j \in \mathbb{R}$, $j=0,\dots,m-1$. Then for any $\varphi_j \in H^{s_j}(\mathbb{R}^n)$, $j=0,\dots,m-1$, there exists a unique solution $u(t,x)$ of boundary value problem (5.82)–(5.83) *such that*

$$\frac{\partial^k u(t,x)}{\partial t^k} \in C^{m-1-k}[(T_1,T_2);H^{\ell_k}(\mathbb{R}^n)],\, k=0,\dots,m-1,$$

and satisfying the estimate

$$\sup_{t\in(T_1,T_2)} \sum_{k=0}^{m-1} \| \frac{\partial^k u}{\partial t^k} | H^{\ell_k} \| \leq C \sum_{j=0}^{m-1} \| \varphi_j | H^{s_j} \|.$$

5.7 Boundary value problems for elliptic operators with a boundary operator of fractional order

In Section 4.5 we briefly studied existence and uniqueness conditions for boundary value problems for elliptic differential equations on the half-space with boundary operators of integer order. However, the exact existence conditions can be obtained only if one extends the class of boundary operators to fractional order boundary operators. In this section we present a detailed analysis of such problems. Important questions arising in this context are "What type of fractional derivatives can be used as a boundary operator?" and "What values of the order of the boundary operator are critical, in terms of changing of orthogonality conditions?" As a boundary operator one can consider fractional derivatives in the sense of Marchaud, Liouville-Weyl, Grünwald-Letnikov, or operators in the form of a hypersingular integral. The exact number of orthogonality conditions for the existence of a solution depends on solution spaces.

Consider the boundary value problem

$$\frac{\partial^2 u}{\partial y^2} + A(D_x)u(y,x) = 0, \quad y>0,\, x\in\mathbb{R}^n, \tag{5.94}$$

$$B^\alpha u(+0,x) = \varphi(x), \quad x\in\mathbb{R}^n, \tag{5.95}$$

$$|u(y,x)| \to 0 \text{ as } y\to\infty, \text{ uniformly for } x\in\mathbb{R}^n, \tag{5.96}$$

where $D_x = -i(\frac{\partial}{\partial x_1},\dots,\frac{\partial}{\partial x_n})$, $A(D_x)$ is an elliptic pseudo-differential operator of order $2m$, $m\in\mathbb{N}$, acting on $u(y,x)$ with respect to the variable x. Its symbol $A(\xi)$ is a smooth function satisfying the condition

$$C_1|\xi|^{2m} \leq -A(\xi) \leq C_2|\xi|^{2m},\, C_1>0,\, C_2>0, \tag{5.97}$$

and B^α is a boundary operator depending on a positive real parameter α acting on $u(y,x)$ with respect to the variable y. As the model operator B^α we consider the one-sided Marchaud fractional derivative $\mathbf{D}_-^\alpha$ defined in (3.116) (see Section "Additional notes" of Chapter 3), defined for $y>0$ as

$$\mathbf{D}_-^\alpha f(y) = \frac{\alpha-k}{\Gamma(1+k-\alpha)} \frac{d^k}{dy^k} \int_0^\infty \frac{f(y)-f(y+h)}{h^{1+\alpha-k}} dh,$$

if $k < \alpha < k+1, k = 0, 1, \ldots$. We recall that in L_p-spaces, $1 \le p < \infty$, the Marchaud derivative coincides with the one-sided Grünwald-Letnikov fractional derivative $\mathscr{D}_-^\alpha$ (see Section 3.10). Moreover, in the class of well-behaved functions both the Marchaud and Grünwald-Letnikov derivatives coincide with the one-sided Liouville-Weyl fractional derivative ${}_{-\infty}D^\alpha$ (see Section 3.6).

We also recall that if $B_\alpha = \mathbf{D}_-^\alpha$, then one has (see Example 3.12)[4]

$$B_\alpha e^{-ay} = (-a)^\alpha e^{-ay}, \quad \forall a > 0. \tag{5.98}$$

This formula plays an essential role in our constructions in this section.

We denote the problem (5.94)–(5.96) by P_α. As particular cases of the problem P_α we obtain the Dirichlet problem, P_0, if $\alpha = 0$ and the Neumann problem, P_1, if $\alpha = 1$. When α is non-integer and $\alpha \in (0,1)$ the problem P_α interpolates these two problems, which are used in description of stationary states. It is well known that the Dirichlet problem is unconditionally solvable, while for solvability of the Neumann problem an additional condition on orthogonality is necessary. We will show that for φ in the Sobolev space $H^s(\mathbb{R})$ (one-dimensional case) the problem P_α preserves unconditional solvability for all $\alpha \in [0, \frac{1}{2})$. At the same time, if $\alpha \in (\frac{1}{2}, 1]$, the corresponding problem P_α is solvable only if φ is orthogonal to 1. We will also show that if α increases, then the number of orthogonality conditions increases as well. A new orthogonality condition appears exactly when α passes through the *critical values*

$$\alpha_j^* = j + \frac{1}{2}, \quad j = 0, 1, 2, \ldots. \tag{5.99}$$

Let $N(\alpha)$ be the number of conditions necessary for solvability of P_α and let $l(\alpha) = [m\alpha - \frac{n}{2} + 1]_+$, where $[s]$ stands for the integer part of s, and $[s]_+ = [s]$, if $[s] > 0$ and $[s]_+ = 0$, if $[s] \le 0$. Then the formula

$$N(\alpha) = \frac{(l(\alpha) + n - 1)!}{(l(\alpha) - 1)! n!}$$

holds. It is easy to see that, if $n = 1$ (one-dimensional case) and $m = 1$, then

$$N(\alpha) = l(\alpha)!/(l(\alpha) - 1)! = l(\alpha) = [\alpha + \frac{1}{2}]_+,$$

justifying the critical values (5.99) in the one-dimensional case. Indeed, if $0 < \alpha < 1/2$, then $N(\alpha) = 0$, if $1/2 \le \alpha < 3/2$, then $N(\alpha) = 1$, and so forth.

Critical values depend on solution spaces. We consider three different types of function spaces. The first one is a subspace of the Sobolev space $H^s(\mathbb{R}^n)$ satisfying some orthogonality conditions. The number of these conditions depends on the order α of the operator B_α in (5.95) and on the order $2m$ of the pseudo-differential operator $A(D)$ in (5.94). This space is the most suitable one in terms of the exact number of conditions necessary for solvability of the problem P_α.

The second type spaces are Lizorkin type spaces. On one hand, these spaces are natural from the point of view of describing the solvability of P_α for any $\alpha \ge 0$.

[4] With the sign correction effected by the definition of $\mathbf{D}_-^\alpha$.

On the other hand, unfortunately, it is not possible to provide the exact number of solvability conditions of P_α in these spaces for a given α.

The third type spaces are some modifications of the spaces $(\Psi_{G,p}(\mathbb{R}^n), \Psi'_{-G,p'}(\mathbb{R}^n))$ introduced in Section 1.10. As a simple corollary we obtain in Section 5.9 that such distributions may be treated as boundary limits of fractional derivatives of harmonic functions in the sense of Marchaud (or Grünwald-Letnikov or Liouville-Weyl).

5.8 Existence theorems for the problem P_α

In this section we will study the existence and uniqueness of a solution of the boundary value problem P_α. A solution of P_α will be understood in the following sense. Let φ be an element of a topological space X. A function $u(y,x) \in L^\infty(y \geq 0; X)$ is said to be a solution of the problem P_α if it satisfies the equation (5.94) in the distributional sense and the boundary conditions (5.95), (5.96) in the sense of the topology of X. Let $C^\ell[y \geq 0; X]$ be the space of X-valued functions $u(y,x)$, having derivatives up to the order ℓ with respect to y continuous up to the boundary $y = 0$. The corresponding meaning is given to $C^\ell[y > 0; X]$. Analogously, we write $C^\infty[y > 0; X]$ for the space of functions infinitely often differentiable with respect to y. As the space X we will use the spaces considered in Section 2. We recall $l(\alpha) = [\alpha m - \frac{n}{2} + 1]_+$.

Theorem 5.20. *Let $\alpha \geq 0$, $\varphi \in H^s_{comp,l(\alpha)}(\mathbb{R}^n)$, $s \geq 0$. Then P_α has a unique solution $u(y,x)$ and for this solution the following inclusions hold:*

(i) $u(y,x) \in C[y \geq 0; H^{s+\alpha m}(\mathbb{R}^n)]$;
(ii) $B^\alpha u(y,x) \in C[y \geq 0; H^s(\mathbb{R}^n)]$;
(iii) $u(y,x) \in C^\infty[y > 0; H^{s+\alpha m}(\mathbb{R}^n)]$.

Proof. Let $0 \leq \alpha < \frac{n}{2m}$, $\varphi \in H^s_{comp}(\mathbb{R}^n)$ and $\operatorname{supp}\varphi \subset Q_N$. By using (5.98) it can be straightforward verified that the solution of the problem P_α has a representation in the form

$$u(y,x) = \Pi_\alpha(y,D)\varphi(x),$$

where $\Pi_\alpha(y,D)$ is the pseudo-differential operator with the symbol

$$\Pi_\alpha(y,\xi) = [-A(\xi)]^{-\alpha} \exp(-i\alpha\pi - y\sqrt{-A(\xi)}), \tag{5.100}$$

Further, since $|\exp(-i\pi\alpha - y\sqrt{-A(\xi)})| \leq 1$, then taking into account (5.97), we have

$$\begin{aligned}\|u(y,x)|H^{s+\alpha m}\|^2 &= \|\Pi_\alpha(y,D)u(y,x)|H^{s+\alpha m}\|^2 \\ &\leq C\int_{\mathbb{R}^n}(1+|\xi|^2)^{s+\alpha m}\frac{|F[\varphi](\xi)|^2}{|\xi|^{2\alpha m}}d\xi \\ &= C\Big(\int_{|\xi|\leq 1}\{\dots\}d\xi+\int_{|\xi|\geq 1}\{\dots\}d\xi\Big)=I_1+I_2, \end{aligned} \tag{5.101}$$

where I_1, I_2 are integrals over $\{|\xi|\leq 1\}$ and $\{|\xi|\geq 1\}$, respectively. Since $2\alpha m < n$, Theorem 1.24 (see Remark 1.10) implies

$$I_1 \leq C_N\|\varphi|H^s\|^2\int_{|\xi|\leq 1}\frac{d\xi}{|\xi|^{2\alpha m}}\leq C_N\|\varphi|H^s\|^2. \tag{5.102}$$

For the second term in the right-hand side of (5.101) we have the estimate

$$I_2 \leq C\int_{|\xi|\geq 1}(1+|\xi|^2)^s|F[\varphi](\xi)|^2 d\xi \leq C\|\varphi|H^s\|^2. \tag{5.103}$$

Further, let $\alpha > \frac{n}{2m}$ be an arbitrary number, $\varphi \in H^s_{comp,l(\alpha)}(\mathbb{R}^n)$ and $\operatorname{supp}\varphi \subset Q_N$. Then, according to Theorem 1.23, there exists a function $v \in H^{s+l(\alpha)}$ such that $\hat{\varphi}(\xi) = |\xi|^{l(\alpha)}\hat{v}(\xi)$ and $\|v|H^{s+l(\alpha)}\| \leq C_N\|\varphi|H^s\|$. We note that in this case the solution of P_α is represented in the form $u(y,x) = \Pi'_\alpha(y,D)v(x)$, where $\Pi'_\alpha(y,D)$ is the pseudo-differential operator with the symbol

$$\Pi_\alpha(y,\xi) = |\xi|^{-\alpha m+l(\alpha)}\exp(-i\alpha\pi - y\sqrt{-A(\xi)})$$

and $\alpha m - l(\alpha) < \frac{n}{2}$. So, one can use the same technique used above in the case $0 \leq \alpha < \frac{n}{2m}$. Hence, we have the estimate

$$\|u(t,x)|H^{s+\alpha m}\| \leq C'_N\|v|H^{s+l(\alpha)}\| \leq C''_N\|\varphi|H^s\|,$$

which, together with (5.102) and (5.103), gives the inclusion (i) in the theorem.

Now we show that the boundary conditions are verified. Indeed,

$$\|B^\alpha u(y,x)-\varphi(x)H^s\|=\|(\exp(-i\alpha\pi-y\sqrt{A(-\xi)})-1)|F[\varphi](\xi)|^2(1+|\xi|^2)^s|L_2\|\to 0$$

as $y \to +0$ according to Lebesgue's dominated convergence theorem.

Further, let $Y > 0$ and $y \geq Y$. Using again the representation

$$F[\varphi](\xi) = |\xi|^{l(\alpha)}F[v](\xi), \quad v \in H^{s+l(\alpha)}(\mathbb{R}^n),$$

we obtain

$$
\begin{aligned}
&|u(y,x)| \\
&= \left| \int_{\mathbb{R}^n} e^{-Y\sqrt{-A(\xi)}} \exp(-i(\alpha\pi - x\xi) - (y-Y)\sqrt{-A(\xi)}) \frac{F[v](\xi)}{(-A(\xi))^{m\alpha - l(\alpha)}} d\xi \right| \\
&\leq C\|v|H^{s+l(\alpha)}\| \left(\int_{\mathbb{R}^n} \frac{e^{-2Y\sqrt{-A(\xi)}}}{|\xi|^{2m\alpha-2l(\alpha)}(1+|\xi|^2)l(\alpha)} d\xi \right)^{\frac{1}{2}} \\
&\leq C\|\varphi|H^s\| \cdot Y^{2m\alpha-2l(\alpha)-n}, \quad y \geq Y,\ x \in \mathbb{R}^n.
\end{aligned}
$$

Since $2m\alpha - 2l(\alpha) < n$, letting $Y \to \infty$, one can see that boundary condition (5.96) is also verified.

Moreover, it is easy to see that $B^\alpha u(y,x) = \Pi_0(y,D)\varphi(x)$, where $\Pi_0(y,\xi) = \exp(-y\sqrt{-A(\xi)})$. Hence, $B^\alpha u(y,x)$ can be represented as a Poisson type integral, that is $B^\alpha u(y,x) = \mathscr{P}(y,x) * \varphi(x)$, where $\mathscr{P}(y,x) = F^{-1}[\Pi_0(y,\xi)](x)$. So, the continuity property of $B^\alpha u(y,x)$ in the variable y is an implication of the continuity property of this integral. Further, since , $|\exp(-y\sqrt{-A(\xi)})| \leq 1$, one has

$$
\|B^\alpha u(y,x)|H^s\| = \|\exp(-y\sqrt{-A(\xi)})|F[\varphi](\xi)|(1+|\xi|^2)^{s/2}|L_2\| \leq C\|\varphi|H^s\|,
$$

proving (ii).

Finally, if $y > 0$, then

$$
\frac{\partial^k}{\partial y^k}\Pi_\alpha(y,\xi) = (-1)^k[A(\xi)]^{\frac{k-\alpha}{2}} \exp(-i\alpha\pi - y\sqrt{A(-\xi)}), \quad k = 1,2,\ldots.
$$

Taking into account (5.97) we have

$$
\left|\frac{\partial^k}{\partial y^k}\Pi_\alpha(y,\xi)\right| = O(e^{-\varepsilon|\xi|^m}), \quad |\xi| \to \infty
$$

with some $\varepsilon > 0$. This yields the assertion (iii).

The next two assertions follow immediately from Theorem 5.20 and from the fact that the total number of n-dimensional monomials up to order K is

$$
\frac{(K+n-1)!}{(K-1)!n!}.
$$

Corollary 5.3. *For the dimensions of the kernel and co-kernel of the operator* P_α, $\alpha \geq 0$, *corresponding to the problem* P_α *the formulas*

$$
dim\,Ker\,P_\alpha = 0, \; dim\,Co\,Ker\,P_\alpha = N(\alpha) = \frac{(l(\alpha)+n-1)!}{(l(\alpha)-1)!n!}
$$

hold.

Corollary 5.4. *Let $\varphi \in C_0^\infty(\mathbb{R}^n)$. Then there exists a unique solution $u(y,x)$ of P_α and for this solution the following inclusions hold true:*

(a) $u(y,x) \in C[y \geq 0; C^\infty(\mathbb{R}^n)]$;
(b) $B^\alpha u(y,x) \in C[y \geq 0; C^\infty(\mathbb{R}^n)]$;
(c) $u(y,x) \in C^\infty[y > 0; C^\infty(\mathbb{R}^n)]$.

Now we consider the problem P_α in the Lizorkin space $\Phi(\mathbb{R}^n)$. First we prove the following auxiliary lemma.

Lemma 5.5. *For every fixed $y \geq 0$ the pseudo-differential operator $\Pi_\alpha(y,D)$ acts continuously from $\Phi(\mathbb{R}^n)$ into $\Phi(\mathbb{R}^n)$.*

Proof. Let $\varphi \in \Phi(\mathbb{R}^n)$. We will first show that $\Pi_\alpha(y,D)\varphi(x) \in \Phi(\mathbb{R}^n)$ as well. Indeed, using the well-known Leibnitz formula, we have

$$D_\xi^\gamma[F(\Pi_\alpha(y,D)\varphi(x))] = D_\xi^\gamma(\Pi_\alpha(y,\xi)F[\varphi](\xi))$$
$$= \sum_\beta \binom{\gamma}{\beta} D_\xi^{\gamma-\beta}\Pi_\alpha(y,\xi)D_\xi^\beta F[\varphi](\xi),$$

where F stands for the Fourier transform operator, D_ξ^β for the derivative with respect to ξ of the "order" $\beta = (\beta_1, \dots, \beta_n)$, and

$$\binom{\gamma}{\beta} = \frac{\gamma!}{\beta!(\gamma-\beta)!}, \quad \beta! = \prod_{i=1}^n \beta_i!.$$

The function $D_\xi^\theta \Pi_\alpha(y,\xi)$ has a singularity of the order $\alpha m + |\theta|$ only at the origin. Since $\varphi \in \Phi(\mathbb{R}^n)$, in accordance with he definition of $\Phi(\mathbb{R}^n)$ (see Section 1.12), we have

$$D_\xi^\gamma(\Pi_\alpha(y,D)\varphi(x))(0) = 0$$

for all multi-indices γ. Furthermore, for arbitrary real number l and multi-index γ, there exists a real number l_1 such that

$$(1+|\xi|^2)^l |D_\xi^\gamma F(\Pi_\alpha(y,D)\varphi(x))| \leq C(1+|\xi|^2)^{l_1}|D_\xi^\gamma \hat{\varphi}(\xi)|.$$

This implies the continuity of $\Pi_\alpha(y,D)$.

Theorem 5.21. *Let $\varphi \in \Phi(\mathbb{R}^n)$. Then for arbitrary $\alpha \geq 0$ there exists a unique solution $u(y,x)$ of P_α and for this solution the following inclusions hold true:*

(i) $u(y,x) \in C[y \geq 0; \Phi(\mathbb{R}^n)]$;
(ii) $B^\alpha u(y,x) \in C[y \geq 0; \Phi(\mathbb{R}^n)]$;
(iii) $u(y,x) \in C^\infty[y > 0; \Phi(\mathbb{R}^n)]$.

Proof. (i) is a simple consequence of Lemma 5.5. (ii) and (iii) can be proved analogously to Theorem 5.21.

Theorem 5.22. *Let $\varphi \in \Phi'(\mathbb{R}^n)$. Then the problem P_α has a solution $u(y,x)$ unique to within an arbitrary additive polynomial, and for this solution the following inclusions hold true:*

(a) $u(y,x) \in C[y \geq 0; \Phi'(\mathbb{R}^n)]$;
(b) $B^\alpha u(y,x) \in C[y \geq 0; \Phi'(\mathbb{R}^n)]$;
(c) $u(y,x) \in C^\infty[y > 0; \Phi'(\mathbb{R}^n)]$.

The proof of Theorem 5.22 is based on the following lemma.

Lemma 5.6. *For every fixed $y \geq 0$ the pseudo-differential operator $\Pi_\alpha(y,-D)$ with the symbol defined by (5.100) acts continuously from $\Phi'(\mathbb{R}^n)$ into $\Phi'(\mathbb{R}^n)$.*

Proof. Let $\varphi \in \Phi'(\mathbb{R}^n)$. Then for arbitrary $v \in \Phi(\mathbb{R}^n)$ the equality

$$< \Pi_\alpha(y,-D)\varphi, v >=< \varphi, \Pi_\alpha(y,D)v >$$

holds. In accordance with Lemma 5.5, $\Pi_\alpha(y,D)v \in \Phi(\mathbb{R}^n)$. The continuity of $\Pi_\alpha(y,-D)$ in $\Phi'(\mathbb{R}^n)$ follows from the continuity of $\Pi_\alpha(y,D)$ in $\Phi(\mathbb{R}^n)$.

Finally, we solve the problem P_α in the spaces $Z_\alpha^\pm(\pm G)$, which was introduced in Section 4.6.4 (see Definition 4.3). Recall that $\varphi \in Z_\alpha^+(G)$, if $\varphi \in \Psi_{G,2}(\mathbb{R}^n) \equiv \Psi_G(\mathbb{R}^n)$ and $|\xi|^{-\beta}|F[\varphi](\xi)| \leq C < \infty$ in the neighborhood of the origin for some $\beta > \alpha - n/2$.

Theorem 5.23. *Let $\varphi \in Z_{m\alpha}^+(G)$. Then there exists a unique solution $u(y,x)$ of P_α and this solution has the properties:*

(i) $u(y,x) \in C[y \geq 0; \Psi_G(\mathbb{R}^n)]$;
(ii) $B^\alpha u(y,x) \in C[y \geq 0; \Psi_G(\mathbb{R}^n)]$;
(iii) $u(y,x) \in C^\infty[y > 0; \Psi_G(\mathbb{R}^n)]$.

The proof of the theorem is based on the following auxiliary result:

Lemma 5.7. *For every fixed $y \geq 0$ and $\alpha \geq 0$ the pseudo-differential operator*

$$\Pi_{m\alpha}(y,D) : Z_{m\alpha}^+(G) \to \Psi_G(\mathbb{R}^n) \tag{5.104}$$

is continuous.

Proof. Let $\varphi \in Z_{m\alpha}^+(G)$. We show that $\Pi_\alpha(y,D)\varphi$ belongs to $\Psi_G(\mathbb{R}^n)$. Indeed, by definition, for $\varphi \in Z_{m\alpha}^+(G)$ there exists a number $\beta > m\alpha - \frac{n}{2}$ and a constant $C > 0$ such that

$$|\xi|^{-\beta}|F[\varphi](\xi)| \leq C < \infty$$

in some neighborhood of the origin. Hence,

$$|\Pi_\alpha(y,\xi)F[\varphi](\xi)| = O(|\xi|^{\beta-\alpha m}), \quad |\xi| \to \infty.$$

Since $m\alpha - \beta < \frac{n}{2}$, then $\Pi_\alpha(y,\xi)\hat{\varphi}(\xi) \in L_2(\mathbb{R}^n)$. In accordance with the Plancherel theorem $\Pi_\alpha(y,D)\varphi(x) \in L_2(\mathbb{R}^n)$. Moreover, $\operatorname{supp} F[\Pi_\alpha(y,D)\varphi] \subset \operatorname{supp} F[\varphi] \subset G$. Thus we have $\Pi_\alpha(y,D)\varphi \in \Psi_G(R^n)$.

Now we show the continuity of the mapping (5.104). Let a sequence of functions $\varphi_k \in Z^+_{m\alpha}(G)$ tend to zero in the topology of $Z^+_{m\alpha}(G)$. This means that there exists a compact set $K \subset G$ and numbers $\beta > m\alpha - \frac{n}{2}$ and $C > 0$ such that:

(i) $\operatorname{supp} F[\varphi_k] \subset K,\ k = 1,2,\ldots;$
(ii) $|\xi|^{-\beta}|F[\varphi_k](\xi)| \leq C < \infty,\ k = 1,2,\ldots;$
(iii) $\|\varphi_k\|_0 \to 0,\ k \to \infty.$

Obviously, $\operatorname{supp} F[\Pi_\alpha(y,D)\varphi_k] \subset K$ for all $k = 1,2,\ldots$. Moreover,

$$\|\Pi_\alpha(y,D)\varphi_k(x)\|_0^2 = (2\pi)^n \int_K |\Pi_\alpha(y,\xi)|^2 |F[\varphi](\xi)|^2 d\xi$$
$$\leq C \int_{|\xi|\leq\varepsilon} \frac{d\xi}{|\xi|^{2(m\alpha-\beta)}} + C_\varepsilon \int_{R^n} |F[\varphi_k]|^2 d\xi,$$

with some positive real numbers ε and C_ε. Let $\delta > 0$ be an arbitrarily small number. Then we can choose ε small enough to ensure that

$$C \int_{|\xi|\leq\varepsilon} \frac{d\xi}{|\xi|^{2(m\alpha-\beta)}} < \frac{\delta}{2}.$$

Further, we can take k large enough to ensure that

$$C_\varepsilon \int_{\mathbb{R}^n} |F[\varphi_k]|^2 d\xi < \frac{\delta}{2}.$$

These inequalities imply $\|\Pi_\alpha(y,D)\varphi_k\|_0 \to 0,\ k \to \infty$.

Theorem 5.24. *Let $\varphi \in \Psi'_{-G}(\mathbb{R}^n)$. Then there exists a unique solution $u(t,x)$ of P_α, and this solution has the properties:*

(a) $u(y,x) \in C[y \geq 0; Z^-_{m\alpha}(-G)]$;
(b) $B^\alpha u(y,x) \in C[y \geq 0; Z^-_{m\alpha}(-G)]$;
(c) $u(y,x) \in C^\infty[y > 0; Z^-_{m\alpha}(-G)]$.

The proof of this theorem is based on the following lemma:

Lemma 5.8. *For every fixed $y \geq 0$ and $\alpha \geq 0$ the pseudo-differential operator*

$$\Pi_\alpha(y,-D) : \Psi'_{-G}(\mathbb{R}^n) \to Z^-_{m\alpha}(-G)$$

is continuous.

Lemma 5.8 follows from Lemma 5.7. Its proof is analogous to that of Lemma 5.6.

Remark 5.8. The obtained results can be easily extended to the more general Sobolev spaces H^s_p with arbitrary $1 < p < \infty$, Besov or Lizorkin-Triebel type spaces. In the paper [Uma98] the case $m = 1$ was considered in these spaces.

5.9 On fractional derivatives of harmonic functions with given traces

In Section 4.6.4 we discussed the problem of the existence of a function $u(t,x)$, harmonic in R_+^{n+1}, whose derivative of integer order $\alpha \in \mathbb{N}_0$ converges to a given hyperfunction $\varphi \in \Psi'_{2,-G}(\mathbb{R}^n)$ in a certain topology. We can extend these results to arbitrary $\alpha \in \mathbb{R}_+$ using the theorems established in the previous section.

Assuming $m = 1$ and $A(D_x) = \Delta$, the Laplace operator, in the problem P_α we have

$$(\frac{\partial^2}{\partial y^2} + \Delta)u(y,x) = 0, \quad y > 0, x \in \mathbb{R}^n,$$

$$B^\alpha u(0,x) = \varphi(x), \quad \lim_{y\to\infty} |u(y,x)| < \infty, \; x \in \mathbb{R}^n.$$

A solution $u(y,x)$ of this problem is a harmonic function in R_+^{n+1} whose fractional derivative of order α has a given trace φ. The following proposition directly follows from Theorem 5.23.

Proposition 5.1. *Let $\varphi \in Z_\alpha^+(G)$. Then there exists a function $u(y,x)$, harmonic on the upper half-space $y > 0$, $x \in \mathbb{R}^n$, such that $\lim_{y\to+0} B^\alpha u(y,x) = \varphi(x)$ in the topology of the space $\Psi_G(\mathbb{R}^n)$.*

Using this proposition we prove the following theorem on a harmonic function whose fractional derivative of order α has a given trace in the space $\Psi'_{-G}(\mathbb{R}^n)$ (a subclass of hyperfunctions).

Proposition 5.2. *Let $\varphi \in \Psi'_{-G}(\mathbb{R}^n)$. Then there exists a function $u(y,x)$, harmonic on the upper half-space $y > 0$, $x \in \mathbb{R}^n$, such that $\lim_{y\to+0} B^\alpha u(y,x) = \varphi(x)$ in the topology of the space $Z_\alpha^-(-G)$.*

Proof. Let $\varphi \in \Psi'_{-G}(\mathbb{R}^n)$. Then the functional $u(y,x) = \Pi_\alpha(y,-D)\varphi(x)$ is a harmonic function in the distributional sense. Here $\Pi_\alpha(y,-D)$ is the pseudo-differential operator with the symbol $\Pi_\alpha(y,\xi) = |\xi|^{-\alpha}\exp(-\alpha\pi - y|\xi|)$. According to Weyl's lemma[5], $u(y,x)$ is an ordinary harmonic function. Moreover, due to Lemma 5.8 $u(y,x)$ is an element of $Z_\alpha^-(-G)$ for any fixed y. Let v be an arbitrary function in $Z_\alpha^+(G)$. We have

$$< B^\alpha u(y,x) - \varphi(x), v(x) >=< \varphi(x), [B^\alpha P(\alpha;y,D) - I]v(x) > .$$

Due to Proposition 5.1 the right-hand side of the latter tends to 0 as $y \to 0$, implying $B^\alpha u(y,x) \to \varphi(x)$ as $y \to +0$.

---

[5] Weyl's lemma [Hor83] states that a distribution $f(x)$, satisfying the equation $\Delta f = 0$ on an open set $\Omega \subset \mathbb{R}^n$ in the weak sense, is an ordinary harmonic function.

5.10 Additional notes

1. Perhaps the first application of fractional integro-differential operators was the Abel's integral equation of first kind (with $\alpha = 1/2$) connected with the famous *tautochrone problem* (see details, for instance, in [OS74], pp. 183–186). Abel published his paper in 1826. In this paper the solution is obtained essentially in the form of the Riemann-Liouville fractional derivative of order 1/2. We note that Abel also solved the problem in the general case, i.e., for arbitrary $\alpha \in (0,1)$ [A26]. A historical perspective on the fractional calculus in linear viscoelasticity is given in Mainardi's book [Mai10]. Recent review paper [RSh10] by Rossikhin and Shitikova contains the novel trends and recent results in applications of the fractional calculus to solid mechanics (containing over 300 citations). Seems Gerasimov [Ger48] was first who modeled movement of a viscous fluid between two moving surfaces using partial differential equations of fractional order in 1948, though earlier Gemant (1936) and Scott-Blair (1944) used fractional order ODEs (of order 1/2) to analyze experimental results obtained from elasto-viscous bodies. Oldham and Spanier, in their book [OS74] published in 1974, discussed fractional generalizations of transport and diffusion equations. Note that in the same year (June, 1974), the University of New Haven hosted the first international conference on fractional calculus and its applications. Proceedings of this conference edited by Ross [Ros75] was published in 1975. Starting from the 1980th applications of fractional calculus to various fields profoundly increased. In particular, Nigmatullin [Nig86] studied anomalous diffusion in a porous media, Wyss [Wis86] and Schneider and Wyss [SW89] fractional diffusion processes, Fujita [Fuj90] investigated fractional diffusion-wave processes, Schneider [Sch90] used fractional differential equations to describe the "grey"-noise. Moreover, applications to cell signaling and protein movement in cell biology [Sax01, SJ97], bioengineering [Mag06], zoology [ScS01], and to many other fields have appeared [MK00, MK04, Lim06, LH02, McC96, RSh10]. These investigations revealed many important intrinsic properties of processes, modeling by fractional equations, including hereditary properties and memory effects, oscillation-relaxation properties, connection with Lévy processes and subordinating processes, and many other properties, which cannot be captured by integer order models. The theoretical background of fractional calculus, its historical development, and various applications can be found in books [SKM87, OS74, MR93, Pod99, Hil00, Mag06, KST06, Mai10]. In Chapter 7 of this book we will discuss further applications of fractional models establishing a triple relationship between fractional Fokker-Planck-Kolmogorov type equations, stochastic differential equations, and their time-changed driving processes.
2. *Diffusion.* Diffusive processes can be classified according to the behavior of their mean square displacement (MSD), $\mathrm{MSD}(t) = \langle X_t^2 \rangle - \langle X_t \rangle^2$, as a function of time t. If the MSD is linear, then the process is classified as normal, otherwise anomalous. For many processes, the MSD satisfies the power-law behavior, $\mathrm{MSD}(t) \sim K_\beta t^\beta$, $t \to \infty$, where K_β is a constant. If $\beta = 1$ the diffusion is normal, if $\beta > 1$ the process is super-diffusive, while if $\beta < 1$ the process is sub-diffusive (see, for instance, [MK00, Zas02]). There are many interesting processes, arising in physics, cell biology, signaling, etc., which do not have the above-mentioned power-law behavior. For example, the ultra-slow diffusion processes studied in [CKS03, MS01, Koc08] have MSD with logarithmic behavior for large t. The MSD of a more complex process with retardation [CGSG03, CGKS8] behaves like t^{β_2} for t small, and t^{β_1} for t large, where $\beta_1 < \beta_2$. We will consider models of such processes in Chapter 6.
3. *The Cauchy and multi-point problems for fractional differential equations.* There is a vast literature on the Cauchy problem for integer order abstract differential-operator equations. The first order evolution equations $u'(t) = Au(t)$ in the spaces of abstract exponential vector-functions of a finite type, $Exp_A(X)$ (and in more general bornological spaces) were studied in [Rad82]. What regards to fractional order differential-operator equations, Kochubei [Koc89] studied existence and uniqueness of a solution to the abstract Cauchy problem $D_*^\alpha u(t) = Au(t)$, $u(0) = u_0$, with Caputo-Djrbashian fractional derivative for $0 < \alpha < 1$ and a closed operator A with a dense domain $\mathscr{D}(A)$ in a Banach space. El-Sayed [ES95] and Bazhlekova [Baz01] investigated Cauchy

problem for $0 < \alpha < 2$. In the papers [Koc89] and [Baz98] the necessary and sufficient conditions for solvability of the abstract Cauchy problem in the case $0 < \alpha < 1$ were given, by extending the conditions of the Hille-Yosida theorem from $\alpha = 1$ to $\alpha \in (0, 1]$. Kostin [Kos93] proved that the abstract initial value problem (Cauchy type problem)

$$D^{\alpha}u(t) = Au(t),\ D^{\alpha-k}u(0) = \varphi_k, \quad k = 1, \ldots, m,$$

where $\alpha \in (m-1, m]$, D^{α} is the Riemann-Liouville derivative of order α, and A is a linear closed operator with a dense domain in a Banach space, is well posed. For more information about recent results on the Cauchy problem for abstract fractional differential-operator equations, we refer the reader to [Baz01, EK04, KJ11, KMSL13].

4. *Fundamental solutions of fractional differential equations.* The fundamental solution of the Cauchy problem for a fractional diffusion-wave equation can be interpreted in terms of Lévy's stable probability distribution [SW89, MPG99]. This fact leads to a range of applications of fractional diffusion-wave equations connected with the description of various stochastic processes arising in science and engineering. The explicit formula for the Green function for the relevant Cauchy problem was given by Mainardi et al. [MPG99]. In the paper [MLP01] the convergent and asymptotic power series forms of fundamental solutions for space-time fractional differential equations were given. The spatial derivative used in this paper is the Riesz-Feller fractional derivative with a skewness parameter. In particular, for the density function $f(\tau)$, $\tau > 0$, of Lévy's stable subordinator the following asymptotic behavior at zero and infinity is obtained (see also [UZ99]):

$$f(\tau) \sim \frac{(\frac{\beta}{\tau})^{\frac{2-\beta}{2(1-\beta)}}}{\sqrt{2\pi\beta(1-\beta)}} e^{-(1-\beta)(\frac{\tau}{\beta})^{-\frac{\beta}{1-\beta}}}, \quad \tau \to 0;$$

$$f(\tau) \sim \frac{\beta}{\Gamma(1-\beta)\tau^{1+\beta}}, \quad \tau \to \infty.$$

In the paper [MPG07] fundamental solutions of some DODEs are obtained in the integral form with the Fox-Wright function involved in the integrand. We note that fundamental solutions to fractional diffusion-wave equations are closely related to the random walk approximation models developed in a series of papers by Gorenflo and Mainardi and their collaborators (see, for example, [GM98-1, GM99, SGM00, GM01, GMM02, GV03, GAR04]). In Chapter 8 we will discuss random walk approximations of time-changed stochastic processes, generalizing some of the results obtained in these series of publications.
5. *Fractional boundary value problems.* As to boundary value problems for partial differential equations of fractional order in bounded and unbounded domains, there are a lot of applications including computer tomography, electrodynamics, electro-statics, and elasticity theory. For mathematical investigations of such problems we refer, for example, to the works by Päivärinta and Rempel [PR92] and Natterer [Nat86], where the equation $\Delta^{\pm 1/2}u = f$ in two dimensions was studied in domains having piecewise smooth boundaries without sharp peaks. Boundary value problems for elliptic and fractional differential equations with boundary operators of fractional order is currently the focus of increasing number of researchers. For instance, in the papers [Nak75, Uma94, TU94, Naz97, Uma98, GLU00a, Goo10, GK10] various boundary value problems with fractional order boundary operators are studied.
6. *Fractional Duhamel principle.* The fractional Duhamel principle is established in the papers [US06, US07] in the case of single time fractional differential equation, and extended for wider classes of distributed fractional order differential equations in [Uma12]. For various applications of the fractional Duhamel principle we refer the reader to papers [ZhX11, Sto13, Ibr14, MN14, KO14, Tat14, WZh14].

Chapter 6
Distributed and variable order differential-operator equations

6.1 Introduction

In Section 5.6 we studied the existence of a solution to the multi-point value problem for a fractional order pseudo-differential equation with m fractional derivatives of the unknown function. This is an example of fractional distributed order differential equations. Our main purpose in this chapter is the mathematical treatment of boundary value problems for general distributed and variable order fractional differential-operator equations. We will study the existence and uniqueness of a solution to initial and multi-point value problems in different function spaces.

In the general setting the distributed time fractional order differential-operator equation has the form

$$\int_0^{\mu} A(r) D_*^r u(t) \Lambda(dr) = Bu(t) + h(t), \, t > 0, \tag{6.1}$$

where $\mu \in (m-1, m]$, $m \in \mathbb{N}$; Λ is a finite Borel measure with $\operatorname{supp} \Lambda \subset [0, \mu]$; $A(r)$ (for a fixed $r \in (0, \mu]$) and B are linear closed operators defined in a certain locally convex topological vector space X, $h(t) \in C(\mathbb{R}_+; X)$; the vector-function $u(t)$ is unknown and belongs to the space $C^{(m)}(0, T; X)$ with some $T > 0$ and D_*^r is the operator of fractional differentiation of order r in the sense of Caputo-Djrbashian (see, Section 3.5). An essential distinctive feature of that model is that integration in (6.1) is performed by the variable r, the order of differentiation. Such models arise naturally in the kinetic theory [CGSG03] when the exact scaling is lacking or in the theory of elasticity [LH02] for description of rheological properties of composite materials. The list of practical applications where distributed order differential equations arise can be continued (see [Cap01, BT00, Lim06, CSK11] and references therein). Mathematical theory of the Cauchy problem for distributed order differential equations was developed in works [BT00, UG05-2, MS06, Koc08].

The equation (6.1) is a generalization of fractional/non-fractional differential equations. To illustrate this consider a few examples.

S. Umarov, *Introduction to Fractional and Pseudo-Differential Equations with Singular Symbols*, Developments in Mathematics 41,
DOI 10.1007/978-3-319-20771-1_6

Example 6.1. 1. Let $\Lambda(dr) = \delta(r-\beta)dr$, where δ is the Dirac delta-function (distribution), $A(r) = I$, the identity operator, and $B = A(D)$ is a ΨDOSS. Let $X = \Psi_{G,p}(\mathbb{R}^n)$ or its dual $X = \Psi'_{-G,p'}(\mathbb{R}^n)$. In this case we have the following fractional differential equation

$$D_*^{\beta} u(t,x) = A(D)u(t,x), \quad t>0,\ x\in\mathbb{R}^n,\ \beta>0,$$

discussed in Chapter 5.

2. Let $\Lambda(dr) = \delta(r-\beta)dr$, $\beta\in(0,2]$, $A = I$, and $B(D) = \mathbb{D}_0^{\alpha}$, $0<\alpha\le 2$. Then we have the space-time fractional differential equation

$$D_*^{\beta} u(t,x) = D_0^{\alpha} u(t,x), \quad t>0,\ x\in\mathbb{R}^n,$$

studied in Section 5.4. As we will see in the next chapter, this equation models non-Gaussian non-Markovian stochastic processes.

3. The next example relates to sub-diffusion equation with retardation, studied in [CGSG03]. Let $\Lambda(dr) = [b_1\delta(r-\beta_1) + b_2\delta(r-\beta_2)]dr$, with $0<\beta_1<\beta_2\le 1, b_1>0, b_2>0, b_1+b_2=1$, $A(r) = I$ and $B(D) = k\frac{\partial^2}{\partial x^2}$ $(n=1)$. Then we have the equation

$$b_1 D_*^{\beta_1} u(t,x) + b_2 D_*^{\beta_2} u(t,x) = k\frac{\partial^2}{\partial x^2} u(t,x),$$

which describes a subdiffusion process with retardation. In [CGSG03] the Cauchy problem for the equation

$$\int_0^1 \tau^{\beta-1} w(\beta) \frac{\partial^{\beta} p}{\partial t^{\beta}} d\beta = k\frac{\partial^2 p}{\partial x^2}$$

referred to as the "normal form" of the distributed order fractional diffusion is also studied. Note that this equation corresponds to the case $\Lambda(dr) = \tau^{r-1}w(r)dr$ with $\tau>0$, $w(r)>0$, $A(r) = I$, and $B(D) = k\frac{\partial^2}{\partial x^2}$.

4. The authors of [LH02] derived the equations of the form

$$\int_0^2 k(q) D^q y(t) dq + F(y) = f(t), \quad t>0,$$

which describes properties of composite materials. Note that $k(q), F(y)$, and $f(t)$ are given functions connected with different characteristics of viscoelastic and viscoinertial materials with rheological properties.

5. Let $A(r) = A(r;D)) = \sum_{k=1}^{m}\delta(r-\beta_k)A_k(D)$ with $k-1<\beta_k\le k$, $A_k(D)$, $k = 0,\dots,m-1$, are pseudo-differential operators with symbols $A_k(\xi)$ continuous in G, and $\Lambda(dr) = dr$. In this case we obtain the equation

$$D_*^{\beta_m} u(t,x) + \sum_{k=1}^{m-1} A_k(D) D_*^{\beta_k} u(t,x) + A_0(D)u(t,x) = 0, \quad t\in(0,T),\ x\in\mathbb{R}^n, \tag{6.2}$$

discussed Section 5.6. In the case $\beta_k \in \mathbb{N},\ k = 1,\dots,m$, the Cauchy problem for this equation with analytic symbols or with symbols having singularities was studied, for example, by Dubinskij [Dub81], Umarov [Uma86], and Tran Duc Van [Van89]. Antipko and Borok [AB92], Borok [Bor71], Ptashnik [Pta84], and Umarov [Uma97, Uma98] (see also references therein) considered multi-point value problems with integer α_k. For fractional α_k the Cauchy and multi-point value problems are studied in [GLU00].

6.2 Distributed order fractional differential-operator equations

In this section we find a representation formula for a solution of the Cauchy problem for the distributed order differential-operator equation

$$\int_0^{\mu} A(r) D_*^r u(t) \Lambda(dr) = Bu(t), \quad t > 0, \tag{6.3}$$

$$\frac{\partial^k u(0)}{\partial t^k} = \varphi_k, \quad k = 0,\dots,m-1. \tag{6.4}$$

To do that we apply a formal operator method. Namely, we assume that $A(r)$ and B complex-valued function and parameter, respectively. Whenever, a formal representation is obtained, we give an informal meaning, depending on the problem being considered.

We split the problem (6.3), (6.4) into m Cauchy problems, one for each index $j \in \{0,1,\dots,m-1\}$, for the same DODE (6.3) with the following Cauchy conditions

$$\begin{aligned} &u(0) = 0,\dots,\frac{\partial^{j-1}u(0)}{\partial t^{j-1}} = 0, \\ &\frac{\partial^j u(0)}{\partial t^j} = \varphi_j, \\ &\frac{\partial^{j+1}u(0)}{\partial t^{j+1}} = 0,\dots,\frac{\partial^{m-1}u(0)}{\partial t^{m-1}} = 0. \end{aligned} \tag{6.5}$$

If one denotes by $u_j(t)$ a solution to (6.3), (6.5) then the general solution to (6.3), (6.4) due to its linearity, has the form

$$u(t) = \sum_{j=0}^{m-1} u_j(t).$$

We rewrite the left-hand side of (6.3) in the form

$$\int_0^{\mu} A(r) D_*^r u(t) \Lambda(dr) = \sum_{k=1}^{m} \int_{k-1}^{k} A(r) D_*^r u(t) \Lambda(dr),$$

and apply the Laplace transform to both sides. Recall that if $k-1<r<k$, then the formula (see (3.55))

$$L[D_*^r f(t)](s) = L[f](s)s^r - \sum_{l=0}^{k-1} f^{(l)}(0)s^{r-l-1},$$

holds. Making use of this formula, we have

$$\int_{k-1}^{k} A(r)L[D_*^r u](s)\Lambda(dr) = L[u](s)\int_{k-1}^{k} s^r A(r)\Lambda(dr) - \sum_{l=0}^{k-1} \frac{u^{(l)}(0)}{s^{l+1}} \int_{k-1}^{k} s^r A(r)\Lambda(dr), \quad (6.6)$$

where $L[u](s)$ is the Laplace transform of u. Now summing up over the indices $k=1,\dots,m$ in equation (6.6), we obtain the following equation for the Laplace transform of u :

$$L[u](s)\int_0^{\mu} A(r)s^r\Lambda(dr) - \sum_{k=0}^{m-1} \frac{\varphi_k}{s^{k+1}} \int_k^m A(r)s^r\Lambda(dr) = BL[u](s). \quad (6.7)$$

Further, let us introduce the functions $\Phi_0(s),\dots,\Phi_m(s)$ by

$$\Phi_0(s)=0; \quad \Phi_j(s) = \int_0^j s^r A(r)\Lambda(dr), \quad j=1,\dots,m. \quad (6.8)$$

Then it follows from (6.7) and the hypothesis that only the j-th Cauchy condition contains nonzero (see (6.5)) right-hand side, that

$$L[u](s) = \frac{\Phi_m(s)-\Phi_j(s)}{s^{1+j}[\Phi_m(s)-B]}\varphi_j, \quad j=0,\dots,m-1. \quad (6.9)$$

Now inverting (6.9), we obtain the j-th solution $u_j(t)=S_j(t)\varphi_j$, through the operator $S_j(t)$, in the form

$$S_j(t) = L^{-1}\left[\frac{\Phi_m(s)-\Phi_j(s)}{s^{1+j}(\Phi_m(s)-B)}\right](t), \quad j=0,\dots,m-1. \quad (6.10)$$

Example 6.2. 1. Let $0<\alpha<1$ and $\Lambda(dr)=\delta_\alpha(r)dr$. Then $\Phi_0(s)=0,\ \Phi_1(s)=s^\alpha$. Hence

$$S(B;t) = L^{-1}\left[\frac{s^{\alpha-1}}{s^\alpha - B}\right](t) = E_\alpha(Bt^\alpha).$$

2. Let $1<\mu<2$ and $0<\beta<1$. Suppose $\Lambda(dr)=[\delta_\mu(r)+a\delta_\beta(r)]dr$, where $a>0$. Then the operator on the left of (6.3) simplifies to $D_*^\mu u(t) + a\,D_*^\beta u(t)$. Moreover, $\Phi_1(s)=a\,s^\beta$, $\Phi_2(s)=s^\mu+as^\beta$. Therefore

$$S_0(B;t) = L^{-1}\left[\frac{s^{\mu-1}+as^{\beta-1}}{s^\mu+as^\beta-B}\right](t), \quad S_1(B;t) = L^{-1}\left[\frac{s^{\mu-1}}{s(s^\mu+as^\beta-B)}\right](t).$$

Using the equality

$$\frac{1}{s^{\mu}+as^{\beta}-B}=\sum_{n=0}^{\infty}\frac{s^{-\beta n-\beta}}{(s^{\mu-\beta}+a)^{n+1}}(-B)^{n}$$

and formula (3.118), one can easily find power series representations of $S_j(B;t)$, $j=0,1$. Leaving the details to the reader, we give the final forms of these series. Namely,

$$S_0(B;t)=S_*(B;t)+at^{\mu-\beta}S_{**}(B;t),\quad S_1(B;t)=JS_*(B;t), \tag{6.11}$$

where

$$S_*(B;t)=\sum_{n=0}^{\infty}\frac{(-Bt^{\mu})^n}{n!}E^{(n)}_{\mu-\beta,\,n\beta+1}(-at^{\mu-\beta}); \tag{6.12}$$

$$S_{**}(B;t)=\sum_{n=0}^{\infty}\frac{(-Bt^{\mu})^n}{n!}E^{(n)}_{\mu-\beta,\,n\beta-\beta+\mu+1}(-at^{\mu-\beta}). \tag{6.13}$$

Remark 6.1. The operators in (6.10) have informal meaning in concrete situations. For example, let $A(r)=A(r,D)$ and $B=B(D)$ be ΨDOSS with respective symbols $A(r,\xi)$ and $B(\xi)$. We assume that these operators are defined on $X=\Psi_{G,p}(\mathbb{R}^n)$ (or on its dual), where $G\subseteq R^n$ is an open set. Let $\xi\in G$ be fixed. Denote by $s_0(\xi)$ the greatest positive root of the equation $\Phi_m(s,\xi)=B(\xi)$, where

$$\Phi_m(s,\xi)=\int_0^m s^rA(r,\xi)\Lambda(dr).$$

If $A(r,\xi)$ preserves its sign for every $0<r<m$, then it follows from the inequality $s_1^r<s_2^r$ for $0<s_1<s_2$ and $r>0$ that $\Phi_m(s,\xi), s>0$, is a monotone function. Hence, the equation $\Phi_m(s,\xi)=B(\xi)$ may have no more than one positive root for every fixed $\xi\in G$. Thus, the function

$$\Psi_j(s,\xi)=\frac{\Phi_m(s,\xi)-\Phi_j(s,\xi)}{s[\Phi_m(s,\xi)-B(\xi)]}, \tag{6.14}$$

where

$$\Phi_j(s,\xi)=\int_0^j s^rA(r,\xi)\Lambda(dr),$$

is well defined for $s>s_0(\xi)$ if the equation $\Phi_m(s,\xi)=B(\xi)$ has a positive root, or for $s>0$ if there is no such a root. Moreover, it is not difficult to verify that the collection of functions $\Phi_0(s,\xi),\dots,\Phi_{m-1}(s,\xi)$ is linearly independent. In this case the solution operators $S_j(t,D), j=0,\dots,m-1$, are ΨDOSS, whose symbols are the Laplace preimages of $s^{-j}\Psi_j(s,\xi)$, that is

$$S_j(t,D)=J^jU_j(t,D),\quad j=0,\dots,m-1, \tag{6.15}$$

where $U_j(t,D)$, $j=0,\dots,m-1$, are ΨDOSS with symbols

$$U_j(t,\xi)=L^{-1}\left[\frac{\Phi_m(s,\xi)-\Phi_j(s,\xi)}{s(\Phi_m(s,\xi)-B(\xi))}\right], \quad j=0,\dots,m-1 \tag{6.16}$$

and J^j is the j-th power of the integration operator with lower limit 0. Using the theorem on uniqueness of the inverse Laplace transform, we can conclude that the collection $S_0(t,\xi),\dots,S_{m-1}(t,\xi)$ is linearly independent as well. Thus for $u_j(t,x)$ we get the representation

$$u_j(t,x)=S_j(t,D)\varphi_j(x),$$

where $S_j(t,D)$ is the pseudo-differential operator with the symbol $S_j(t,\xi)$. Here, for the solution of (6.3), (6.4) we have the representation

$$u(t,x)=\sum_{j=0}^{m-1} S_j(t,D)\varphi_j(x), \tag{6.17}$$

Now returning to the abstract case we have the representation for the solution

$$u(t)=\sum_{j=0}^{m-1} S_j(t)\varphi_j, \tag{6.18}$$

where $S_j(t)$ is defined in (6.10). We note that in the abstract case also we can write $S_j(t)=J^jU_j(t)$, where

$$U_j(t)=L^{-1}\left[\frac{\Phi_m(s)-\Phi_j(s)}{s(\Phi_m(s)-B)}\right], \quad j=0,\dots,m-1$$

Remark 6.2. The obtained representation formula is useful both from mathematical and physics point of views. This representation is obtained as the action of the operators $J^jU_j(t,\xi)$, $j=0,\dots,m-1$, called *j-th solution operator*, to the given functions. These operators have the same structure. The formula (6.18) says that no matter how many different fractional order derivatives in the sense of Caputo-Djrbashian are there between two consecutive integers, only the initial data with integer order derivatives define the solution. Moreover from this formula it can be derived that if the maximal order of the derivatives in the equation is not greater than $m-1$, then the Cauchy problem with m given data becomes ill-posed. Indeed if $A(r)=\delta(r-\alpha), \alpha\le m-1$, rewriting (6.14) for $j=m-1$ in the form

$$\Psi_{m-1}(s,\xi)=\frac{\int_{m-1}^{m} s^rA(r,\xi)dr}{s[\Phi_m(s,\xi)-B(\xi)]}$$

we have $S_{m-1}(t,\xi)\equiv 0$.

Further, to describe solution spaces of Cauchy problem (6.3)–(6.4) we need to study properties of symbols of solution operators. Below we establish necessary properties of symbols of solution operators. Let

$$F_a(k;f) := \int_0^a e^{kt} f(t)\Lambda(dt), \quad k \in \mathbb{R}, \tag{6.19}$$

where a is a fixed positive real number, $\Lambda(dt)$ is a finite measure on the interval $[0,a]$, and f is a distribution with *supp* $f \subset [0,a]$.

Lemma 6.1. *1. For a regular distribution $f(t)$ with $suppf \subset [0,\mu], \quad d \leq a$,*

$$|F_a(k;f)| = O(e^{\mu k}),\ k \to \infty;$$

2. For a singular generalized function $f(t)$ with $suppf = \{d\}$

$$|F_a(k;f)| = o(e^{(\mu+\varepsilon)k}), \quad k \to \infty;$$

where $\varepsilon > 0$ is arbitrarily small.

Proof. Let first $f \in L_\infty(0,a)$, $|f(t)| \leq M < \infty$ and $suppf \subset [0,\mu]$. Then

$$|F_a(k;f)| = |\int_0^\mu e^{kt} f(t)dt| \leq Me^{\mu t}\Lambda([0,a]) = Ce^{k\mu}, \quad k \to \infty.$$

For a regular generalized function f with $suppf \subset [0,\mu]$ there is a sequence $f_m \in L_\infty(0,a)$, all supported in $[0,\mu]$ and with common constant M bounding above, and such that $f_m \to f, m \to \infty$ in the weak sense. For f_m we have $|F_a(k;f_m)| \leq Ce^{k\mu}$ with positive constant C. Letting $m \to \infty$ we obtain the desired result.

If f is a singular generalized function with support $supp\ f = \{\alpha\}, 0 < \alpha < a$, then due to Proposition 1.14, f is a finite linear combination of $\delta^{(j)}(t-\alpha)$, where δ is the Dirac function. Substituting this linear combination to (6.19), one has $|F_a(k;f)| = O(k^N e^{\alpha k})$ for some N, and hence, $|F_a(k;f)| = o(e^{(\alpha+\varepsilon)k})$ for any $\varepsilon > 0$ as $k \to \infty$.

Lemma 6.2. *For $\Phi_j(s), j = 0,\ldots,m$, defined in (6.8) the following assertions hold:*

1. *If $A(r)$ is a regular distribution with $suppA(r) \subset [0,\mu]$, then $\Phi_j(s) = O(s^\nu)$, $s \to \infty$, where $\nu = min\{\mu, j\}$;*
2. *If $A(r)$ is a singular distribution with $suppA(r) = \{\mu\}$, then $\Phi_j(s) = o(s^{\mu+\varepsilon})$, where ε is arbitrarily small, $s \to \infty$, in the case $\mu \leq j$ and $\Phi_j(s) = 0$ when $\mu > j$.*

Proof. The function $\Phi_j(s)$ can be reduced to $F_a(k;f)$. In fact,

$$\Phi_j(s) = \int_0^j s^r A(r)\Lambda(dr) = \int_0^j e^{r\ln s} A(r)\Lambda(dr) =: F_j(\ln s; A(r)), \quad s > \sigma(B) > 0,$$

where $\sigma(B)$ is a positive number depending on the operator B. Now it is an easy exercise to apply Lemma 6.1 and obtain the asymptotics in cases 1) and 2).

Corollary 6.1 *Let*

$$\Psi_j(s) = \frac{\Phi_m(s) - \Phi_j(s)}{s[\Phi_m(s) - B]}, \quad j = 0, \dots, m. \tag{6.20}$$

Then $\Psi_j(s) = O(\frac{1}{s}), s \to \infty$, *for each* $j = 0, \dots, m$.

Lemma 6.3. *For every* $j = 0, \dots, m$, *and* $k = 0, 1, \dots, j-1$, *the equality*

$$\Phi_j^{(k)}(s) = \frac{F_j\left(\ln s; \frac{\Gamma(r+1)r^k}{\Gamma(r-k+1)} A(r)\right)}{s^k}, \quad s > \sigma_0(B) > 0,$$

holds.

Proof. Computing first, second, etc. derivatives consecutively, we have

$$\Phi_j'(s) = \int_0^j rs^{r-1} A(r)\Lambda(dr) = \frac{1}{s}\int_0^j rs^r A(r)\Lambda(dr) = \frac{F_j(\ln s; rA(r))}{s},$$

$$\Phi_j''(s) = \int_0^j r(r-1)s^{r-2} A(r)\Lambda(dr) = \frac{1}{s^2}\int_0^j r(r-1)s^r A(r)\Lambda(dr)$$

$$= \frac{F_j(\ln s; r(r-1)A(r))}{s^2} = \frac{F_j\left(\ln s; \frac{\Gamma(r+1)}{\Gamma(r-1)} A(r)\right)}{s^2}.$$

Similarly for the k-th derivative,

$$\Phi_j^{(k)}(s) = \int_0^j r(r-1)\dots(r-k+1)s^{r-k} A(r)\Lambda(dr)$$

$$= \frac{1}{s^k}\int_0^j \frac{\Gamma(r+1)}{\Gamma(r-k+1)} s^r A(r)\Lambda(dr) = \frac{F_j(\ln s; \frac{\Gamma(r+1)r^k}{\Gamma(r-k+1)} A(r))}{s^k}.$$

Corollary 6.2 *For* $j = 0, \dots, m-1$, $\Psi_j^{(k)}(s) = O(\frac{1}{s^{k+1}}), s \to \infty$, $k = 0, \dots, j-1$.

Lemma 6.4. *For* $U_j(t) = L^{-1}_{s \to t}\Psi_j(s), j = 0, \dots, m-1$, *the following assertions hold:*

1. $U_j(t) \to 1$ *as* $t \to +0$, $j = 0, \dots, m-1$;
2. $U_j^{(\ell)}(t) \to 0$ *as* $t \to +0$, $\forall j = 0, \dots, m-\ell-1$, $\ell = 1, \dots, m-1$.

Proof. It follows from the representation (6.20) for $\Psi_j(s)$ and the fact that $\Psi_j(s) = O(1/s)$ for large s (see Corollary 6.1), that $\Psi_j(s)$ is Laplace invertible. Thus $U_j(t)$ exists for all $j = 0, \dots\dots, m-1$. Further, we use the following relation [DP65], [Wid46]

$$\lim_{s \to \infty} sL[f](s) = \lim_{\varepsilon \to +0} \frac{1}{\varepsilon}\int_0^\varepsilon f(t)dt, \tag{6.21}$$

which is an implication of tauberian theorems, valid for functions bounded below, meaning that if one of these limits exists, then the other limit also exists and the equality holds. It is easy to see that Corollary 6.1 implies $s\Psi_j(s) \to 1$, or the same $sL[U_j](s) \to 1$, as $s \to \infty$. Hence, due to (6.21), we have $U(t)$ is continuous near zero and $U_j(t) \to 1$, as $t \to +0$. Part 1) of the lemma is proved.

Further, the Laplace transform of $U_j'(t)$ is

$$L[U'](s) = s\Psi_j(s) - U_j(+0) = \frac{\Phi_m(s) - \Phi_j(s)}{\Phi_m(s) - B} - 1 = \frac{-\Phi_j(s) + B}{\Phi_m(s) - B}, \quad s > \sigma(B),$$

which is $O(1/s^{\mu-j})$ for large s, where $\mu = \sup \operatorname{supp} A(r)$. Note that $\mu > m-1$. Otherwise the Cauchy problem is meaningless (see Remark 6.2). Hence, $sL[U'](s) = O(1/s^{\mu-j-1})$, which due to relation (6.21) implies $U_j'(+0) = 0$ if $\mu - j - 1 > 0$, or for all $j = 0, \ldots, m-2$. Thus, we have

$$U_j'(+0) = 0, \quad j = 0, \ldots, m-2.$$

Similarly, the Laplace transform of $U_j''(t)$ is

$$L[U''](s) = s^2\Psi_j(s) - U_j'(+0) - sU_j(+0) = s[s\Psi_j(s) - 1] = O(\frac{1}{s^{\mu-j-1}}), \quad s \to \infty$$

which implies $sL[U''](s) = O(1/s^{\mu-j-2})$, $s \to \infty$. Hence, due to relation (6.21), we have $U''(+0) = 0$ for all $j = 0, \ldots, m-3$.

Continuing this process for the Laplace transforms of derivatives $U_j^{(\ell)}(t)$, $\ell = 3, 4, \ldots, m-1$, by induction we obtain

$$L[U_j^{(\ell)}](s) = s^{l-1}[s\Psi_j(s) - U_j(+0)] = O(1/s^{\mu-j-l+1}), \quad s \to \infty. \tag{6.22}$$

Consequently, we have $sL[U_j^{(\ell)}](s) = O(1/s^{\mu-j-l})$, $s \to \infty$. Using this and relation (6.21) we obtain 2). The proof is complete.

Lemma 6.4 can be reformulated in the following more convenient form:

Lemma 6.5. *For $U_j(t) = L^{-1}_{s\to t}[\Psi_j(s)](t)$, $j = 0, \ldots, m-1$, the following assertions hold:*

1. $U_j(t) \to 1$ as $t \to +0$, $j = 0, \ldots, m-1$;
2. $U_j^{(\ell)}(t) \to 0$ as $t \to +0$, $\forall j = 0, \ldots, m-1$, $\ell = 1, \ldots, m-j-1$.

Lemma 6.6. *For every $j = 0, \ldots, m-1$ the following assertions hold:*

1. $U_j(t) \in C^{m-j-1}[0, \infty)$;
2. If the upper bound of $\operatorname{supp} A(r) = \mu = m$, then $U_j^{(m-j)}(t)$, $j = 0, \ldots, m-1$, exists for almost all $t \in (0, \infty)$.

Proof. Proving the previous lemma we have noticed that the $U_j(t)$, $j = 0, \ldots, m-1$, exist. Now we will check their differentiability properties. It is known that if for given $f(t)$ its Laplace transform $L[f](s)$ additionally to (6.21) satisfies the condition $sL[f](s) \to 0$, when $s \to \infty$, then f is a continuous function. This fact directly follows from the inverse Laplace transform formula. It follows from (6.22) that

$$sL[U_j^{(\ell)}](s) = O(1/s^{\mu-l-j}), \quad s \to \infty. \tag{6.23}$$

Let $l = m - j - 1$. Then (6.23) takes the form $sL[U_j^\ell](s) = O(1/s^{\mu-m+1})$, $s \to \infty$. Hence, $sL[U_j^\ell](s)$ vanishes as $s \to \infty$, since $\mu > m-1$. Thus, $U^{(m-j-1)}(t)$ is continuous on $[0,\infty)$. Now assume that $\mu = m$ and $\ell = m-j$. Then it follows from (6.23) that $L[U_j^{(m-j)}](s) = O(1/s)$, as $s \to \infty$, which implies $U_j^{(m-j)}(t)$, $j = 0,\ldots,m-1$, exists for a.e. $t \in (0,\infty)$.

Remark 6.3. If $m-1 < \mu < m$ one can show that $D_*^{\mu-j} S_j(t,\xi)$, $j = 0,\ldots,m-1$, exists and bounded for a.e. t. Compare with Lemma 7 in [GLU00].

Lemma 6.7. *For $S_j(t) = J^j U_j(t)$, $j = 0,\ldots,m-1$, the following assertions hold:*

1. $S_j^{(j)}(t) \to 1$ *as* $t \to +0$, $\quad j = 0,\ldots,m-1$;
2. $S_j^{(\ell)}(t) \to 0$ *as* $t \to +0$, $\forall j,\ell = 0,\ldots,m-1$, $\ell \neq j$;
3. $S_j(t) \in C^{m-1}[0,\infty)$.

Proof. Since $S_j^{(j)}(t) = U_j(t)$, the first statement immediately follows from Lemma 6.5. Further, if $1 \le \ell < j$, then obviously, $S_j^\ell(t) = J^{j-\ell} U_j(t) \to 0$, as $t \to 0+$. If $j < \ell \le m-1$, then $S_j^\ell(t) = U_j^{\ell-j}(t) \to 0$, as $t \to 0+$, due to Part 2 of Lemma 6.5. Finally, the third statement is a simple implication of Part 1 of Lemma 6.6.

The established properties of solution operators (of course, in the formal level) play an important role in the description of solution spaces for boundary value problems (6.3)–(6.4). In particular, these abstract results combined with properties of ΨDOSS, studied in Chapter 2, can be applied for the Cauchy problem for distributed order fractional pseudo-differential equations with singular symbols. Consider the following distributed order time-fractional differential equation with spatial pseudo-differential operators

$$\int_0^m A(r;D) D_*^r u(t,x)dr = B(D)u(t,x), \quad t > 0,\, x \in \mathbb{R}^n, \tag{6.24}$$

with the Cauchy conditions

$$\frac{\partial^k u(0,x)}{\partial t^k} = \varphi_k(x), \quad x \in \mathbb{R}^n,\, k = 0,\ldots,m-1, \tag{6.25}$$

where the φ_k, $k = 0,\ldots,m-1$, are given functions in certain spaces described later, $D = (D_1,\ldots,D_n)$, $D_j = -i\frac{\partial}{\partial x_j}$, $j = 1,\ldots,n$, $A(r;D)$ (for every fixed value of the parameter $r \in [0,m]$) and $B(D)$ are pseudo-differential operators with the symbols $A(r;\xi)$ and $B(\xi)$ in $CS_p(G)$, respectively.

The following two corollaries, which immediately follow from Lemmas 6.7 and 6.6, will be used in the next section.

Corollary 6.3 *For symbols* $S_j(t,\xi) = J^j U_j(t,\xi), j = 0,\dots,m-1,$ *of operators* $S_j(t,D)$ *defined in* (6.15), *for any fixed* $\xi \in G$, *the following assertions hold:*

1. $S_j(t,\xi) \to 1$ *for* $t \to +0,\ j = 0,\dots,m-1;$
2. $\frac{\partial^\ell S_j(t,\xi)}{\partial t^\ell} \to 0$ *for* $t \to +0,\ j,\ \ell = 0,\dots,m-1,\ \ell \neq j.$

Corollary 6.4 *For every* $j = 0,\dots,m-1$ *the following assertions hold:*

1. $S_j(t,\xi) \in C^{m-1}[t \geq 0; C(G)];$
2. *If the upper bound of* $supp A(r,\xi) = m$, *then* $U_j^{(m-j)}(t,\xi)$ *exists for almost all* $t \in (0,\infty)$.

6.3 Solution of the Cauchy problem for distributed order pseudo-differential equations

In this section we describe solution spaces using the properties of solution operators and their symbols established in the previous section. Consider the following Cauchy problem for distributed order fractional differential operator

$$L^\Lambda[u] \equiv \int_0^\mu f(\alpha,A) D_*^\alpha u(t) d\Lambda(\alpha) = B(A)u, \quad t > 0, \tag{6.26}$$

$$u^{(k)}(0) = \varphi_k, \quad k = 0,\dots,m-1, \tag{6.27}$$

where $\mu \in (m-1,m]$ and φ_k, $k = 0,\dots,m-1$, are elements of a locally convex topological vector space $\mathrm{Exp}_{A,G}(X)$, defined in Section 2.8. The operator $B(A)$ and the family of operators $f(\alpha,A)$ are defined through the symbol $B(z)$ and the family of symbols $f(\alpha,z)$, that are continuous in the variable $\alpha \in [0,\mu]$, and analytic in the variable $z \in G \subset \mathbb{C}$; see definition in Section 2.8. The operator A is a closed linear operator in a reflexive Banach space X with a dense domain $\mathscr{D}(A)$. The measure Λ is finite and defined on $[0,\mu]$.

The strong and weak solutions of the Cauchy problem (6.3), (6.4) are understood in the following sense.

Definition 6.1. A function $u(t)$ is called a strong solution to the Cauchy problem (6.3), (6.4) if the following conditions are verified:

1. $u(t) \in C^{m-1}[t \geq 0; X]$;
2. $u^{(m)}(t) \in X$ exists for almost all $t > 0$; and
3. it satisfies the equation (6.3) for almost all $t \in (0,\infty)$ and initial conditions (6.4) pointwise.

Definition 6.2. A function $u(t)$ is called a weak solution to the Cauchy problem (6.3), (6.4) (replacing A by A^*, the adjoint of A) if

1. $u(t) \in C^{m-1}[t \geq 0; X']$ (X' is the dual space to X);
2. $u^{(m)}(t) \in X'$ for almost all $t > 0$ and
3. the following equalities hold true for arbitrary $v \in X$

$$\int_0^m < D_*^{\alpha} u(t), f(\alpha, A)v > \Lambda(d\alpha) \; = < u(t), B(A)v >,$$

for almost all $t \in (0, \infty)$ and

$$lim_{t \to +0} < u^{(k)}(t), v > = < \varphi_k, v >, \quad k = 0, \ldots, m-1.$$

Theorem 6.5. *Let G be a domain of continuity of the symbols $A(r,z)$ (r fixed) and $B(z)$. Let $\varphi_j \in \mathrm{Exp}_{A,G}(\mathrm{X})$, $j = 0, \ldots, m-1$. Then the Cauchy problem (6.26), (6.27) has a unique strong solution. This solution is given by the following representation (cf. (6.18))*

$$u(t) = \sum_{j=0}^{m-1} S_j(t,A)\varphi_j, \tag{6.28}$$

where $S_j(t,A)$ is the operator with the symbol $S_j(t,z) = J^j U_j(t,z)$, $U_j(t,z) = L^{-1}_{s \to t}[\Psi_j(s,z)](t)$, and $\Psi_j(s,z)$, $j = 0, \ldots, m-1$, are defined in (6.14).

Proof. Let $G \subset \mathbb{C}$ be a domain of continuity of the symbols $A(r,z)$ and $B(z)$ and $\varphi_j \in X$, $j = 0, \ldots, m-1$. By construction of the representation (6.28) each of its term satisfies the equation (6.26) and, by virtue of Corollary 6.3 (part 1), the conditions (6.27). Moreover, due to Corollary 6.4, for every $j = 0, \ldots, m-1$, the inclusion $J^j U_j(t,z) \in C^{m-1}(t \geq 0; C(G))$ holds and $D_*^d J^j U_j(t,z)$ is bounded for almost every $t \in (0, \infty)$. Lemma 2.3 in Section 2.8 yields $J^j U_j(t,A)\varphi_j \in \mathrm{Exp}_{A,G}(\mathrm{X})$, $j = 0, \ldots, m-1$, for every fixed $t > 0$. Hence, $u(t,x) \in C^{m-1}[t \geq 0; \mathrm{Exp}_{A,G}(\mathrm{X})]$, and $u(t,x)$ is a strong solution of the Cauchy problem (6.26), (6.27). Its uniqueness follows from the representation formula (6.28) and the $\mathrm{Exp}_{A,G}(\mathrm{X})$-continuity of operators $S_j(t,A)$ with continuous symbols $S_j(t,z) = J^j U_j(t,z)$ (see Lemma 2.3).

Theorem 6.6. *Let $\varphi_j \in \mathrm{Exp}'_{A^*,G^*}(\mathrm{X}^*)$, $j = 0, \ldots, m-1$. Then the Cauchy problem (6.26)–(6.27) has a unique weak solution. This solution is given by*

$$u(t,x) = \sum_{j=0}^{m-1} S_j^w(t,A^*)\varphi_j, \tag{6.29}$$

where $S_j^w(t,A^*)$, $j = 1, \ldots, m$, is the j-th solution operator with the symbol $S_j(t,z)$.

Proof. Let $\varphi_j \in \Psi'_{-G,q}(\mathbb{R}^n)$, $j = 0, \ldots, m-1$. It follows from Lemma 2.3 that every term on the right-hand side of (6.29), namely, $u_j(t) = J^j S_j^w(t,A^*)\varphi_{k-1}$, $j = 0, \ldots, m-1$, is a functional from the space $\mathrm{Exp}'_{A^*,G^*}(\mathrm{X}^*)$. To prove the theorem we have to show that $u_j(t)$, $j = 0, \ldots, m-1$, satisfies the equation (6.26) and the initial conditions (6.27) in the weak sense. Let $v \in \mathrm{Exp}_{A,G}(\mathrm{X})$ be an arbitrary element. Then for $u_j(t) = J^j S_j^w(t,A^*)\varphi_j$ we have

$$\int_0^\mu < D_*^\alpha u_j(t), f(\alpha,A)v > \Lambda(d\alpha) \; - \; < u_j(t), B(A)v >$$
$$= < \int_0^\mu D_*^\alpha S_j^w(t,A^*)\varphi_j, A(\alpha,A)v > \Lambda(d\alpha) \; - \; < S_j^w(t,A^*)\varphi_j, B(A)v >$$
$$= < \varphi_j, \left[\int_0^\mu D_*^\alpha A(\alpha,A) S_j(t,A)\Lambda(d\alpha) \; - B(A)S_j(t,A) \right] v > . \qquad (6.30)$$

Since, by construction, $S_j(t,A)$ for each $j = 0,\dots,m-1$, satisfies equation (6.26), we conclude that the expression in the square brackets on the right of equation (6.30) is zero for a. e. $t \in (0,\infty)$. Moreover, Corollary 6.3 yields

$$\lim_{t\to+0} < u_j^{(k)}(t), v > = < \delta_{j,k}\varphi_k, v >, \quad j,k = 0,\dots,m-1.$$

Hence, $u(t)$ defined by (6.29) satisfies Cauchy problem (6.26)-(6.27) in the weak sense.

The following two theorems follow immediately from Theorems 6.5 and 6.6, respectively.

Theorem 6.7. *Let G be a domain of continuity of the symbols $A(r,\xi)$ (r fixed) and $B(\xi)$. Let $\varphi_j \in \Psi_{G,p}(\mathbb{R}^n)$, $j = 0,\dots,m-1$. Then the Cauchy problem (6.24)-(6.25) has a unique strong solution. This solution is given by the representation (6.17).*

Theorem 6.8. *Let $\varphi_j \in \Psi'_{-G,q}(\mathbb{R}^n)$, $j = 0,\dots,m-1$. Then the Cauchy problem (6.24)-(6.25) has a unique weak solution. This solution is given by*

$$u(t,x) = \sum_{j=0}^{m-1} S_j(t,-D)\varphi_j(x),$$

where $S_j(t,-D)$, $j = 1,\dots,m$, is the j-th solution operator with the symbol $S_j(t,\xi) = J^j U_j(t,\xi)$ defined in equations(6.15) *and* (6.16).

6.4 Abstract Duhamel principle for DODE with fractional Caputo-Djrbashian derivative

Let X be a reflexive Banach space and $A : \mathscr{D} \to X$ be a closed linear operator with a domain $\mathscr{D} \subset X$. In Section 2.8 we introduced a Frechét type topological vector space $\mathrm{Exp}_{A,G}(\mathrm{X})$ and its dual, where G is an open subset of the complex plain $\mathbb{C}$. We also introduced a functional calculus $f(A)$, where f is an analytic function defined on G. The function f is called the symbol of the operator $f(A)$.

The goal of this section is to generalize the Duhamel principle for the Cauchy problem for general inhomogeneous fractional distributed order differential-operator equations of the form

$$L^{\Lambda}[u] \equiv \int_0^{\mu} f(\alpha,A)D_*^{\alpha}u(t)d\Lambda(\alpha) = B(A)u(t) + h(t), \quad t > 0,$$
$$u^{(k)}(0) = \varphi_k, \quad k = 0,\dots,m-1,$$

where $\mu \in (m-1,m]$ and $h(t)$ and φ_k, $k = 0,\dots,m-1$, are given X-valued vector-functions. The family of operators $f(\alpha,A)$ is defined through the family of symbols $f(\alpha,z)$ that are continuous in the variable $\alpha \in [0,\mu]$, and analytic in the variable $z \in G \subset \mathbb{C}$. The measure Λ is defined on $[0,\mu]$, and such that the DODO L^{Λ} is well defined (see, Section 3.11). The integrals are understood in the sense of Bochner if the integrand is a vector-function with values in some topological-vector space.

Let $\Lambda = \delta_{\mu} + \lambda$, where μ is a number such that $m-1 < \mu < m$, and λ is a finite measure with $supp\,\lambda \subset [0,m-1]$. Consider the operator

$$_{\tau}L^{(\mu,\lambda)}[u](t) \equiv {_{\tau}D_*^{\mu}}u(t) + \int_0^{m-1} f(\alpha,A)\,{_{\tau}D_*^{\alpha}}u(t)\lambda(d\alpha), \tag{6.31}$$

acting on m-times differentiable vector-functions $u(t), t \geq \tau \geq 0$. If $\tau = 0$, then instead of $_0L^{(\mu,\lambda)}[u](t)$ we write $L^{(\mu,\lambda)}[u](t)$.

Consider the Cauchy problem for the inhomogeneous equation

$$L^{(\mu,\lambda)}[u](t) = h(t), \quad t > 0, \tag{6.32}$$

with the homogeneous Cauchy conditions

$$u^{(k)}(0) = 0, \quad k = 0,\dots,m-1. \tag{6.33}$$

The fractional Duhamel principle establishes a connection between the solutions of this problem and the Cauchy problem for the homogeneous equation

$$_{\tau}L^{(\mu,\lambda)}[V(\cdot,\tau)](t) = 0, \quad t > \tau, \tag{6.34}$$

$$\frac{\partial^k V}{\partial t^k}(t,\tau)|_{t=\tau+0} = 0, \quad k = 0,\dots,m-2, \tag{6.35}$$

$$\frac{\partial^{m-1} V}{\partial t^{m-1}}(t,\tau)|_{t=\tau+0} = D_+^{m-\mu}h(\tau), \tag{6.36}$$

where $h(t)$ is a given vector-function and D_+^{γ} is the Riemann-Liouville fractional derivative of order γ. In Theorem 6.9 we assume that the vector-functions $h(t), t \geq 0$, and $V(t,\tau), t \geq \tau \geq 0$, are $\mathrm{Exp}_{A_{\lambda},G}(\mathrm{X})$-, or $\mathrm{Exp}'_{A^*,G^*}(\mathrm{X}^*)$-valued, $h(t)$ is differentiable, $V(t,\tau)$ is an m times differentiable with respect to the variable t, and the derivatives $\frac{\partial^j V(t,\tau)}{\partial t^j}, 0 \leq j \leq k-1$, are jointly continuous in the topology of $\mathrm{Exp}_{A_{\lambda},G}(\mathrm{X})$, or of $\mathrm{Exp}'_{A^*,G^*}(\mathrm{X}^*)$, respectively.

We start with the following lemma proved in Section 3.3, Example 3.6.

Lemma 6.8. *Let $h(t)$ be a continuously differentiable function for all $0 \leq t < T < \infty$. Then the equation $J^{\alpha}u(t) = h(t), 0 < t < T$, where $0 < \alpha < 1$, has a unique continuous solution given by the formula*

$$u(t) = D_+^{\alpha}h(t), \quad 0 < t < T. \tag{6.37}$$

Lemma 6.8 is essentially the well-known result on a solution of Abel's integral equation of first kind. Tonelli [Ton28] sowed that if h is absolutely continuous on $[0,T]$, then a unique solution to $J^{\alpha}u(t) = h(t)$ is given by (6.37) and $u \in L_1(0,T)$. If h is in a Hölder class $C^{\gamma}[0,T]$, $\alpha < \gamma < 1$, then $u \in C^{\beta}[0,T]$ for some $\beta < \gamma - \alpha$. See [GV91, SKM87] for further details.

Theorem 6.9. *Suppose that* $V(t,\tau)$ *is a solution of the Cauchy problem* (6.34)–(6.36). *Then Duhamel's integral*

$$u(t) = \int_0^t V(t,\tau)d\tau \tag{6.38}$$

solves the Cauchy problem (6.32), (6.33).

Proof. First notice that since $m-1 < \mu < m$, or $0 < m-\mu < 1$, due to Lemma 6.8, the equation $J^{m-\mu}g(t) = h(t)$ has a unique solution

$$g(t) = D_+^{m-\mu}h(t). \tag{6.39}$$

Let $V(t,\tau)$ as a function of the variable t be a solution to Cauchy problem (6.34)-(6.36) for any fixed τ. We verify that $u(t) = \int_0^t V(t,\tau)d\tau$ satisfies equation (6.32), and conditions (6.33). Splitting the interval $(0,m-1]$ into subintervals $[0,1]$, $(1,2]$, $\ldots,(m-2,m-1]$, we have

$$L^{(\mu,\lambda)}[u](t) = D_*^{\mu}u(t) + \sum_{k=1}^{m-1}\int_{(k-1,k]} f(\alpha,A)D_*^{\alpha}u(t)\lambda(d\alpha). \tag{6.40}$$

For $\alpha \in (k-1,k)$, $k = 1,\ldots,m-1$, using Definition (3.3) of D_*^{α}, we have

$$D_*^{\alpha}u(t) = \frac{1}{\Gamma(k-\alpha)}\int_0^t (t-s)^{k-\alpha-1}\frac{d^k}{ds^k}\int_0^s V(s,\tau)d\tau ds.$$

Lemma 4.4 and conditions (6.35) imply that

$$\frac{d^k}{ds^k}\int_0^s V(s,\tau)d\tau = \int_0^s \frac{\partial^k}{\partial s^k}V(s,\tau)d\tau, \quad k = 1,\ldots,m-1. \tag{6.41}$$

Hence,

$$D_*^{\alpha}u(t) = \int_0^t \frac{1}{\Gamma(k-\alpha)}\int_{\tau}^t (t-s)^{k-\alpha-1}\frac{\partial^k}{\partial s^k}V(s,\tau)dsd\tau. \tag{6.42}$$

Again due to Lemma 4.4 and condition (6.36),

$$\begin{aligned}\frac{d^m}{ds^m}\int_0^s V(s,\tau)d\tau &= \frac{\partial^{m-1}}{\partial s^{m-1}}V(s,\tau)_{|\tau=s} + \int_0^s \frac{\partial^m}{\partial s^m}V(s,\tau)d\tau \\ &= D_+^{m-\mu}h(s) + \int_0^s \frac{\partial^m}{\partial s^m}V(s,\tau)d\tau.\end{aligned}$$

Therefore the first term on the right-hand side of (6.40) takes the form

$$D_*^{\mu} u(t) = \frac{1}{\Gamma(m-\mu)} \int_0^t (t-s)^{m-\mu-1} \frac{d^m}{ds^m} \int_0^s V(s,\tau) d\tau ds$$
$$= \frac{1}{\Gamma(m-\mu)} \int_0^t (t-s)^{m-\mu-1} \left(D_+^{m-\mu} h(s) + \int_0^s \frac{\partial^m}{\partial s^m} V(s,\tau) d\tau \right) ds. \quad (6.43)$$

Furthermore, by virtue of (6.39),

$$\frac{1}{\Gamma(m-\mu)} \int_0^t (t-s)^{m-\mu-1} D_+^{m-\mu} h(s) = J^{m-\mu} D_+^{m-\mu} h(t) = h(t). \quad (6.44)$$

Now equations (6.40), (6.42), (6.43), and (6.44) imply that

$$L^{(\mu,\lambda)}[u](t) = h(t) + \frac{1}{\Gamma(m-\mu)} \int_0^t (t-s)^{m-\mu-1} \int_0^s \frac{\partial^m}{\partial s^m} V(s,\tau) d\tau ds$$
$$+ \sum_{k=1}^{m-1} \int_{k-1}^{k} f(\alpha,A) \frac{1}{\Gamma(k-\alpha)} \int_0^t (t-s)^{k-\alpha-1} \int_0^s \frac{\partial^k}{\partial s^k} V(s,\tau) d\tau ds \lambda(d\alpha). \quad (6.45)$$

Changing the order of integration (Fubini is allowed) in (6.45) we get

$$L^{(\mu,\lambda)}[u](t) = h(t) + \int_0^t \int_\tau^t \frac{1}{\Gamma(m-\mu)} (t-s)^{m-\mu-1} \frac{\partial^m}{\partial s^m} V(s,\tau) ds d\tau$$
$$+ \sum_{k=1}^{m-1} \int_0^t \int_{k-1}^{k} f(\alpha,A) \int_\tau^t \frac{1}{\Gamma(k-\alpha)} (t-s)^{k-\alpha-1} \frac{\partial^k}{\partial s^k} V(s,\tau) ds \lambda(d\alpha) d\tau$$
$$= h(t) + \int_0^t {}_\tau D_*^{\alpha_m} V(t,\tau) d\tau + \int_0^t \int_0^{m-1} f(\alpha,A) \; {}_\tau D_*^{\alpha} V(t,\tau) \lambda(d\alpha) d\tau$$
$$= h(t) + \int_0^t {}_\tau L^{(\mu,\lambda)}[V(\cdot,\tau)](t) d\tau = h(t).$$

Finally, using the relations (6.41) it is not hard to verify that $u(t)$ in (6.38) satisfies initial conditions (6.33) as well.

If the vector-function h satisfies the additional condition $h(0) = 0$ then condition (6.36), in accordance with the relationship (7.41), can be replaced by

$$\frac{\partial^{m-1} V}{\partial t^{m-1}}(t,\tau)|_{t=\tau} = D_*^{m-\mu} h(\tau),$$

with the Caputo-Djrbashian derivative $D_*^{m-\mu}$ of order $m-\mu$. As a consequence the formulation of the fractional Duhamel principle takes the form:

Theorem 6.10. *Suppose that for all $\tau : 0 < \tau < t$ a function $V(t,\tau)$, is a solution to the Cauchy problem for the homogeneous equation*

$$
\begin{aligned}
&{}_\tau L^{(\mu,\lambda)}[V(\cdot,\tau)](t)=0, \quad t>\tau,\\
&\frac{\partial^k V}{\partial t^k}(t,\tau)|_{t=\tau+0}=0, \quad k=0,\dots,m-2,\\
&\frac{\partial^{m-1} V}{\partial t^{m-1}}(t,\tau)|_{t=\tau+0}=D_*^{m-\mu}h(\tau),
\end{aligned}
$$

where $h(t)$ is a given differentiable vector-function such that $h(0)=0$. Then Duhamel's integral $u(t)=\int_0^t V(t,\tau)d\tau$ solves the Cauchy problem for the inhomogeneous equation (6.32), (6.33).

Remark 6.4. In Theorems 6.9 and 6.10 we assumed that $f(\mu,A)$ is the identity operator (see equation (6.31)). In the general case, with appropriate selection of G, we can assume that the inverse operator $[f(\mu,A)]^{-1}$ exists. Then with the condition

$$
\frac{\partial^{m-1}V(t,\tau)}{\partial t^{m-1}}|_{t=\tau+}=[f(\mu,A)]^{-1}D_+^{m-\mu}h(\tau)
$$

instead of (6.36), Theorems 6.9 and 6.10 remain valid.

6.5 Abstract Duhamel principle for DODE with fractional Riemann-Liouville derivative

The operator ${}_\tau L^\Lambda$ in Theorem 6.9 is defined via the fractional derivative in the sense of Caputo-Djrbashian. A fractional generalization of the Duhamel principle is also possible when this operator is defined via the Riemann-Liouville fractional derivative. In this section we prove the fractional Duhamel principle in the case $\Lambda(d\alpha)=[\delta_\mu(\alpha)+\Sigma_{k=1}^{m-1}\delta_{\mu-k}(\alpha)]d\alpha$, where $m-1<\mu\le m$. Namely, consider the following Cauchy type problem for a nonhomogeneous differential-operator equation

$$
L_+^\Lambda[u]=D^\mu u(t)+\sum_{k=1}^{m-1}B_kD^{\mu-k}u(t)+B_0u(t)=h(t), \quad t>0, \tag{6.46}
$$

$$
D^{\mu-j}u(t)\big|_{t=0+}=0, \quad j=1,\dots,m-1, \tag{6.47}
$$

$$
J^{m-\mu}u(t)\big|_{t=0+}=0. \tag{6.48}
$$

where B_k, $k=0,\dots,m-1$, are linear closed operators independent of the variable t, and with domains dense in X.

The following lemma will be used in the proof of the Duhamel principle for Cauchy type problem (6.46)–(6.48).

Lemma 6.9. *Suppose $V(t,\tau)$ is a X-valued function defined for all $t \geq \tau \geq 0$ and jointly continuous in the X-norm. Then for any $\beta > 0$ the equality*

$$J^{\beta}\left(\int_0^t V(t,\tau)d\tau\right) = \int_0^t {}_{\tau}J_t^{\beta} V(t,\tau)d\tau$$

holds.

Proof. We have

$$\begin{aligned}
J^{\beta}\left(\int_0^t V(t,\tau)d\tau\right) &= \frac{1}{\Gamma(\beta)}\int_0^t (t-s)^{\beta-1}\left(\int_0^t V(t,\tau)d\tau\right)ds \\
&= \frac{1}{\Gamma(\beta)}\int_0^t \int_{\tau}^t (t-s)^{\beta-1} V(t,\tau)dsd\tau \\
&= \int_0^t \left(\frac{1}{\Gamma(\beta)}\int_{\tau}^t (t-s)^{\beta-1} V(s,\tau)ds\right)d\tau \\
&= \int_0^t {}_{\tau}J_t^{\beta} V(t,\tau)d\tau.
\end{aligned}$$

Theorem 6.11. *Suppose that $V(t,\tau)$ is a solution of the following Cauchy type problem:*

$$ {}_{\tau}L_+^{\Lambda}[V(t,\tau)] = {}_{\tau}D_t^{\mu} V(t,\tau) + \sum_{k=1}^{m-1} B_k \, {}_{\tau}D_t^{\mu-k} V(t,\tau) + B_0 V(t,\tau) = 0, \quad t > \tau, \tag{6.49}$$

$$ {}_{\tau}D_t^{\mu-1} V(t,\tau)_{|_{t=\tau+}} = h(\tau), \tag{6.50}$$

$$ {}_{\tau}D_t^{\mu-j} V(t,\tau)_{|_{t=\tau+}} = 0, \quad j = 2,\ldots,m-1, \tag{6.51}$$

$$ {}_{\tau}J_t^{m-\mu} V(t,\tau)_{|_{t=\tau+}} = 0. \tag{6.52}$$

Then the Duhamel integral

$$u(t) = \int_0^t V(t,\tau)d\tau \tag{6.53}$$

solves the Cauchy type problem (6.46)–(6.48).

Proof. Let $V(t,\tau)$ as a function of the variable t be a solution to Cauchy type problem (6.49)-(6.52) for any fixed τ. We show that $u(t) = \int_0^t V(t,\tau)d\tau$ satisfies equation (6.46), and conditions (6.47) and (6.48). We set $\alpha = \mu - k$, $k = 0,1,\ldots,m-1$. Then $m-k-1 < \alpha < m-k$. Using the definition $D^{\alpha} = D^k J^{k-\alpha}$ of D^{α}, and Lemma 6.9, we have

$$D^{\alpha}u(t) = D^{m-k}J^{m-k-\alpha}\left(\int_0^s V(s,\tau)d\tau\right) = \frac{d^{m-k}}{dt^{m-k}}\int_0^t {}_{\tau}J_t^{m-k-\alpha} V(t,\tau)d\tau. \tag{6.54}$$

For a solution $V(t,\tau)$ of problem (6.49)-(6.52) the function ${}_\tau J_t^{k-\alpha}V(t,\tau)$ satisfies the conditions of Lemma 4.4. Using Lemma 4.4, we have

$$\frac{d^{m-k}}{dt^{m-k}}\int_0^t {}_\tau J_t^{m-k-\alpha}V(t,\tau)d\tau = \sum_{j=0}^{m-k-1}\frac{d^j}{dt^j}\left[\frac{\partial^{m-k-1-j}}{\partial t^{m-k-1-j}}{}_\tau J_t^{m-k-\alpha}V(t,\tau)_{|\tau=t}\right]$$
$$+\int_0^t \frac{\partial^{m-k}}{\partial t^{m-k}}{}_\tau J_t^{m-k-\alpha}V(t,\tau)d\tau. \qquad (6.55)$$

Conditions (6.51) and (6.52) imply that for all $k=1,\dots,m-1$

$$\frac{\partial^{m-k-1-j}}{\partial t^{m-k-1-j}}{}_\tau J_t^{m-k-\alpha}V(t,\tau)_{|\tau=t}=0, \quad j=0,\dots,k-1. \qquad (6.56)$$

Indeed, since $\alpha\in(m-k-1,m-k)$, we have $0<m-k-\alpha<1$. Therefore,

$$\frac{\partial^{m-k-1-j}}{\partial t^{m-k-1-j}}{}_\tau J_t^{m-k-\alpha}V(t,\tau)_{|\tau=t} = D^{m-k-j-1}\,{}_\tau J_t^{m-k-j-1-(\alpha-j-1)}V(t,\tau)_{|\tau=t}$$
$$= {}_\tau D_t^{\alpha-j-1}V(t,\tau)_{|\tau=t} = {}_\tau D_t^{\mu-k-j-1}V(t,\tau)_{|\tau=t}=0. \qquad (6.57)$$

The special case is $k=0$. In this case, when $j=0$, in equation (6.56) due to condition (6.50) we have

$$\frac{\partial^{m-1}}{\partial t^{m-1}}{}_\tau J_t^{m-\alpha}V(t,\tau)_{|\tau=t} = D^{m-1}\,{}_\tau J_t^{m-1-(\mu-1)}V(t,\tau)_{|\tau=t}\,\tau$$
$$={}_\tau D_t^{\mu-1}V(t,\tau)_{|\tau=t}=h(t). \qquad (6.58)$$

Taking into account (6.57) and (6.58), it follows from (6.54) and (6.55) that

$$D^{\mu-k}u(t)=\int_0^t {}_\tau D_t^{\mu-k}V(t,\tau)d\tau, \quad k=1,\dots,m-1, \qquad (6.59)$$

$$D^{\mu}u(t)=\int_0^t {}_\tau D_t^{\mu}V(t,\tau)d\tau+h(t). \qquad (6.60)$$

Substituting (6.59) and (6.60) into equation (6.46), we have

$$L_+^{\Lambda}[u]=h(t)+\int_0^t {}_\tau L_+^{\Lambda}[V(t,\tau)]d\tau=h(t).$$

Thus, the Duhamel integral $u(t)$ in (6.53) satisfies equation (6.46). Finally, letting $t\to 0+$ in (6.59) one can see that $u(t)$ satisfies initial conditions (6.47). Evidently, $u(t)$ satisfies condition (6.48), as well.

Corollary 6.1. *Suppose that $V(t,\tau)$ is a solution of the following Cauchy type problem:*

$$\begin{aligned} {}_\tau D_t^\mu V(t,\tau) + B_0 V(t,\tau) &= 0, \quad t > \tau, \\ {}_\tau D_t^{\mu-1} V(t,\tau)_{|t=\tau+} &= h(\tau), \\ {}_\tau D_t^{\mu-j} V(t,\tau)_{|t=\tau+} &= 0, \quad j = 2,\ldots,m-1, \\ {}_\tau J_t^{m-\mu} V(t,\tau)_{|t=\tau+} &= 0. \end{aligned}$$

Then the Duhamel integral

$$u(t) = \int_0^t V(t,\tau)d\tau$$

solves the Cauchy type problem

$$\begin{aligned} D^\mu u(t) + B_0 u(t) &= h(t), \quad t > 0, \\ D^{\mu-1} u(t)_{|t=0+} &= 0, \\ D^{\mu-j} u(t)_{|t=0+} &= 0, \quad j = 2,\ldots,m-1, \\ J^{m-\mu} u(t)_{|t=0+} &= 0. \end{aligned}$$

Remark 6.5. Theorem 6.11 represents a fractional generalization of the Duhamel principle for integer order differential-operator equations proved in Theorem 4.12 in Section 4.7, recovering it in the case $\mu = m$.

6.6 Applications of the fractional Duhamel principles

Theorems 6.9, 6.10, and 6.11 can be applied to general boundary value problems for distributed order differential equations, in terms of analysis of the existence and uniqueness of a solution. Below we demonstrate this in two different cases, namely in the case of an abstract Cauchy problem for DODEs determined by the Caputo-Djrbashian derivative, and in the case of a Cauchy type problem for DODEs determined by the Riemann-Liouville derivative.

Case I. Let L^Λ be the distributed fractional order abstract differential operator defined in (6.31), that is

$${}_\tau L^{(\mu,\lambda)}[u](t) \equiv {}_\tau D_*^\mu u(t) + \int_0^{m-1} f(\alpha,A)\, {}_\tau D_*^\alpha u(t) \lambda(d\alpha),$$

with the lower endpoint of the time interval $\tau = 0$. The characteristic function of this operator is

$$\Delta(s,z) = s^\mu + \int_0^{m-1} f(\alpha,z) s^\alpha d\lambda,$$

where μ is a fixed number in the interval $(m-1,m]$, λ is a finite measure with $\operatorname{supp}\lambda \subset [0,m-1]$, and $f(\alpha,z)$ is a function continuous in α and analytic in $z \in G \subset \mathbb{C}$. Denote by $\tilde{v}(s) = \mathscr{L}[v](s)$, the Laplace transform of a vector-function $v(t)$, namely

$$\mathscr{L}[v](s) = \int_0^\infty e^{-st} v(t)dt, \quad s > s_0,$$

where $s_0 \geq 0$ is a real number. It is not hard to verify that if $v(t) \in Exp_{A,G}(X)$ for each $t \geq 0$ and satisfies the condition $\|v(t)\| \leq Ce^{\gamma t}, t \geq 0$, with some constants $C > 0$ and γ, then $\tilde{v}(s)$ exists and the inequality

$$\|A^k \tilde{v}(s)\| \leq \frac{C_s}{s-\gamma} \nu^k, \quad s > \gamma,$$

holds, implying $\tilde{v}(s) \in Exp_{A,G}(X)$ for each fixed $s > \gamma$. The lemma below gives a formal representation formula for a solution of the general abstract Cauchy problem

$$L^\Lambda[u](t) = h(t), \quad t > 0, \tag{6.61}$$

$$u^{(k)}(0+) = \varphi_k, \quad k = 0, \dots, m-1. \tag{6.62}$$

Let $\delta_{j,k}$ denote the Kronecker delta, that is $\delta_{j,k} = 1$ if $j = k$, and $\delta_{j,k} = 0$, if $j \neq k$.

Lemma 6.10. *Let $c_\beta(t,z) = \mathscr{L}^{-1}[\frac{s^\beta}{\Delta(s,z)}](t)$, $z \in G \subset \mathbb{C}$, where $\mathscr{L}^{-1}$ stands for the inverse Laplace transform, and*

$$S_k(t,z) = c_{\mu-k-1}(t,z) + \int_k^{m-1} f(\alpha,z) c_{\alpha-k-1}(t,z)\lambda(d\alpha), \, k = 0, \dots, m-1. \tag{6.63}$$

Then $S_k(t,A)\varphi_k$ solves the Cauchy problem

$$L^\Lambda[u] = 0, \quad u^{(j)}(0) = \delta_{j,k}\varphi_j, \; j = 0, \dots, m-1.$$

Proof. Applying formula (3.55) we have

$$\widetilde{L^\Lambda[u]}(s) = s^\mu \tilde{u}(s) - \sum_{i=0}^{m-1} u^i(0) s^{\mu-i-1}$$
$$+ \sum_{k=1}^{m-1} \int_{k-1}^k f(\alpha,A)(s^\alpha \tilde{u}(s) - \sum_{j=0}^{k-1} u^j(0) s^{\alpha-j-1})\lambda(d\alpha) = 0.$$

Due to the initial conditions $u^j(0) = \delta_{j,k}\varphi_j$, $j = 0, \dots, m-1$, the latter reduces to

$$\Delta(s,z)\tilde{u}(s) = \varphi_k \left(s^{\mu-k-1} + \int_k^{m-1} f(\alpha,z) s^{\alpha-k-1} \lambda(d\alpha) \right).$$

Now it is easy to see that the solution in this case is represented as $u_k = S_k(t,A)\varphi_k$, $k = 0, \dots, m-1$.

Corollary 6.12 *Let $S_k(t,A), k=0,\dots,m-1$, be the collection of solution operators with the symbols $S_k(t,z)$ defined in Lemma 6.10. Then the solution of the Cauchy problem*

$$L^{\Lambda}[u]=0, \quad u^{(j)}(0)=\varphi_j, \quad j=0,\dots,m-1. \tag{6.64}$$

is given by the following representation formula

$$u(t)=\sum_{k=0}^{m-1} S_k(t,A)\varphi_k.$$

Remark 6.6. 1. Corollary 6.12 can easily be extended to the operator ${}_{\tau}L^{\Lambda}$ in (6.64) as well with the initial conditions $u^j(\tau)=\varphi_j, j=0,\dots,m-1$. In this case the symbols of solution operators depend on τ and have the form $S_k(t,\tau,z)=S_k(t-\tau,z), k=0,\dots,m-1$, where $S_k(t,z)$ is defined in (6.63).
2. A particular case of Lemma 6.10 when $\Lambda=\sum_{k=0}^{m}\delta_{\alpha_k}, k-1<\alpha_k<k$, is proved in [GLU00].

A vector-function $u(t)\in C^{(m)}[t>0; Exp_{A,G}(X)]\cap C^{(m-1)}[t\geq 0; Exp_{A,G}(X)]$ is called a solution of the problem (6.61), (6.62) if it satisfies the equation (6.61) and the initial conditions (6.62) in the topology of $Exp_{A,G}(X)$.

Theorem 6.9 and Corollary 6.12 imply the following results.

Theorem 6.13. *Let $\varphi_k\in Exp_{A,G}(X)$, $k=0,\dots,m-1$, and for any $T>0$ the vector-function $h(t)\in AC[0\leq t\leq T; Exp_{A,G}(X)]$. Then the Cauchy problem (6.61), (6.62) has a unique solution. This solution is given by*

$$u(t)=\sum_{k=0}^{m-1} S_k(t,A)\varphi_k+\int_0^t S_{m-1}(t-\tau,A)D_+^{m-\mu}h(\tau)d\tau. \tag{6.65}$$

Proof. We split the Cauchy problem (6.61),(6.62) into two Cauchy problems

$$L^{\Lambda}[U](t)=0, \quad t>0, \tag{6.66}$$

$$U^{(k)}(0+)=\varphi_k, \quad k=0,\dots,m-1, \tag{6.67}$$

and

$$L^{\Lambda}[v](t)=h(t), \quad t>0, \tag{6.68}$$

$$v^{(k)}(0+)=0, \quad k=0,\dots,m-1. \tag{6.69}$$

Due to Corollary 6.12 the unique solution to (6.66),(6.67) is given by

$$U(t)=\sum_{k=0}^{m-1} S_k(t,A)\varphi_k. \tag{6.70}$$

Lemma 2.3 implies that $U(t) \in C^m[t > 0; Exp_{A,G}(X)]$. For the Cauchy problem (6.68)–(6.69), in accordance with the fractional Duhamel's principle (Theorem 6.9), it suffices to solve the Cauchy problem for the homogeneous equation:

$$
\begin{aligned}
&{}_\tau L^\Lambda[V(t,\tau)](t) = 0, \quad t > \tau,\\
&\frac{\partial^k V(t,\tau)}{\partial t^k}_{|t=\tau+} = 0, \quad k = 0,\dots,m-2,\\
&\frac{\partial^{m-1} V(t,\tau)}{\partial t^{m-1}}_{|t=\tau+} = D_+^{m-\mu} h(\tau).
\end{aligned}
$$

The solution of this problem, again using Corollary 6.12 (with the note in Remark 6.6), has the representation

$$V(t,\tau) = S_{m-1}(t-\tau,A) D_+^{m-\mu} h(\tau).$$

Again it follows from Lemma 2.3 that $V(t,\tau) \in C^m[t > \tau; Exp_{A,G}(X)]$ for all $\tau \geq 0$, as well as its Duhamel integral. Thus, the Duhamel integral of $V(t,\tau)$ and representation (6.70) lead to formula (6.65). The uniqueness of a solution also follows from the obtained representation (6.65).

By duality we immediately obtain the following theorem.

Theorem 6.14. *Let $\varphi_k^* \in Exp'_{A^*,G^*}(X^*)$, $k = 0,\dots,m-1$, and the vector-functional $h^*(t) \in AC[t \leq t \leq T; Exp'_{A^*,G^*}(X^*)]$. Assume also that $Exp_{A,G}(X)$ is dense in X. Then the Cauchy problem (6.61), (6.62) (with A switched to A^*) is meaningful and has a unique weak solution. This solution is given by*

$$u^*(t) = \sum_{k=0}^{m-1} S_k(t,A^*)\varphi_k^* + \int_0^t S_{m-1}(t-\tau,A^*) D_+^{m-\mu} h^*(\tau) d\tau.$$

Assume that $Exp_{A,G}(X)$ is densely embedded into X. Besides, let the solution operators $S_k(t,A)$ for each $k = 0,\dots,m-1$, satisfy the estimates

$$\|S_k(t,A)\varphi\| \leq C\|\varphi\|, \quad \forall t \in [0,T], \tag{6.71}$$

where $\varphi \in Exp_{A,G}(X)$, and $C > 0$ does not depend on φ. Then there exists a unique closure $\bar{S}_k(t)$ to X of the operator $S_k(t,A)$ which satisfies the estimate $\|\bar{S}_k(t)u\| \leq C\|u\|$ for all $u \in X$. Using the standard technique of closure (see Section 2.3), we can prove the following theorem.

Theorem 6.15. *Let $\varphi_k \in X$, $k = 0,\dots,m-1$, $h(t) \in AC[0 \leq t \leq T; X]$ for any $T > 0$, and $D_+^{m-\mu} h(t) \in C[0 \leq t \leq T; X]$. Further let $Exp_{A,G}(X)$ be densely embedded into X, and the estimates (6.71) hold for solution operators $S_k(t,A)$,*

$k = 0, \ldots, m-1$. *Then the Cauchy problem (6.61), (6.62) has a unique solution* $u(t) \in C^m[0 < t \leq T; X]$. *This solution is given by*

$$u(t) = \sum_{k=0}^{m-1} \bar{S}_k(t)\varphi_k + \int_0^t \bar{S}_{m-1}(t-\tau) D_+^{m-\mu} h(\tau) d\tau.$$

Case II. Consider the following Cauchy type problem

$$D^{3/2}u(t) + bD^{1/2}u(t) + Au(t) = h(t), \quad t > 0, \tag{6.72}$$

$$(D^{1/2}u)(0+) = \varphi_1 \tag{6.73}$$

$$(J^{1/2}u)(0+) = \varphi_2. \tag{6.74}$$

where $b \in \mathbb{R}$, D^α is the Riemann-Liouville derivative of order α, and A is a closed operator defined on a Hilbert space $\mathscr{H}$, has a discrete spectrum $\lambda_1, \lambda_2, \ldots$, with corresponding eigenvectors $\phi_1, \phi_2, \ldots$, which form a base in $\mathscr{H}$. We assume that $\varphi_k \in Exp_{A,G}(X)$, $k = 1, 2$, and $h(t) \in C[t \geq 0; Exp_{A,G}(X)]$, where $G \subseteq \mathbb{R}^n$.

Remark 6.7. Equation (6.72) belongs to the family of fractional differential equations (also interpreted as a fractional diffusion-wave equation)

$$aD^\alpha u(t) + bD^\beta u(t) + Au(t) = f(t),$$

where $0 < \beta \leq 1 < \alpha \leq 2$, a and b are constants, and A is a linear operator. The case $\alpha = 2$, $\beta = 1/2$, and $A \equiv d^4/dx^4$ is studied by Agraval [Agr04], and the case $\alpha = 2$, $\beta = 3/2$, and A is constant, by Bagley and Torvik in [BT00].

Let us first solve the Cauchy type problem for homogeneous equation (6.72) with nonhomogeneous conditions, namely

$$D^{3/2}u(t) + bD^{1/2}u(t) + Au(t) = 0, \quad t > 0, \tag{6.75}$$

$$(D^{1/2}u)(0+) = \varphi_1 \tag{6.76}$$

$$(J^{1/2}u)(0+) = \varphi_2. \tag{6.77}$$

Applying the Fourier transform to (6.75), we have

$$M(s)L[u](s) = (s+b)\psi + \varphi, \quad s > \eta_0,$$

where $M(s) = s^{3/2} + bs^{1/2} + A$, and $\eta_0 > 0$ is a number such that $M(s) \neq 0$ for all $s > \eta_0$.[1] Let $S_1(t,A) = L^{-1}[1/M(s)]$ and $S_2(t,A) = L^{-1}[(s+b)/M(s)]$. The power series representations of operators $S_j(t,A)$ can be obtained similar to (6.11)–(6.13). Then the solution to problem (6.75)–(6.77) has the representation

$$u(t) = S_1(t,A)\varphi_1 + S_2(t,A)\varphi_2,$$

[1] We assume that such η_0 exists.

where

$$S_k(t,A)\varphi_k = \sum_{n=1}^{\infty} S_k(t,\lambda_n)\varphi_{k,n}\phi_n, \quad k=1,2,$$

Here $\varphi_{k,n} = (\varphi_k, \phi_n)$, Fourier coefficients of elements φ_k, $k=1,2$.

Now, we return to the nonhomogeneous problem (6.72)–(6.74); however, we assume now that $\varphi_k = 0$, $k=1,2$. To solve this we apply the fractional Duhamel principle. In accordance with Theorem 6.11 the solution is given by the Duhamel integral $\int_0^t V(t,\tau)d\tau$, where $V(t,\tau)$ is a solution to the following problem:

$$_\tau D_t^{3/2} V(t,\tau) + b\,_\tau D_t^{1/2} V(t,\tau) + Au(t) = 0, \quad t>\tau, \tag{6.78}$$

$$_\tau D_t^{1/2} V(\tau+,\tau) = h(\tau), \tag{6.79}$$

$$_\tau J_t^{1/2} V(\tau+,\tau) = 0. \tag{6.80}$$

It is not hard to verify that the solution of the Cauchy type problem (6.78)–(6.80) can be obtained from the solution of problem (6.75)–(6.77) by translation $t \to t+\tau$. Namely, $V(t,\tau) = S_1(t-\tau,A)h(\tau)$. Hence, the solution to the given problem in (6.72)–(6.74) has the representation

$$u(t) = S_1(t,A)\varphi_1 + S_2(t,A)\varphi_2 + \int_0^t S_1(t-\tau,\, A)h(\tau)d\tau.$$

6.7 The Cauchy problem for variable order differential equations

In this section we study the Cauchy problem for variable order differential equations with a piecewise constant order function $\beta(t) = \sum_{k=0}^{N} \mathscr{I}_k \beta_k$, where $\mathscr{I}_k$ is the indicator function of $[T_k, T_{k+1})$, and $0 < \beta_k \le 1$, $k = 0,\dots,N$. We assume that the diffusion mode change times $T_1 < T_2, \dots, T_N$ are known, and set $T_0 = 0$, $T_{N+1} = \infty$. We assume that the solution will stay continuous at the diffusion mode change times. Hence, the Cauchy problem for variable order differential equations has the form

$$\mathscr{D}_{*\{\mu,\nu\}}^{\beta(t)} u(t,x) = \mathscr{A}(D)u(t,x), \quad t>0, x\in\mathbb{R}^n, \tag{6.81}$$

$$u(0,x) = \varphi(x), \tag{6.82}$$

$$u(T_k-0,x) = u(T_k+0,x), \quad k=1,\dots,T_N, x\in\mathbb{R}^n. \tag{6.83}$$

where $\mathscr{D}_{*\{\mu,\nu\}}^{\beta(t)}$ is the Caputo-Djrbashian type (μ,ν)-VODO, defined as (see Definition 3.7)

$$\mathscr{D}_{*\mu,\nu}^{\beta(t)} f(t) = \int_0^t \mathscr{K}_{\mu,\nu}^{\beta(t)}(t,\tau) \frac{df(\tau)}{d\tau} d\tau,$$

with the kernel function

$$\mathscr{K}^{\beta}_{\mu,\nu}(t,\tau) = \frac{1}{\Gamma(1-\beta(\mu t+\nu\tau))(t-\tau)^{\beta(\mu t+\nu\tau)}}, \quad 0<\tau<t.$$

The parameters ν and μ belong to the following *causality LH-parallelogram*: $\Pi = \{(\mu,\nu)\in\mathbb{R}^2 : 0\le\mu\le 1, -1\le\nu\le +1, 0\le\mu+\nu\le 1\}$ (see Figure 3.1).

An interesting phenomenon related to variable order fractional differential equations is that an internal memory quantified as an inhomogeneous term in the equation may be generated. Consider a simple example, which demonstrates how such an inhomogeneous term arises in a single change of diffusion mode. Assume the function $\beta(t)$ takes only two values β_1, if $0<t<T$ and β_2, if $t>T$. In other words, the diffusion mode changes at time $t=T$ from a sub-diffusive mode β_1 to a sub-diffusive mode β_2. Since the first mode is sub-diffusive, a non-Markovian memory arises, which effects on the actual change of diffusion mode occurring at time $T_* \ge T$. Here T_* depends on the parameters μ and ν; see Section 3.12, where the value of T_* is calculated. For simplicity, suppose $\nu=0$ and $\mu=1$. In this case $T_*=T$, and we assume the following continuity condition at the change of mode time $t=T$:

$$u(T) = u(T-0). \tag{6.84}$$

For $0<t<T$, equation (6.81) is a fractional equation of order β_1, so a solution to the Cauchy problem (6.81)–(6.82) can be found by standard methods in this interval (see, Section 5.3). If $t>T$, then one has

$$\mathscr{D}^{\beta(t)}_{*} u(t) = \int_0^T \mathscr{K}^{\beta_1}_{1,0}(t,\tau)\frac{du(\tau)}{d\tau}d\tau + \int_T^t \mathscr{K}^{\beta_2}_{1,0}(t,\tau)\frac{du(\tau)}{d\tau}d\tau.$$

Hence, equation (6.81) takes the form

$${}_T\mathscr{D}^{\beta_2}_{*,t} u(t) = \mathscr{A} u(t) + h(t), \quad t>T, \tag{6.85}$$

with the initial condition (6.84). Equation (6.85) is no longer homogeneous, due to the nonhomogeneous term

$$h(t) = -\int_0^T \mathscr{K}^{\beta_1}_{1,0}(t,\tau)\frac{du(\tau)}{d\tau}d\tau.$$

Therefore, the Duhamel principle developed for fractional order differential equations play an important role in the theory of the Cauchy problem for variable order fractional differential equations.

We note that, since the order $\beta(t)$ depends on the variable t, the fractional integration operator becomes

$$J^{\beta(t)} f(t) = J^{\beta(t)}_{\mu,\nu} f(t) = \sum_{k=0}^{N} J_k f(t),$$

where

$$J_k f(t) = \int_0^t \frac{\mathscr{I}_k(\mu t + \nu\tau)(t-\tau)^{\beta_k(\mu t+\nu\tau)-1} f(\tau)}{\Gamma(\beta_k(\mu t + \nu\tau))} d\tau, \quad k = 0, \ldots, N.$$

Lemma 6.11. *Suppose that* $min_{0\le j\le N}\{\beta_j\} = \beta_* > 0$ *and* $[k\beta_*]$ *is the integer part of* $k\beta_*$. *Let* $v(t)$ *be a function continuous in* $[0,\infty)$. *Then for arbitrary* $T > 0$ *and every* $k = 1, 2, \ldots$ *the estimate*

$$max_{0\le t\le T}|J^{\beta(t)k} v(t)| \le \frac{\psi^{k-1}(T)}{\Gamma(k\beta_*)} max_{0\le t\le T}|v(t)| \tag{6.86}$$

holds with

$$\psi(\tau) = \begin{cases} \tau^{\beta_*}, & \text{if } 0 < \tau < 1, \\ \tau, & \text{if } \tau \ge 1. \end{cases}$$

Proof. Let $v(t)$ be a function continuous in $[0,\infty)$. For k large enough, such that $\beta_* k \ge 2$, we have $min\Gamma(k\beta(\mu t + \nu\tau)) = \Gamma(k\beta_*)$. Taking this into account, for all such k and for all $t \in (0, T]$ we obtain the estimate

$$|J^{\beta(t)k} v(t)| = \left| \int_0^t \frac{(t-\tau)^{k\beta(\mu t+\nu\tau)-1} v(\tau) d\tau}{\Gamma(k\beta(\mu t + \nu\tau))} \right| \le \frac{\psi^{k-1}(T)}{\Gamma(k\beta_*)} \max_{0\le t\le T} |v(t)|,$$

and hence, the estimate in equation (6.86).

Let $t_{cr,j} = T_j/(\mu + \nu)$, $j = 1, \ldots, N$, be critical points corresponding to diffusion mode change times T_j, $j = 1, \ldots, N$. We accept the conventions $t_{cr,0} = 0$, $t_{cr,N+1} = \infty$. Let $E_\beta(z)$ be the Mittag-Leffler function with parameter $\beta \in (0, 1]$. Now we introduce the symbols which play an important role in the representation of a solution. Let

$$S_j(t,\xi) = E_{\beta_j}((t - t_{cr,j})^{\beta_j} A(\xi)), \quad t \ge t_{cr,j},\ j = 0, \ldots, N, \tag{6.87}$$

and

$$M_k(t,\xi) = S_k(t - t_{cr,k}, \xi) \prod_{j=0}^{k-1} S_j(t_{cr,j+1} - t_{cr,j}, \xi), t \ge t_{cr,k}, \quad k = 1, \ldots, N. \tag{6.88}$$

Further, we define recurrently the symbols

$$\mathscr{R}_1(t,\xi) = -\frac{1}{\Gamma(1-\beta_1)} \int_0^{t_{cr,1}} \frac{\frac{\partial}{\partial\tau} S_0(\tau,\xi)}{(t-\tau)^{\beta_1}} d\tau$$
$$= \frac{-\beta_0 A(\xi)}{\Gamma(1-\beta_1)} \int_0^{t_{cr,1}} \frac{E'_{\beta_0}(\tau^{\beta_0} A(\xi)) d\tau}{\tau^{1-\beta_0}(t-\tau)^{\beta_1}},$$
$$P_{-1}(t,\xi) \equiv 0, \quad P_0(t,\xi) \equiv 1,$$
$$P_1(t,\xi) = \int_{t_{cr_1}}^t S_j(t + t_{cr,1} - \tau, \xi)\, {}_{t_{cr,1}}D_\tau^{1-\beta_1} \mathscr{R}_1(\tau,\xi) d\tau,$$

and if

$$P_j(t,\xi) = \int_{t_{cr\,j}}^{t} S_j(t+t_{cr,j}-\tau,\xi)\, {}_{t_{cr,j}}D_\tau^{1-\beta_j}\mathscr{R}_j(\tau,\xi)d\tau,$$

is defined for $t \geq t_{cr,j}$ and for all $j \leq k-1$, then for $t \geq t_{cr,k}$,

$$\mathscr{R}_k(t,\xi)$$
$$= -\frac{1}{\Gamma(1-\beta_k)}\sum_{j=0}^{k-1}\int_{t_{cr,j}}^{t_{cr,j+1}} \frac{\frac{\partial}{\partial\tau}[M_j(\tau,\xi)+S_j(\tau-t_{cr,j},\xi)P_{j-1}(t_{cr,j},\xi)+P_j(\tau,\xi)]}{(t-\tau)^{\beta_k}}d\tau, \tag{6.89}$$

for $k = 2,\ldots,N$.

The case $\nu = 0$**.** First we solve problem (6.81)–(6.83) in the particular case $\nu = 0$.

Theorem 6.16. *Assume* $\nu = 0$ *and* $\varphi \in \Psi_{G,p}(\mathbb{R}^n)$*. Then Cauchy problem (6.81)-(6.83) has a unique solution* $u(t,x) \in C([0,T],\Psi_{G,p}(\mathbb{R}^n))$*,* $T < \infty$*, which is represented in the form* $u(t,x) = \mathscr{S}(t,D)\varphi(x)$*, where* $\mathscr{S}(t,D)$ *is the pseudo-differential operator with the symbol*

$$\mathscr{S}(t,\xi) = \mathscr{I}_0' S_0(t,\xi) + \sum_{k=1}^{N}\mathscr{I}_k'(t)\Big\{M_k(t,\xi)$$
$$+S_k(t,\xi)\int_{t_{cr,k-1}}^{t_{cr,k}} S_{k-1}(t_{cr,k}+t_{cr,k-1}-\tau,\xi)\ {}_{t_{cr,k-1}}D_\tau^{1-\beta_k}\mathscr{R}_k(\tau,\xi)d\tau$$
$$+\int_{t_{cr,k}}^{t} S_k(t+t_{cr,k}-\tau,\xi)\ {}_{t_{cr,k}}D_\tau^{1-\beta_k}\mathscr{R}_k(\tau,\xi)d\tau\Big\}. \tag{6.90}$$

Here $\mathscr{I}_k' = \mathscr{I}_{[t_{cr,k},t_{cr,k+1})}(t)$*,* $k = 0,\ldots,N$*, are indicator functions of the intervals* $[t_{cr,k},t_{cr,k+1})$*,* $k = 0,\ldots,N$*;* $S_j(t,\xi)$*,* $j = 0,\ldots,N$*,* $M_k(t,\xi)$ *and* $\mathscr{R}_k(t,\xi)$*,* $k = 1,\ldots,N$*, are defined in (6.87), (6.88) and (6.89), respectively.*

Proof. It is not hard to verify that

$$J_{\{\mu,0\}}^{\beta(t)}\mathscr{D}_{*\{\mu,0\}}^{\beta(t)}u(t,x) = \sum_{k=0}^{N}\mathscr{I}_k' J^{\beta_k}D_*^{\beta_k}u(t,x) = u(t,x) - \sum_{k=0}^{N}\mathscr{I}_k' u(t_{cr,k},x) + g(t,x), \tag{6.91}$$

where

$$g(t,x) = \sum_{k=1}^{N}\mathscr{I}_k'\frac{t^{\beta_k}}{\Gamma(1+\beta_k)}\sum_{j=0}^{k-1} {}_{t_{cr,j}}\mathscr{D}_{*\{\mu,0\}}^{\beta_k}u(t_{cr,j+1},x).$$

Multiplying both sides of equation (6.81) by $J_{\{\mu,0\}}^{\beta(t)}$ and applying the formula (6.91), we obtain

$$u(t,x) - \sum_{k=0}^{N}\mathscr{I}_k' J^{\beta_k}\mathscr{A}(D)u(t,x) = \sum_{k=0}^{N}\mathscr{I}_k' u(t_{cr,k},x) - g(t,x). \tag{6.92}$$

Let $t \in (0, t_{cr,1})$. Then $\beta(\mu t) = \beta_0$ and $g(t,x) \equiv 0$. In this case taking into account the initial condition (6.82), we can rewrite equation (6.92) in the form

$$u(t,x) - J^{\beta_0} \mathscr{A}(D) u(t,x) = \varphi(x), \quad 0 < t < t_{cr,1}. \tag{6.93}$$

The obtained equation can be solved by using the iteration method. Determine the sequence of functions $\{u_0(t,x), \dots, u_m(t,x)\}$ in the following way. Let $u_0(t,x) = \varphi(x)$ and by iteration

$$u_m(t,x) = J^{\beta_0} \mathscr{A}(D) u_{m-1}(t,x) + \varphi(x), \quad m = 1, 2, \dots \tag{6.94}$$

We show that this sequence is convergent in the topology of the space $C[0,T; \Psi(\mathbb{R}^n)]$ and its limit is a solution to the Cauchy problem (6.81)-(6.82). Moreover, this solution can be represented in the form of functional series

$$u(t,x) = \sum_{k=0}^{\infty} J^{\beta_0 k} \mathscr{A}^k(D) \varphi(x). \tag{6.95}$$

Indeed, it follows from the iteration process (6.94) that

$$u_m(t,x) = J^{\beta(t)m} \mathscr{A}^m(D) \varphi(x) + J^{\beta(t)(m-1)} \mathscr{A}^{m-1}(D) \varphi(x) + \dots + \varphi(x). \tag{6.96}$$

Now we estimate $u_m(t,x)$ applying Lemma 6.11 term by term in the right-hand side of (6.96). Indeed, let $N \in \mathbb{N}$. Then taking into account the fact that the Fourier transform in x commutes with $J^{\beta(t)}$, we have

$$\max_{[0,T]} p_N \left(J^{\beta_0 k} \mathscr{A}^k(D) \varphi(x) \right) \leq \frac{\psi^{k-1}(T)}{\Gamma(k\beta_0)} p_N(\mathscr{A} \varphi(x)).$$

Further, since $A(\xi)$ is continuous on G there exists a constant $C_N > 0$, such that $max_{\xi \in supp\, \kappa_N} |A(\xi)| \leq C_N$, or, by induction $max_{\xi \in supp\, \kappa_N} |A^k(\xi)| \leq C_N^k$. Hence, for every $N \in \mathbb{N}$ we have

$$p_N(J^{\beta(t)k} \mathscr{A}^k(D) \varphi(x)) \leq \|\varphi\|_p \frac{C_N^{k-1} \psi^{k-1}(T)}{\Gamma(k\beta_0)}. \tag{6.97}$$

It follows from (6.97) that

$$\begin{aligned} \max_{[0,T]} p_N(u_m(t,x)) &\leq \|\varphi(x)\|_p \sum_{k=0}^{m} \frac{C_N^k \psi^k(T)}{\Gamma(\beta_0 k + 1)} \\ &\leq C \|\varphi(x)\|_p E_{\beta_0}(C_N \psi(T)), \quad N = 1, 2, \dots, \end{aligned}$$

where $E_{\beta_0}(\tau)$ is the Mittag-Leffler function corresponding to β_0. Since the right-hand side of the latter does not depend on m, we conclude that $u_m(t,x)$ defined in (6.96) is convergent. Again making use of Lemma 6.11 we have

$$p_N(u(t,x) - u_m(t,x)) \leq \|\varphi(x)\|_p \sum_{k=m+1}^{\infty} \frac{C_N \psi^k(T)}{\Gamma(\beta_0 k + 1)}, \quad N = 1, 2, \dots. \tag{6.98}$$

The function

$$\mathscr{R}_m(\eta) = \sum_{k=m+1}^{\infty} \frac{\eta^k}{\Gamma(\beta_0 k+1)}$$

on the right side of equation (6.98) is the reminder of the (convergent) power series representation of the Mittag-Leffler function $E_{\beta_0}(\eta)$, and, hence, $\mathscr{R}_m(\eta) \to 0$, when $m \to \infty$ for any real (or even complex) η. Consequently, $u_m(t,x) \to u(t,x)$ for every $N = 1,2,\ldots,$, that is in the inductive topology of the space $C\left[t \geq 0; \Psi_{G,p}(\mathbb{R}^n)\right]$. Thus, $u(t,x) \in C\left[t \geq 0; \Psi_{G,p}(\mathbb{R}^n)\right]$ is a solution. Moreover, it is readily seen that $u(t,x)$ in (6.95) in the interval $(0,t_{cr,1})$ can be represented through the pseudo-differential operator $S(t,D)$ with the symbol $S_0(t,\xi) = E_{\beta_0}(t^{\beta_0}\mathscr{A}(\xi))$ in the form

$$u(t,x) = u_1(t,x) = S_0(t,D)\varphi(x), \quad t \in (0,t_{cr,1}). \tag{6.99}$$

By construction, the solution $u(t,x)$ is unique and continuous in t. So, the limit

$$\lim_{t \to t_{cr,1}-0} u_1(t,x) = E_{\beta_0}\left(t_{cr,1}^{\beta_0}\mathscr{A}(D)\right)\varphi(x)$$

exists in $\Psi_{G,p}(\mathbb{R}^n)$. Further we extend $u_1(t,x)$ to $[t_{cr,1},t_{cr,2})$. We denote this extension by $u_2(t,x)$. Equation (6.81) in the interval $(t_{cr,1},t_{cr,2})$ reads

$$D_*^{\beta_1} u(t,x) = \mathscr{A}(D)u(t,x), \quad t \in (t_{cr,1},t_{cr,2}).$$

Splitting the integration interval $(0,t)$ on the left-hand side of the last equation into subintervals $(0,t_{cr,1})$ and $(t_{cr,1},t)$, we can rewrite it in the form

$${}_{t_{cr,1}}D_*^{\beta_1} u(t,x) = \mathscr{A}(D)u(t,x) + F_1(t,x), \quad t \in (t_{cr,1},t_{cr,2}),$$

where

$$F_1(t,x) = -\frac{1}{\Gamma(1-\beta_1)} \int_0^{t_{cr,1}} \frac{\frac{\partial}{\partial \tau}u_1(\tau,x)}{(t-\tau)^{\beta_1}} d\tau,$$

Taking into account relation (6.99), it is not hard to see that $F_1(t,x) = \mathscr{R}_1(t,D)\varphi(x)$, where

$$\mathscr{R}_1(t,\xi) = \frac{-\beta_0 A(\xi)}{\Gamma(1-\beta_1)} \int_0^{t_{cr,1}} \frac{E'_{\beta_0}(\tau^{\beta_0}A(\xi))}{\tau^{1-\beta_0}(t-\tau)^{\beta_1}} d\tau.$$

Due to continuity condition (6.83), we have also

$$u(t_{cr,1}-0,x) = u_1(t_{cr,1}+0,x) = E_{\beta_0}\left(t_{cr,1}^{\beta_0}\mathscr{A}(D)\right)\varphi(x).$$

In the general case, assuming that solutions $u_1(t,x),\ldots,u_k(t,x)$ are found in the respective intervals $[0,t_{cr,1}),\ldots,$ $[t_{cr,k-1},t_{cr,k})$, we have the following inhomogeneous Cauchy problem for the interval $(t_{cr,k},t_{cr,k+1})$:

$$_{t_{cr,k}}D_*^{\beta_k}u(t,x) = \mathscr{A}(D)u(t,x) + F_k(t,x), \quad t \in (t_{cr,k}, t_{cr,k+1}), \tag{6.100}$$

$$u(t_{cr,k},x) = u_k(t_{cr,k},x), \tag{6.101}$$

where

$$F_k(t,x) = -\frac{1}{\Gamma(1-\beta_k)} \sum_{j=0}^{k-1} \int_{t_{cr,j}}^{t_{cr,j+1}} \frac{\frac{\partial}{\partial\tau}u_j(\tau,x)}{(t-\tau)^{\beta_k}} d\tau.$$

The solution of this problem we denote by $u_{k+1}(t,x)$. It is not hard to verify that $F_k(t,x)$ can be represented in the form $\mathscr{R}_k(t,D)\varphi(x)$ with a pseudo-differential operator $\mathscr{R}_k(t,D)$, whose symbol is given in (6.89). A unique solution to (6.100),(6.101) can be found by applying the fractional Duhamel principle (see Section 5.5)

$$u_{k+1}(t,x) = S_k(t,D)u_k(t_{cr,k},x) + \int_{t_{cr,k}}^{t} S_k(t-(\tau-t_{cr,k}),D) \; _{t_{cr,k}}D_\tau^{1-\beta_k}F_k(\tau,x)d\tau,$$
$$k = 1,\dots,N, \quad t_{cr,k} < t < t_{cr,k+1}.$$

Now taking into account the equality

$$u_k(t_{cr,k},x) = \left[\prod_{j=0}^{k} S_j(t_{cr,j+1}-t_{cr,j},D)\right] \varphi(x)$$
$$+ \int_{t_{cr,k-1}}^{t_{cr,k}} S_{k-1}(t_{cr,k}-(\tau-t_{cr,k-1}),D)_{t_{cr,k-1}}D_\tau^{1-\beta_{k-1}}F_{k-1}(\tau,x)d\tau,$$

we obtain a solution $u(t,x) = \mathscr{S}(t,D)\varphi(x)$ through the solution operator $\mathscr{S}(t,D)$, whose symbol is given by equation (6.90).

Remark 6.8. Assume in equation (6.81) $\beta(t) = \beta$, where β is a constant in $(0,1]$. Then the representation formula in (6.90) is reduced to

$$u(t,x) = E_\beta(t^\beta \mathscr{A}(\mathscr{D}))\varphi(x),$$

which coincides with the result obtained in Section 5.3.

Applying the technique used in Section 5.3 and the duality of the spaces $\Psi_{G,p}(\mathbb{R}^n)$ and $\Psi'_{-G,q}(\mathbb{R}^n)$ one can prove the following theorem.

Theorem 6.17. *Assume $\nu = 0$ and $\varphi \in \Psi'_{-G,q}(\mathbb{R}^n)$. Then the Cauchy problem (6.81)–(6.83) (with '-D' instead of 'D') has a unique weak solution $u(t,x) \in C[[0,T], \Psi'_{-G,q}(\mathbb{R}^n)]$, $T < \infty$, which is represented in the form*

$$u(t,x) = \mathscr{I}_0' S_0(t,-D)\varphi(x) + \sum_{k=1}^{N} \mathscr{I}_k'(t) \Big\{ M_k(t,-D)\varphi(x)$$

$$+ S_k(t,-D) \int_{t_{cr,k-1}}^{t_{cr,k}} S_{k-1}(t_{cr,k} + t_{cr,k-1} - \tau, -D) \ {}_{t_{cr,k-1}}D_\tau^{1-\beta_k} \mathscr{R}_k(\tau,-D)\varphi(x) d\tau$$

$$+ \int_{t_{cr,k}}^{t} S_k(t + t_{cr,k} - \tau, -D) \ {}_{t_{cr,k}}D_\tau^{1-\beta_k} \mathscr{R}_k(\tau,-D)\varphi(x) d\tau \Big\},$$

Corollary 6.2. *If $\nu = 0$, then the fundamental solution of equation (6.81) with the continuity conditions in (6.83) is represented in the form*

$$U(t,x) = \mathscr{I}_0'(t) \frac{1}{(2\pi)^n} \int_{R^n} E_{\beta_0}(t^{\beta_0} \mathscr{A}(-\xi)) d\xi$$

$$+ \sum_{k=1}^{N} \mathscr{I}_k'(t) \frac{1}{(2\pi)^n} \int_{R^n} \Big\{ E_{\beta_k}((t-t_k)^{\beta_k} \mathscr{A}(-\xi)) \prod_{j=0}^{k-1} E_{\beta_j}((t_{j+1}-t_j)^{\beta_j} \mathscr{A}(-\xi))$$

$$+ E_{\beta_k}((t-t_k)^{\beta_k} A(-\xi)) \int_{t_{k-1}}^{t_k} E_{\beta_{k-1}}((t_k-\tau)^{\beta_{k-1}} A(-\xi)) \ {}_{t_{k-1}}D_\tau^{1-\beta_{k-1}} \mathscr{R}_{k-1}(\tau,-\xi) d\tau$$

$$+ \int_{t_k}^{t} E_{\beta_k}((t-\tau)^{\beta_k} A(-\xi)) \ {}_{t_k}D_\tau^{1-\beta_k} \mathscr{R}_k(\tau,-\xi) d\tau \Big\} e^{ix\xi} \, d\xi,$$

where $t_j = t_{cr,j}$. Moreover, $U(t,x) \in \Psi'_{-G,q}(\mathbb{R}^n)$ for every fixed $t > 0$.

The case $-1 < \nu \le 1$. Now we derive asymptotic behaviors for large and small t of the solution of Cauchy problem (6.81)–(6.83) in the general case of ν, that is $-1 < \nu \le 1$.

The solution $u(t,x) = \mathscr{S}(t,D)\varphi(x)$ obtained in Theorem 6.16 in the case $\nu = 0$ has the structure $u(t,x) = \Psi_1(t,D)\varphi(x) + \Psi_2(t,D)\varphi(x)$, where $\Psi_1(t,D)$ and $\Psi_2(t,D)$ are operators with symbols

$$\Psi_1(t,\xi) = \mathscr{I}_0' S_0(t,\xi) + \sum_{k=1}^{N} \mathscr{I}_k'(t) M_k(t,\xi),$$

and

$$\Psi_2(t,\xi)$$

$$= \sum_{k=1}^{N} \mathscr{I}_k'(t) \Big[S_k(t,\xi) \int_{t_{cr,k-1}}^{t_{cr,k}} S_{k-1}(t_{cr,k} + t_{cr,k-1} - \tau, \xi) \ {}_{t_{cr,k-1}}D_\tau^{1-\beta_k} \mathscr{R}_k(\tau,\xi) d\tau$$

$$+ \int_{t_{cr,k}}^{t} S_k(t + t_{cr,k} - \tau, \xi) \ {}_{t_{cr,k}}D_\tau^{1-\beta_k} \mathscr{R}_k(\tau,\xi) d\tau \Big]$$

The term $v(t,x) = \Psi(t,D)\varphi(x)$ reflects the effect of diffusion modes, while the term $w(t,x) = \Psi_2(t,D)\varphi(x)$ reflects the memory of past. We note that this structure remains valid in the general case $\nu \in (-1,1]$ also; however, the symbols of solution operators get further restructuring, depending on the intervals of mixture of (two or more) modes. The theorems below concern time intervals free of mixed modes, that is time intervals, where a new diffusion mode is established, or not yet started to effect.

Theorem 6.18. *Assume $\mu \neq 0$, $\mu + \nu \neq 0$ and $\varphi \in \Psi_{G,p}(\mathbb{R}^n)$. Then there exists a number $T^* > 0$ and pseudo-differential operators $\mathscr{P}^*(D)$ and $\mathscr{R}^*(t,D)$ with continuous symbols, such that for $t > T^*$ the solution of the Cauchy problem (6.81)-(6.83) coincides with the solution of the Cauchy problem*

$$
{}_{T^*}D_*^{\beta_N} u(t,x) = \mathscr{A}(D)u(t,x) + f^*(t,x), \quad t > T^*, x \in \mathbb{R}^n, \tag{6.102}
$$

$$
u(T^*,x) = \varphi^*(x), \quad x \in \mathbb{R}^n. \tag{6.103}
$$

where $f^*(t,x) = \mathscr{R}^*(t,D)\varphi(x)$ and $\varphi^*(x) = \mathscr{P}^*(D)\varphi(x)$.

Proof. Without loss of generality one can assume that $\nu > 0$. Then as it follows from Theorem 3.7 that the actual mode changes occur at times $T_j^* = T_j/\mu$ and $t_j^* = T_j/(\mu+\nu)$, $j = 1,\dots,N$, if diffusion modes change at times $T_j, j = 1,\dots,N$. Obviously, $t_1^* < \dots < t_N^*$ and $T_1^* < \dots < T_N^*$ if $T_1 < \dots < T_N$. The order function $\beta(\mu t + \nu\tau)$ under the integral in $D_{*\{\mu,\nu\}}^{\beta(t)}$ takes the value β_N for all $t > T_N^*$ and $\tau > 0$. Hence, the variable order operator on the left side of (6.81) becomes $D_*^{\beta_N}$ if $t > T_N^*$. Analogously it follows from Theorem 3.8 that if $\nu < 0$, then $\beta(\mu t + \nu\tau)$ takes the value β_N for all $t > t_N^*$ and $\tau > 0$. Thus, if $\nu \neq 0$, then for all $t > T^* = max\{T_N^*, t_N^*\}$ and $0 < \tau < t$ we have $\beta(\mu t + \nu\tau) = \beta_N$. Similar to the case $\nu = 0$, splitting the interval $(0,t)$, $t > T^*$, into subintervals, we can represent the equation (6.81) in the form (6.102). Further, from the continuity condition (6.83) we have $u(T^*,x) = \lim_{t\to T^*-0} v(t,x)$, where $v(t,x)$ is a solution to the Cauchy problem for fractional order pseudo-differential equations in sub-intervals of the interval $[0,T_N^*)$ constructed by continuation. Therefore there exists an operator $S^*(t,D)$, such that $v(t,x) = S^*(t,D)\varphi(x)$. Denote $\mathscr{P}^*(D) = S^*(T^*,D)$. Then $u(T^*,x) = \mathscr{P}^*(D)\varphi(x)$. This means that for $t > T^*$ solutions of problems (6.81)-(6.83) and (6.102),(6.103) coincide. If $\nu = 0$, then the statement follows from Theorem 6.16.

Theorem 6.19. *Assume $\varphi \in \Psi_{G,p}(\mathbb{R}^n)$. Then there exists a number $t^* > 0$, such that for $0 < t < t^*$ the solution of the Cauchy problem (6.81)–(6.83) coincides with the solution of the Cauchy problem*

$$
D_*^{\beta_0} u(t,x) = \mathscr{A}(D)u(t,x), \quad t > 0, \quad x \in \mathbb{R}^n, \tag{6.104}
$$

$$
u(0,x) = \varphi(x), \quad x \in \mathbb{R}^n. \tag{6.105}
$$

Proof. It follows from Theorems 3.7 and 3.8 that the order function $\beta(\mu t + \nu\tau)$ under the integral in $D_{*\{\mu,\nu\}}^{\beta(t)}$ takes the value β_0 for all $t < t^* = min\{t_1^*, T_1^*\}$ and $0 < \tau < t$. Hence, the variable order operator in (6.81) becomes $D_*^{\beta_0}$ if $0 < t < t^*$. The order β_1 (or diffusion mode $\{\beta_1,\mu,\nu\}$) has no influence in this interval. For $t > t^*$ two diffusion modes $\{\beta_0,\mu,\nu\}$ and $\{\beta_1,\mu,\nu\}$ are present. If $t < min\{t_2^*, T_2^*\}$, then for all $\tau > 0$ we have $\mu t + \nu\tau < T_2$. That is, there is no influence of the mode $\{\beta_2, \mu, \nu\}$ if $t < min\{t_2^*, T_2^*\}$. In the same manner the other values of β have no influence in the interval $0 < t < t^*$. This means that for $0 < t < t^*$ solutions of problems (6.81)–(6.83) and (6.104)–(6.105) coincide.

6.8 Additional notes

1. *Models with DODEs.* A distributed fractional order differential operator was first considered in the paper [Cap67] by Michele Caputo in 1967 in connection with modeling of linear dissipation. In Equation (3) of this paper, generalizing a stress–strain relation used earlier by Knopoff, he writes the following relation:

$$\tau_{rs} = \int_{a_1}^{b_1} f_1(r,z)\frac{d^z}{dt^z}(g^{hi}g_{rs}e_{hi})dz + 2\int_{a_1}^{b_1} f_2(r,z)\frac{d^z}{dt^z}(e_{rs})dz, \tag{$*$}$$

where $\frac{d^z}{dt^z}$ in this expression is understood in the sense of Caputo-Djrbashian derivative D_*^z, the definition of which is given in Equation (5) on the same page. Both integrals on the right of equation $(*)$ are distributed order differential operators in the sense of Definition 3.5. Namely, the first term corresponds to $\mu(d\alpha) = f_1(r,\alpha)d\alpha$, and the second term to $\mu(d\alpha) = 2f_2(r,\alpha)d\alpha$. In the sense of Definition 3.5 differential equations with a finite number of fractional derivatives in the sum are also qualified as DODEs. In this case μ is defined as $\mu(d\alpha) = \sum_{j=1}^{m} C_j \delta_{\alpha_j}(\alpha)d\alpha$. An example is the Bagley-Torvik equation [Pod99]

$$\frac{d^2u}{dt^2} + aD_+^{3/2}u + cu = h(t), \tag{6.106}$$

arising in the theory of viscoelastic materials and corresponding to the case $\mu(d\alpha) = [\delta_2(\alpha) + a\delta_{3/2}(\alpha) + c\delta_0(\alpha)]d\alpha$. The solution to equation (6.106) satisfying the homogeneous initial conditions can be represented in the form $u(t) = (G * h)(t)$, where the function G is defined as

$$G(t) = \sum_{k=0}^{\infty} \frac{(-1)^k c^k t^{2k+1}}{k!} E^{(k)}_{1/2,\,2+3k/2}(-a\sqrt{t}),$$

with the generalized Mittag-Leffler function $E_{\mu,\nu}(z)$. Agraval [Agr04] used the model

$$\frac{d^2u}{dt^2} + aD_+^{1/2}u + Au = h(t),$$

which corresponds to the measure $\mu(d\alpha) = [\delta_2(\alpha) + a\delta_{1/2}(\alpha) + c\delta_0(\alpha)]d\alpha$, to describe fractionally damped beam. Podlubny [Pod99] investigated general m-term equations of the form

$$\sum_{k=0}^{m} C_k D^{\alpha_k} u = h(t),$$

and found the corresponding Green function assuming $\alpha_m > \alpha_{m-1} > \cdots > \alpha_0$. DODEs are broadly used by many researchers to model various processes arising in modern science and engineering, see, e.g., [Pod99, Cap01, LH02, CSK11, JCP12, AUS06, KAN10, HKU10]. Numerical methods of solution of DODEs are discussed in [DF09, DF01, Kat12]. See also papers [BT00, UG05-2, MS06, Koc08], where abstract properties of DODEs, not tied to any physical model, are presented. DODEs also arise in the theory of factional Fokker-Planck-Kolmogorov equations discussed in Chapter 7.

2. *Models with VODEs.* The theory of variable order differential equations is a relatively new branch of fractional differential equations. In the last few decades a number of publications appeared where VODEs were used to model processes with changing in time diffusion exponents. We recall the classification of diffusion processes through their mean square displacement (MSD), which is a function of time: MSD$(t) \sim K_\beta t^\beta, t \to, \infty$, where the exponent β indicates the actual diffusion mode. For instance, if a diffusion process has the normal mode, then $\beta = 1$, if it has a sub-diffusive (slower) mode, then $\beta < 1$, and if the process has a super-diffusive mode, then $\beta > 1$. In many applications the exponent β turns out to depend on time, that is $\beta = \beta(t)$, or some other parameters of the model. In the paper by Lorenzo et al. [LH02] a number of examples of such a dependence arising naturally in applied sciences are given. Chechkin et al. [CGS05] modeled the evolution of a composite system with different sub-diffusion exponents using a space-variable fractional differential equation. Papers [SCK05, RC10] used VODE for analysis of viscoelastic oscillators.
   The mathematical theory of initial and boundary value problems for VODEs is not yet satisfactorily developed. The Cauchy problem studied in Section 6.7 partially shows difficulties arising in such an analysis. Even for piecewise order functions at every mode change time an inhomogeneous term emerges, making analysis complicated. It is not clear into what form turns this phenomenon in the case of more general, for instance, continuous order functions.
3. *Fractional Duhamel principle for DODEs and VODEs.* the fractional Duhamel principle for general DODEs formed with the help of a bounded measure Λ is discussed in the paper [Uma12]. In this paper the basic fractional derivative is the Caputo-Djrbashian derivative. In this case Λ can be rather general measure. However, in the case of DODEs defined with the Riemann-Liouville derivative, the problem on the fractional Duhamel principle is more challenging. In Section 6.5 we proved the fractional Duhamel principle only in the case of measures of the form
$$\Lambda(d\alpha) = \left[\delta_\mu(\alpha) + \sum_{k=1}^{m-1} c_k \delta_{\mu-k}(\alpha)\right] d\alpha.$$
   Note that there is another way to prove Theorem 6.11. Namely, applying the operator $J^{\mu-m+1}$ to both sides, one can reduce it to an integer order integro-differential equation (differential equation if $B_0 = 0$) for which the classic Duhamel principle is applicable. Is the Duhamel principle valid for more general measures of the form
$$\Lambda(d\alpha) = \left[\delta_\mu(\alpha) + \sum_{k=1}^{m-1} c_k \delta_{\alpha_k}(\alpha)\right] d\alpha,$$
   with arbitrary $\alpha_k \in [k-1,k)$? If the answer is "Yes," then in what form? This is a challenging open question, as well as the Duhamel principle in the case of variable order differential equations.

Chapter 7
Fractional order Fokker-Planck-Kolmogorov equations and associated stochastic processes

7.1 Introduction

This chapter discusses the connection between pseudo-differential and fractional order differential equations considered in Chapters 2–6 with some random (stochastic) processes defined by stochastic differential equations. We assume that the reader is familiar with basic notions of probability theory and stochastic processes, such as a random variable, its density function, mathematical expectation, characteristic function, etc. Since we are interested only in applications of fractional order ΨDOSS, we do not discuss in detail facts on random processes that are already established and presented in other sources. For details of such notations and related facts we refer the reader to the book by Applebaum [App09] (or [IW81, Sat99]). We only mention some basic notations directly related to our discussions on fractional Fokker-Planck-Kolmogorov equations.

Fokker-Planck-Kolmogorov (FPK) equations are partial differential equations introduced first by Fokker (1913) and Planck (1917) for time evolution of the density function of the velocity of a minute substance diffusing in a white noise environment. Later Kolmogorov (1931) developed a mathematical theory [of a broad class of such equations] obtaining forward and backward Kolmogorov equations. FPK equations are closely related to stochastic differential equations. Historically, first stochastic differential equation was introduced and studied by Langevin (1908), three years after a theoretical explanation of Brownian motion by Einstein (1905). Wiener (1927) showed that Brownian motion is nowhere differentiable and its path has infinite total variation over an arbitrary time interval. Due to these facts, mathematically, both differential and integral versions of stochastic differential equations were not justified. In the second half of the 1940s Itô developed a strict mathematical theory of stochastic differentials and integrals, which now is referred to as an Itô stochastic calculus. Section 7.4 discusses briefly Brownian motion, Itô's stochastic differential equations driven by Brownian motion, and associated FPK equations.

S. Umarov, *Introduction to Fractional and Pseudo-Differential Equations with Singular Symbols*, Developments in Mathematics 41,
DOI 10.1007/978-3-319-20771-1_7

Fractional FPK equations model sub- and super-diffusion and other complex stochastic processes revealing subtle intrinsic properties of such processes. For instance, in the last few decades, fractional FPK equations have appeared as an essential tool for the study of dynamics of various complex stochastic processes arising in anomalous diffusion in physics, finance, hydrology, cell biology, etc. Complexity includes phenomena such as the presence of weak or strong correlations, simultaneous presence of different sub- or super-diffusive modes, and various types of jump effects, which occur in various real world processes. Consider one example from cell biology. Experimental studies of the motion of proteins and other macromolecules in the cell membrane show apparent subdiffusive motion (see details in [Sax01]). Moreover, these experiments show that several diffusive modes simultaneously affect the motion. An experiment describing such a phenomenon is provided in [GW94], which recorded that approximately 50 % of case measurements on the LDL receptor labeled with *diILDL* show subdiffusive motion, with diffusion mode parameter β between 0.2 and 0.9. Subdiffusive motion with $0.1 < \beta < 0.9$ or with $0.22 < \beta < 0.48$ were found in [GW94, WEKN04], depending on a type of macromolecules and cells. Protein molecules diffuse 5 to 100 times slower [Edi97, GW94, Sax01, SJ97] than free Brownian motion at different times and regions. Here, the smaller the parameter β, the more slowly the particles scatter, whereas the case $\beta = 1$ corresponds to the classical diffusion. Examples can be drawn from numerous other fields.

Boundary value problems for fractional order differential equations discussed in Chapters 5 and 6 can be used to model random processes driven by time-changed stochastic processes. A deeper relationship between processes modeled by stochastic differential equations driven by a time-changed stochastic process and their associated deterministic fractional order differential equations (fractional Fokker-Planck-Kolmogorov equations) is the main subject of study in this chapter.

The SDEs associated with fractional FPK equations are driven by Brownian motion subordinated to a special time-change process, the first hitting time of a Lévy's stable subordinator. Section 7.6 presents a description of Lévy's stable subordinators of stability index $\beta \in (0,1)$ and their inverse (first hitting time) processes.

Driving processes of stochastic differential equations play a key role in modeling of complex stochastic processes. Therefore, understanding of the properties of the driving process elucidate many properties of the process itself. Driving processes can be approximated by random walks. For instance, as is shown in the next chapter, Brownian motion without drift (the definition is given in Section 7.2) can be approximated by a simple random walk. Moreover, the transition probabilities $p_t(x,y)$ from a point x to a point y at a time t satisfies the following initial value problem:

$$\frac{\partial p_t(x,y)}{\partial t} = \frac{1}{2}\frac{\partial^2 p_t(x,y)}{\partial y^2}, t > 0, \tag{7.1}$$

$$p_0(x,y) = \delta_x(y). \tag{7.2}$$

Equation (7.1) is the Fokker-Planck-Kolmogorov equation associated with Brownian motion. The stochastic processes associated with fractional order Fokker-Planck-Kolmogorov equations are usually driven by complex processes. Even in the simplest case of the fractional equation

$$D_t^\beta u = \frac{1}{2}\frac{\partial^2 u}{\partial y^2},$$

where D^β is a fractional derivative of order $0 < \beta < 1$, the driving process belongs to the class of non-Markovian semimartingales (see Section 7.16, "Additional notes"). We note that semimartingales form the largest class of processes for which Itô's stochastic calculus is valid. Driving processes play a key role for processes defined by stochastic differential equations. As we will see, driving processes of fractional Fokker-Planck-Kolmogorov equations are time-changed processes. These time-changed processes are scaling limits of, so-called, continuous time random walks (CTRW). The connection between CTRW and fractional order differential equations will be discussed in detail in Chapter 8. Thus, in the theory of fractional Fokker-Planck-Kolmogorov equations a triple relationship between a driving process, the corresponding stochastic differential equation, and the associated Fokker-Planck-Kolmogorov equation, is apparent. This triple relationship for SDEs driven by a time-changed Lévy processes was recently studied in the papers [HKU10, HKU11, HU11].

For additional comments and historical notes see Section 7.16.

7.2 Itô's stochastic calculus and stochastic differential equations

In this section, for the reader's convenience, we introduce some basic notions related to stochastic processes, Itô's calculus, and Itô's stochastic differential equations (SDEs). In Probability Theory, one always assumes that random variables (vectors) under consideration are given in a probability space consisting of a triple $(\Omega, \mathscr{F}, \mathbb{P})$, which depends on an underlying model. Here Ω is a set of elementary events (sample set), $\mathscr{F}$ is a σ-algebra of subsets of Ω, and $\mathbb{P}$ is a measure defined on $\mathscr{F}$, such that $\mathbb{P}(\Omega) = 1$. By definition, an n-dimensional random vector X is a mapping $X : \Omega \to \mathbb{R}^n$, such that $X^{-1}(A) \in \mathscr{F}$ for any Borel set $A \subset \mathbb{R}^n$. We denote the density function of X by $f_X(x)$ and the mathematical expectation of X by $\mathbb{E}(X)$, which, by definition, is

$$\mathbb{E}(X) = \int_\Omega X d\mathbb{P}, \tag{7.3}$$

or if the density function f_X is known, then

$$\mathbb{E}(X) = \int_\mathbb{R} x f_X(x) dx. \tag{7.4}$$

A real (complex) function $g(X)$ of the random variable X defines naturally as a mapping $g: A \times \Omega \to \mathbb{R}$ ($A \times \Omega \to \mathbb{C}$), where $A \subset \mathbb{R}^n$ is in the range of X. Note that the expectation of $g(X)$ is

$$\mathbb{E}[g(X)] = \int_\Omega g(X) d\mathbb{P} = \int_\mathbb{R} g(x) f_X(x) dx,$$

generalizing (7.3) and (7.4). For given events A, B and a random variable X we frequently use notations $\mathbb{P}(A|B)$ and $\mathbb{E}[X|B]$, called a conditional probability and conditional expectation, respectively, and meaning the probability of the event A and the expectation of the random variable X, under the condition B.

Definition 7.1. Let $\mathbb{T}$ be a set of indices. A family of random variables $\{X_t, t \in \mathbb{T}\}$, is called a stochastic process.

It follows from this definition that for each fixed $t \geq 0$ the stochastic process X_t is a random variable defined on $(\Omega, \mathscr{F}_t, \mathbb{P})$. Frequently one needs to consider a stochastic process adopted to a certain filtration. By definition, a filtration $\mathscr{F}_t$ is an increasing family of sub-σ-algebras of $\mathscr{F}$, that is $\mathscr{F}_s \subset \mathscr{F}_t$, if $0 \leq s \leq t$. A filtration $\mathscr{F}_t$ is called right-continuous, if $\mathscr{F}_{t+} = \cap_{\varepsilon>0} \mathscr{F}_{t+\varepsilon} = \mathscr{F}_t$ for each $t \geq 0$. We say that the stochastic process X_t is adapted to the filtration $\mathscr{F}_t$, or $\mathscr{F}_t$-adapted if X_t is $\mathscr{F}_t$-measurable for each $t \geq 0$.

Definition 7.2. Brownian motion B_t is a stochastic process $\{B_t, t \geq 0\}$, such that

1. $B_0 = 0$;
2. for all nonoverlapping intervals (t_1, t_2) and (s_1, s_2) random variables (increments) $B_{t_2} - B_{t_1}$ and $B_{s_2} - B_{s_1}$ are independent;
3. for arbitrary $0 \leq s < t$ the random variable $B_t - B_s$ has the density

$$f_{B_t - B_s}(x) = p(t-s, x) = \frac{1}{\sqrt{2\pi(t-s)}} e^{-\frac{x^2}{2(t-s)}}, \tag{7.5}$$

   that is $B_t - B_s$ is normal with mean 0 and variance $t - s$;
4. B_t has a continuous path.

N. Wiener [Wie28] proved that B_t is nowhere differentiable and has the infinite total variation over arbitrary interval (t_1, t_2). Hence, the integral $\int_0^t f(s) dB_s$ is meaningless if one understands it in the Lebesgue-Stieltjes sense. Itô developed a special stochastic calculus in the frame of which the above integral becomes meaningful. Below we briefly reproduce the definition of the Itô integral and its properties without proofs. For details we refer the reader to nicely written book [IW81].

Introduce the space $\mathscr{L}_2$ of $\mathscr{F}_t$-measurable stochastic processes X_t, such that for an arbitrary $T > 0$

$$\|X|\mathscr{L}_2\|_T^2 = \mathbb{E}\left[\int_0^T X_s^2 ds\right] < \infty. \tag{7.6}$$

$\mathscr{L}_2$ is a complete metric space with respect to the metric

$$\rho(X,Y) = \sum_{k=1}^{\infty} \frac{\|(X-Y)|\mathscr{L}_2\|_k \wedge 1}{2^k},$$

where the symbol $a \wedge b$ means $\min\{a,b\}$. A random variable Y is said to belong to $L(\mathbb{P})$, if $\mathbb{E}[|Y|] < \infty$. Hence, equation (7.6) means that $\int_0^T X_s^2 ds \in L(\mathbb{P})$ for each fixed $T > 0$. We also introduce a class $L^\infty[0,T;L(\mathbb{P})]$ of stochastic processes Y_t, such that $Y_t \in L(\mathbb{P})$ for each fixed $t \in [0,T]$, and $\sup_{t\in[0,T]} \mathbb{E}[Y_t] < \infty$.

Let $\mathscr{L}_0$ be a subspace of $\mathscr{L}_2$ containing stochastic processes X_t with the following property: there exists a partition $0 = t_0 < t_1 < \cdots < t_n < \ldots$, and a sequence of $\mathscr{F}_{t_k}$-measurable random variables f_k, $k = 0,1,\ldots$, with $\sup_k \|f_k|L_\infty\| < \infty$, such that

$$X_t = f_0 \mathscr{I}_{t=0}(t) + \sum_{k=0}^{\infty} f_k \mathscr{I}_{(t_k,t_{k+1}]}(t).$$

Here $\mathscr{I}_{(a,b)}$ is the indicator function of the interval (a,b). Elements of $\mathscr{L}_0$ are called simple processes.

Proposition 7.1. *1. $\mathscr{L}_2$ is a complete metric space;*
2. $\mathscr{L}_0$ is dense in $\mathscr{L}_2$.

Define an operator I in $\mathscr{L}_0$ by

$$I(X_t) = \sum_{k=0}^{n-1} f_k (B_{t_{k+1}} - B_{t_k}) + f_n (B_t - B_{t_n})$$

for $t \in [t_n, t_{n+1}]$, $n = 0,1,\ldots$. $I(X_t)$ is called a stochastic integral of a simple process $X_t \in \mathscr{L}_0$ with respect to Brownian motion B_t, and denoted by

$$I(X_t) = \int_0^t X_s dB_s.$$

Using the independence of increments of Brownian motion and the fact that the variance $\mathbb{E}(B_t - B_s)^2 = t - s$, one has

$$\mathbb{E}[I(X_t)^2] = \sum_{k=0}^{n-1} \mathbb{E}[f_k^2](t_{k+1} - t_k) + \mathbb{E}[f_n^2](t - t_n) = \mathbb{E}\left[\int_0^t X_s^2 ds\right].$$

Hence we have the following equality

$$\mathbb{E}\left[\left(\int_0^t X_s dB_s\right)^2\right] = \mathbb{E}\left[\int_0^t X_s^2 ds\right], \quad \forall X_t \in \mathscr{L}_0. \tag{7.7}$$

Now let $X_t \in \mathscr{L}_2$ be an arbitrary process. Due to Proposition 7.1 there exists a sequence $X_{tn} \in \mathscr{L}_0$, such that $X_{tn} \to X_t$ in the sense $\rho(X_{tn}, X_t) \to 0$, when $n \to \infty$. We set

$$I(X_t) = \int_0^t X_s dB_s = \lim_{n\to\infty} \int_0^t X_{sn} ds = \lim_{n\to\infty} I(X_{tn}).$$

Using Lebesgue's dominated convergence theorem it follows from (7.7) that

$$\mathbb{E}[I(X_t)^2] = \mathbb{E}\left[\left(\int_0^t X_s dB_s\right)^2\right] = \mathbb{E}\left[\int_0^t X_s^2 ds\right], \quad \forall X_t \in \mathscr{L}_2, \tag{7.8}$$

which is called the Itô identity. This identity shows that the mapping

$$I : \mathscr{L}_2 \to L^\infty[0,T;L((P))],$$

is continuous for each fixed $T > 0$.

Thus, for each $X_t \in \mathscr{L}_2$ the stochastic process defined by the integral

$$\int_0^t X_s dB_s$$

is well defined and is called *Itô's stochastic integral*. The proposition below contains some important properties of Itô's stochastic integral.

Proposition 7.2. *1. (Linearity)* $\int_0^t [\alpha X_s + \beta Y_s] dB_s = \alpha \int_0^t X_s dB_s + \beta \int_0^t Y_s dB_s$, *where* $\alpha, \beta \in \mathbb{C}$, *and* $X_t, Y_t \in \mathscr{L}_2$;
2. (Itô's identity) $\mathbb{E}\left[\left(\int_0^t X_s dB_s\right)^2\right] = \mathbb{E}\left[\int_0^t X_s^2 ds\right]$;
3. $\mathbb{E}\left[\int_0^t X_s dB_s\right] = 0$ *for all* $t \geq 0$;

Proof. The linearity property immediately follows from the definition of Itô's integral. Itô's identity is proved above. The third property (called a martingale property) follows from the fact that $\mathbb{E}(B_t) = 0$ for each $t \geq 0$ and from the independence of Brownian motion B_t and the stochastic process X_t.

Let $b(x)$ and $\sigma(x)$ be Lipschitz continuous functions with linear growth[1] for large $x \in \mathbb{R}$. Consider a stochastic differential equation (SDE)

$$dX_t = b(X_t)dt + \sigma(X_t)dB_t, \tag{7.9}$$

with the initial condition

$$X_{t=0} = X_0, \tag{7.10}$$

where X_0 is a random variable independent of B_t. The meaning of SDE (7.9) with initial condition (7.10) is

$$X_t = X_0 + \int_0^t b(X_s)ds + \int_0^t \sigma(X_s)dB_s, \tag{7.11}$$

where the first integral is a usual integral and the second one is in the sense of Itô stochastic integral.

[1] The Lipschitz and linear growth conditions in n-dimensional case are given in (7.12) and (7.13), respectively.

Further, suppose B_t is an m-dimensional Brownian motion, that is $B_t = (B_{1t},\dots,B_{mt})$, where $B_{1t},\dots,B_{mt}$ are independent Brownian motions in the sense of Definition 7.2. Then n-dimensional analog of SDE (7.9) takes the following component-wise form

$$dX_{jt} = b_j(X_{1t},\dots,X_{nt})dt + \sum_{k=1}^{m} \sigma_{jk}(X_{1t},\dots,X_{nt})dB_{kt}, \quad j = 1,\dots,n,$$

where the mappings $b = (b_1,\dots,b_n) : \mathbb{R}^n \to \mathbb{R}^n$, $\sigma = \{\sigma_{jk}\}_{j=1,k=1}^{n,\,m} : \mathbb{R}^n \to \mathbb{R}^{n\times m}$ satisfy the conditions:

$$\text{Lipschitz:} \quad \|b(x)-b(y)\| + |||\sigma(x)-\sigma(y)||| \le C\|x-y\|, \quad x,y \in \mathbb{R}^n, \tag{7.12}$$

$$\text{Linear growth:} \quad \|b(x)\| + |||\sigma(x)||| \le C(1+\|x\|), \quad x \in \mathbb{R}^n. \tag{7.13}$$

Here the vector-norm $\|b\|$ and the matrix-norm $|||\sigma|||$ are defined

$$\|b\|^2 = \sum_{j=1}^{n} |b_j|^2, \quad |||\sigma|||^2 = \sum_{j=1}^{n}\sum_{k=1}^{m} |\sigma_{jk}|^2,$$

respectively.

In the theory of Ito's stochastic differential equations Ito's formula plays an important role. Below we formulate Ito's formula in a particular case. In the proposition below we assume that $b(t)$ and $\sigma(t)$ are $\mathscr{F}_t$-adapted stochastic processes, where $\mathscr{F}_t$ is a filtration associated with B_t.

Proposition 7.3. *(Itô's formula) Let $f \in C_0^2(\mathbb{R})$ and X_t be a stochastic process of the form*

$$X_t = X_0 + \int_0^t b(s)ds + \int_0^t \sigma(s)dB_s.$$

Then

$$f(X_t) = f(X_0) + \int_0^t \left[b(s)f'(X_s) + \frac{1}{2}\sigma^2(s)f''(X_s) \right] ds + \int_0^t \sigma(s)f'(X_s)dB_s. \tag{7.14}$$

For the proof see, e.g., [IW81]. See Section "Additional Notes" for more general forms of Itô's formula.

7.3 Connection between stochastic and deterministic descriptions of random processes: Fokker-Planck-Kolmogorov equations

In Section 8.2 we will show that the limiting stochastic process of the so-called *simple* random walk can be described with the help of the diffusion equation (7.1), whose solution satisfying the initial condition (7.2) is the Gaussian density function

(7.5) (with $s = 0$), evolved in time. This limiting stochastic process is Brownian motion without drift. Thus, Brownian motion B_t and the partial differential equation (7.1) for the density of B_t are related. Equation (7.1) is the Fokker-Planck equation associated with B_t. Such a relationship holds for a wide class of stochastic processes.

The aim of this section is to give a precise formulation of the connection between certain classes of stochastic processes and their associated deterministic partial differential equations (FPK equations). Let $(\Omega, \mathscr{F}, \mathbb{P})$ be a probability space with a complete right-continuous filtration $(\mathscr{F}_t)$. Consider an example $Y_t = \sigma B_t + bt$ of an n-dimensional Brownian motion (with constant drift bt, $b \in \mathbb{R}^n$, and a constant covariance $(n \times m)$-matrix σ) defined on this filtered probability space. This example is useful for understanding of the general form of operators appearing in partial differential equations associated with stochastic processes. Let $A \subset \mathbb{R}^n$ be a Borel set and $P^Z(t,x,A) = \mathbb{P}(Z_t \in A | Z_0 = x)$ be the transition probability (from a point $x \in \mathbb{R}^n$ to A) of the process Z_t with density $p^Z(t,x,y)$, i.e., $p^Z(t,x,y)dy = P^Z(t,x,dy)$. Then $p^Z(t,x,y)$ satisfies (in the weak sense) the following partial differential equation (see Section 7.4)

$$\frac{\partial p^Z(t,x,y)}{\partial t} = -\sum_{j=1}^{n} b_j \frac{\partial p^Z(t,x,y)}{\partial y_j} + \frac{1}{2} \sum_{i,j=1}^{n} a_{ij} \frac{\partial^2 p^Z(t,x,y)}{\partial y_i \partial y_j}, \quad t > 0,\, x, y \in \mathbb{R}^n,$$

with the additional condition $p^Z(0,x,y) = \delta_x(y)$, where δ_x is the Dirac delta function with mass on x. Here, the matrix a_{ij} is a square $(n \times n)$-matrix equal to the product of σ by its transpose σ^T.

A deep generalization of this relationship between a stochastic process and its associated partial differential equation is expressed through the Fokker-Planck-Kolmogorov forward and backward equations. As we will show in Section 7.4 that this concept is based on the relationship between two main components:

(I) the Cauchy problem

$$\frac{\partial u(t,x)}{\partial t} = \mathscr{A} u(t,x), \quad t > 0, x \in \mathbb{R}^n, \tag{7.15}$$

$$u(0,x) = \varphi(x), \quad x \in \mathbb{R}^n; \tag{7.16}$$

where $\mathscr{A}$ is a differential operator

$$\mathscr{A} = \sum_{j=1}^{n} b_j(x) \frac{\partial}{\partial x_j} + \frac{1}{2} \sum_{i,j=1}^{n} a_{i,j}(x) \frac{\partial^2}{\partial x_i \partial x_j}, \tag{7.17}$$

with coefficients $b_j(x)$ and $a_{i,j}(x)$ satisfying some mild regularity conditions; and

(II) the associated class of Itô SDEs given by

$$dX_t = b(X_t)dt + \sigma(X_t)dB_t, \quad X_0 = x, \tag{7.18}$$

where B_t is an m-dimensional Brownian motion. Here X_t is a solution, and the coefficients are connected with the coefficients of the operator $\mathscr{A}$ as follows: $b(x) = (b_1(x), \ldots, b_n(x))$ and $a_{i,j}(x)$ is the (i,j)-th entry of the product of the $n \times m$ matrix $\sigma(x)$ with its transpose $\sigma^T(x)$. Finally, x is a random variable, independent of B_t and with the density $\varphi(x)$.

Example 7.1. Let in equation (7.18) the coefficients $b(x) = 0$, identically zero vector, and $\sigma(x) = I$, the identity matrix. In this case, $X_t = B_t$, and as follows from the definition of B_t and equations (8.4) and (8.5) that the density of B_t satisfies the Cauchy problem

$$\frac{\partial p}{\partial t} = \frac{1}{2}\Delta p, \quad t > 0, x \in \mathbb{R}^n, \tag{7.19}$$

$$p(0,x) = \varphi(x), \quad x \in \mathbb{R}^n. \tag{7.20}$$

That is, in this case $\mathscr{A} = \Delta$.

One mechanism for establishing the relationship between (i) and (ii) is via semigroup theory of linear operators. Let $\mathscr{X}$ be a Banach space and T_t, $t \geq 0$, be a one-parameter family of linear operators mapping $\mathscr{X}$ to itself. The family T_t is called a *strongly continuous semigroup* if

1. $T_0 = I$, the identity operator;
2. $T_t T_s = T_{t+s}$, for all $t, s \geq 0$; and
3. $T_t \varphi \to T_{t_0} \varphi$ for all $\varphi \in \mathscr{X}$ in the norm of $\mathscr{X}$ as $t \to t_0$.

A linear operator A defined as

$$A\varphi = \lim_{t \to 0+} \frac{T_t \varphi - \varphi}{t}, \tag{7.21}$$

provided the limit exists, is called an *infinitesimal generator* of the semigroup T_t. In fact, the set of elements $\varphi \in \mathscr{X}$ for which the limit (7.21) exists is a dense subset of $\mathscr{X}$ and is the domain of the operator A. We will denote the domain of A by $\mathrm{Dom}(A)$.

Returning to our discussion on the connection of the stochastic process X_t defined by SDE (7.18) and the operator $\mathscr{A}$ in (7.19), we notice that the operator $\mathscr{A}$ is recognized as the infinitesimal generator of the semigroup $T_t \varphi(x) := \mathbb{E}[\varphi(X_t)|X_0 = x]$ (defined, for instance, on the Banach space $C_0(\mathbb{R}^n)$ with sup-norm), i.e., $\mathscr{A}\varphi(x) = \lim_{t \to 0} (T_t - I)\varphi(x)/t$, $\varphi \in \mathrm{Dom}(\mathscr{A})$, the domain of $\mathscr{A}$. A unique solution to (7.15)–(7.16) for $\mathscr{A}$ in (7.17) is represented by $u(t,x) = (T_t \varphi)(x)$ (see details, e.g., in [App09]).

Example 7.2. Suppose again $b(x) = 0$ and $\sigma(x) = I$. Then

$$T_t \varphi(x) = \mathbb{E}[\varphi(B_t)|B_0 = x] = \frac{1}{(\sqrt{2\pi t})^n} \int_{\mathbb{R}^n} e^{-\frac{|y-x|^2}{2t}} \varphi(y) dy. \tag{7.22}$$

Due to relationships (2.11) and (2.12) with $\kappa = 1/2$ the latter can be written in the form of ΨDOSS

$$T_t\varphi(x) = \frac{1}{(2\pi)^n}\int_{\mathbb{R}^n} e^{-\frac{t}{2}|\xi|^2 - ix\xi} F[\varphi](\xi)d\xi = e^{t\left(\frac{1}{2}\Delta\right)}\varphi(x).$$

In terms of the semigroup theory the latter is advantageous, since it shows the direct connection between the semigroup T_t and its infinitesimal generator $\mathscr{A} = \frac{1}{2}\Delta$, i.e., $T_t = e^{t\left(\frac{1}{2}\Delta\right)}$. Recall also that the unique solution of (7.19)–(7.20) is given by the right-hand side of (7.22) (see (2.12) of Section 2.2 with $\kappa = 1/2$): $u(t,x) = T_t\varphi(x)$.

The relationship between (I) and (II) says that the equation given in (7.15), with the first order time derivative on the left and the operator $\mathscr{A}$ on the right defined in equation (7.17), is related to SDE in (7.18) driven by a Brownian motion with drift, as long as the coefficients satisfy appropriate conditions. In such cases we say that the deterministic partial differential equation in (7.15) (or Cauchy problem (7.15)–(7.16)) is associated with SDE (7.18), or vice versa, SDE (7.18) is associated with deterministic equation (7.15).

The mechanism for establishing the relationship reveals that the transition probabilities $P^X(t,x,dy) = \mathbb{P}(X_t \in dy | X_0 = x)$ of a solution X_t to (7.18) satisfy in the weak sense the following partial differential equations (this is discussed in the next section):

$$\frac{\partial P^X(t,x,dy)}{\partial t} = \mathscr{A}\, P^X(t,x,dy), \quad (\mathscr{A} \text{ acts on the variable } x) \tag{7.23}$$

$$\frac{\partial P^X(t,x,dy)}{\partial t} = \mathscr{A}^*\, P^X(t,x,dy), \quad (\mathscr{A}^* \text{ acts on the variable } y) \tag{7.24}$$

where $\mathscr{A}^*$ is the formal adjoint to $\mathscr{A}$. Equation (7.23), in which $\mathscr{A}$ acts on the backward variable x, is called a *backward Kolmogorov equation*. Equation (7.24), where $\mathscr{A}^*$ acts on the forward variable y, is called a *forward Kolmogorov equation* or, in the Physics literature, a Fokker-Planck equation. We call them *Fokker-Planck-Kolmogorov* equations, of for short, *FPK equations*.

In subsequent sections we will establish FPK equations, including space and time fractional, as well as DODE FPK equations, associated with various classes of SDEs.

7.4 FPK equations associated with SDEs driven by Brownian motion

There are different ways of derivation of FPK equations. Below we show the derivation of FPK equations (7.23) and (7.24) based on Itô's formula (7.14).

Suppose a process X_t solves the stochastic differential equation

$$dX_t = b(X_t)dt + \sigma(X_t)dB_t, \quad X_0 = x. \tag{7.25}$$

Consider the conditional expectation

$$u(t,x) = \mathbb{E}^x[f(X_t)] = \mathbb{E}[f(X_t)|X_0 = x], \tag{7.26}$$

of $f(X_t)$ given $X_0 = x$. Here $f(y)$, $y \in \mathbb{R}^n$, is an arbitrary twice differentiable function with compact support. If $p(t,y;x)$ is the density of X_t (in the variable y), given that $X_0 = x$ (transition probability from x to y during the time period t), then we have

$$u(t,x) = \int_{\mathbb{R}} f(y)p(t,y;x)dy.$$

We will show that $u(t,x)$ satisfies the following Cauchy problem

$$\frac{\partial u}{\partial t} = -\frac{\partial}{\partial x}\Big(b(x)u(t,x)\Big) + \frac{1}{2}\frac{\partial^2}{\partial x^2}\Big(\sigma^2(x)u(t,x)\Big), \quad u(0,x) = f(x).$$

Applying Ito's formula (Proposition 7.3) to the stochastic process X_t, that is

$$X_t = X_0 + \int_0^t b(X_s)ds + \int_0^t \sigma(X_s)dB_s,$$

we have

$$f(X_t) = f(X_0) + \int_0^t \Big[f'(X_s)b(X_s) + \frac{1}{2}f''(X_s)\sigma^2(X_s)\Big]ds + \int_0^t f'(X_s)\sigma(X_s)dB_s.$$

Then, the expectation of $f(X_t)$ under the condition $X_0 = x$ becomes

$$\begin{aligned}\mathbb{E}[f(X_t)|X_0 = x] &= f(x) + \mathbb{E}[\int_0^t [f'(X_s)b(X_s) + \frac{1}{2}f''(X_s)\sigma^2(X_s)]ds] \\ &= \int_0^t \int_{\mathbb{R}} [f'(y)b(y) + \frac{1}{2}f''(y)\sigma^2(y)]p(s,y;x)dyds, \end{aligned} \tag{7.27}$$

since

$$\mathbb{E}[\int_0^t f'(X_s)\sigma(X_s)dB_s] = 0,$$

due to the martingale property of Itô's stochastic integral. Therefore, differentiating (7.27) with respect to the variable t, we have

$$\int_{\mathbb{R}} f(y)\frac{\partial p(t,y;x)}{\partial t}dy = \int_{\mathbb{R}} f(y)A^*p(t,y;x)dy,$$

where

$$A^*\varphi(y) = -\frac{\partial}{\partial y}\Big(b(y)\varphi(y)\Big) + \frac{1}{2}\frac{\partial^2}{\partial y^2}\Big(\sigma^2(y)\varphi(y)\Big).$$

Due to arbitrariness of f it follows from the latter that

$$\frac{\partial p(t,y;x)}{\partial t} = A^* p(t,y;x), \tag{7.28}$$

in the weak sense, and $p(t,y;x)$ as the density of X_t with the condition $X_0 = x$, satisfies the initial condition

$$p(0,y;x) = \delta_x(y),$$

where $\delta_x(y)$ is the Dirac delta function concentrated at x. Equation (7.28) is the Fokker-Planck, or Kolmogorov forward equation associated with stochastic differential equation (7.25).

Now let A be the formal adjoint operator to A^*, that is

$$A\phi(x) = b(x)\frac{\partial \phi(x)}{\partial x} + \frac{\sigma^2(x)}{2}\frac{\partial^2 \phi(x)}{\partial x^2}.$$

Then, as an adjoint equation to (7.28), we obtain the backward Kolmogorov equation in the form

$$\frac{\partial p(t,y;x)}{\partial t} = Ap(t,y;x),$$

with the initial condition

$$p(0,y;x) = \delta_y(x).$$

In the n-dimensional case the operator A takes the form (c.f. (7.17))

$$A\phi(x) = \sum_{j=1}^{n} b_j(x)\frac{\partial \phi(x)}{\partial x_j} + \frac{1}{2}\sum_{i,j=1}^{n} a_{ij}(x)\frac{\partial^2 \phi(x)}{\partial x_i \partial x_j}, \tag{7.29}$$

where $a_{ij}(x), i,j = 1,\dots,n$, are entries of the matrix obtained by multiplying the matrix $\sigma(x)$ by its transpose $\sigma(x)^T$.

7.5 Lévy processes and Lévy stable subordinators

Fractional Fokker-Planck-Kolmogorov equations are connected with the SDEs driven by a specific time-changed stochastic process. A time-change process is the inverse to, so-called, a stable Lévy subordinator, which is a Lévy process. Below we introduce Lévy processes and stable subordinators. Lévy processes form a wide class of stochastic processes. For us particular Lévy processes, namely, Lévy stable subordinators, and symmetric stable Lévy processes will be of interest.

Definition 7.3. By definition, a *Lévy process* $L_t \in \mathbb{R}^n$, $t \geq 0$, is an adapted stochastic process satisfying the following conditions:

1. $L_0 = 0$;
2. has independent stationary increments;
3. for all $\varepsilon, t > 0$, $\lim_{s \to t} \mathbb{P}(|L_t - L_s| > \varepsilon) = 0$.

Comparing with Definition 7.2 of Brownian motion, we notice that a Lévy process is not required to have the density (7.5), and the path wise continuity condition is weakened to the "continuity in probability." Hence, Lévy processes may have jumps in a countable number of points. To study jumps it is convenient to introduce a class of càdlàg processes. A càdlàg process, by definition, is right continuous with left limits. Any Lévy process has a càdlàg modification, which is again Lévy process [App09]. Moreover, any Lévy process is a semimartingale. (See the definition in Section 7.16 "Additional notes").

Lévy processes are characterized by three parameters (b, Σ, ν), called a *characteristic triple*, where $b \in \mathbb{R}^n$, Σ is a nonnegative definite $(n \times n)$-matrix, and ν is a measure defined on $\mathbb{R}^n \setminus \{0\}$, such that

$$\int_{\mathbb{R}^n} \min(1, |x|^2) d\nu < \infty. \tag{7.30}$$

The measure ν is called a *Lévy measure.* The Lévy-Khintchine formula characterizes a Lévy process (as an infinitely divisible process) in terms of its characteristic function

$$\Phi_t(\xi) = \mathbb{E}(e^{i\xi L_t}) = e^{t\Psi(\xi)}, \tag{7.31}$$

with

$$\Psi(\xi) = i(b, \xi) - \frac{1}{2}(\Sigma\xi, \xi) + \int_{\mathbb{R}^n \setminus \{0\}} (e^{i(w,\xi)} - 1 - i(w, \xi)\chi_{(|w| \leq 1)}(w))\nu(dw). \tag{7.32}$$

The function Ψ is called the *Lévy symbol* of L_t.

Another characterization of Lévy processes is given by the Lévy-Itô decomposition theorem, which states that

$$L_t = b_0 t + \sigma B_t + \int_{|w|<1} w\tilde{N}(t, dw) + \int_{|w| \geq 1} wN(t, dw), \tag{7.33}$$

where $b_0 \in \mathbb{R}^n$, σ is an $n \times m$-matrix such that $\sigma\sigma^T = \Sigma$, B_t is an m-dimensional Brownian motion, and $N(t, dw)$ is a Poisson random measure and $\tilde{N}(t, dw) = N(t, dw) - t\nu(dw)$ is a compensated Poisson martingale-valued measure; see "Additional notes" for the definition.

The first two terms in equations (7.32) and (7.33) characterize a Brownian component of the Lévy process and the other terms are responsible for jumps. In Lévy-Ito's decomposition (7.33) small and large jumps are classified by the third and fourth terms.

Consider some examples of Lévy processes. Two important subclasses of Lévy processes (examples 2 and 3 below), called *symmetric α-stable Lévy processes* and *β-stable Lévy subordinators* will essentially be used in our further considerations. Symmetric α-stable processes will be used as alternatives to Brownian motion, and inverses to stable subordinators as time-change processes.

Example 7.3. 1. Let in the triple (b,Σ,ν) the Lévy measure $\nu \equiv 0$. Then the corresponding Lévy process is Brownian motion, that is $L_t = B_t$, with the drift b and the correlation matrix Σ. This is the only class of Lévy processes with no jump components.

2. *Lévy's symmetric α-stable processes.* Let $b = 0$, $\Sigma = 0$ in the triple (b,Σ,ν), and the Lévy measure ν is defined so that the Lévy symbol is

$$\Psi(\xi) = -|\xi|^\alpha, \quad 0 < \alpha < 2. \tag{7.34}$$

In this case the corresponding Lévy process is called a symmetric α-stable process. Let, for example, the Lévy measure ν depend only on the radial variable $r = \sqrt{x_1^2 + \cdots + x_n^2}$, and be defined as

$$\nu(dx) = \frac{C_\alpha dx}{|x|^{\alpha+n}},$$

where C_α is the normalizing constant specified below. This measure satisfies the condition (7.30) if $0 < \alpha < 2$. One can show that in this case the Lévy symbol has the form (7.34). Indeed, it follows from (7.32) that

$$\Psi(\xi) = C_\alpha \int\limits_{|w|\le 1} \frac{e^{i(w,\xi)} - 1 - i(w,\xi)}{|w|^{\alpha+n}} dw + C_\alpha \int\limits_{|w|>1} \frac{e^{i(w,\xi)} - 1}{|w|^{\alpha+n}} dw.$$

The substitution $w = x/|\xi|$, $\xi \neq 0$, leads

$$\Psi(\xi) = C_\alpha |\xi|^\alpha \left(\int\limits_{|x|\le|\xi|} \frac{e^{i(x,\theta)} - 1 - i(x,\theta)}{|x|^{\alpha+n}} dx + \int\limits_{|x|>|\xi|} \frac{e^{i(x,\theta)} - 1}{|x|^{\alpha+n}} dx \right),$$

where $|\theta|$ is a point on the unit sphere in $\mathbb{R}^n$ with the center at the origin, and therefore, the expression in parentheses does not depend on θ. Taking into account the equality

$$\int_{\mathbb{R}^n} \frac{(x,\theta)(I_{|x|\le 1} - I_{|x|\le|\xi|})}{|x|^{n+\alpha}} dx = 0,$$

one has

$$\Psi(\xi) = C_\alpha |\xi|^\alpha \int\limits_{\mathbb{R}^n} \left(e^{i(x,\theta)} - 1 - i(x,\theta) I_{|x|\le 1} \right) \frac{dx}{|x|^{\alpha+n}} dx = -|\xi|^\alpha,$$

where we set

$$C_\alpha = -\left[\int\limits_{\mathbb{R}^n} \left(e^{i(x,\theta)} - 1 - i(x,\theta) I_{|x|\le 1} \right) \frac{dx}{|x|^{\alpha+n}} dx \right]^{-1}.$$

We denote Lévy's m-dimensional symmetric α-stable process by $\mathbb{L}_{\alpha,t}$. A stochastic differential equation driven by $\mathbb{L}_{\alpha,t}$ we write in the form

$$dX_t = g(X_{t-})d\mathbb{L}_{\alpha,t}, \quad X_{t=0} = X_0,$$

where $g(x)$ is $(n \times m)$-matrix valued function satisfying the Lipschitz continuity and growth conditions.

3. *Lévy's β-stable subordinators.* One-dimensional nonnegative, nondecreasing Lévy processes are called *subordinators.* This implies Σ to be the zero matrix, $b \geq 0$, $\nu(-\infty,0) = 0$ and $\int \min(1,|x|)\nu(dx) < \infty$. We will not consider in this book subordinators in such a general form. We will be interested only in the subclass of Lévy's *stable* subordinators. For $\beta \in (0,1)$, a *β-stable subordinator* is a strictly increasing subordinator W_t, which is self-similar, i.e. $W_t = t^{1/\beta}W_1$ in the sense of finite-dimensional distributions, and with the Lévy symbol $\Psi(s) = -s^\beta$, $s \geq 0$. Due to equations (7.31) and (7.32), the latter can be written as

$$\mathbb{E}[e^{-sW_1}] = e^{-s^\beta}, \quad s \geq 0. \tag{7.35}$$

This in terms of the density function $f_{W_1}(\tau)$, $\tau \geq 0$, of the random variable W_1 takes the form

$$\mathbb{E}[e^{-sW_1}] = \int_0^\infty e^{-s\tau} f_{W_1}(\tau)d\tau = L[f_{W_1}](s) = e^{-s^\beta},$$

where $L[f_{W_1}](s)$ is the Laplace transform of $f_{W_1}(\tau)$. Since the Laplace transform of $f_{W_1}(\tau)$ decays exponentially at infinity, it follows from the general theory of Laplace transforms and Watson's lemma that $f_{W_1}(\tau)$ is infinitely differentiable on $(0,\infty)$, and vanishes at zero at an exponential rate. In fact, $f_{W_1}(\tau)$ has the following asymptotic behavior at zero and infinity: [MLP01, UZ99]:

$$f_{W_1}(\tau) \sim \frac{(\frac{\beta}{\tau})^{\frac{2-\beta}{2(1-\beta)}}}{\sqrt{2\pi\beta(1-\beta)}} e^{-(1-\beta)(\frac{\tau}{\beta})^{-\frac{\beta}{1-\beta}}}, \quad \tau \to 0; \tag{7.36}$$

$$f_{W_1}(\tau) \sim \frac{\beta}{\Gamma(1-\beta)\tau^{1+\beta}}, \quad \tau \to \infty. \tag{7.37}$$

The Lévy-Ito decomposition (7.33), in fact, gives a clue how the stochastic differential equation (7.11) driven by Brownian motion can be extended to SDEs driven by a Lévy process L_t. Namely, by SDE driven by a Lévy process we understand the equation

$$X_t = X_0 + \int_0^t b(X_s)ds + \int_0^t \sigma(X_s)dB_s + \int_0^t \int_{|w|<1} H(X_{-s}, w)\tilde{N}(ds, dw)$$
$$+ \int_0^t \int_{|w|\geq 1} K(X_{-s}, w)N(ds, dw), \tag{7.38}$$

where the continuous mappings $b(x) : \mathbb{R}^n \to \mathbb{R}^n$, $\sigma(x) : \mathbb{R}^n \to \mathbb{R}^{n\times m}$, and $G(x,w) = \chi_{(|w|<1)}(w)H(x,w) + \chi_{(|w|\geq 1)}(w)K(x,w) : \mathbb{R}^n \times \mathbb{R}^n \to \mathbb{R}^n$ satisfy the following Lipschitz and growth conditions, respectively:

$$|b(x)-b(y)|^2 + \|\sigma(x)-\sigma(y)\|^2 + \int_{|w|<1} |H(x,w)-H(y,w)|^2 \nu(dw)$$

$$\leq C_1|x-y|^2, \quad \forall x,y \in \mathbb{R}^n; \tag{7.39}$$

$$\int_{|w|<1} |H(x,w)|^2 \nu(dw) \leq C_2(1+|x|^2), \quad \forall x \in \mathbb{R}^n. \tag{7.40}$$

7.6 Inverse processes to Lévy's stable subordinators

Let E_t be the first hitting time process for a stable subordinator W_t with stability index $\beta \in (0,1)$. The process E_t is also called an inverse to W_t. The relation between E_t and W_t can be expressed as $E_t = \min\{\tau : W_\tau > t\}$. Since W_t is strictly increasing, its inverse process E_t is continuous and nondecreasing, but not a Lévy process. Likewise the time-changed process B_{E_t} is also not a Lévy process (see details in [HKU10]).

We denote by $f_t(\tau)$ the density function of E_t. If $f_{W_1}(t)$ is the density function of W_1, then

$$\begin{aligned} f_t(\tau) &= \frac{\partial}{\partial\tau}P(E_t \leq \tau) = \frac{\partial}{\partial\tau}(1 - P(W_\tau < t)) \\ &= -\frac{\partial}{\partial\tau}P(W_1 < \frac{t}{\tau^{1/\beta}}) = -\frac{\partial}{\partial\tau}[Jf_{W_1}](\frac{t}{\tau^{1/\beta}}) \\ &= -\frac{\partial}{\partial\tau}\int_0^{\frac{t}{\tau^{1/\beta}}} f_{W_1}(u)du = \frac{t}{\beta\tau^{1+\frac{1}{\beta}}} f_{W_1}(\frac{t}{\tau^{\frac{1}{\beta}}}), \quad \tau \geq 0. \end{aligned} \tag{7.41}$$

Since $f_{W_1}(u) \in C^\infty(0,\infty)$, it follows from representation (7.41) that $f_t(\tau) \in C^\infty(\mathbb{R}^2_+)$, where $\mathbb{R}^2_+ = (0,\infty)\times(0,\infty)$. Further properties of $f_t(\tau)$ are represented in the following lemma.

Lemma 7.1. *Let $f_t(\tau)$ be the function given in* (7.41). *Then*

(a) $\lim_{t\to+0} f_t(\tau) = \delta_0(\tau)$ *in the sense of the topology of the space of Schwartz distributions* $\mathcal{D}'(\mathbb{R})$;

(b) $\lim_{\tau\to+0} f_t(\tau) = \frac{t^{-\beta}}{\Gamma(1-\beta)}$, $t > 0$;

(c) $\lim_{\tau\to\infty} f_t(\tau) = 0$, $t > 0$;

(d) $L_{t\to s}[f_t(\tau)](s) = s^{\beta-1}e^{-\tau s^\beta}$, $s > 0$, $\tau \geq 0$,

where $L_{t\to s}$ denotes the Laplace transform with respect to the variable t.

Proof. (a) Let $\psi(\tau)$ be an infinitely differentiable function with compact support. We have to show that $\lim_{t\to+0} < f_t, \psi >= \psi(0)$. Here $< f_t, \psi >$ denotes the value of $f_t \in \mathscr{D}'(\mathbb{R})$ on ψ. We have

$$\lim_{t\to+0} < f_t(\tau), \psi(\tau) > = \lim_{t\to+0} \int_0^\infty f_t(\tau)\psi(\tau)d\tau = \lim_{t\to+0} \int_0^\infty f_{W_1}(u)\psi((\frac{t}{u})^\beta)du$$
$$= \psi(0) \int_0^\infty f_{W_1}(u)du = \psi(0).$$

Parts (b) and (c) follow from asymptotic relations (7.37) and (7.36), respectively. Part (d) is straightforward. One needs just to compute the Laplace transform of $f_t(\tau)$ using the representation $f_t(\tau) = -\frac{\partial}{\partial\tau}[Jf_{W_1}](\frac{t}{\tau^{1/\beta}})$. Indeed,

$$L_{t\to s}[f_t(\tau)](s) = L_{t\to s}\left[-\frac{\partial}{\partial\tau}[Jf_{W_1}](\frac{t}{\tau^{1/\beta}})\right](s)$$
$$= -\frac{1}{s}\frac{\partial}{\partial\tau}L_{t\to s}\left[f_{W_1}(\frac{t}{\tau^{1/\beta}})\right](s)$$
$$= -\frac{1}{s}\frac{\partial}{\partial\tau}\left(e^{-\tau s^\beta}\right) = s^{\beta-1}e^{-\tau s^\beta}.$$

Due to part (b) of Lemma 7.1, $f_t \in C^\infty(0,\infty)$ for each fixed $\tau \geq 0$. Hence, the fractional derivative $D^\beta_{*,t}f_t(\tau)$ in the variable t is meaningful and is a generalized function of the variable τ. Notice also that Part (d) means

$$L^{-1}_{s\to t}\left[s^{\beta-1}e^{-\tau s^\beta}\right](t) = f_t(\tau).$$

Lemma 7.2. *The function $f_t(\tau)$ defined in* (7.41) *for each $t > 0$ satisfies the equation*

$$D^\beta_{*,t}f_t(\tau) = -\frac{\partial}{\partial\tau}f_t(\tau) - \frac{t^{-\beta}}{\Gamma(1-\beta)}\delta_0(\tau). \tag{7.42}$$

Remark 7.1. Equality (7.42) is understood in the sense of distributions. The fractional derivative $D^\beta_{*,t}$ and the partial derivative $\frac{\partial}{\partial\tau}$ in this equation are in the usual sense, since $f_t(\tau) \in C^\infty(\mathbb{R}^2_+)$.

Proof. The Laplace transform (in variable t) of $D^\beta_{*,t}f_t(\tau)$, due to formula (3.55) in the case $0 < \beta < 1$, and using Parts (a) and (d) of Lemma 7.1, equals

$$L_{t\to s}[D^\beta_{*,t}f_t(\tau)](s) = s^\beta L_{t\to s}[f_t(\tau)](s) - s^{\beta-1}\lim_{t\to+0} f_t(\tau)$$
$$= s^{2\beta-1}e^{-\tau s^\beta} - s^{\beta-1}\delta_0(\tau), \quad s > 0.$$

On the other hand, computing the inverse Laplace transform of both sides of this equality, one obtains

$$
\begin{aligned}
D^{\beta}_{*,t} f_t(\tau) &= L^{-1}_{s\to t}\left[s^{2\beta-1}e^{-\tau s^{\beta}} - s^{\beta-1}\delta_0(\tau)\right](t) \\
&= L^{-1}_{s\to t}\left[-s^{\beta-1}\frac{\partial}{\partial\tau}e^{-\tau s^{\beta}}\right](t) - \delta_0(\tau)L^{-1}_{s\to t}\left[\frac{1}{s^{1-\beta}}\right](t) \\
&= -\frac{\partial}{\partial\tau}L^{-1}_{s\to t}[s^{\beta-1}e^{-\tau s^{\beta}}](t) - \delta_0(\tau)\frac{t^{-\beta}}{\Gamma(1-\beta)} \\
&= -\frac{\partial}{\partial\tau}f_t(\tau) - \frac{t^{-\beta}}{\Gamma(1-\beta)}\delta_0(\tau),
\end{aligned}
$$

completing the proof.

Proposition 7.4. *The function $f_t(\tau)$ defined in* (7.41) *for each $t > 0$ satisfies the equation*

$$
f_t(\tau) = -\frac{\partial}{\partial\tau}J^{\beta}_t f_t(\tau). \tag{7.43}
$$

Proof. Applying the fractional integration operator J^{β} to equation (7.42), we have

$$
f_t(\tau) - \lim_{t\to 0+} f_t(\tau) = -\frac{\partial}{\partial\tau}J^{\beta}_t f_t(\tau) - \frac{\delta_0(\tau)}{\Gamma(1-\beta)}J^{\beta}_t t^{-\beta},
$$

in the sense of distributions. Due to part (a) of Lemma 7.1 we have $\lim_{t\to 0+} f_t(\tau) = \delta_0(\tau)$. This fact together with the equation $J^{\beta}t^{-\beta} = \Gamma(1-\beta)$ implies (7.43).

7.7 Fractional FPK equations

Suppose $X_t \in \mathbb{R}^n$ is a solution to the stochastic differential equation

$$
dX_t = b(X_t)dt + \sigma(X_t)dB_t, \quad X_{t=0} = x,
$$

where vector-functions $b : \mathbb{R}^n \to \mathbb{R}^n$ and $\sigma : \mathbb{R}^n \to \mathbb{R}^{n\times m}$ satisfy the Lipschitz and linear growth conditions, B_t is an m-dimensional Brownian motion, and $x \in \mathbb{R}^n$ is a fixed point. We have seen above (Section 7.3) that in this case the associated FPK is

$$
\frac{\partial u(t,x)}{\partial t} = \mathscr{A}u(t,x), \quad t > 0, x \in \mathbb{R}^n, \tag{7.44}
$$

$$
u(0,x) = \varphi(x), \quad x \in \mathbb{R}^n; \tag{7.45}
$$

where $\mathscr{A}$ is a differential operator defined in (7.17). The solution $u(t,x)$ to Cauchy problem (7.44)–(7.45) is connected with the stochastic process X_t via the relationship $u(t,x) = \mathbb{E}[\varphi(X_t)|X_0 = x]$.

Below we will show that the FPK equation associated with the time-changed process X_{E_t}, where E_t is the process inverse to the Lévy's stable subordinator with the stability index β, has the form

$$
D^{\beta}_{*}v(t,x) = \mathscr{A}v(t,x), \quad t > 0, x \in \mathbb{R}^n, \tag{7.46}
$$

with the initial condition

$$v(0,x) = \varphi(x), \tag{7.47}$$

where D_*^β is the fractional derivative in the sense of Caputo-Djrbashian. We note that solutions to equations (7.44) and (7.46) are connected via a certain relationship. Namely, a solution $v(t,x)$ to equation (7.46) satisfying the initial condition (7.47) can be represented through the solution $u(t,x)$ to equation (7.44), satisfying the same initial condition (7.45), by the formula

$$v(t,x) = \int_0^\infty f_t(\tau)u(\tau,x)d\tau,$$

where $f_t(\tau)$ is the density function of E_t for each fixed $t > 0$. Indeed, conditioning on the event $[E_t = \tau]$, $\tau \in (0,\infty)$,

$$\begin{aligned} v(t,x) = \mathbb{E}^x(\varphi(X_{E_t})) &= \int_0^\infty \mathbb{E}^x\Big(\varphi(X_{E_t})|E_t = \tau\Big)\mathbb{P}(E_t \in d\tau) \\ &= \int_0^\infty u(\tau,x)f_t(\tau)d\tau. \end{aligned}$$

Using Lemma 7.2, we have

$$\begin{aligned} D_{*,t}^\beta v(t,x) &= \int_0^\infty D_{*,t}^\beta f_t(\tau)u(\tau,x)d\tau \\ &= -\int_0^\infty \left[\frac{\partial}{\partial\tau}f_t(\tau) + \frac{t^{-\beta}}{\Gamma(1-\beta)}\delta_0(\tau)\right]u(\tau,x)d\tau \\ &= -\lim_{\tau\to\infty}[f_t(\tau)u(\tau,x)] + \lim_{\tau\to 0}[f_t(\tau)u(\tau,x)] \\ &\quad + \int_0^\infty f_t(\tau)\frac{\partial}{\partial\tau}u(\tau,x)d\tau - \frac{t^{-\beta}}{\Gamma(1-\beta)}u(0,x). \end{aligned}$$

Due to Lemma 7.1, part (c) implies the first term vanishes since $u(\tau,x)$ is bounded, while part (b) implies the second and last terms cancel. Taking into account (7.44),

$$D_{*,t}^\beta v(t,x) = \int_0^\infty f_t(\tau)\mathscr{A}u(\tau,x)d\tau = \mathscr{A}v(t,x).$$

Moreover, by property (a) of Lemma 7.1,

$$\lim_{t\to+0} v(t,x) = <\delta_0(\tau),u(\tau,x)> = u(0,x) = \varphi(x).$$

Thus, we proved the following theorem:

Theorem 7.1. *Let $u(t,x)$ be a solution of Cauchy problem* (7.44)–(7.45). *Then the function $v(t,x) = \int_0^\infty f_t(\tau)u(\tau,x)$, where $f_t(\tau)$ is the density function of E_t, satisfies the Cauchy problem for fractional order differential equations* (7.46)–(7.47).

7.8 Mixed time-changed processes

Let W_t be an $(\mathscr{F}_t)$-adapted strictly increasing càdlàg process, or equivalently, a strictly increasing $(\mathscr{F}_t)$-semimartingale, and $E_t = \inf\{\tau \geq 0 : W_\tau > t\}$, the first hitting time process. Then it follows from the definition of E_t that E_t is a continuous $(\mathscr{F}_t)$-time-change and $\mathbb{P}(E_t \leq \tau) = \mathbb{P}(W_\tau > t)$. If $W_t = W_{1,t} + W_{2,t}$, where $W_{1,t}$ and $W_{2,t}$ are independent $(\mathscr{F}_t)$-adapted strictly increasing càdlàg processes, then W_t also possesses the same property and its inverse process, E_t, satisfies $\mathbb{P}(E_t \leq \tau) = 1 - (F_\tau^{(1)} * F_\tau^{(2)})(t)$, where for $k = 1,2$, $F_\tau^{(k)}(t) = \mathbb{P}(W_{k,\tau} \leq t)$ with density $f_\tau^{(k)}$, and $*$ denotes convolution of cumulative distribution functions or densities, whichever is required. For notational convenience, if $a, b > 0$, let

$$\left[F_1^{(1)}\left(\frac{\cdot}{a}\right) * F_1^{(2)}\left(\frac{\cdot}{b}\right)\right](t) := \int_{s=0}^{s=t} F_1^{(1)}\left(\frac{t-s}{a}\right) dF_1^{(2)}\left(\frac{s}{b}\right),$$

which through the density functions can also be written as

$$\left[F_1^{(1)}\left(\frac{\cdot}{a}\right) * F_1^{(2)}\left(\frac{\cdot}{b}\right)\right](t) = \frac{1}{b}\int_{s=0}^{s=t} (Jf_1^{(1)})\left(\frac{t-s}{a}\right) f_1^{(2)}\left(\frac{s}{b}\right) ds,$$

where J is the usual integration operator.

Lemma 7.3. *Let $W_t = c_1 W_{1,t} + c_2 W_{2,t}$, where c_1, c_2 are positive constants and $W_{1,t}$ and $W_{2,t}$ are independent stable subordinators with respective indices β_1 and β_2 in $(0,1)$. Then the inverse E_t of W_t satisfies*

$$\mathbb{P}(E_t \leq \tau) = 1 - \left[F_1^{(1)}\left(\frac{\cdot}{c_1 \tau^{\frac{1}{\beta_1}}}\right) * F_1^{(2)}\left(\frac{\cdot}{c_2 \tau^{\frac{1}{\beta_2}}}\right)\right](t) \tag{7.48}$$

and has density

$$f_{E_t}(\tau) = -\frac{\partial}{\partial \tau}\left\{\frac{1}{c_2 \tau^{\frac{1}{\beta_2}}}\left[(Jf_1^{(1)})\left(\frac{\cdot}{c_1 \tau^{\frac{1}{\beta_1}}}\right) * f_1^{(2)}\left(\frac{\cdot}{c_2 \tau^{\frac{1}{\beta_2}}}\right)\right](t)\right\}. \tag{7.49}$$

Proof. Since $W_{1,\tau}$ and $W_{2,\tau}$ are independent and self-similar processes,

$$\begin{aligned}\mathbb{P}(E_t \leq \tau) &= \mathbb{P}(W_\tau > t) = 1 - \mathbb{P}\left(c_1 \tau^{\frac{1}{\beta_1}} W_{1,1} + c_2 \tau^{\frac{1}{\beta_2}} W_{2,1} \leq t\right) \\ &= 1 - \left[F_1^{(1)}\left(\frac{\cdot}{c_1 \tau^{\frac{1}{\beta_1}}}\right) * F_1^{(2)}\left(\frac{\cdot}{c_2 \tau^{\frac{1}{\beta_2}}}\right)\right](t),\end{aligned}$$

from which (7.49) follows immediately upon differentiating with respect to τ.

The following lemma provides an estimate for the density function $f_{E_t}(\tau)$.

Lemma 7.4. *For any $t < \infty$, the density $f_{E_t}(\tau)$ in (7.49) is bounded and there exist a number $\beta \in (0,1)$ and positive constants C, k, not depending on τ, such that*

$$f_{E_t}(\tau) \le C\exp\left(-k\tau^{\frac{1}{1-\beta}}\right) \tag{7.50}$$

for τ large enough.

Proof. Suppose for clarity that $0 < \beta_1 < \beta_2 < 1$ in representation (7.49). It follows that $f_{E_t}(\tau) = I_1 + I_2 + I_3$, where

$$I_1 = \frac{1}{\beta_2 c_2 \tau^{1+\frac{1}{\beta_2}}} \int_0^t (Jf_1^{(1)})\left(\frac{s}{c_1\tau^{\frac{1}{\beta}_1}}\right) f_1^{(2)}\left(\frac{t-s}{c_2\tau^{\frac{1}{\beta}_2}}\right) ds,$$

$$I_2 = \frac{1}{\beta_1 c_1 c_2 \tau^{1+\frac{1}{\beta_1}+\frac{1}{\beta_2}}} \int_0^t s\cdot f_1^{(1)}\left(\frac{s}{c_1\tau^{\frac{1}{\beta}_1}}\right) f_1^{(2)}\left(\frac{t-s}{c_2\tau^{\frac{1}{\beta}_2}}\right) ds,$$

and

$$I_3 = \frac{1}{c_2\tau^{\frac{1}{\beta_2}}} \int_0^t s\cdot (Jf_1^{(1)})\left(\frac{s}{c_1\tau^{\frac{1}{\beta}_1}}\right) (f_1^{(2)})'\left(\frac{t-s}{c_2\tau^{\frac{1}{\beta}_2}}\right) ds.$$

It is easy to see that integration by parts reduces I_3 to the sum of integrals of types I_1 and I_2, namely, $I_3 = \beta_2 c_2 \tau^{1+\frac{1}{\beta}_2} I_1 + \beta_1 \tau I_2$. Therefore, it suffices to estimate I_1 and I_2. First notice that both functions $f_1^{(1)}$, $f_1^{(2)}$ are continuous on $[0,\infty)$, and $Jf_1^{(1)}(t) \le 1$. Consequently, in accordance with the mean value theorem, there exist numbers $s_*, s_{**} \in (0,t)$ such that

$$I_1 \le \frac{t}{\beta_2 c_2 \tau^{1+\frac{1}{\beta_2}}} f_1^{(2)}\left(\frac{s_*}{c_2\tau^{\frac{1}{\beta}_2}}\right), \tag{7.51}$$

and

$$I_2 = \frac{t s_{**}}{\beta_1 c_1 c_2 \tau^{1+\frac{1}{\beta_1}+\frac{1}{\beta_2}}} f_1^{(1)}\left(\frac{s_{**}}{c_1\tau^{\frac{1}{\beta}_1}}\right) f_1^{(2)}\left(\frac{t-s_{**}}{c_2\tau^{\frac{1}{\beta}_2}}\right). \tag{7.52}$$

For τ small enough, (7.37) implies

$$I_1 \le C_1, \; I_2 \le C_2\tau \text{ and } I_3 \le C_3\tau^2,$$

where C_1, C_2, and C_3 are constants not depending on τ. These estimates and continuity of convolution imply boundedness of $f_{E_t}(\tau)$ for any $\tau < \infty$.

Now suppose that τ is large enough. Then taking into account (7.36) in (7.51) and (7.52), it is not hard to verify that

$$I_1 \le \frac{C_3}{\tau^{\frac{1-2\beta_2}{2(1-\beta_2)}}} \exp\left(-k_1\tau^{\frac{1}{1-\beta_2}}\right),$$

and

$$I_2 \le \frac{C_4}{\tau^{1-\frac{\beta_1}{2(1-\beta_1)}-\frac{\beta_2}{2(1-\beta_2)}}} \exp\left(-k_2(\tau^{\frac{1}{1-\beta_1}}+\tau^{\frac{1}{1-\beta_2}})\right),$$

where C_3, C_4 and k_1, k_2 are positive constants not depending on τ. Selecting $\beta = \beta_1 = \min(\beta_1, \beta_2)$, $C = \max(C_3, C_4)$, and $k = \min(k_1, 2k_2) - \varepsilon$, where $\varepsilon \in (0, \min(k_1, 2k_2))$, yields (7.50).

Two lemmas proved above can be extended to weighted averages of an arbitrary number of independent stable subordinators. It is easy to verify that the process $W_t = \sum_{k=1}^N c_k W_{k,t}$ satisfies

$$\ln \mathbb{E}\left[e^{-sD_t}\right]\Big|_{t=1} = -\sum_{k=1}^{N} c_k^{\beta_k} s^{\beta_k}, \ s \ge 0. \tag{7.53}$$

The function on the right-hand side of (7.53) can be expressed as the integral $-\int_0^1 s^\beta d\mu(\beta)$, with μ the finite atomic measure,

$$d\mu(\beta) = \sum_{k=1}^{N} c_k^{\beta_k} \delta_{\beta_k}(\beta) d\beta.$$

Definition 7.4. Let μ be a finite measure defined on the interval $[0,1]$. Let $\mathbb{S}$ designate the class of strictly increasing $(\mathcal{F}_t)$-semimartingales V_t, $V_0 = 0$, whose Laplace transform is given by

$$\ln L[f_{V_t}(\tau)](s) = \ln \mathbb{E}\left[e^{-sV_t}\right] = -t \int_0^1 s^\beta d\mu(\beta), \ s \ge 0,$$

where $f_{V_t}(\tau)$, $\tau > 0$, is the density function of the process V_t. This class obviously contains stable subordinators and all mixtures of finitely many independent stable subordinators. By construction, $V_0 = 0$ a.s., and V_t can be considered as a weighted mixture of independent stable subordinators. For the process $V_t \in \mathbb{S}$ corresponding to a finite measure μ, we use the notation $V_t = W_t^\mu$ to indicate this correspondence. In particular, if $d\mu(\beta) = a(\beta)d\beta$, where a is a positive continuous function on $[0,1]$, we write $V_t = W_t^a$.

Remark 7.2. Lemma 7.4 remains valid for the inverse E_t^μ of any mixture W_t^μ of independent Lévy's stable subordinators with a mixing measure μ whose support $\operatorname{supp}\mu \subset (0,1]$.

7.9 Distributed order FPK equations

The technique used in Section 7.7 extends to the more general case when the time-change process is the first hitting time for an arbitrary mixture of independent stable subordinators. Let $\phi(s) = \int_0^1 s^\beta d\mu(\beta)$, where μ is a finite measure with $supp\,\mu \subset (0,1]$. Let W_t^μ be a nonnegative stochastic process satisfying $\mathbb{E}(e^{-sW_t^\mu}) = e^{-t\phi(s)}$, and $E_t^\mu = \min\{\tau : W_\tau^\mu > t\}$. The process W_t^μ represents a mixture of independent stable subordinators with a mixing measure μ.

Theorem 7.2. *Let $u(t,x)$ be a solution of the Cauchy problem*

$$\frac{\partial u(t,x)}{\partial t} = \mathscr{A} u(t,x), \quad t > 0,\ x \in \mathbb{R}^n, \tag{7.54}$$

$$u(0,x) = \varphi(x), \quad x \in \mathbb{R}^n. \tag{7.55}$$

Then the function $v(t,x) = \int_0^\infty f_t^\mu(\tau) u(\tau,x)$, where $f_t^\mu(\tau)$ is the density function of E_t^μ, satisfies the initial value problem for the distributed order differential equation

$$D_\mu v(t,x) \equiv \int_0^1 D_{*,t}^\beta v(t,x) d\mu(\beta) = \mathscr{A} v(t,x), \quad t > 0,\ x \in \mathbb{R}^n, \tag{7.56}$$

$$v(0,x) = \varphi(x),\ x \in \mathbb{R}^n. \tag{7.57}$$

The proof of this theorem requires two lemmas which generalize Lemmas 7.1 and 7.2. Define the function

$$\Phi_\mu(t) = \int_0^1 \frac{t^{-\beta}}{\Gamma(1-\beta)} d\mu(\beta), \quad t > 0. \tag{7.58}$$

Lemma 7.5. *Let $f_t^\mu(\tau)$ be the function defined in Theorem 7.1. Then*

(a) $\lim_{t\to+0} f_t^\mu(\tau) = \delta_0(\tau),\ \tau \geq 0;$

(b) $\lim_{\tau\to+0} f_t^\mu(\tau) = \Phi_\mu(t),\ t > 0;$

(c) $\lim_{\tau\to\infty} f_t^\mu(\tau) = 0,\ t > 0;$

(d) $\mathscr{L}_{t\to s}[f_t^\mu(\tau)](s) = \frac{\phi(s)}{s} e^{-\tau\phi(s)},\ s > 0,\ \tau \geq 0.$

Proof. First, notice that $f_t^\mu(\tau) = f_{E_t^\mu}(\tau) = -\frac{\partial}{\partial\tau}[J f_{W_\tau^\mu}](t)$, where J is the usual integration operator. The proofs of parts $(a)-(c)$ are similar to the proofs of parts $(a)-(c)$ of Lemma 7.1. Further, using the definition of W_t^μ,

$$\mathscr{L}_{t\to s}[f_t^\mu(\tau)](s) = -\frac{1}{s}\frac{\partial}{\partial\tau}\mathscr{L}_{t\to s}[f_{W_\tau^\mu}(t)](s) = \frac{\phi(s)}{s} e^{-\tau\phi(s)}, \quad s > 0,$$

which completes the proof.

Lemma 7.6. *The function $f_t^\mu(\tau)$ defined in Theorem 7.1 satisfies for each $t > 0$ the following equation*

$$D_{\mu,t} f_t^\mu(\tau) = -\frac{\partial}{\partial\tau} f_t^\mu(\tau) - \delta_0(\tau)\Phi_\mu(t), \tag{7.59}$$

in the sense of tempered distributions.

Proof. Integrating both sides of the equation

$$\mathscr{L}_{t\to s}\left[D^{\beta}_{*,t} f^{\mu}_{t}(\tau)\right] = s^{\beta}\mathscr{L}_{t\to s}[f^{\mu}_{t}(\tau)](s) - s^{\beta-1}\delta_0(\tau),$$

and taking into account part (d) of Lemma 7.5, yields

$$\mathscr{L}_{t\to s}\left[D_{\mu,t} f^{\mu}_{t}(\tau)\right] = \frac{\phi^2(s)}{s}e^{-\tau\phi(s)} - \frac{\phi(s)}{s}\delta_0(\tau).$$

It is easy to verify that the latter coincides with the Laplace transform of the right-hand side of (7.59).

Proof (of Theorem 7.2). Using Lemma 7.6, we have

$$\begin{aligned} D_{\mu,t}v(t,x) &= \int_0^\infty D_{\mu,t} f^{\mu}_{t}(\tau)u(\tau,x)d\tau \\ &= -\lim_{\tau\to\infty}[f^{\mu}_{t}(\tau)u(\tau,x)] + \lim_{\tau\to 0}[f^{\mu}_{t}(\tau)u(\tau,x)] \\ &+ \int_0^\infty f^{\mu}_{t}(\tau)\frac{\partial}{\partial\tau}u(\tau,x)d\tau - \Phi_{\mu}(t)u(0,x) = \int_0^\infty f^{\mu}_{t}(\tau)\frac{\partial}{\partial\tau}u(\tau,x)d\tau, \end{aligned}$$

since the limit $\lim_{\tau\to\infty}[f^{\mu}_{t}(\tau)u(\tau,x)] = 0$ due to parts (c) of Lemma 7.5, and $\lim_{\tau\to 0}[f^{\mu}_{t}(\tau)u(\tau,x)] = \Phi_{\mu}(t)u(0,x)$ due to part (b) of Lemma 7.5. Now taking into account equation (7.54),

$$D_{\mu,t}v(t,x) = \int_0^\infty f^{\mu}_{t}(\tau)\mathscr{A}u(\tau,x)d\tau = \mathscr{A}v(t,x).$$

The initial condition (7.55) is also verified by using property (a) of Lemma 7.5:

$$\lim_{t\to+0} v(t,x) =< \delta_0(\tau), u(\tau,x) >= u(0,x) = \varphi(x),$$

which completes the proof.

Corollary 7.1. *Let the Cauchy problem* (7.54)–(7.55) *represent the FPK equation associated with stochastic differential equation* $dX_t = b(X_t)dt + \sigma(X_t)dB_t$ *with the initial condition* X_0. *Then the fractional FPK equation associated with the time-changed stochastic process* $X_{E^{\mu}_{t}}$ *is given by Cauchy problem* (7.56)–(7.57).

Unfortunately, this method does not provide any information about stochastic differential equations associated with fractional order FPK equations. To establish the connection between fractional FPK equations and their associated stochastic differential equations we further need to study properties of densities of time-change processes, and to establish some auxiliary results. This is done in the next section.

7.10 Connection with semigroups and their infinitesimal generators

In Example 7.2 we saw that Brownian motion B_t is associated with a semigroup $T_t : C_0(\mathbb{R}^n) \to C_0(\mathbb{R}^n)$ (or $T_t : L_2(\mathbb{R}^n) \to L_2(\mathbb{R}^n)$) of linear bounded operators with the infinitesimal generator $\mathscr{A} = \frac{1}{2}\Delta$, which is a closed operator with the domain $\mathscr{D}(\mathscr{A}) = C_0^2(\mathbb{R}^n)$ (or $\mathscr{D}(\mathscr{A}) = H^1(\mathbb{R}^n)$).

This relationship is true for wide class of (Markovian) stochastic processes. For every Lévy process (see, e.g., [App09]) there is an associated semigroup $\{T_t, t \geq 0\}$ defined on $C_0(\mathbb{R}^n)$ with infinitesimal generator $\mathscr{A}$ whose domain contains $C_0^2(\mathbb{R}^n)$. The infinitesimal generator of the Lévy process with characteristics (b, Σ, ν) is a pseudo-differential operator $\mathscr{A} = \Psi(D_x)$ with the symbol $\Psi(\xi)$ defined in (7.32). The explicit form of this operator with the domain $C_0^2(\mathbb{R}^n)$ is

$$\Psi(D_x)\varphi(x) = \sum_{j=1}^n b_j \frac{\partial \varphi}{\partial x_j} + \frac{1}{2} \sum_{ij=1}^n \sigma_{ij} \frac{\partial^2 \varphi}{\partial x_i \partial x_j} + \int_{\mathbb{R}^n \setminus \{0\}} \left[\varphi(x-w) - \varphi(x) - \chi_{(|w| \leq 1)}(w) \sum_{j=1}^n w_j \frac{\partial \varphi}{\partial x_j} \right] \nu(dw). \quad (7.60)$$

Indeed, due to definition (7.32) of the symbol $\Psi(\xi)$, one has

$$\begin{aligned}
\Psi(D_x)\varphi(x) &= \frac{1}{(2\pi)^n} \int_{\mathbb{R}^n} e^{-i(x,\xi)} \Psi(\xi) F[\varphi](\xi) d\xi \\
&= \frac{1}{(2\pi)^n} \int_{\mathbb{R}^n} e^{-i(x,\xi)} \Big[i(b,\xi) - \frac{1}{2}(\Sigma\xi,\xi) \\
&+ \int_{\mathbb{R}^n \setminus \{0\}} \left(e^{i(w,\xi)} - 1 - i(w,\xi)\chi_{(|w|\leq 1)}(w) \right) \nu(dw) \Big] F[\varphi](\xi) d\xi \\
&= \sum_{j=1}^n b_j \frac{\partial \varphi}{\partial x_j} + \frac{1}{2} \sum_{ij=1}^n \sigma_{ij} \frac{\partial^2 \varphi}{\partial x_i \partial x_j} \\
&+ \int_{\mathbb{R}^n \setminus \{0\}} \Big[\frac{1}{(2\pi)^n} \int_{\mathbb{R}^n} e^{-i(x-w,\xi)} F[\varphi](\xi) d\xi - \frac{1}{(2\pi)^n} \int_{\mathbb{R}^n} e^{-i(x,\xi)} F[\varphi](\xi) d\xi \\
&- \chi_{(|w|\leq 1)}(w) \sum_{j=1}^n w_j \frac{\partial \varphi}{\partial x_j} \Big] \nu(dw)
\end{aligned}$$

Now changing the order of integration, valid for functions $\varphi \in \mathbb{C}_0^2(\mathbb{R}^n)$, implies (7.60).

We note that $\{T_t, t \geq 0\}$, associated with the Lévy process L_t, is also a semigroup on $L^2(\mathbb{R}^n)$ and the domain of its infinitesimal generator $\Psi(D_x)$ is the anisotropic Sobolev space $W(\mathbb{R}^n) = \{\varphi \in L^2(\mathbb{R}^n) : \int_{\mathbb{R}^n} |\Psi(\xi)|^2 |\hat{\varphi}(\xi)|^2 d\xi < \infty\}$ (see [Jac01]). If L_t is a symmetric α-stable process, then $W(\mathbb{R}^n)$ coincides with the Sobolev space $H^\alpha(\mathbb{R}^n)$.

Further, if X_t solves SDE (7.38), then $(T_t\varphi)(x) = \mathbb{E}[\varphi(X_t)|X_0 = x]$ is a strongly continuous contraction semigroup defined on the Banach space $C_0(\mathbb{R}^n)$. Moreover, its infinitesimal generator is the operator [App09, Sit05]

$$\mathscr{L}(x,D_x)\varphi(x) = \sum_{j=1}^{n} b_j(x)\frac{\partial \varphi}{\partial x_j} + \frac{1}{2}\sum_{ij=1}^{n} \sigma_{ij}(x)\frac{\partial^2 \varphi}{\partial x_i \partial x_j}$$
$$+ \int\limits_{\mathbb{R}^n\setminus\{0\}} \left[\varphi(x - G(x,w)) - \varphi(x) - i\chi_{(|w|<1)}(w)\sum_{j=1}^{n} G_j(x,w)\frac{\partial \varphi(x)}{\partial x_j}\right] \nu(dw), \tag{7.61}$$

with $G(x,w) = H(x,w)$ if $|w| < 1$, and $G(x,w) = K(x,w)$ if $|w| \geq 1$. This is a pseudo-differential operator with the symbol

$$\Psi(x,\xi) = i(b(x),\xi) - \frac{1}{2}(\Sigma(x)\xi,\xi)$$
$$+ \int_{\mathbb{R}^n\setminus\{0\}} (e^{i(G(x,w),\xi)} - 1 - i(G(x,w),\xi)\chi_{(|w|<1)}(w))\nu(dw), \tag{7.62}$$

Indeed, for $\varphi \in C_0^2(\mathbb{R}^n) \subset \mathfrak{D}(\mathscr{L}(x,D))$, one has

$$\mathscr{L}(x,D_x)\varphi(x) = \frac{1}{(2\pi)^n}\int\limits_{\mathbb{R}^n} e^{-i(x,\xi)}\Psi(x,\xi)F[\varphi](\xi)d\xi$$
$$= \frac{1}{(2\pi)^n}\int\limits_{\mathbb{R}^n} e^{-i(x,\xi)}\Big[i(b(x),\xi) - \frac{1}{2}(\Sigma(x)\xi,\xi)$$
$$+ \int_{\mathbb{R}^n\setminus\{0\}} (e^{i(G(x,w),\xi)} - 1 - i(G(x,w),\xi)\chi_{(|w|<1)}(w))\nu(dw)\Big]F[\varphi](\xi)d\xi$$
$$= \sum_{j=1}^{n} b_j(x)\frac{\partial \varphi}{\partial x_j} + \frac{1}{2}\sum_{ij=1}^{n} \sigma_{ij}(x)\frac{\partial^2 \varphi}{\partial x_i \partial x_j}$$
$$+ \frac{1}{(2\pi)^n}\int\limits_{\mathbb{R}^n} e^{-i(x,\xi)}\Big[\int\limits_{\mathbb{R}^n\setminus\{0\}} (e^{i(G(x,w),\xi)} - 1 - i(G(x,w),\xi)\chi_{(|w|<1)}(w))\nu(dw)\Big]F[\varphi](\xi)d\xi$$

Changing the order of integration in the last line, which is valid for any $\varphi \in C_0^2(\mathbb{R}^n)$, we reduce it to

$$\int_{\mathbb{R}^n\setminus\{0\}} \Big[\frac{1}{(2\pi)^n}\int\limits_{\mathbb{R}^n} e^{-i(x-G(x,w),\xi)}F[\varphi](\xi)d\xi - \frac{1}{(2\pi)^n}\int\limits_{\mathbb{R}^n} e^{-i(x,\xi)}F[\varphi](\xi)d\xi$$
$$- \chi_{(|w|<1)}(w)\sum_{j=1}^{n} G_j(x,w)\frac{\partial \varphi(x)}{\partial x_j}\Big]\nu(dw).$$

Further, using properties of Fourier transform, the expression under the outer integral in the latter can be written in the form

$$\varphi\big(x - G(x,w)\big) - \varphi(x) - \chi_{(|w|<1)}(w) \sum_{j=1}^{n} G_j(x,w) \frac{\partial \varphi(x)}{\partial x_j}.$$

Thus, the infinitesimal generator $\mathscr{L}(x,D_x)$ of the solution of SDE (7.38) has the form (7.61). Moreover, if vector-functions $b(x)$, $\Sigma(x)$, and $G(x,w)$ satisfy Lipschitz and linear growth conditions, then the mapping $\mathscr{L}(x,D_x) : C_0^2(\mathbb{R}^n) \to C_0(\mathbb{R}^n)$ is continuous, that is $C_0^2(\mathbb{R}^n) \subset \mathrm{Dom}\big(\mathscr{L}(x,\mathbf{D}_x)\big)$.

A time-changed stochastic process, in general, does not have an associated semigroup. This is the case when the time-change process E_t is the inverse to Lévy's β-stable subordinator. However, if a stochastic process X_t has an associated strongly continuous semigroup T_t with an infinitesimal generator $\mathscr{A}$, then this information can be effectively used in description of the time-changed process X_{E_t} or $X_{E_t^{\mu}}$, where μ is a mixing measure. Below we establish two important abstract theorems in this context which are required for the main results of Section 7.11.

Let $\{T_t, t \geq 0\}$ be a strongly continuous semigroup defined on a Banach space $\mathscr{X}$ with norm $\|\cdot\|$, such that the estimate

$$\|T_t \varphi\| \leq M \|\varphi\| e^{\omega t} \tag{7.63}$$

is valid for some constants $M > 0$ and $\omega \geq 0$. This assumption implies that any number s with $Re(s) > \omega$ belongs to the resolvent set $\rho(\mathscr{A})$ of the infinitesimal generator $\mathscr{A}$ of T_t and the resolvent operator is represented in the form $R(s,\mathscr{A}) = \int_0^\infty e^{-st} T_t \, dt$ [EN99].

Theorem 7.3. *Define the process $W_t = c_1 W_{1,t} + c_2 W_{2,t}$, where $W_{1,t}$ and $W_{2,t}$ are independent stable subordinators with respective indices $\beta_1, \beta_2 \in (0,1)$ and constants $c_1 > 0$, $c_2 > 0$. Let E_t be the inverse process to W_t. Suppose T_t is a strongly continuous semigroup in a Banach space $\mathscr{X}$ satisfies (7.63), and has infinitesimal generator $\mathscr{A}$ with $Dom(\mathscr{A}) \subset \mathscr{X}$. Then, for each fixed $t \geq 0$, the integral $\int_0^\infty f_{E_t}(\tau) T_\tau \varphi \, d\tau$ exists and the vector-function $v(t) = \int_0^\infty f_{E_t}(\tau) T_\tau \varphi \, d\tau$, where $\varphi \in Dom(\mathscr{A})$, satisfies the abstract Cauchy problem for the distributed order fractional differential equation*

$$C_1 D_*^{\beta_1} v(t) + C_2 D_*^{\beta_2} v(t) = \mathscr{A} v(t), \quad t > 0, \tag{7.64}$$

$$v(0) = \varphi, \tag{7.65}$$

where D_*^{β} is the fractional derivative of order β in the sense of Caputo-Djrbashian, and $C_1 = c_1^{\beta_1}$ and $C_2 = c_2^{\beta_2}$.

Proof. First, define a vector-function $p(\tau) = T_\tau \varphi$, where $\varphi \in \mathrm{Dom}(\mathscr{A})$. In accordance with the conditions of the theorem, $p(\tau)$ satisfies the abstract Cauchy problem

$$\frac{\partial p(\tau)}{\partial \tau} = \mathscr{A} p(\tau), \quad p(0) = \varphi, \tag{7.66}$$

where the operator $\mathscr{A}$ is the infinitesimal generator of T_τ. Now consider the integral $\int_0^\infty f_{E_t}(\tau)T_\tau\varphi\,d\tau$. It follows from Lemma 7.4 and condition (7.63) that

$$\left\|\int_0^\infty f_{E_t}(\tau)T_\tau\varphi\,d\tau\right\| \leq \int_0^\infty f_{E_t}(\tau)\|T_\tau\varphi\|\,d\tau \qquad (7.67)$$
$$\leq C\|\varphi\|\int_0^\infty e^{-(k\tau^{\frac{1}{1-\beta}}-\omega\tau)}\,d\tau < \infty,$$

where $\beta \in (0,1)$ and $C,\, k > 0$ are constants. Hence, the integral $\int_0^\infty f_{E_t}(\tau)T_\tau\varphi\,d\tau$ exists in the sense of Bochner for each fixed $t \geq 0$. Denote this vector-function by

$$v(t) = \int_0^\infty f_{E_t}(\tau)T_\tau\varphi\,d\tau.$$

It follows immediately from the definition of the semigroup T_t that

$$v(0) = \lim_{t\to 0+}\int_0^\infty f_{E_t}(\tau)T_\tau\varphi\,d\tau = T_0\varphi = \varphi,$$

in the norm of $\mathscr{X}$. By (7.49),

$$v(t) = -\int_0^\infty \frac{\partial}{\partial\tau}\left\{\frac{1}{c_2\tau^{\frac{1}{\beta_2}}}\left[(Jf_1^{(1)})\left(\frac{\cdot}{c_1\tau^{\frac{1}{\beta_1}}}\right) * f_1^{(2)}\left(\frac{\cdot}{c_2\tau^{\frac{1}{\beta_2}}}\right)\right](t)\right\}T_\tau\varphi\,d\tau.$$

Since

$$L\left[\frac{1}{b}\,(Jf_1^{(1)})\left(\frac{t}{a}\right) * f_1^{(2)}\left(\frac{t}{b}\right)\right](s) = \frac{1}{b}\frac{1}{as}\left(a\widetilde{f_1^{(1)}}(as)\right)\left(b\widetilde{f_1^{(2)}}(bs)\right)$$
$$= \frac{1}{s}\widetilde{f_1^{(1)}}(as)\widetilde{f_1^{(2)}}(bs),$$

using (7.35), the Laplace transform of $v(t)$ takes the form

$$\tilde{v}(s) = -\int_0^\infty \frac{\partial}{\partial\tau}\left\{\frac{1}{s}e^{-\tau c_1^{\beta_1}s^{\beta_1}}e^{-\tau c_2^{\beta_2}s^{\beta_2}}\right\}T_\tau\varphi\,d\tau \qquad (7.68)$$
$$= \left(c_1^{\beta_1}s^{\beta_1-1} + c_2^{\beta_2}s^{\beta_2-1}\right)\int_0^\infty e^{-\tau(c_1^{\beta_1}s^{\beta_1}+c_2^{\beta_2}s^{\beta_2})}T_\tau\varphi\,d\tau$$
$$= (C_1s^{\beta_1-1} + C_2s^{\beta_2-1})\,\widetilde{[T_\tau\varphi]}(C_1s^{\beta_1} + C_2s^{\beta_2})$$
$$= (C_1s^{\beta_1-1} + C_2s^{\beta_2-1})\,\tilde{p}\,(C_1s^{\beta_1} + C_2s^{\beta_2}),$$

which is well defined for all s such that $C_1s^{\beta_1} + C_2s^{\beta_2} > \omega$, where $C_k = c_k^{\beta_k}$, $k = 1,2$. On the other hand it follows from (7.66) that

$$(s - \mathscr{A})\tilde{p}(s) = \varphi, \quad \forall s > \omega. \qquad (7.69)$$

Let $\omega_0 \geq 0$ be a number such that $s > \omega_0$ iff $C_1 s^{\beta_1} + C_2 s^{\beta_2} > \omega$. Then (7.68) and (7.69) together yield

$$[C_1 s^{\beta_1} + C_2 s^{\beta_2} - \mathscr{A}]\tilde{v}(s) = (C_1 s^{\beta_1 - 1} + C_2 s^{\beta_2 - 1})\varphi, \quad s > \omega_0.$$

Writing this in the form

$$C_1[s^{\beta_1}\tilde{v}(s) - s^{\beta_1 - 1}v(0)] + C_2[s^{\beta_2}\tilde{v}(s) - s^{\beta_2 - 1}v(0)] = \mathscr{A}\tilde{v}(s), \quad s > \omega_0, \tag{7.70}$$

recalling the formula $L[D_*^\beta v(t)](s) = s^\beta L[v](s) - s^{\beta-1}v(0)$ for $0 < \beta < 1$ (see Proposition 3.11), and applying the inverse Laplace transform to both sides of equation (7.70), we obtain

$$C_1 D_*^{\beta_1} v(t) + C_2 D_*^{\beta_2} v(t) = \mathscr{A} v(t).$$

Hence $v(t)$ satisfies the Cauchy problem (7.64)–(7.65).

This theorem can easily be extended to the linear combination of a finite number of processes $W_{k,t}$, $k = 1, \ldots, N$. The proof has no essential difference.

Theorem 7.4. *Define the process $W_t = \sum_{k=1}^N c_k W_{k,t}$, where $W_{k,t}$, $k = 1, \ldots, N$, are independent stable subordinators with respective indices $\beta_k \in (0,1)$ and constants $c_k > 0$. Let E_t be the inverse process to W_t. Suppose T_t is a strongly continuous semigroup in a Banach space $\mathscr{X}$, satisfies (7.63), and has infinitesimal generator $\mathscr{A}$ with $Dom(\mathscr{A}) \subset \mathscr{X}$. Then, for each fixed $t \geq 0$, the integral $\int_0^\infty f_{E_t}(\tau) T_\tau \varphi \, d\tau$ exists and the vector-function $v(t) = \int_0^\infty f_{E_t}(\tau) T_\tau \varphi \, d\tau$, where $\varphi \in Dom(\mathscr{A})$, satisfies the abstract Cauchy problem for the distributed order fractional differential equation*

$$\sum_{k=1}^N C_k D_*^{\beta_k} v(t) = \mathscr{A} v(t), \quad t > 0,$$
$$v(0) = \varphi,$$

where D_*^β is the fractional derivative of order β in the sense of Caputo-Djrbashian, and $C_k = c_k^{\beta_k}$, $k = 1, \ldots, N$.

The next theorem provides an extension of Theorem 7.4 to an arbitrary time-change process $W_t^\mu \in \mathbb{S}$; see Definition 7.4.

Theorem 7.5. *Assume that $W_t^\mu \in \mathbb{S}$ where μ is a positive finite measure with $supp\, \mu \subset (0,1)$, and let E_t^μ be the inverse process to W_t^μ. Then the vector-function $v(t) = \int_0^\infty f_{E_t^\mu}(\tau) T_\tau \varphi \, d\tau$, where T_t and φ are as in Theorem 7.4, exists and satisfies the abstract Cauchy problem*

$$D_\mu v(t) = \int_0^1 D_*^\beta v(t) d\mu(\beta) = \mathscr{A} v(t), \quad t > 0, \tag{7.71}$$
$$v(0) = \varphi. \tag{7.72}$$

Proof. We briefly sketch the proof, since the idea is similar to the proof of Theorem 7.4. Since $supp\,\mu \subset (0,1)$, the density $f_{W_t^\mu}(\tau), \tau \geq 0$, exists and has asymptotics (7.36) with some $\beta = \beta_0 \in (0,1)$ and (7.37) with some $\beta = \beta_1 \in (0,1)$. This implies existence of the vector-function $v(t)$. Further, one can readily see that

$$v(t) = -\int_0^\infty \frac{\partial}{\partial \tau}\{Jf_{W_\tau^\mu}(t)\}(T_\tau \varphi)d\tau.$$

Now it follows from the definition of W_t^μ that the Laplace transform of $v(t)$ satisfies

$$\begin{aligned} L[v](s) &= \frac{\int_0^1 s^\beta d\mu(\beta)}{s}\int_0^\infty e^{-\tau \int_0^1 s^\beta d\mu(\beta)}(T_\tau\varphi)d\tau \\ &= \frac{\phi(s)}{s}L[p](\phi(s)), \quad s > \bar{\omega}, \end{aligned} \tag{7.73}$$

where $\phi(s) = \int_0^1 s^\beta d\mu(\beta)$, $p(t)$ is a solution to the abstract Cauchy problem (7.66), and $\bar{\omega} > 0$ is a number such that $s > \bar{\omega}$ if $\rho(s) > \omega$ ($\bar{\omega}$ is uniquely defined, since $\phi(s)$ is a strictly increasing function). Combining (7.73) and (7.69),

$$(\phi(s) - \mathscr{A})\tilde{v}(s) = \varphi\frac{\eta(s)}{s}, \quad s > \bar{\omega}. \tag{7.74}$$

Applying the Laplace transform to (7.71) yields (7.74), as desired.

Remark 7.3. If $\omega = 0$ in (7.63), that is the semigroup T_t satisfies the inequality $\|T_t\| \leq M$, then the condition $supp\,\mu \subset (0,1)$ in Theorem 7.5 can be replaced by $supp\,\mu \subset [0,1)$.

Example 1. Time-changed Lévy process. The operator $\mathscr{A}$ associated with the Lévy process L_t with characteristics (b,Σ,ν) is a pseudo-differential operator with the symbol $\Psi(\xi)$ given in (7.32). The corresponding Cauchy problem takes the form

$$\frac{\partial u(t,x)}{dt} = \mathscr{A}(D_x)u(t,x), \quad u(0,x) = \varphi(x).$$

Theorem 7.5 implies that if E_t^μ is the first hitting time of the process W_t^μ defined in this theorem and if E_t^μ is independent of L_t, then the Cauchy problem associated with the time-changed Lévy process $L_{E_t^\mu}$ is the initial value problem for the time-fractional distributed order pseudo-differential equation

$$\begin{aligned} D_\mu u(t,x) &= \mathscr{A}(D_x)u(t,x), \quad t>0, x \in \mathbb{R}^n, \\ u(0,x) &= \varphi(x), \quad x \in \mathbb{R}^n. \end{aligned}$$

7.11 Fractional Fokker-Planck-Kolmogorov equations associated with SDEs driven by a time-changed Lévy process

Suppose L_t is a Lévy process and E_t is a continuous time-change process, both with respect to a filtration $\mathscr{F}_t$. Consider the following SDE driven by the time-changed Lévy process L_{E_t}:

$$X_t = x + \int_0^t b(X_{s-})dE_s + \int_0^t \sigma(X_{s-})dB_{E_s} + \int_0^t \int_{|w|<1} H(X_{s-},w)\tilde{N}(dE_s,dw)$$
$$+ \int_0^t \int_{|w|\geq 1} K(X_{s-},w)N(dE_s,dw), \tag{7.75}$$

where the mappings $b(x) : \mathbb{R}^n \to \mathbb{R}^n$, $\sigma(x) : \mathbb{R}^n \to \mathbb{R}^{n\times m}$, and $H(x,w)$, $K(x,w) : \mathbb{R}^n \times \mathbb{R}^n \to \mathbb{R}^n$ satisfy the same conditions as in SDE (7.38). SDE (7.75) is obtained from SDE (7.38) upon replacing its driving process L_t by a time-changed process L_{E_t}. It is known [Jac79] that, if L_t is an $(\mathscr{F}_t)$-semimartingale and E_t is a continuous time-change process, then L_{E_t} is an $(\mathscr{F}_{E_t})$-semimartingale. Thus, (7.75) is the integral form of an SDE driven by an $(\mathscr{F}_{E_t})$-semimartingale. We use the following shorthand differential form of SDE (7.75):

$$dX_t = F(X_{t-}) \odot dL_{E_t}, \quad X_0 = x, \tag{7.76}$$

where $F(x) = (b(x), \sigma(x), G(x,\cdot))$ indicates the triple of coefficients controlling the drift, Brownian, and jump terms, respectively. Similarly, the SDE in (7.76) we write in the following shorthand differential form (to avoid confusion with SDE (7.76) we use letter Y for the unknown process and τ for the time variable):

$$dY_\tau = F(Y_{\tau-}) \odot dL_\tau, \quad Y_0 = x. \tag{7.77}$$

SDE (7.77) has been a focus of many researchers (see, [Sit05, App09] and the references therein). In particular, the following theorem is proved.

Theorem 7.6. ([Sit05, App09]) *If $F(x) = (b(x), \sigma(x), G(x,\cdot))$ satisfies the Lipschitz and growth conditions* (7.39) *and* (7.40)*, respectively, then SDE (7.77) has a unique strong solution with càdlàg paths.*

Theorem 7.7. *Let W_t be a $(\mathscr{F}_t)$-adapted strictly increasing càdlàg process and E_t be its inverse. Suppose a stochastic process Y_τ satisfies SDE (7.77). Then $X_t = Y_{E_t}$ is an $(\mathscr{F}_{E_t})$-semimartingale and satisfies SDE (7.76).*

Proof. Since W_t is a strictly increasing $(\mathscr{F}_t)$-adapted process, its inverse E_t is a continuous $(\mathscr{F}_t)$-time-change. Suppose Y_τ satisfies SDE (7.77) and let $X_t = Y_{E_t}$. Then due to well-known time-change formula [Jac79], we have

$$X_t = x + \int_0^{E_t} F(s, Y_{s-}) \odot dL_s = x + \int_0^t F(E_s, Y_{E(s)-}) \odot dL_{E_s}. \tag{7.78}$$

X_t will satisfy SDE (7.76) provided $X_{s-} = (Y \circ E)_{s-}$ can replace $Y_{E(s)-}$ in (7.78). The equality $Y_{E(s)-} = (Y \circ E)_{s-}$ fails only when $s > 0$ and E is constant on some closed interval $[s-\varepsilon, s] \subset (0,t]$ with $\varepsilon > 0$. However, the integrator $L \circ E$ on the right-hand side of (7.78) is constant on this interval. Hence, the difference between the two values $Y_{E(s)-}$ and $X_{s-} = (Y \circ E)_{s-}$ does not affect the value of the integral. Consequently, (7.78) is valid with X_{s-} in place of $Y_{E(s)-}$. Thus, X_t satisfies SDE (7.76), as desired.

Remark 7.4. There is a general duality between the classes of SDEs (7.76) and SDE (7.77) studied in detail in [Kob11]. Theorem 7.7 is an adopted case to to our special case.

Theorems 7.7 and 7.6 together yield

Corollary 7.2. *If $F(u,x)$ satisfies the Lipschitz and growth conditions* (7.39) *and* (7.40)*, respectively, then SDE (7.76) has a unique strong solution with càdlàg paths.*

Now we are ready to prove the following theorem, which generalizes Theorem 7.3.

Theorem 7.8. *Let $W_{1,t}$ and $W_{2,t}$ be independent stable subordinators of respective indices β_1, $\beta_2 \in (0,1)$. Define $W_t = c_1 W_{1,t} + c_2 W_{2,t}$, with positive constants c_1 and c_2, and let E_t be its inverse. Suppose that a stochastic process Y_τ satisfies the SDE* (7.77) *driven by a Lévy process L_t. Let $X_t = Y_{E_t}$. Then*

1) X_t satisfies the SDE (7.76) *driven by the time-changed Lévy process L_{E_t}.*
2) if Y_τ is independent of E_t, then the function $u(t,x) = \mathbb{E}[\varphi(X_t)|X_0 = x]$ satisfies the following Cauchy problem

$$C_1 D_*^{\beta_k} u(t,x) + C_1 D_*^{\beta_k} u(t,x) = \mathscr{L}(x, D_x) u(t,x), \quad t > 0, x \in \mathbb{R}^n, \tag{7.79}$$
$$u(0,x) = \varphi(x), \tag{7.80}$$

where $\varphi \in C_0^2(\mathbb{R}^n)$, $C_k = c_k^{\beta_k}$, $k = 1,2$, and the pseudo-differential operator $\mathscr{L}(x,D_x)$ is as in (7.61) with symbol in (7.62).

Proof. The proof of part 1) easily follows from Theorem 7.7. Notice that since W_t is a linear combination of stable subordinators, which are càdlàg and strictly increasing, it follows that W_t is also càdlàg and strictly increasing. Hence, $X_t = Y_{E_t}$ satisfies SDE (7.76).

2) Consider $T_\tau^Y \varphi(x) = \mathbb{E}[\varphi(Y_\tau)|Y_0 = x]$, where Y_τ is a solution of SDE (7.77). Then T_τ^Y is a strongly continuous contraction semigroup in the Banach space $C_0(\mathbb{R}^n)$ (see [App09]) which satisfies (7.63) with $\omega = 0$, has infinitesimal generator given by the pseudo-differential operator $\mathscr{L}(x,D_x)$ with symbol $\Psi(x,\xi)$ defined in (7.62), and $C_0^2(\mathbb{R}^n) \subset \mathrm{Dom}(\mathscr{L}(x,D_x))$. So the function $p^Y(\tau,x) = T_\tau^Y \varphi(x)$ with $\varphi \in C_0^2(\mathbb{R}^n)$ satisfies the Cauchy problem

$$\frac{\partial p^Y(\tau,x)}{\partial \tau} = \mathscr{L}(x,D_x) p^Y(\tau,x), \quad p^Y(0,x) = \varphi(x).$$

Furthermore, consider $p^X(t,x) = \mathbb{E}[\varphi(X_t)|X_0 = x] = \mathbb{E}[\varphi(Y_{E_t})|Y_0 = x]$ (recall that $E_0 = 0$). Using independence of the processes Y_τ and E_t,

$$p^X(t,x) = \int_0^\infty \mathbb{E}[\varphi(Y_\tau)|E_t = \tau, Y_0 = x] f_{E_t}(\tau)d\tau = \int_0^\infty f_{E_t}(\tau)T_\tau^Y \varphi(x)d\tau. \quad (7.81)$$

Now, in accordance with Theorem 7.4, $p^X(t,x)$ satisfies the Cauchy problem (7.79)-(7.80).

Theorem 7.9. *([HKU10]) Let $W_{k,t}$, $k = 1,\dots,N$ be independent stable subordinators of respective indices $\beta_k \in (0,1)$. Define $W_t = \sum_{k=1}^N c_k W_{k,t}$, with positive constants c_k, and let E_t be its inverse. Suppose that a stochastic process Y_τ satisfies the SDE* (7.77) *driven by a Lévy process L_t. Let $X_t = Y_{E_t}$. Then*

1) X_t satisfies the SDE (7.76) *driven by the time-changed Lévy process L_{E_t}.*
2) if Y_τ is independent of E_t, then the function $u(t,x) = \mathbb{E}[\varphi(X_t)|X_0 = x]$ satisfies the following Cauchy problem

$$\sum_{k=1}^N C_k D_*^{\beta_k} u(t,x) = \mathcal{L}(x,D_x)u(t,x), \quad t > 0, x \in \mathbb{R}^n,$$
$$u(0,x) = \varphi(x),$$

where $\varphi \in C_0^2(\mathbb{R}^n)$, $C_k = c_k^{\beta_k}$, $k = 1,\dots,N$, and the pseudo-differential operator $\mathcal{L}(x,D_x)$ is as in (7.61) with symbol in (7.62).

Theorem 7.10. *([HKU10]) Assume that $W(\mu;t) \in \mathbb{S}$, where μ is a positive finite measure with $\operatorname{supp}\mu \subset [0,1)$, and let E_t be its inverse. Suppose that a stochastic process Y_τ satisfies SDE (7.77), and let $X_t = Y_{E_t}$. Then*

1) X_t satisfies SDE (7.76);
2) if Y_τ is independent of E_t, then the function $u(t,x) = \mathbb{E}[\varphi(X_t)|X_0 = x]$ satisfies the following Cauchy problem for the time-fractional distributed order pseudo-differential equation

$$D_\mu u(t,x) = \mathcal{L}(x,D_x)u(t,x), \quad t > 0, x \in \mathbb{R}^n,$$

and the initial condition

$$u(0,x) = \varphi(x).$$

Proof. The proof of part 1) again follows from Theorem 7.7. Part 2) follows from Theorem 7.5 in a manner similar to the proof of part 2) of Theorem 7.9.

Remark 7.5. Theorems 7.9 and 7.10 reveal the class of SDEs which are associated with the wide class of time fractional distributed order pseudo-differential equations. Each SDE in this class is driven by a semimartingale which is a time-changed Lévy process, where the time-change is given by the inverse of a mixture of independent stable subordinators. Therefore, these SDEs cannot be represented as classical SDEs driven by a Brownian motion or a Lévy process. The general extensions provided by these two theorems were motivated by their requirement in many applications, such as the cell biology example considered in the introduction.

Corollary 7.3. *Let the coefficients b, σ, H, K of the pseudo-differential operator $\mathscr{L}(x, D_x)$ defined in (7.61) with symbol in (7.62) be continuous, bounded, and satisfy Lipschitz and growth condition. Suppose $\varphi \in C_0^2(\mathbb{R}^n)$. Then the Cauchy problem for the time-fractional DODE*

$$D_\mu u(t,x) = \mathscr{L}(x, D_x)u(t,x), \quad t > 0, x \in \mathbb{R}^n,$$
$$u(0,x) = \varphi(x), \quad x \in \mathbb{R}^n,$$

has a unique solution $u(t,x) \in C_0^2(\mathbb{R}^n)$ for each $t > 0$.

Proof. The result follows from the representation (7.81) in conjunction with estimate (7.67).

Example 7.4. Time-changed α-stable Lévy process. Let $L_\alpha(t)$ be a symmetric n-dimensional α-stable Lévy process, which is a pure jump process. If $p^L(t,x) = E[\varphi(L_\alpha(t))|L_\alpha(0) = x]$, where $\varphi \in C_0^2(\mathbb{R}^n)$ (or, $\varphi \in H^\alpha(\mathbb{R}^n)$, the Sobolev space of order α), then $p^L(t,x)$ satisfies in the strong sense the Cauchy problem

$$\frac{\partial p^L(t,x)}{\partial t} = -\kappa_\alpha(-\Delta)^{\alpha/2} p^L(t,x), \quad t > 0, \ x \in \mathbb{R}^n, \tag{7.82}$$
$$p^L(0,x) = \varphi(x), \quad x \in \mathbb{R}^n, \tag{7.83}$$

where κ_α is a constant depending on α and $(-\Delta)^{\alpha/2}$ is a fractional power of the Laplace operator. The operator on the right-hand side of (7.82) can be represented as a pseudo-differential operator with the symbol $\psi(\xi) := |\xi|^\alpha$. It can also be represented as a hyper-singular integral (Section 3.8), which is more convenient in random walk approximation of α-stable Lévy processes (see Chapter 8).

Example 7.5. Let $\mathbb{SS}$ be the set of Lévy processes X_t^ρ, such that

$$\ln \mathbb{E}(e^{iX^\rho \xi}) = \Psi(\xi) = -\int_0^2 |\xi|^\alpha d\rho(\alpha)$$

Evidently, if $\rho(d\alpha) = \delta_{\alpha_0}(\alpha)d\alpha$, then $X_t^\rho = L_{\alpha_0}(t)$. Hence, the set $\mathbb{SS}$ contains all the symmetric α-stable Lévy processes. Moreover, if $L_{\alpha_1}(t)$ and $L_{\alpha_2}(t)$ are independent symmetric α_1- and α_2-stable Lévy processes, respectively, then $AL_{\alpha_1}(t) + BL_{\alpha_2}(t) \in \mathbb{SS}$ with $\rho(d\alpha) = [A\delta_0(\alpha - \alpha_1) + B\delta_0(\alpha - \alpha_2)]d\alpha$.

Now let $X_{E_t^\mu}^\rho$ be the time-changed process, where E_t^μ is the inverse to a process $W(\mu;t) \in \mathbb{S}$. Then due to Theorem 7.10, the FPK equation, associated with $X_{E_t^\mu}^\rho$, has the form

$$D_\mu u(t,x) = \int_0^2 \mathbb{D}_0^\alpha u(t,x)d\rho(\alpha),$$
$$u(0,x) = \varphi(x).$$

Now suppose Y_t solves SDE

$$dY_t = g(Y_{t-})dL_\alpha(t), \quad Y_0 = x,$$

where $g(x)$ is a function satisfying the growth and Lipschitz conditions, and such that $g(x) \neq 0, x \in \mathbb{R}^n$. In other words each for each component $Y_j(t)$ of the n dimensional stochastic process Y_t, we have SDE

$$dY_j(t) = g(Y_j(t-))dL_{\alpha j}(t),\ Y_j(0) = x_j, \quad j = 1,\dots,n,$$

where $L_{\alpha j}(t)$ is j-th component of the process $L_\alpha(t)$. In this case, the forward FPK equation takes the form

$$\frac{\partial p^Y(t,x)}{\partial t} = -\kappa_\alpha(-\Delta)^{\alpha/2}\{|g(x)|^\alpha p^Y(t,x)\}, \quad t>0,\ x\in\mathbb{R}^n. \tag{7.84}$$

In order to prove this statement we recall that the forward FPK equation uses the adjoint operator $\mathscr{A}^*$ (see (7.24)). In our case $L_\alpha(t)$ is $S\alpha S$-process, and therefore, the operator $\mathscr{A} = \Psi(x,D)$, due to formula (7.61), is

$$\Psi(x,D)\varphi(x) = \int\limits_{\mathbb{R}^n\setminus\{0\}} [\varphi(x-g(x)w) - \varphi(x) - g(x)(w,\nabla\varphi(x))\chi_{|w|\le 1}(w)]\frac{dw}{|w|^{n+\alpha}},$$

where $\varphi \in \mathscr{D}(\Psi(x,D)) = H^1(\mathbb{R}^n)$. Using the substitution $y_j = -g(x)w_j,\ j=1,\dots,n$, in the latter integral, one has

$$\Psi(x,D)\varphi(x) = \int\limits_{\mathbb{R}^n\setminus\{0\}} [\varphi(x+y) - \varphi(x) + (y,\nabla\varphi(x))\chi_{|y|\le|g(x)|}(y)]\frac{|g(x)|^\alpha dy}{|y|^{n+\alpha}}.$$

Therefore, for arbitrary $v \in H^1(\mathbb{R}^n)$,

$$\begin{aligned}
&\Big(\Psi(x,D)\varphi(x), v(x)\Big)\\
&= \int\limits_{\mathbb{R}^n}\int\limits_{\mathbb{R}^n\setminus\{0\}} v(x)[\varphi(x+y) - \varphi(x) + (y,\nabla\varphi(x))\chi_{|y|\le|g(x)|}(y)]\frac{|g(x)|^\alpha dy}{|y|^{n+\alpha}}dx\\
&= \int\limits_{\mathbb{R}^n}\int\limits_{\mathbb{R}^n\setminus\{0\}} \varphi(x)[(|g|^\alpha v)(x-y) - (|g|^\alpha v)(x) - (y,\nabla(|g|^\alpha v)(x))\chi_{|y|\le 1}(y)]\frac{dy}{|y|^{n+\alpha}}dx\\
&= \Big(\varphi(x), \Psi(x,D)\Big(|g(x)|^\alpha v(x)\Big)\Big) = (\varphi(x), \Psi^*(x,D)v(x)).
\end{aligned}$$

Hence, the adjoint operator is $\Psi^*(x,D)v(x) = \Psi(x,D)\Big(|g(x)|^\alpha v(x)\Big)$, and we obtain (7.84).

Application of Theorem 7.10 implies that $X_t = Y_{E_t}$ satisfies the SDE

$$dX_t = g(X_{t-})dL_{\alpha,E_t}, \quad X_0 = x, \tag{7.85}$$

where E_t is the first hitting time of the process $D(\mu;t)$ described in this theorem. Moreover, if E_t is independent of Y_t, then the corresponding forward Kolmogorov equation becomes a {time-fractional DODE/pseudo-differential} equation

$$D_\mu p^X(t,x) = -\kappa_\alpha(-\Delta)^{\alpha/2}\{[g(x)]^\alpha p^X(t,x)\}, \quad t>0,\, x\in\mathbb{R}^n, \tag{7.86}$$

where $\mathscr{D}_\mu$ is the operator defined in (7.71). When the SDE in (7.85) is driven by a nonsymmetric α-stable Lévy process, an analogue of (7.86) holds using instead of (7.84) its analogue appearing in [SLDYL01].

Example 7.6. Fractional analogue of the Feynman-Kac formula. Suppose Y_t is a strong solution of SDE (7.77). Let $\bar{Y}\in\mathbb{R}^n$ be a fixed point, which we call a terminal point. Let q be a nonnegative continuous function. Consider the process

$$Y_t^q = \begin{cases} Y_t, & \text{if } 0\le t<\mathscr{T}_q, \\ \bar{Y}, & \text{if } t\ge\mathscr{T}_q, \end{cases}$$

where $\mathscr{T}_q$ is an $(\mathscr{F}_t)$-stopping time satisfying

$$\mathbb{P}(\mathscr{T}_q>t|\mathscr{F}_t) = \exp\left(-\int_0^t q(Y_s)ds\right).$$

The process Y_t^q is a Feller process with associated semigroup (see [App09])

$$(T_t^q\varphi)(y) = \mathbb{E}\left[\exp\left(-\int_0^t q(Y_s)ds\right)\varphi(Y_t)\Big|Y_0=y\right], \tag{7.87}$$

and infinitesimal generator $\mathscr{L}_q(x,D_x) = -q(x)+\mathscr{L}(x,D_x)$, where $\mathscr{L}(x,D_x)$ is the pseudo-differential operator defined in (7.61). Let E_t be the inverse to a β-stable subordinator independent of Y_t. Then it follows from Theorem 7.9 with $N=1$ that the transition probabilities of the process $X_t = Y_{E_t}$ solve the Cauchy problem for the fractional order equation

$$\begin{aligned} D_*^\beta u(t,x) &= [-q(x)+\mathscr{L}(x,D_x)]u(t,x), \quad t>0,\, x\in\mathbb{R}^n, \\ u(0,x) &= \varphi(x), \quad x\in\mathbb{R}^n. \end{aligned}$$

Consequently, (7.87), with $X_t = Y_{E_t}$ replacing Y_t, and dE_t replacing dt, represents a fractional analogue of the Feynman-Kac formula.

7.12 Fractional Brownian motion

Brownian motion does not adequately model stochastic processes having correlations arising in a range of diverse applied fields, including finance, biology, hydrology, solar physics, turbulence, etc. A better mathematical model for such processes is achieved using fractional Brownian motion (fBM). Our next goal is to derive fractional Fokker-Planck-Kolmogorov type equations, associated with stochastic differential equations driven by a time-changed fBM.

By definition, a one-dimensional fBM B_t^H is a zero-mean Gaussian process with continuous paths and covariance function

$$R_H(s,t) = E(B_s^H B_t^H) = \frac{1}{2}(s^{2H} + t^{2H} - |s-t|^{2H}),$$

where the parameter H, called *a Hurst parameter,* takes values in the interval $(0,1)$. If $H = \frac{1}{2}$, then B_t^H coincides with the standard Brownian motion. In this case, obviously, $R_{1/2}(s,t) = \min(s,t)$, which is a well known property of Brownian motion.

Fractional Brownian motion, as a driving process for SDEs, does not satisfy conditions required for Itô's calculus,[2] unless $H = \frac{1}{2}$ (see Section "Additional notes"). Nevertheless, there are several approaches [Ben03, BHOZ08, DU98, Nua06] to a stochastic calculus in order to interpret in a meaningful way as an SDE of the form

$$X_t = X_0 + \int_0^t b(X_s)ds + \int_0^t \sigma(X_s)dB_s^H, \tag{7.88}$$

driven by an m-dimensional fBM B_t^H, where mappings $b : \mathbb{R}^n \to \mathbb{R}^n$ and $\sigma : \mathbb{R}^n \to \mathbb{R}^{n\times m}$ are Lipschitz continuous; X_0 is a random variable independent of B_t^H. We do not discuss here these approaches referring the interested reader to [BHOZ08, DU98, Nua06]. Instead, we focus our attention on the FPK equation associated with SDE (7.88) driven by fBM whose generic form is given by

$$\frac{\partial u(t,x)}{\partial t} = B(x,D_x)u(t,x) + Ht^{2H-1}A(x,D_x)u(t,x), \tag{7.89}$$

where

$$B(x,D_x) = \sum_{j=1}^{n} b_j(x)\frac{\partial}{\partial x_j}, \tag{7.90}$$

a first order differential operator, and $A(x,D_x)$ is a second order elliptic differential operator

$$A(x,D_x) = \sum_{j,k=1}^{n} a_{jk}(x)\frac{\partial^2}{\partial x_j \partial x_k}. \tag{7.91}$$

Functions $a_{jk}(x),\ j,k = 1,\dots,n$ are entries of the matrix $\mathscr{A}(x) = \sigma(x) \times \sigma^T(x)$, where $\sigma^T(x)$ is the transpose of matrix $\sigma(x)$. By definition $\mathscr{A}(x)$ is positive definite:

---

[2] It is not a semimartingale

for any $x \in \mathbb{R}^n$ and $\xi \in \mathbb{R}^n$ one has $\sum_{j,k=1}^n a_{jk}(x)\xi_j\xi_k \geq C|\xi|^2$, where C is a positive constant. The operator $A(x,D_x)$ can also be given in the divergent form

$$A(x,D_x) = \sum_{j,k=1}^{n} \frac{\partial}{\partial x_j}\Big(a_{jk}(x)\frac{\partial}{\partial x_k}\Big). \tag{7.92}$$

The right-hand side of (7.89) depends on the time variable t, which, in fact, reflects the presence of correlation. Additionally, $u(t,x)$ in equation (7.89) satisfies the initial condition

$$u(0,x) = \varphi(x), \quad x \in \mathbb{R}^n, \tag{7.93}$$

where $\varphi(x)$ belongs to some function space, or is a generalized function. In the particular case of FPK equation associated with SDE (7.88), $\varphi(x) = f_{X_0}(x)$, the density function of X_0. If $X_0 = x_0 \in \mathbb{R}^n$, then $\varphi(x) = \delta_{x_0}(x)$, Dirac's delta with mass on x_0. In this case the solution to the FPK equation is understood in the weak sense.

In the one-dimensional case with $H \in (\frac{1}{4}, 1)$, as is shown in [BC07], the function $u(t,x) = E_x[\varphi(X_t)]$ solves the equation (7.89) with initial condition (7.93) when X_t solves SDE (7.88) with $b = 0$ and a stochastic integral in the sense of Stratanovich. The operator $A(x,D_x)$ appearing in (1.3) is expressed in the divergence form (7.92).

7.13 Abstract theorem

In this section we prove an abstract theorem for a class of differential operator equations, containing (7.89) as a particular case. Let A and B be linear closed operators with $\mathscr{D}(A) \subset \mathscr{D}(B) \subset X$, where X is a Banach space. Introduce the operator

$$L_\gamma(t) = B + \frac{\gamma+1}{2}t^\gamma A, \quad t > 0, \tag{7.94}$$

where $\gamma \in (-1,1)$. The parameter γ is related to the Hurst parameter H through $\gamma = 2H - 1$. The introduction of γ is made so that the operators G_γ arising below (see (7.99)) will have the semigroup property.

Our starting point is the differential-operator equation

$$\frac{du(t)}{dt} = L_\gamma(t)u(t), \quad t > 0, \tag{7.95}$$

with an initial condition

$$u(0) = u_0 \in X. \tag{7.96}$$

If $\gamma = 0$, or equivalently $H = \frac{1}{2}$, and operators $B = B(x,D_x)$ and $A = A(x,D_x)$ are defined in (7.90) and (7.91), respectively, then the operator $L_0(t) = L_0(t,x,D_x)$ has a form with coefficients not depending on t:

$$L_0(t,x,D_x) \equiv L(x,D_x) = B(x,D_x) + \frac{1}{2}A(x,D_x),$$

and equation (7.95) coincides with the FPK equation associated with the SDE driven by Brownian motion (see Section 7.4)

$$\frac{\partial u(t,x)}{\partial t} = L(x,D_x)u(t,x) \quad t>0, x\in\mathbb{R}^n.$$

As before, integrals below are understood in the sense of Bochner, if integrands are vector-functions with values in a topological vector space.

Theorem 7.11. *Let $u(t)$ be a solution to initial value problem (7.95)–(7.96). Let $f_t(\tau)$ be the density function of the process inverse to a Lévy's stable subordinator of index β. Then the vector function $v(t)=\int_0^\infty f_t(\tau)u(\tau)d\tau$ satisfies the following initial value problem for a fractional order differential-operator equation*

$$D_*^\beta v(t) = Bv(t) + \frac{\gamma+1}{2}AG_{\gamma,t}v(t), \quad t>0, \tag{7.97}$$

$$v(0) = \varphi, \tag{7.98}$$

where the operator $G_{\gamma,t}$ is defined through $u(t)$ by

$$G_{\gamma,t}v(t) = \int_0^\infty f_t(\tau)\tau^\gamma u(\tau)d\tau.$$

Moreover, for $G_{\gamma,t}$ the following explicit representation holds:

$$G_{\gamma,t}v(t) = \beta\Gamma(\gamma+1)J_t^{1-\beta}\mathscr{L}_{s\to t}^{-1}\left[\frac{1}{2\pi i}\int_{C-i\infty}^{C+i\infty}\frac{L[v](z)}{(s^\beta - z^\beta)^{\gamma+1}}dz\right](t), \tag{7.99}$$

where $0<C<s$, and $z^\beta = e^{\beta Ln(z)}$, $Ln(z)$ being the principal value of the complex $\ln(z)$ *with cut along the negative real axis.*

Proof. Let $v(t)=\int_0^\infty f_t(\tau)u(\tau)d\tau$, where $u(t)$ satisfies initial value problem (7.95)–(7.96). Using relation (7.42) valid for $f_t(\tau)$, we have

$$\begin{aligned}D_{*,t}^\beta v(t) &= \int_0^\infty D_{*,t}^\beta f_t(\tau)u(\tau)d\tau = -\int_0^\infty\left[\frac{\partial}{\partial\tau}f_t(\tau) + \frac{t^{-\beta}}{\Gamma(1-\beta)}\delta_0(\tau)\right]u(\tau)d\tau\\ &= -\lim_{\tau\to\infty}[f_t(\tau)u(\tau)] + \lim_{\tau\to 0}[f_t(\tau)u(\tau)]\\ &\quad + \int_0^\infty f_t(\tau)\frac{du(\tau)}{d\tau}d\tau - \frac{t^{-\beta}}{\Gamma(1-\beta)}u(0).\end{aligned}$$

Due to Lemma 7.1, part (c) the first term vanishes since $u(\tau,x)$ is bounded, and due to part (b) of the same lemma the second and last terms cancel. Moreover, taking into account (7.95), one has

$$D^{\beta}_{*,t}v(t) = \int_0^{\infty} f_t(\tau)L_{\gamma}(\tau)u(\tau)d\tau = \int_0^{\infty} f_t(\tau)\left[Bu(\tau) + \frac{\gamma+1}{2}\tau^{\gamma}Au(\tau)\right]d\tau$$
$$= Bv(t) + \frac{\gamma+1}{2}AG_{\gamma,t}v(t),$$

where

$$G_{\gamma,t}v(t) = \int_0^{\infty} f_t(\tau)\tau^{\gamma}u(\tau)d\tau. \tag{7.100}$$

It follows from the definition of $v(t)$ and equation (7.100) that if $\gamma = 0$, then $G_{0,t} = \int_0^{\infty} f_t(\tau)u(\tau)d\tau = v(t)$, that is the identity operator. To show representation (7.99) in the case $\gamma \neq 0$, we find the Laplace transform of $G_{\gamma,t}v(t)$. In accordance with the property (d) of Lemma 7.1, we have

$$L[G_{\gamma,t}v(t)](s) = s^{\beta-1}\int_0^{\infty} e^{-\tau s^{\beta}}\tau^{\gamma}u(\tau)d\tau = s^{\beta-1}L[\tau^{\gamma}u(\tau)](s^{\beta}).$$

Obviously, if $\gamma = 0$, then $L[G_{0,t}v(t)](s) = s^{\beta-1}\tilde{u}(s^{\beta})$, which implies $\tilde{v}(s) = s^{\beta-1}$ $\tilde{u}(s^{\beta})$. If $\gamma \neq 0$, then

$$L[t^{\gamma}u(t)](s) = L[t^{\gamma}](s) * \tilde{u}(s) = \frac{1}{2\pi i}\int_{c-i\infty}^{c+i\infty}\frac{\Gamma(\gamma+1)}{(s-z)^{\gamma+1}}L[u](z)dz, \tag{7.101}$$

where $*$ stands for the convolution of Laplace images of two functions and $0 < c < s$. Now using the substitution $z = e^{\beta \mathrm{Ln}(\zeta)}$, with $\mathrm{Ln}(\zeta)$ the principal part of the complex function $\ln(\zeta)$, the right-hand side of (7.101) reduces to

$$L[t^{\gamma}u(t)](s) = \frac{\beta}{2\pi i}\int_{C-i\infty}^{C+i\infty}\frac{\Gamma(\gamma+1)}{(s-\zeta^{\beta})^{\gamma+1}}\zeta^{\beta-1}L[u](\zeta^{\beta})d\zeta \tag{7.102}$$
$$= \frac{\beta}{2\pi i}\int_{C-i\infty}^{C+i\infty}\frac{\Gamma(\gamma+1)}{(s-\zeta^{\beta})^{\gamma+1}}L[v](\zeta)d\zeta.$$

The last equality uses the relation $\tilde{v}(\zeta) = \zeta^{\beta-1}\tilde{u}(\zeta^{\beta})$. Further, replacing s by s^{β} and taking the inverse Laplace transform in (7.102) yields the desired representation (7.99) for the operator $G_{\gamma,t}$ since $\mathscr{L}[J^{1-\beta}f](s) = s^{\beta-1}\tilde{f}(s)$. In accordance with part (a) of Lemma 7.1 we have $v(0,x) = u(0,x)$ as well, which completes the proof.

In the more general case when the time-change process E_t^{μ} is the inverse to W_t^{μ}, the mixture of stable subordinators with the mixing measure μ, a representation for the abstract fractional FPK equation is given in the following theorem.

Theorem 7.12. *Let $u(t)$ be a solution to initial value problem (7.95)–(7.96). Let $f_t^{\mu}(\tau)$ be the density function of the process inverse to W_t^{μ}. Then the vector-function $v(t) = \int_0^{\infty} f_t^{\mu}(\tau)u(\tau)d\tau$ satisfies the following initial value problem for a fractional order differential equation*

$$D_\mu v(t) = Bv(t) + \frac{\gamma+1}{2} AG^\mu_{\gamma,t} v(t), \quad t > 0, \tag{7.103}$$

$$v(0) = \varphi \in X. \tag{7.104}$$

The operator $G^\mu_{\gamma,t}$ acts on the variable t and is defined by

$$G^\mu_{\gamma,t} v(t) = \int_0^\infty f_t^\mu(\tau)\tau^\gamma u(\tau)d\tau.$$

Moreover, for $G_{\gamma,t}$ the following explicit representation holds:

$$G^\mu_{\gamma,t} v(t) = \Phi_\mu(t) * \mathscr{L}^{-1}_{s\to t}\Big[\frac{\Gamma(\gamma+1)}{2\pi i}\int_{C-i\infty}^{C+i\infty} \frac{m_\mu(z)\tilde{v}(z)}{(\rho(s)-\rho(z))^{\gamma+1}}dz\Big](t), \tag{7.105}$$

where $*$ denotes the usual convolution of two functions, $0 < C < s$, $\Phi_\mu(t)$ is defined in (7.58)*, and*

$$\rho(z) = \int_0^1 e^{\beta Ln(z)} d\mu(\beta), \quad m_\mu(z) = \frac{\int_0^1 \beta z^\beta d\mu(\beta)}{\rho(z)}.$$

Proof. The proof is similar to the proof of Theorem 7.11. We only sketch how to obtain representation (7.105) for the operator

$$G^\mu_{\gamma,t} v(t) = \int_0^\infty f_t^\mu(\tau)\tau^\gamma u(\tau)d\tau.$$

The Laplace transform of $G^\mu_{\gamma,t} v(t)$, due to part (d) of Lemma 7.5, is

$$L_{t\to s}\left[G^\mu_{\gamma,t} v(t)\right](s) = \frac{\rho(s)}{s} L[t^\gamma u(t)](\rho(s)), s > 0.$$

Since $L[\Phi_\mu](s) = \frac{\rho(s)}{s}, s > 0$, we have

$$G^\mu_{\gamma,t} v(t) = \Phi_\mu(t) * L^{-1}_{s\to t}\left[L[t^\gamma u(t)](\rho(s))\right](t).$$

Further, replacing s by $\rho(s)$ in (7.101), followed by the substitution $z = \rho(\zeta) = \int_0^1 e^{\beta Ln(\zeta)} d\mu(\beta)$ in the integral on the right side of (7.101), yields the form (7.105).

The following theorem represents the general case when the time-change process E_t is not necessarily the first hitting time process for a stable subordinator or their mixtures.

Theorem 7.13. *Let $\gamma \in (-1,1)$. Let E_t be a time-change process and assume that its density $K(t,\tau) = f_{E_t}(\tau)$ satisfies the hypotheses:*

i) $\lim_{\tau\to+0}\left[K(t,\tau)\tau^{-\gamma}\right] < \infty$ *for all $t > 0$;*

ii) $\lim_{\tau\to\infty}[K(t,\tau)\tau^{-\gamma}u(\tau,x)] = 0$ *for all $t > 0$ and $x \in \mathbb{R}^n$,* *where $u(t)$ is a solution to the initial value problem (7.95)–(7.96). Let H_t be an operator acting in the variable t such that*

$$H_t K(t,\tau) = -\frac{\partial}{\partial \tau}\Big[K(t,\tau)(\frac{t}{\tau})^{\gamma}\Big] - \delta_0(\tau) \lim_{\tau \to +0}\Big[(\frac{t}{\tau})^{\gamma} K(t,\tau)\Big].$$

Then the function $v(t) = \int_0^\infty K(t,\tau)u(\tau)d\tau$ satisfies the initial value problem

$$H_t v(t) = t^{\gamma} \bar{G}_{-\gamma,t} B v(t) + \frac{\gamma+1}{2} t^{\gamma} A v(t), \quad \tau > 0, \tag{7.106}$$

$$v(0) = u(0), \tag{7.107}$$

where $\bar{G}_{-\gamma,t} v(t) = \int_0^\infty K(t,\tau)\tau^{-\gamma}u(\tau)d\tau$.

Remark 7.6. Obviously, if $\gamma \neq 0$, then H_t cannot be a fractional derivative in the sense of Caputo (or Riemann-Liouville). A representation of H_t in cases when E_t is the inverse to a stable subordinator, is given below in Corollary 7.4.

Proof. We have

$$\begin{aligned} H_t v(t) &= \int_0^\infty H_t K(t,\tau)u(\tau)d\tau \\ &= -\int_0^\infty \Big\{\frac{\partial}{\partial \tau}\Big[K(t,\tau)(\frac{t}{\tau})^{\gamma}\Big] + \delta_0(\tau)\lim_{\tau\to+0}\Big[(\frac{t}{\tau})^{\gamma}K(t,\tau)\Big]\Big\}u(\tau)d\tau \\ &= -t^{\gamma}\lim_{\tau\to\infty}[K(t,\tau)\tau^{-\gamma}u(\tau)] + t^{\gamma}\lim_{\tau\to 0+}[K(t,\tau)\tau^{-\gamma}u(\tau)] \\ &+ \int_0^\infty K(t,\tau)(\frac{t}{\tau})^{\gamma}\frac{du(\tau)}{d\tau}d\tau - \lim_{\tau\to+0}\Big[(\frac{t}{\tau})^{\gamma}K(t,\tau)\Big]u(0). \end{aligned} \tag{7.108}$$

The first term on the right of (7.108) is zero by hypothesis *ii*) of the theorem. The sum of the second and last terms, which exist by hypothesis *i*), also equals zero. Now taking equation (7.95) into account, we have

$$H_t v(t) = t^{\gamma} B \int_0^\infty K(t,\tau)\tau^{-\gamma}u(\tau,x)d\tau + \frac{\gamma+1}{2}t^{\gamma}Av(t).$$

Further, since $E_0 = 0$ it follows that

$$\lim_{t\to 0} v(t) = \int_0^\infty \delta_0(\tau)u(\tau)d\tau = u(0),$$

which completes the proof.

Let Π_γ denote the operator of multiplication by t^γ, i.e. $\Pi_\gamma h(t) = t^\gamma h(t)$, $h \in C(0,\infty)$. Applying Theorem 7.13 to the case $K(t,\tau) = f_t(\tau)$ in conjunction with Theorem 7.11, we obtain the following corollary.

Corollary 7.4. *Let $\gamma \in (-1,0]$ and $K(t,\tau) = f_t(\tau)$, where $f_t(\tau)$ is defined in (7.41). Then (i) $G_{-\gamma,t} = G_{\gamma,t}^{-1}$; (ii) $H_t = \Pi_\gamma G_{-\gamma,t} D_*^\beta$.*

This corollary yields an equivalent form for equation (7.97) in the case when E_t is the inverse to the stable subordinator with index β and $\gamma \in (-1,0]$:

$$H_t v(t) = t^{\gamma} G_{-\gamma,t} B v(t) + \frac{\gamma+1}{2} t^{\gamma} A v(t), \tag{7.109}$$

with H_t as in Corollary 7.4.

Suppose the operator in the "drift" term $B=0$. Then equation (7.109) takes the form

$$H_t v(t) = \frac{\gamma+1}{2} t^{\gamma} A(x, D_x) v(t). \tag{7.110}$$

Notice that equation (7.109) is valid for $\gamma \in (0,1)$ as well. Indeed, part (*ii*) of Corollary 7.4 can be rewritten in the form $G_{\gamma,t} = G_{-\gamma,t}^{-1}$ for $\gamma > 0$. For $\gamma < 0$ part (*ii*) of Corollary 7.4 also implies $(G_{\gamma,t}^{-1})^{-1} = G_{-\gamma,t}^{-1} = G_{\gamma,t}$. Now applying operators $G_{-\gamma,t}$ and Π_{γ} consecutively to both sides of (7.97) we obtain (7.109) for all $\gamma \in (-1,1)$.

Analogously, the fractional order equation (7.103) obtained in Theorem 7.12 with the mixing measure μ can be represented in its equivalent form as

$$H_t^{\mu} v(t) = t^{\gamma} G_{-\gamma,t}^{\mu} B v(t) + \frac{\gamma+1}{2} t^{\gamma} A v(t), \quad t>0,\ \tau>0, \tag{7.111}$$

where $H_t^{\mu} = \Pi_{\gamma} G_{\gamma,t}^{\mu} D_{\mu}$. We leave verification of the details to the reader as an exercise.

The equivalence of equations (7.97) and (7.109) and the equivalence of equations (7.103) and (7.111) are obtained by means of Theorem 7.13. This fact can also be established with the help of the semigroup property of the family of operators $\{G_{\gamma}, -1 < \gamma < 1\}$:

$$G_{\gamma} g(t) = \int_0^{\infty} f_t(\tau) \tau^{\gamma} h(\tau) d\tau = \mathscr{F}_{\gamma} h(t), \tag{7.112}$$

where $h \in C^{\infty}(0,\infty)$ is a nonnegative bounded function. Denote the class of such functions by U. Functions g and h in (7.112) are connected through the relation $g(t) = \int_0^{\infty} f_t(\tau) h(\tau) d\tau = \mathscr{F} h(t)$. It follows from the behavior of $f_t(\tau)$ as a function of t, that $g \in C^{\infty}(0,\infty)$. On the other hand, obviously, operator $\mathscr{F}$ is bounded, $\|\mathscr{F} h\| \le \|h\|$ in the sup-norm, and one-to-one due to positivity of $f_t(\tau)$. Therefore, the inverse $\mathscr{F}^{-1} : \mathscr{F}U \to U$ exists. Let a distribution $H(t,\tau)$ with $supp H \subset \mathbb{R}_+^2$ be such that $\mathscr{F}^{-1} g(t) = \int_0^{\infty} H(t,\tau) g(\tau) d\tau$. Since $f_t(\tau) \in \mathscr{F}U$ as a function of t for each $\tau > 0$, for an arbitrary $h \in U$ one has

$$\begin{aligned} h(t) = \mathscr{F}^{-1}\mathscr{F} h(t) &= \int_0^{\infty} H(t,s) \Big(\int_0^{\infty} f_s(\tau) h(\tau) d\tau \Big) ds \\ &= \int_0^{\infty} h(\tau) \Big(\int_0^{\infty} H(t,s) f_s(\tau) ds \Big) d\tau = < \int_0^{\infty} H(t,s) f_s(\tau) ds, h >_{\tau}. \end{aligned}$$

We write this relation between $H(t,\tau)$ and $f_t(\tau)$ in the form

$$\int_0^{\infty} H(t,s) f_s(\tau) ds = \delta_t(\tau). \tag{7.113}$$

Proposition 7.5. *Let* $-1<\gamma<1, -1<\alpha<1,$ *and* $-1<\gamma+\alpha<1$. *Then* $G_\gamma \circ G_\alpha = G_{\gamma+\alpha}$.

Proof. The proof uses the following two relations:

(1) $G_\gamma g(t) = \int_0^\infty \mathscr{F}_{\gamma,t}H(t,s)g(s)ds, \ \gamma \in (-1,1)$;

(2) $\int_0^\infty \mathscr{F}_{\gamma,t}H(t,s)\mathscr{F}_{\alpha,s}H(s,\tau)ds = \mathscr{F}_{\gamma+\alpha,t}H(t,\tau)$, with $-1<\gamma,\alpha<1$, and $-1<\gamma+\alpha<1$.

Indeed, using (7.112) and changing the order of integration, we obtain the first relation

$$G_\gamma g(t) = \int_0^\infty f_t(\tau)\tau^\gamma\Big(\int_0^\infty H(\tau,s)g(s)ds\Big)d\tau \qquad (7.114)$$
$$= \int_0^\infty g(s)\Big(\int_0^\infty f_t(\tau)H(\tau,s)\tau^\gamma d\tau\Big)ds = \int_0^\infty \mathscr{F}_{\gamma,t}H(t,s)g(s)ds.$$

It is readily seen that the internal integral in the second line of (7.114) is meaningful, since $f_t(\tau)$ is a function of exponential decay when $\tau\to\infty$, which follows from (7.36). Further, in order to show the second relation, we have

$$\int_0^\infty \mathscr{F}_{\gamma,t}H(t,s)\mathscr{F}_{\alpha,s}H(s,\tau)ds = \int_0^\infty\Big(\int_0^\infty f_t(p)H(p,s)p^\gamma dp\Big)\Big(\int_0^\infty f_s(q)H(q,\tau)q^\alpha dq\Big)ds$$
$$= \int_0^\infty\int_0^\infty f_t(p)H(q,\tau)p^\gamma q^\alpha\Big(\int_0^\infty H(p,s)f_s(q)ds\Big)dp\,dq.$$

Due to (7.113), this equals

$$\int_0^\infty f_t(p)p^\gamma\Big(\int_0^\infty H(q,\tau)q^\alpha\delta_p(q)dq\Big)dp = \int_0^\infty H(p,\tau)p^\alpha f_t(p)p^\gamma dp = \mathscr{F}_{\gamma+\alpha,t}H(t,\tau).$$

Now we are ready to prove the claimed semigroup property. Making use of the two proved relations,

$$(G_\gamma\circ G_\alpha)g(t) = G_\gamma\big[G_\alpha g(t)\big]$$
$$= G_\gamma\Big[\int_0^\infty \mathscr{F}_{\alpha,t}H(t,s)g(s)ds\Big] = \int_0^\infty \mathscr{F}_{\gamma,t}H(t,s)\Big[\int_0^\infty \mathscr{F}_{\alpha,s}H(s,\tau)g(\tau)d\tau\Big]ds$$
$$= \int_0^\infty g(\tau)\int_0^\infty \mathscr{F}_{\gamma,t}H(t,s)\mathscr{F}_{\alpha,s}H(s,\tau)ds\,d\tau = \int_0^\infty \mathscr{F}_{\gamma+\alpha,t}H(t,\tau)g(\tau)d\tau = G_{\gamma+\alpha}g(t),$$

which completes the proof.

Similarly one can prove the semigroup property of the family of operators $G^\mu_{\gamma,t}$.

Proposition 7.6. *The operator* $G^\mu_{\gamma,t}$ *possesses the semigroup property. Namely, for any* $\gamma,\delta\in(-1,1), \gamma+\delta\in(-1,1)$, *one has* $G^\mu_{\gamma,t}\circ G^\mu_{\delta,t} = G^\mu_{\gamma+\delta,t} = G^\mu_{\delta,t}\circ G^\mu_{\gamma,t}$, *where* "$\circ$" *denotes the composition of two operators.*

Remark 7.7.

1. FPK equations associated with SDEs driven by Brownian motion and time-changed Brownian motion had very simple connection. Namely, retain the right-hand side of FPK equation corresponding to SDE driven by Brownian motion and change the left-hand side to a fractional derivative, to obtain FPK equation corresponding to SDE driven by the time-changed Brownian motion. Equation (7.110) shows that this drastically changes in the case of fBM. Moreover, if a fractional derivative is desired on the left-hand side in the time-changed case, then (3.4) shows that the right-hand side must be a different operator from that in the non-time-changed case.
2. Proposition 7.5 immediately implies that $G_\gamma^{-1} = G_{-\gamma}$ for arbitrary $\gamma \in (-1,1)$. Indeed, $G_\gamma \circ G_{-\gamma} = G_0 = I$, as well as $G_{-\gamma} \circ G_\gamma = I$, where I is the identity operator. Thus, the statement in Corollary 7.4 is valid for all $\gamma \in (-1,1)$.
3. Theorem 7.12 generalizes Theorem 7.5. In fact, if $B = 0$ and $\gamma = 0$, then $G^\mu_{0,t} \equiv I$, where I is the identity operator, so Theorem 7.12 represents Theorem 7.5 in a slightly disguised formulation. Notice that Theorem 7.12 does not use the semi-group structure. The Cauchy problem (7.95)-(7.96) is important from the applications point of view too. Indeed, if $B = 0$, $\gamma = 2H - 1$ and $A = \Delta$, then (7.95) is the FPK equation associated with the fractional Brownian motion with the Hurst parameter $H \in (0,1)$.

7.14 Applications of the abstract theorem

7.14.1 Fractional FPK equations associated with time-changed fBM

Now let us focus on the FPK equation associated with SDE driven by a time-changed fBM $B^H_{E_t}$:

$$X_t = X_0 + \int_0^t b(X_s)dE_s + \int_0^t \sigma(X_s)dB^H_{E_s},$$

where E_t is the inverse process to the Lévy's stable subordinator of the stability index $\beta \in (0,1)$. Recall that the FPK equation associated with an SDE driven by an fBM (without time-change) has the form

$$\frac{\partial u(t,x)}{\partial t} = L_\gamma(t,x,D_x)u(t,x),$$

where $L_\gamma(t,x,D_x)$ is defined in (7.94) with operators B and A defined in (7.90) and (7.91), respectively. Here the Hurst parameter H is connected with γ via $2H - 1 = \gamma$. Again for simplicity, we first consider a time-change process E_t inverse to a single stable subordinator W_t, and then E^μ_t, the inverse to a mixture of stable subordinators with mixing measure μ.

Theorem 7.14. *([HKU11]) Let $u(t,x)$ be a solution to the initial value problem*

$$\frac{\partial u(t,x)}{\partial t} = B(x,D_x)u(t,x) + \frac{\gamma+1}{2} t^{\gamma} A(x,D_x)u(t,x), \quad t>0,\ x\in\mathbb{R}^n, \tag{7.115}$$

$$u(0,x) = \varphi(x),\ x\in\mathbb{R}^n. \tag{7.116}$$

Let $f_t(\tau)$ be the density function of the process inverse to a stable subordinator of index β. Then $v(t,x) = \int_0^\infty f_t(\tau)u(\tau,x)d\tau$ satisfies the following initial value problem for a fractional order differential equation

$$D_*^{\beta} v(t,x) = B(x,D_x)v(t,x) + \frac{\gamma+1}{2} G_{\gamma,t} A(x,D_x)v(t,x), \quad t>0,\ x\in\mathbb{R}^n, \tag{7.117}$$

$$v(0,x) = \varphi(x), \quad x\in\mathbb{R}^n, \tag{7.118}$$

where the operator $G_{\gamma,t}$ acts on the variable t and is defined by (7.99).

Theorem 7.15. *[HKU11] Let $u(t,x)$ be a solution to the initial value problem (7.115)–(7.116). Let $f_t^{\mu}(\tau)$ be the density function of the process inverse to W_t^{μ}. Then*

$$v(t,x) = \int_0^\infty f_t^{\mu}(\tau)u(\tau,x)d\tau$$

satisfies the following initial value problem for a fractional order differential equation

$$D_{\mu} v(t,x) = B(x,D_x)v(t,x) + \frac{\gamma+1}{2} G_{\gamma,t}^{\mu} A(x,D_x)v(t,x), \quad t>0,\ x\in\mathbb{R}^n, \tag{7.119}$$

$$v(0,x) = \varphi(x),\ x\in\mathbb{R}^n. \tag{7.120}$$

The operator $G_{\gamma,t}^{\mu}$ acts on the variable t and is defined by (7.105).

Proof. The proofs of these theorems follow immediately from abstract Theorems 7.11 and 7.12, respectively.

Example 7.7. 1. If $H = 1/2$, then B_t^H is Brownian motion. In this case Theorem 7.11 implies the fractional FPK equation obtained in (7.46)–(7.47). Similarly, Theorem 7.12 reduces to Theorem 7.1.

2. Consider the following equation ($0 < H < 1$):

$$\frac{\partial h}{\partial t}(t,x) = 2Ht^{2H-1} a \frac{\partial^2 h}{\partial x^2}(t,x).$$

This equation was obtained in the paper [MNX09] as a governing equation for fBM. Theorem 7.11 and Proposition 7.5 imply that the governing equation for the corresponding time-changed fBM is either of the following equivalent forms:

$$D_*^{\beta} h(t,x) = 2HG_{2H-1,t}\, a \frac{\partial^2 h}{\partial x^2}(t,x),$$

$$G_{1-2H,t} D_*^{\beta} h(t,x) = 2Ha \frac{\partial^2 h}{\partial x^2}(t,x).$$

Remark 7.8. The formula $v(t,x) = \mathscr{F}u(t,x)$ for a solution of FPK equations associated with time-changed fBM provides a useful tool for analysis of properties of a solution to initial value problem (7.97)–(7.98), (7.103)–(7.104), and (7.106)–(7.107).

7.14.2 Fractional FPK equation for LFSM

The method used above can be applied for derivation of the fractional FPK equation for SDEs driven by time-changed Lévy's stable processes and linear fractional stable motions (LFSM).

The Lévy symbol of a one-dimensional Lévy's α-stable (not necessarily symmetric) process L_t for $t = 1$ is given by

$$\psi(\xi) = ia\xi - b|\xi|^{\alpha}\{1 - i\beta\frac{\xi}{|\xi|}\omega(\xi,\alpha)\}, \tag{7.121}$$

where a, b, α, β are constants, a is real, $b > 0$, $0 < \alpha \leq 2$, $-1 < \beta < 1$ and

$$\omega(\xi,\alpha) = \begin{cases} tan(\frac{\pi}{2}\alpha), & \text{if } \alpha \neq 1; \\ \frac{2}{\pi}log|\xi|, & \text{if } \alpha = 1. \end{cases}$$

One can easily see that the case $a = 0, \beta = 0$ corresponds to the symmetric distribution. The corresponding FPK equation in the 1-D case has the form

$$\frac{\partial u}{\partial t} = -a\frac{\partial u}{\partial x} + Dq\frac{\partial^{\alpha} u}{\partial(-x)^{\alpha}} + Dp\frac{\partial^{\alpha} u}{\partial(x)^{\alpha}}, \quad t > 0,\ x \in \mathbb{R},$$

where D is some constant depending on α, $p \geq 0$, $q \geq 0$ and $p + q = 1$, and in multi-dimensional case has the form [MBB01]

$$\frac{\partial u(t,x)}{\partial t} = -\mathbf{a}\nabla u(t,x) + D\nabla_M^{\alpha} u(t,x), \quad t > 0,\ x \in \mathbb{R}^n,$$

where $\mathbf{a} \in \mathbb{R}^n$ and ∇^{α} is the pseudo-differential operator with the symbol

$$\int_{|\theta|=1} (-i\xi,\theta)^{\alpha} M(d\theta)$$

with $M(d\theta)$, a probability measure on the unit sphere. If $a = 0$ and

$$M(d\theta) = \frac{\Gamma(1+\frac{n}{2})d\theta}{n\pi^{n/2}},$$

then we get the symmetric case considered above. Setting $\gamma = 0$, $B = -a\nabla$, and $A = D\nabla_M^{\alpha}$ in Theorem 7.12, we obtain the following assertion.

Theorem 7.16. *Let L_t be an n-dimensional Lévy's α-stable process. Let E_t be the inverse to the mixture of Lévy's stable subordinators with a mixing measure μ, and independent of L_t. Then the density function $p(t,x)$ of the time-changed process L_{E_t} satisfies the following initial value problem*

$$D_\mu p(t,x) = -a\nabla p(t,x) + D\nabla_M^\alpha p(t,x), \quad t > 0,\ x \in \mathbb{R}^n,$$
$$p(0,x) = \delta_0(x), \quad x \in \mathbb{R}^n.$$

Theorem 7.17 below is an application of Theorem 7.11 to linear LFSM. Let $L_{\alpha,H}$, $0 < \alpha < 2$, $0 < H < 1$, be a LFSM. Then its density solves the following equation [MNX09]

$$\frac{\partial u(t,x)}{\partial t} = \alpha H t^{\alpha H-1}[ap\partial_x^\alpha u(t,x) + aq\partial_{-x}^\alpha u(t,x)] \tag{7.122}$$

where $p+q=1$, ∂_x^α and ∂_{-x}^α are space-fractional order derivatives in the sense of Liouville. Denoting $\gamma = \alpha H - 1$ and $A = 2a[ap\partial_x^\alpha + aq\partial_{-x}^\alpha]$, one can rewrite equation (7.122) in the form (7.95) with $B = 0$ and the initial condition

$$u(0,x) = \varphi(x), \quad x \in (-\infty,\infty). \tag{7.123}$$

Theorem 7.17. *Let $u(t,x)$ be a solution to the Cauchy problem (7.122), (7.123). Let $f_t^\mu(\tau)$ be the density function of the process W_t^μ. Then $v(t,x) = \int_0^\infty f_t^\mu(\tau)u(\tau,x)d\tau$ satisfies the following initial value problem for a fractional order differential equation*

$$D_\mu v(t,x) = \alpha H t^{\alpha H-1} G_{\gamma,t}^\mu [ap\partial_x^\alpha u(t,x) + aq\partial_{-x}^\alpha u(t,x)]v(t,x),$$
$$t > 0, x \in (-\infty,\infty),$$
$$v(0,x) = \varphi(x), \quad x \in (-\infty,\infty),$$

where the operator $G_{\gamma,t}^\mu$, $\gamma = \alpha H - 1$, acts in the variable t, and is defined in (7.105).

7.14.3 Fractional FPK equation associated with time-changed infinite-dimensional Wiener process

Theorem 7.12 can be applied to fractional FPK equations in the infinite dimensional case. Below we consider only the simplest case. Let H be an infinite dimensional separable Hilbert space, and Q be a positive definite trace operator on H. In this section we suppose B_t is the infinite dimensional Wiener process associated with the operator Q. We refer the reader to [DPZ02] for the definition and properties of the infinite dimensional Wiener process. Then the corresponding FPK equation has the form (see [DPZ02])

$$\frac{\partial u(t,x)}{\partial t} = \frac{1}{2}Tr[QD^2u(t,x)], \quad t > 0, x \in H, \tag{7.124}$$

where Tr stands for the trace, and D^2 is the second order Fréchet derivative. Note that equation (7.124) is the FPK equation associated with the simplest Itô SDE $dX_t = dB_t$. Denote $A(\cdot) = \frac{1}{2}Tr[QD^2 \cdot]$ with $Dom(A) = UC_b^2(H)$, the space of functions $u : H \to R$ such that D^2u is uniformly continuous and bounded. Then applying Theorem 7.12 with $B = 0$ and $\gamma = 0$, one obtains the following theorem.

Theorem 7.18. *Let $u(t,x)$ be a strong solution to equation (7.124) with the initial condition $u(0,x) = \varphi(x)$, $\varphi \in UC_b(H)$. Let $f_t^\mu(\tau)$ be the density function of the process W_t^μ. Then $v(t,x) = \int_0^\infty f_t^\mu(\tau)u(\tau,x)d\tau$ is a strong solution to the following initial value problem for the infinite dimensional time-fractional distributed order differential equation*

$$D_\mu v(t,x) = \frac{1}{2}Tr[QD^2 v(t,x)], \quad t > 0, x \in H,$$

$$v(0,x) = \varphi(x), \quad x \in H.$$

Remark 7.9. The associated stochastic process is, obviously, the time-changed Wiener process $X_t = B_{W_t}$.

7.15 Filtering problem: fractional Zakai equation

The filtering problem is a wide generalization of the concept discussed in this chapter. Namely, in the filtering problem one is interested in a stochastic process under additional information obtained from observation/measurement. The FPK equations correspond to the particular case, when additional information consists of only the initial condition. The additional information obtained through certain measurements form a sigma-algebra of events. Given this sigma algebra one needs to optimize the state process. As a result, the FPK counterpart of the filtering process is not deterministic, but is a stochastic partial differential equation. The latter is called a Zakai equation, which was first derived by Zakai [Zak69] in 1969.

In this section we are interested in the fractional Zakai type equations, which describe filtering problems whose state and observation processes are driven by a time-changed Brownian motion (or other standard driving processes). Below we will show a derivation of the fractional Zakai equation, in which the time-change process is the inverse to a Lévy stable subordinator with the stability index $\beta \in (0,1)$ and discuss existence and uniqueness of a solution, as well as some methods of solution.

We have seen above that if one is interested in a solution of an SDE conditioned on the value at the initial time $t = 0$, then the associated FPK equation is a deterministic PDE (Section 7.4). In the filtering problem one has information of the past for all times s, $0 \le s < t$, coming from observations (measurements). Suppose

$$Z_t = \int_0^t h(X_s)ds + W_t, \tag{7.125}$$

are R^m-valued measurements, or *observations* related to the process X_t in the noisy environment. Let $\mathscr{Z}_t$ be a σ-algebra generated by the measurement process Z_t. One of the formulations of the filtering problem is to find the best estimation of X_t at time t in the mean square sense, given $\mathscr{Z}_t$. Namely, to find a stochastic process X_t^* such that

$$E[\|X_t - X_t^*\|^2] = \inf_{\{Y_t\}} E[\|X_t - Y_t\|^2],$$

where inf is taken over all stochastic processes $Y_t \in L^2(\mathbb{P})$ under the condition that the sigma-algebra $\mathscr{Z}_t$ is given. It follows from the abstract theory of functional analysis that X_t^* is the projection of X_t onto the space of stochastic processes $\mathscr{L}(Z_t) = \{Y \in L^2(\mathbb{P}) : \textit{given } \mathscr{Z}_t\}$. The latter can be written in the form $X_t^* = \mathbb{E}[f(X_t)|\mathscr{Z}_t]$, generalizing (7.26) from the initial condition $X_0 = x$ to the entire history $\mathscr{Z}_t$. Hence, the filtering problem comprises of SDEs

$$dY_t = b(Y_t)dt + \sigma(Y_t)dB_t, \quad Y_{t=0} = X_0, \tag{7.126}$$

called a state process, and

$$dZ_t = h(Y_t)dt + dW_t, \quad Z_0 = 0, \tag{7.127}$$

called an observation process obtained from (7.125) by differentiating. Brownian motion W_t is assumed to be independent of B_t and the initial random variable X_0.

This problem was first posed and solved in the linear case by Kalman and Bucy [KB61] in 1961. The filtering problem is still under active development due to its significant applications. In the linear case Kalman and Bucy [KB61] reduced the filtering problem to a linear SDE and a deterministic Riccati type differential equation. In the case of *nonlinear filtering* Kushner [Kus67], Lipster and Shiryaev [LS02], and Fujisaki, Kallianpur and Kunita [FKK72] obtained a nonlinear infinite dimensional stochastic differential equations for the posterior conditional density of X_t given $\mathscr{Z}_t$. However, two issues arise:

(1) it is not easy to solve these equations, and
(2) it is computationally 'expensive' due to the two-stage calculation procedure (prediction and correction) in the real time.

In 1969 Zakai [Zak69] suggested a simpler approach, reducing the solution of the filtering problem to a partial stochastic differential equation for the posterior unnormalized conditional density $\Phi(t,x) = p(t,x|\mathscr{Z}_t)$ for X_t. Below we briefly sketch this method. Introduce the process

$$\rho(t) = \exp\{-\sum_{k=1}^{m} \int_0^t h_k(Y_s)dW_s - \frac{1}{2}\int_0^t |h(Y_s)|^2 ds\}$$

and the probability measure $d\mathbb{P}_0 = \rho(t)d\mathbb{P}$. Further, let

$$\Lambda_t = \frac{d\mathbb{P}}{d\mathbb{P}_0}\Big|_{\mathscr{Z}_t},$$

and $\hat{E}$ be the expectation under the reference measure $\mathbb{P}_0$. Then, as is known, the optimal solution of the filtering problem (7.126), (7.127) is given by the following Kallianpur-Striebel's formula (see, e.g., [Roz90])

$$\mathbb{E}[f(Y_t)|\mathscr{Z}_t] = \frac{\hat{\mathbb{E}}[f(Y_t)\Lambda_t|\mathscr{Z}_t]}{\hat{\mathbb{E}}[\Lambda_t|\mathscr{Z}_t]}. \tag{7.128}$$

Moreover, under some mild conditions the unnormalized filtering measure $p_t(f) = \hat{\mathbb{E}}[f(Y_t)\Lambda_t|\mathscr{Z}_t]$ satisfies the following stochastic differential equation, called the Zakai equation:

$$p_t(f) = p_0(f) + \int_0^t p_s(Af)ds + \sum_{k=1}^m \int_0^t p_s(h_k f)dZ_s^k, \tag{7.129}$$

where A is a second order elliptic differential operator given by equation (7.29). Further, introducing the filtering density $U(t,x)$ through

$$p_t(f) = \int_{\mathbb{R}^n} f(x)U(t,x)dx,$$

one can show that $U(t,x)$ solves the following partial stochastic differential equation (called an adjoint Zakai equation)

$$dU(t,x) = A^*U(t,x)dt + \sum_{k=1}^m h_k(x)U(t,x)dZ_k(t), \tag{7.130}$$

with the initial condition $U(0,x) = p_0(x)$. Here A^* is the adjoint operator of A defined in (7.29). Thus, if one has a solution of equation (7.130), then one will be able to establish a solution to the original filtering problem using Kallianpur-Striebel's formula (7.128). Equation (7.130) reduces to FPK equation (7.28) if the observation process Z_t stays constant in time, which means no additional information is obtained from measurement/observation.

Now consider a filtering problem with the state process and observation process driven by time-changed Brownian motions. Namely, suppose the state process is

$$dX_t = f(X_t)dT_t + \sigma(X_t)dB_{T_t}, \quad X_{t=0} = X_0, \tag{7.131}$$

and the observation process is

$$dZ_t = h(t,X_t)dT_t + dW_{T_t}, \quad Z_0 = 0, \tag{7.132}$$

where T_t is the inverse of the Lévy stable subordinator with the stability index $\beta \in (0,1)$, and independent of B_t and W_t. We show that the Zakai equation corresponding to this problem has the form

$$\Phi(t,x) = p_0(x) + \int_0^t A^*\Phi(s,x)dT_s + \int_0^t h_s(x)\Phi(s,x)dZ_{T_s}. \tag{7.133}$$

A few remarks before deriving this equation. The stochastic integral in equation (7.133) is well defined in the sense of Itô's integral. If $\beta \to 1$, then we recover the classical Zakai equation, since $T_t = t$ in this case. Hence, equation (7.133) generalizes the classic Zakai equation (7.130) for the case of filtering problem with time-changed driving processes. Note also that the time-changed process B_T is not Markovian and has no independent increments. Therefore, the model (7.131), (7.132) can be applied to a class of correlated state processes. An important question is the existence and uniqueness (in an appropriate sense) of a solution for this new Zakai equation. We will discuss this question in this section as well as some solution methods useful from the application point of view.

The fractional, or time-changed version of the Zakai equation in the general case of time-changed Lévy processes is obtained in the paper [UDN14] for filtering problems driven by Lévy processes. For completeness, we demonstrate the derivation of the fractional Zakai equation in our particular case of filtering problem (7.131)–(7.132). We assume that the following conditions on the input data of the filtering problem:

(C1) the vector-functions $f(x)$, $h(x)$, and $n \times m$-matrix-function $\sigma(x)$ satisfy the Lipschitz and linear growth conditions:

$$\|f(x)-f(y)\|^2 + \|h(x)-h(y)\|^2 + |||\sigma(x)-\sigma(y)|||^2 \le C_1\|x-y\|^2, \quad \forall x,y \in \mathbb{R}^n;$$
$$\|b(x)\|^2 + \|h(x)\|^2 + |||\sigma(x)|||^2 \le C_2(1+\|x\|^2), \quad \forall x \in \mathbb{R}^n,$$

where $\|\cdot\|$ and $|||\cdot|||$ are vector- and matrix-norms, respectively.

(C2) the time-change process T_t and Brownian motions B_t and W_t are independent processes;

(C3) the initial random vector X_0 is independent of processes B_t, W_t, and T_t and has an infinite differentiable density function $p_0(x)$ decaying at infinity faster than any power of $|x|$.

Theorem 7.19. *Let the conditions (C1)-(C3) be verified. Then the filtering density $\Phi(t,x)$ associated with the filtering measure $\phi_t(f) = \hat{\mathbb{E}}[f(X_t)\Lambda_{T_t}|\mathscr{V}_t]$, where $\mathscr{V}_t$ is the filtration generated by $V_t = Z_{T_t}$, satisfies the following Zakai equation*

$$\Phi(t,x) - \Phi(0,x) = \int_0^t A^*\Phi(s,x)dT_s + \sum_{k=1}^m \int_0^t h_k(x)\Phi(s,x)dZ_{T_s}^{(k)}. \tag{7.134}$$

Proof. Let conditions (C1)-(C3) be verified. Then, in particular, the conditions for the existence of an unnormalized filtering distribution $p_t(f) = \hat{\mathbb{E}}[f(Y_t)\Lambda_t|\mathscr{Z}_t]$ which solves the Zakai equation (7.129), are also verified. Here Y_t is a solution to stochastic differential equation (7.126). According to Theorem 3.3 in [HKU10] the time-changed process $X_t = Y_{T_t}$ solves stochastic differential equation (7.131).

The connection $X_t = Y_{T_t}$ between the state processes X_t and Y_t implies the connection $V_t = Z_{T_t}$ between the observation processes V_t and Z_t. Indeed, letting $T_t = \tau$, or the same $D_\tau = t$, one obtains from the relation $dV_t = h(Y_{T_t})dT_t + dW_{T_t}$ and from (7.127) that $Z_\tau = V_{D_\tau}$, or the same $V_t = Z_{T_t}$. It follows that the filtration $\mathscr{V}_t$ coincides with the filtration $\mathscr{Z} \circ \mathscr{T}_t \equiv \mathscr{Z}_{\mathscr{T}_t}$ generated by the time-changed observation process Z_{T_t}. Hence, the unnormalized filtering distribution $\phi_t(f) = \hat{\mathbb{E}}[f(X_t)\Lambda_{T_t}|\mathscr{Z}\circ\mathscr{T}_t]$ corresponding to the filtering problem (7.131), (7.132) is the time-changed process

$$\phi_t(f) = p_{T_t}(f). \tag{7.135}$$

Therefore, due to equation (7.19) the process $\phi_t(f)$ satisfies

$$\phi_t(f) = p_{T_t}(f) = p_0(f) + \int_0^{T_t} p_s(Af)ds + \sum_{k=1}^{m}\int_0^{T_t} p_s(h_k f)dZ_s^{(k)}. \tag{7.136}$$

Further, using the change of variable formula (see [Jac79], Proposition 10.21) $\int_0^{T_t} H_s dS_s = \int_0^t H_{T_s-}dS_{T_s}$, for stochastic integrals driven by a semimartingale S_t, we obtain

$$\begin{aligned}\int_0^{T_t} p_s(Af)ds &= \int_0^t \hat{\mathbb{E}}[Af(Y_{T_s})\Lambda_{T_s}|\mathscr{Z}_{T_s}]dZ_{T_s}^{(k)} = \int_0^t \hat{\mathbb{E}}[Af(X_s)\Lambda_{T_s}|\mathscr{Z}_{T_s}]dZ_{T_s}^{(k)} \\ &= \int_0^t \phi_s(Af)dZ_{T_s}^{(k)}.\end{aligned} \tag{7.137}$$

and

$$\begin{aligned}\sum_{k=1}^{m}\int_0^{T_t} p_s(h_k f)dZ_s^{(k)} &= \sum_{k=1}^{m}\int_0^t \hat{\mathbb{E}}[h_k(Y_{T_s})f(Y_{T_s})\Lambda_{T_s}|\mathscr{Z}_{T_s}]dZ_{T_s}^{(k)} \\ &= \sum_{k=1}^{m}\int_0^t \hat{\mathbb{E}}[h_k(X_s)f(X_s)\Lambda_{T_s}|\mathscr{Z}_{T_s}]dZ_{T_s}^{(k)} \\ &= \sum_{k=1}^{m}\int_0^t \phi_s(h_k f)dZ_{T_s}^{(k)}.\end{aligned} \tag{7.138}$$

Equations (7.136), (7.137), and (7.138) imply the desired equation (7.134).

Let T_t be the inverse to a stable Lévy subordinator D_t of a stability index $\beta \in (0,1)$ and let the stochastic processes $\Pi_t(f)$ and $\Pi_{t,Z}(f)$ are defined by

$$\Pi_t(f) = Ap_t(f) = \int_0^\infty g_t(\tau)p_\tau(f)d\tau, \tag{7.139}$$

$$\Pi_{t,Z}(f) = Cp_t(f) = \int_0^\infty g_t(\tau)p_\tau(f)dZ_\tau. \tag{7.140}$$

where $g_t(\tau)$ is the density function of the process T_t and $p_t(f)$ is the unnormalized filtering distribution of the Zakai equation (7.129) corresponding to the filtering model (7.126)–(7.127). Then it follows from equation (7.134) that the following stochastic relation holds:

$$\Pi_t(f) - p_0(f) = J_t^\beta\Big(\Pi_t(Af) + \sum_{k=1}^{m}\Pi_{t,Z^{(k)}}(h_k f)\Big), \tag{7.141}$$

where J_t^β is the fractional integration operator of order β.

Let B map the class of stochastic processes $\Pi_t(f)$ to the class of processes $\Pi_{t,Z}(f)$, that is $\Pi_{t,Z}(f) = B\Pi_t(f)$. One can verify easily that the operator B can be expressed with the help of operators A and C in equation (7.139). Namely,

$$B = CA^{-1}.$$

Using $L^2(\mathbb{P})$-norm and calculus of stochastic processes one can show that A is a one-to-one bounded linear operator and C is a bounded linear operator. Therefore, it follows that operator B is well defined bounded linear operator. We note that equation (7.141) can be written in the form

$$\Pi_t(f) - p_0(f) = J_t^{\beta}\left(\Pi_t(Af) + \sum_{k=1}^{m} B_k\Pi_t(h_k f)\right),$$

where

$$B_k\Pi_t(f) = \Pi_{t,Z^{(k)}}(f).$$

The differential form of (7.141) involves a fractional derivative in the Riemann-Liouville sense

$$d\Pi_t(f) = \mathscr{D}_t^{1-\beta}\Pi_t(Af)dt + \sum_{k=1}^{m}\mathscr{D}_t^{1-\beta}B_k\Pi_t(h_k f)dt,\ \Pi_{t=0}(f) = p_0(f). \quad (7.142)$$

The latter in terms of unnormalized densities associated with the process $\Pi_t(f)$ can be represented in the form

$$D_*^{\beta}U(t,x) = A^*U(t,x) + \sum_{k=1}^{m} h_k(x)B_kU(t,x),\ U(0,x) = f(x), \quad (7.143)$$

where D_*^{β} is the fractional derivative in the sense of Caputo. Equation (7.143) generalizes the forward version (that is A^* instead of A) of fractional FPK equation (7.46) to the case of fractional adjoint Zakai equation.

Theorem 7.20. *Let the conditions (C1)-(C3) be verified. Then there exists a unique filtering density $\Phi(t,x)$ satisfying the fractional Zakai equation* (7.134). *Moreover, there exist uniquely defined stochastic processes in equation* (7.139) *satisfying initial value problem* (7.142).

Proof. Suppose there are two filtering densities $\Phi_1(t,x)$ and $\Phi_2(t,x)$ such that both satisfy equation (7.134). Then the process $\Psi(t,x) = \Phi_2(t,x) - \Phi_1(t,x)$ satisfies the following equation

$$\Psi(t,x) = \int_0^t A^*\Psi(s,x)dT_s + \sum_{k=1}^{m}\int_0^t h_k(x)\Psi(s,x)dZ_{T_s}^{(k)}.$$

This is an SDE driven by time-changed processes T_s given Z_{T_s}. Its counterpart with non-time-changed process has the form

$$V(t,x) - V_0(x) = \int_0^t A^* V(s,x) ds + \sum_{k=1}^{m} \int_0^t h_k(x) V(s,x) dZ_s^{(k)}, \tag{7.144}$$

with $V_0(x) \equiv 0$. The solutions $V(t,x)$ and $\Psi(t,x)$ are related through $\Psi(t,x) = V(T_t,x)$. Equation (7.144) has a unique solution (see, e.g., [Roz90]). Since the initial condition is $V(0,x) = 0$, then the corresponding solution $V(t,x) \equiv 0$ in the sense of $L^2(\mathbb{P})$. This implies $\Psi(t,x) \equiv 0$, or the same, $\Phi_1(t,x) = \Phi_2(t,x)$ in the sense of $L^2(\mathbb{P})$. The latter, in turn, implies that the process defined in equation (7.139) is unique.

How to solve a fractional filtering problem? Knowing a solution of a nonlinear filtering problem one can use it for solution of the associated fractional nonlinear filtering problem. Two approaches to the solution of nonlinear filtering problems are commonly used. Namely,

(1) direct solution of filtering problem (7.131)–(7.132).
(2) solution of the Zakai equation followed by the Kallianpur-Striebel formula.

For the filtering problem with no time-changed driving processes both approaches are well studied; see, e.g., works [Ku90, Roz90, Da87, IX00] for the first approach, and [Zak69, BGR90, BK96, LMR97] for the second approach. Both type of Zakai equations (7.134) and (7.142) (or its adjoint form (7.143)) are of great interest in various applications of fractional filtering problems. For solutions of these equations, due to relations (7.135) and (7.140), the following formulas are important:

$$\Phi_t(f) = p_{T_t}(f)$$

and

$$\Pi_t(f) = \int_0^\infty g_t(\tau) p_\tau(f) d\tau,$$

where $p_t(f)$ is the solution of non-time-changed filtering problem and $g_t(\tau)$ is the density function of the time-changed process T_t. Therefore, in the first step one needs to find the stochastic process $p_t(f)$ which solves the classical Zakai equation. Then using the above formulas one can find solutions to fractional Zakai equations. Like the non-fractional case, one can develop analytic and numerical methods, and methods for solution of filtering problem directly, or through the associated Zakai equation. Accordingly, in the fractional case the methods can be developed for solution of the three following situations:

(a) Direct solution of the fractional filtering problem;
(b) Solution of the fractional filtering problem through the Zakai equation (7.134);
(c) Solution of the fractional filtering problem through the adjoint Zakai equation (7.143).

7.16 Additional notes

1. *Brownian motion.* The term "Brownian motion" was coined in one of Albert Einstein's *Annus Mirabilis* 1905 papers, titled "Über die von der molekularkinetischen Theorie der Wärme geforderte Bewegung von in ruhenden Flüssigkeiten suspendierten Teilchen" ("On the Motion of Small Particles Suspended in a Stationary Liquid, as Required by the Molecular Kinetic Theory of Heat"). In this paper A. Einstein first provided theoretical explanation of Brownian motion from the point of view of thermal diffusion. A little earlier (in 1900) Bachelier published his doctoral dissertation "Théorie de la spéculation" ("The Theory of Speculation") modeling Brownian motion from the economics point of view, founding financial mathematics. In 1908 Langevin published his work with a stochastic differential equation which was "understood mathematically" only after a stochastic calculus was introduced by Itô in 1944–48. The Fokker-Planck equation, a deterministic form of describing of the dynamics of a random process in terms of transition probabilities, was invented in 1913–17. Its complete "mathematical understanding" become available after the appearance of the distribution (generalized function) theory (Sobolev, 1938; Schwartz, 1951) and was embodied in Kolmogorov's backward and forward equations (1931). Deep mathematical properties of Brownian motion, like nowhere differentiability and infinite total variation over arbitrary time interval, were studied by N. Wiener in his paper [Wie28] published in 1927.
2. *Fractional FPK equation.* The classic FPK equation establishes a relationship between Itô's stochastic differential equation driven by Brownian motion and its associated partial differential equation. In fact, this is a triple relationship. Indeed, changing the driving process, one gets a different FPK equation. In the paper [MGZ14] a fractional Fokker-Planck equation is obtained in the form
$$\frac{\partial u}{\partial t} = \left[-\frac{\partial F(t,x)}{\partial x} + \frac{1}{2}\frac{\partial^2 D(t,x)}{\partial x^2} \right] D^{1-\alpha} u, \quad t > 0,\ x \in \mathbb{R}, \tag{7.145}$$
with the initial condition $u(0,x) = \delta_0(x)$. Here $D^{1-\alpha}$ is the Riemann-Liouville fractional derivative, $F(t,x)$ and $D(t,x)$ are the drift and diffusion coefficients, respectively. If $F(t,x) = F(x)$ and $D(t,x) = D(x)$, i.e., do not depend on the time variable t, then one can easily verify using Proposition 3.1 that (7.145) is equivalent to the forward version of the fractional FPK equation in (7.46). Equation (7.145) in the case $F(x,t) = F(x)$ and $D(t,x) = const$ was first established using CTRW approach in [MBK99], in the case $F(x,t) = F(t)$ and $D(t,x) = const$ in [SK06], and in the case $F(x,t) = F(x)f(t)$ and $D(t,x) = D(x)d(t)$ in [LQR12]. Fractional FPK equations associated with SDEs with time-independent coefficients and driven by a time-changed Lévy processes are studied in the papers [HKU10, HKU11, HU11, HKRU11]. Another approach to the theory of FPK is based on the Tsallis entropy, which leads to a nonlinear equation of the form [Tsa09]
$$\frac{\partial u^{\mu}}{\partial t} = -\frac{\partial}{\partial x}[F(x)u^{\mu}] + D\frac{\partial^2 [u^{\nu}]}{\partial x^2}, \quad t > 0,\ x \in \mathbb{R},$$
where $(\mu,\nu) \in \mathbb{R}^2$, $D > 0$ is a diffusion constant, and $F(x)$ is a drift coefficient. The solution of this equation under certain conditions is given by Tsallis' q-Gaussian.
3. *Feynman-Kac formula.* The relationship between the stochastic process X_t in (7.18) and another associated partial differential equation
$$\frac{\partial w}{\partial t} = \mathscr{A} w - qw,\ w(0,x) = \varphi(x),$$
for a nonnegative continuous function q, is given by the Feynman-Kac formula:
$$w(t,x) = \mathbb{E}\left[\exp\left(-\int_0^t q(X_s)ds \right) \varphi(X_t) \middle| X_0 = x \right].$$

4. *Martingales, Poisson processes and random measures.* Brownian motion B_t adapted to a filtration $\mathscr{F}_t$ possess the following property: $\mathbb{E}[B_t|\mathscr{F}_s] = B_s$. This property is called a martingale property of Brownian motion. A stochastic process X_t adapted to a filtration $\mathscr{F}_t$ is called a *martingale,* if it satisfies the following conditions:

   a. $\mathbb{E}[|X_t|] < \infty$ for all $t \geq 0$;
   b. $\mathbb{E}[X_t|\mathscr{F}_s] = X_s$, for all $0 \leq s \leq t$.

   One of the general properties of martingales is $\mathbb{E}[X_t]$ does not depend on t, that is constant, if X_t is a martingale. For example, for Brownian motion $\mathbb{E}[B_t] = 0$ for all $t \geq 0$.
   A process $N(t)$ with values on $\mathbb{N}_0$, defined on a probability space $(\Omega, \mathscr{F}, \mathbb{P})$, and adapted to a filtration $\mathscr{F}_t$ is a *Poisson process* with an intensity $\lambda > 0$, if $N(0) = 0$, $N(t) - N(s)$ and $N(b) - N(a)$ are independent for any nonoverlapping intervals (s,t) and (a,b) of the semiaxis $\mathbb{R}_+$, has stationary increments, and

   $$\mathbb{P}(N(t) = n) = \frac{(\lambda t)^n}{n!} e^{-\lambda t}, \quad n \in \mathbb{N}_0, \quad \forall t \geq 0.$$

   It is easy to see that the characteristic function of N_t is

   $$\mathbb{E}[\exp(i\xi N(t))] = e^{\lambda t(e^{i\xi}-1)}. \tag{7.146}$$

   The Poisson process $N(t)$ is not a martingale. Indeed, its expected value $\mathbb{E}[N(t)] = \lambda t$, that is t-dependent. However, the process $\tilde{N}(t) = N(t) - t\lambda$, called a *compensated Poisson process,* is a martingale (try to show this!). Let Y_i be a sequence of independent and identically distributed random variables independent of the Poisson process $N(t)$. Then the process

   $$X_t = \sum_{i=1}^{N(t)} Y_i$$

   is called a compound Poisson process.
   Let μ be a measure defined on a measurable space $(S, \mathscr{E})$, that is for each set $A \in \mathscr{E}$ its measure $\mu(A)$ is defined. Let $(\Omega, \mathscr{F}, \mathbb{P})$ be a probability space. By Poisson random measure we mean a (random) measure $N : (\Omega, \mathscr{E}) \to \mathbb{R}_+$, such that

   a. For each fixed ω, $N(\omega, A)$ is a measure on $(S, \mathscr{E})$, and for any fixed $A \in \mathscr{E}$ the random variable $N(\omega, A)$ is $\mathbb{P}$-measurable;
   b. $N(\omega, \emptyset) = 0$;
   c. For arbitrary disjoint collection of sets $A_1, \ldots, A_n \in \mathscr{E}$ random variables $N(A_1), \ldots, N(A_n)$ are independent;
   d. For all $A \in \mathscr{E}$ with $\mu(A) < \infty$, the random variable $N(\omega, A)$ has the Poisson distribution with the intensity parameter $\mathbb{E}[N(A)] = \mu(A)$.

   The first integral on the right of equation (7.33) is performed with respect to the compensated Poisson random measures $\tilde{N}(t, dw) = N(t, \nu(dw)) - t\nu(dw)$, where ν is a Lévy measure. The random measure $\tilde{N}(t, dw)$ is a martingale-valued measure. The second integral in this representation is performed with respect to the Poisson random measure $N(t, \nu(dw))$, and this integral represents a compound Poisson process [Sat99, App09].
5. *Lévy processes.* Lévy processes were introduced in the 1930th by Paul Lévy. Today there are many books on Lévy processes among which we would like to indicate Bertoin [Ber96], Sato [Sat99], Barndorff-Nielsen, Mikosch, Resnick (Editors) [BMR01], Rong [Sit05], and Applebaum [App09]. One of the important properties of Lévy processes is they are infinite divisible. Such processes are characterized by Lévy-Khinchin representation. Namely, for a Lévy process L_t its characteristic function $\mathbb{E}[e^{i\xi L_t}] = e^{t\Psi(\xi)}$, where $\Psi(\xi)$ has the form given in (7.32):

   $$\Psi(\xi) = i(b, \xi) - \frac{1}{2}(\Sigma\xi, \xi) + \int_{\mathbb{R}^n \setminus \{0\}} (e^{i(w,\xi)} - 1 - i(w, \xi)\chi_{(|w| \leq 1)}(w))\nu(dw). \tag{7.34}$$

The function $\Psi(\xi)$, called a Lévy symbol, is continuous, Hermitian, conditionally positive definite, and satisfies $\Psi(0) = 0$ [App09]. The triplet (b, Σ, ν) uniquely defines the corresponding Lévy process. If $b = 0$ and $\Sigma = \Theta$, then the Levy process is a purely jump process. Comparing (7.146) with (7.32) one can be convinced that any Poisson process is a purely jump Lévy process.

6. *Semigroup.* The semigroup theory and boundary value problems for ΨDOSS are powerful tools for the study of Markovian diffusion processes. Many researchers have contributed to the development of the interrelation of semigroups, pseudo-differential operators, and Markov processes; see [Bo55, Co65, Tai91, FOT94, Jac01, App09, JSc02, Hoh00] and the references therein. In the fractional case solution operators do not possess the semigroup property. Bazhlekova [Baz98] studied strong continuity and analyticity of abstract solution operators to the Cauchy problem for fractional order differential-operator equations of orders α and $\beta, 0 < \alpha < \beta \leq 2$, with a closed operator A on the right, satisfying some conditions. In particular, she established the following relationship, called a subordination principle:
$$S_\alpha(t) = \int_0^\infty \phi_{t,\gamma}(s) S_\beta(s) ds, \quad t > 0, \tag{7.147}$$
where $\gamma = \beta/\alpha$, and $\phi_{t,\gamma}(s) = t^{-\gamma} M_\gamma(st^{-\gamma})$. Here $M_\gamma(z)$ is the M-Wright or Mainardi function discussed in Section 3.13. In the case $\alpha = 1$ and $\beta = 2$ relationship (7.147) implies the known fact $S_1(t) = e^{tA}$, which is a semigroup with the infinitesimal generator A. Relation (7.147) also allows to get a solution of the Cauchy problem for a fractional order α equation, through the solution of the second order equation ($\beta = 2$) with $A = k^2 d^2/dx^2$:
$$S_\alpha f(x) = \frac{1}{2kt^{\alpha/2}} \int_{\mathbb{R}} M_{\alpha/2}\left(\frac{|s|}{kt^{\alpha/2}}\right) f(x-s) ds.$$
This representation was obtained earlier in [Mai96].

7. *Càdlàg and semimartingales.* A càdlàg process, by definition, is right continuous with left limits. Any Lévy process has a càdlàg modification, which is again a Lévy process [App09]. Lévy processes are an important subclass of the so-called semimartingale processes. By definition, an $(\mathscr{F}_t)$*-semimartingale* is a càdlàg process Z_t which allows a decomposition
$$Z_t = Z_0 + M_t + A_t, \tag{7.148}$$
where Z_0 is $\mathscr{F}_0$-measurable, M_t is an $(\mathscr{F}_t)$-local martingale, and A_t is an $(\mathscr{F}_t)$-adapted càdlàg process of finite variation on compact sets. Rearranging the Lévy-Itô decomposition given in (7.33) of a Lévy process L_t, one can write it in the form $L_t = M_t + A_t$, where
$$M_t = \sigma B_t + \int_{|w|<1} w\tilde{N}(t, dw), \quad A_t = b_0 t + + \int_{|w|\geq 1} wN(t, dw), \tag{7.149}$$
confirming that L_t has form (7.148) with $Z_0 = 0$ and M_t and A_t in (7.149). Hence, L_t is a semimartingale. Semimartingales are the widest class of integrators for which the stochastic calculus in the sense of Itô can be extended; see details, for instance, in [Pro91].

8. *Time-change process. Itô's formula.* In our discussions in this chapter semimartingales have appeared in the context of time-changed Brownian motion or time-changed Lévy processes. By definition, an $(\mathscr{F}_t)$*-time-change* is a càdlàg, nondecreasing family of $(\mathscr{F}_t)$-stopping times. Let E_t be an $(\mathscr{F}_t)$-time-change and define a new filtration $(\mathscr{G}_t)$ by $\mathscr{G}_t := \mathscr{F}_{E_t}$. Then the right continuity of $\mathscr{F}_t$ and E_t yields that of $\mathscr{G}_t$. The following statement justifies that all the stochastic integrals, considered in this chapter and driven by a time-changed Brownian motion or time-changed Lévy process, are meaningful.

Proposition 7.7. ([Jac79]) *Let Z_t be an $(\mathscr{F}_t)$-semimartingale and let E_t be an $(\mathscr{F}_t)$-time-change. Then the time-changed process $Z_{E_t} = (Z \circ E)_t$ is a $(\mathscr{G}_t)$-semimartingale, where $\mathscr{G}_t = \mathscr{F}_{E_t}$.*

The Itô's formula plays an essential role in the theory of SDEs driven by semimartingales. It states that if Z_t is a semimartingale and $f \in C^2(\mathbb{R})$, then $f(Z_t)$ is again a semimartingale and

$$f(Z_t) - f(0) = \int_0^t f'(Z_{s-})dZ_s + \frac{1}{2}\int_0^t f''(Z_{s-})d[Z,Z]_s^c + \sum_{0<s\leq t} \{f(Z_s) - f(Z_{s-}) - f'(Z_{s-})\Delta Z_s\}.$$

We also note that Kobayashi [Kob11] proved an analog of Itô's formula for time-changed stochastic processes and applied it to solution of various SDEs driven by a time-changed process and to derivation of fractional FPK type equations.

Theorem 7.21. (Time-changed Itô Formula [Kob11]) *Let Z_t be an $(\mathcal{F}_t)$-semimartingale. Let D_t be a strictly increasing $(\mathcal{F}_t)$-semimartingale with $\lim_{t\to\infty} D_t = \infty$ and let E_t denote the inverse of D_t. Define a filtration $(\mathcal{G}_t)$ by $\mathcal{G}_t = \mathcal{F}_{E_t}$. Let X_t be a process defined by*

$$X_t := \int_0^t A_s ds + \int_0^t F_s dE_s + \int_0^t G_s dZ_{E_s}$$

where $A_t \in L(m,\mathcal{G}_t)$, $F_t \in L(E,\mathcal{G}_t)$, $G_t \in L(Z \circ E,\mathcal{G}_t)$, and m is the identity map on $\mathbb{R}$ corresponding to Lebesgue measure. If $f \in C^2(\mathbb{R})$, then $f(X_t)$ is a $(\mathcal{G}_t)$-semimartingale and with probability one, for all $t \geq 0$,

$$\begin{aligned} f(X_t) - f(0) &= \int_0^t f'(X_{s-})A_s ds + \int_0^{E_t} f'(X_{D(s-)-})F_{D(s-)}ds \\ &+ \int_0^{E_t} f'(X_{D(s-)-})G_{D(s-)}dZ_s + \frac{1}{2}\int_0^{E_t} f''(X_{D(s-)-})\{G_{D(s-)}\}^2 d[Z,Z]_s^c \\ &+ \sum_{0<s\leq t} \{f(X_s) - f(X_{s-}) - f'(X_{s-})\Delta X_s\}. \end{aligned}$$

Here $L(\mu_t,\mathcal{F}_t)$ is the set of $\mathcal{F}_t$-adapted processes H_t for which the stochastic integral $\int_0^t H_t d\mu_s$ exists.

9. *Fractional Brownian motion.* Fractional Brownian motion was introduced by Kolmogorov [Kol40] in 1940. Mandelbrot and Van Ness in their paper [MVN68] coined the name "fractional Brownian motion" and studied its self-similarity, path continuity, dependent increments, and other properties, and indicate various applications of fBM B_t^H. The parameter H is called the Hurst exponent, due to British hydrologist E.H. Hurst, who studied statistics of water levels of Nile river. Fractional Brownian motion, like standard Brownian motion, has nowhere differentiable sample-paths and stationary increments, but the increments over nonoverlapping intervals are no longer independent. Namely, the covariance between increments over nonoverlapping intervals is positive, if $\frac{1}{2} < H < 1$, and negative, if $0 < H < \frac{1}{2}$. In particular, when $\frac{1}{2} < H < 1$, increments of B_t^H exhibit long range dependence. B_t^H has the integral representation

$$B_t^H = \int_0^t K_H(t,s)dB_s,$$

where B_t is a Brownian motion. We refer the reader to [ST94, Nua06, BHOZ08] for details of the above properties, including various representations for $K_H(t,s)$. Fractional Brownian motion is not a semimartingale [BHOZ08, Nua06], unless $H = \frac{1}{2}$, and hence, as a driving process for SDEs, does not satisfy conditions required for Itô's calculus, However, it works [Ben03, BHOZ08, DU98, Nua06] several different approaches are used to develop an appropriate stochastic calculus in order to interpret in a meaningful way SDEs of the form

$$X_t = X_0 + \int_0^t b(X_s)ds + \int_0^t \sigma(X_s)dB_s^H, \tag{7.114}$$

driven by fBM B_t^H. What concerns FPK equation associated with fractional Brownian motion, it is derived with different methods in [BC07, BHOZ08, MNX09]. In the general setting, a FPK equation associated with SDE (7.88), to author's best knowledge, is not yet known. In the paper [BC07], in the case $b = 0$ for $H \in (1/4, 1)$, it is shown that $u(t,x) = \mathbb{E}[\varphi(X_t)|X_0 = x]$ solves the equation of the form $\partial_t u = Ht^{2H-1}A(x,D)u$, when X_t solves SDE (7.88) and the stochastic integral in the sense of Stratanovich. Fractional Black-Scholes partial differential equations are obtained in [Mag09] when the driving process is Brownian motion, and in [LW12] when the driving process is fBM. In the paper [LW12] also a fractional Fokker-Planck equation associated with the fractional geometric Brownian motion is derived.

Chapter 8
Random walk approximants of mixed and time-changed Lévy processes

8.1 Introduction

Random walks are used to model various random processes in different fields. In this chapter we are only interested in random walks as approximating processes of some basic driving processes of stochastic differential equations discussed in the previous chapter. There is a vast literature (see, e.g., [GK54, Don52, Bil99, Taq75, GM98-1, GM01, MS01]) devoted to approximation of various basic stochastic processes like Brownian motion, fractional Brownian motion, Lévy processes, and their time-changed counterparts. In the context of approximation, the question in what sense a random walk approximates (or converges to) an associated stochastic process becomes important. We will be interested only in the convergence in the sense of finite-dimensional distributions, which is equivalent to the locally uniform convergence of corresponding characteristic functions (see, e.g., [Bil99]).

We start our discussion with the model case - a simple random walk (Section 8.2), which approximates Brownian motion. The idea of convergence of random walks to associated mixed and time-changed stochastic processes, used in subsequent sections, is given here in the simplest case. In Section 8.3 we construct random walks approximating symmetric α-stable Lévy processes $\mathbb{L}_{\alpha,t}$, which was used as driving processes for SDEs in Section 7.5. Random walks approximating stable Lévy processes (not necessarily symmetric) in the one-dimensional case were studied in a series of papers [GM98-1, GM99, GM01], and symmetric multi-dimensional case in [UG05-1]. In Section 8.4 we construct random walks approximating mixed symmetric Lévy processes with some mixing measure. And finally, in Section 8.5 we will construct continuous time random walk (CTRW) approximants of time-changed mixed symmetric Lévy processes and develop an analytic method for the convergence.

Note that CTRW was first introduced by Montroll and Weiss in their paper [MW65] in 1965, and by definition, is a random walk subordinated to a renewal process. More precisely, this means that CTRW comprises of two i.i.d. sequences

S. Umarov, *Introduction to Fractional and Pseudo-Differential Equations with Singular Symbols*, Developments in Mathematics 41,
DOI 10.1007/978-3-319-20771-1_8

of random variables (vectors), one expressing jumps, and another one expressing waiting times between successive jumps. The mathematical definition of CTRW and some of its properties are provided in Section 8.5.

8.2 Simple random walk as an approximant

Suppose a variable X takes randomly two values, $\pm h$, with probability $1/2$ each, i.e., $\mathbb{P}(X=h)=1/2$ and $\mathbb{P}(X=-h)=1/2$. Here $\mathbb{P}(A)$ means the probability of a random event A. With the help of the variable X one can model random movement of a particle on the one-dimensional uniform lattice $x_j = jh$, $j \in \mathbb{Z}$. Indeed, the position of the particle, initially located at some point x_{j_0} ($Y_0 = x_{j_0} = 0$), after n moves is $Y_n = X_1 + \ldots + X_n$, where each X_k is the same as X, and each movement is independent of other movements. In probability theory X_k, $k=1,\ldots,n$, are called an independent and identically distributed (i.i.d.) random variables, which represent "n independent copies of X," and the sequence $Y_0, Y_1, \ldots$, is called a (simple) *random walk.* Suppose the probability of the particle being at x_j in n-th movement is y_j^n, that is $P(Y_n = x_j) = y_j^n$. Since the particle can arrive at x_j only from two neighboring points x_{j-1} and x_{j+1}, with probability $1/2$, then for the $(n+1)$-st movement we have

$$y_j^{n+1} = \frac{1}{2} y_{j-1}^n + \frac{1}{2} y_{j+1}^n, \; y_0^0 = 1, \quad n \in \mathbb{N}_0, \; j \in \mathbb{Z}. \tag{8.1}$$

From this recursive equation one gets a unique solution $\{y_j^n, \; j \in \mathbb{Z}, \, n \in \mathbb{N}_0\}$ for the probability distribution of the above random walk model.

Simple random walk models serve as discrete approximations of, the so-called, Gaussian stochastic processes (diffusion process in physics terminology). Therefore, if one is interested in the limiting process of the above random walk, then a natural question would be what is a "continuous" version of equation (8.1) and its solution? In fact, the answer to this question can be obtained by letting $h \to 0$ in equation (8.1). For this purpose we subtract y_j^n from both sides of equation (8.1) to obtain

$$y_j^{n+1} - y_j^n = \frac{1}{2}\left(y_{j+1}^n - 2y_j^n + y_{j-1}^n\right).$$

Dividing by h^2 and assuming that the time step $\tau = t_{n+1} - t_n$ for each movement equals h^2, that is $\tau = h^2$, we have

$$\frac{y_j^{n+1} - y_j^n}{\tau} = \frac{1}{2} \frac{y_{j+1}^n - 2y_j^n + y_{j-1}^n}{h^2}, \tag{8.2}$$

Let $p(t,x)$ be the probability of being the particle at the position x at time t. Then, taking into account the equality $y_j^n = p(x_j, t_n)$, one can rewrite equation (8.2) in the form

$$\frac{p(t_n+\tau, x_j) - p(t_n, x_j)}{\tau} = \frac{1}{2} \frac{p(t_n, x_j+h) - 2p(t_n, x_j) + p(t_n, x_j - h)}{h^2}.$$

Now letting $h \to 0$, for $(t,x) = (t_n, x_j)$ one obtains the equation

$$\frac{\partial p(t,x)}{\partial t} = \frac{1}{2} \frac{\partial^2 p(t,x)}{\partial x^2}. \tag{8.3}$$

Due to arbitrariness of the pair (t_n, x_j), in fact, equation (8.3) is valid for all $t > 0$, $x \in \mathbb{R}$. The initial condition $Y_0 = 0$ implies $p(0,x) = \delta_0(x)$.

In the d-dimensional case[1] the simple random walk is constructed with the help of the random vector $\mathbf{X}$ taking $2d$ values $(\pm h, 0, \ldots, 0), \ldots, (0, \ldots, 0, \pm h)$, with probability $\frac{1}{2d}$. Then $\mathbf{Y}_n = \mathbf{X}_1 + \ldots + \mathbf{X}_n$, $\mathbf{Y}_0 = 0$, where random vectors $\mathbf{X}_j$, $j = 1, \ldots, n$, have the same distribution as $\mathbf{X}$ and independent, represent a random walk on the d-dimensional uniform lattice $h\mathbb{Z}^d = \{hk = (hk_1, \ldots, hk_d) : (k_1, \ldots, k_d) \in \mathbb{Z}^d\}$. Repeating the above arguments, for the probability $p(t,x)$ of finding the particle at $x \in \mathbb{R}^d$ at time $t > 0$, one has the equation

$$\frac{\partial p(t,x)}{\partial t} = \frac{1}{2} \Delta p(t,x), \quad t > 0, x \in \mathbb{R}^d, \tag{8.4}$$

where Δ is the d-dimensional Laplace operator, with the initial condition

$$p(0,x) = \delta_0(x). \tag{8.5}$$

The unique solution to this initial value problem is given by the following function[2]

$$G_2(t,x) = \frac{1}{(2\pi t)^{d/2}} e^{\frac{-|x|^2}{2t}}, \quad t > 0, x \in \mathbb{R}^d. \tag{8.6}$$

We have seen (see Section 1.5.3) that $G_2(t,x) \to \delta_0(x)$, as $t \to 0$, in the weak sense.

These heuristic calculations show that the simple random walk introduced above converges to a Brownian motion, whose density for each fixed t is given by (8.6). Below we show that in fact the random walk $\mathbf{Y}_n$ converges as $n \to \infty$ to a d-dimensional Brownian motion $B_t = (B_{1t}, \ldots, B_{dt})$ in the sense of finite-dimensional distributions.

The characteristic function $\hat{y}_n(\xi) = \mathbb{E}\left[e^{i\mathbf{Y}_n \xi}\right]$ of $\mathbf{Y}_n$ has the form

$$\hat{y}_n(\xi) = \mathbb{E}\big(e^{i[(\mathbf{X}_1,\xi)+\cdots+(\mathbf{X}_n,\xi)]}\big) = \prod_{j=1}^{n} \mathbb{E}(e^{i\mathbf{X}_j \xi}) = [\hat{y}(\xi)]^n, \quad \xi \in \mathbb{R}^d,$$

where

$$\hat{y}(\xi) = \mathbb{E}(e^{i\mathbf{X}\xi}) = \prod_{j=1}^{d} \left[\frac{1}{2d} e^{-ih\xi_j} + \frac{1}{2d} e^{ih\xi_j}\right] = \prod_{j=1}^{d} \frac{\cos(h\xi_j)}{d}.$$

Therefore, taking into account the relation $t = n\tau = nh^2$, we have

$$\ln[\hat{y}_n(\xi)] = n \sum_{j=1}^{d} \ln\left[\frac{\cos(h\xi_j)}{d}\right] = t \sum_{j=1}^{d} \frac{\ln\left[\frac{\cos(h\xi_j)}{d}\right]}{h^2}.$$

---

[1] For the dimension in this chapter we use the letter d, since n is overloaded.

[2] The Gaussian density function with mean 0 and correlation matrix I evolving in time.

Now, applying L'Hôpital's rule, one has $\lim_{h\to 0} h^{-2}\ln\left[\frac{\cos(h\xi_j)}{d}\right] = -\xi_j^2/2$. Hence,

$$\lim_{h\to 0}\hat{y}_n(h\xi) = e^{-t\frac{|\xi|^2}{2}} = F\left[\frac{1}{(2\pi t)^{d/2}}e^{\frac{-|x|^2}{2t}}\right] = \mathbb{E}\left[e^{iB_t\xi}\right],$$

with $B_t\xi = \xi_1 B_{1t} + \cdots + \xi_d B_{dt}$. Thus, for each fixed $t > 0$ the characteristic function of the random walk $\mathbf{Y}_n$ locally uniformly converges as $n \to \infty$ to the characteristic function of the standard Brownian motion B_t with mean zero and variance t. The latter convergence is equivalent to the convergence in distributions (in law). Now, since t is an arbitrary fixed number, it follows the convergence in the sense of finite-dimensional distributions, as well. Concluding, we have that the simple random walk on the lattice $h\mathbb{Z}^d$ approximates Brownian motion.

8.3 Random walk approximants for Lévy-Feller processes

Now consider the fractional order differential equation

$$\frac{\partial u(t,x)}{\partial t} = \frac{1}{2}D_0^\alpha u(t,x), \quad t > 0, x \in \mathbb{R}^d, \tag{8.7}$$

which generalizes equation (8.4), and where D_0^α, $0 < \alpha < 2$, is the pseudo-differential (hyper-singular) operator

$$D_0^\alpha f(x) = b(\alpha)\int_{\mathbb{R}^d}\frac{\Delta_y^2 f(x)}{|y|^{n+\alpha}}dy, \tag{8.8}$$

defined in Section 3.8, with the symbol $-|\xi|^\alpha$. See equation (3.75) for $b(\alpha) = b_d(\alpha)$ in (8.8). In accordance with $\lim_{\alpha\to 2}|\xi|^\alpha = |\xi|^2$ one can accept $D_0^2 = \Delta$, where Δ is the Laplace operator. In the general case we have formally $D_0^\alpha = -(-\Delta)^{\alpha/2}$ (see Section 3.9). A weak solution, namely a distribution $G_\alpha(t,x)$, which satisfies (8.7) and the condition

$$u(0,x) = \delta_0(x), \quad x \in \mathbb{R}^d, \tag{8.9}$$

with the Dirac delta-function $\delta_0(x)$, concentrated at 0, in the sense of distributions, is called a fundamental solution of the Cauchy problem (8.7), (8.9).

It is clear that in the case $\alpha = 2$ we have equation (8.4) whose fundamental solution is the Gaussian probability density function evolving in time, given by (8.6). In the case $\alpha = 1$ the corresponding fundamental solution is given by the Cauchy-Poisson probability density function evolving in time (c.f. (1.119)):

$$G_1(t,x) = \frac{\Gamma(\frac{d+1}{2})}{2\pi^{(d+1)/2}}\frac{t}{\left(|x|^2 + (t/2)^2\right)^{(d+1)/2}}.$$

It is well known that for the Fourier transforms of the functions $G_q(t,x), q = 1,2,$ the relations (cf. formulas (1.16) and (1.17))

$$\hat{G}_2(t,\xi) = e^{-\frac{t}{2}|\xi|^2} \quad \text{and} \quad \hat{G}_1(t,\xi) = e^{-\frac{t}{2}|\xi|}$$

hold. For other values of α, $0 < \alpha < 2$, applying the Fourier transform to equation (8.7), and then the inverse Fourier transform to the solution of the equation in the Fourier domain, one can find the fundamental solution to the Cauchy problem (8.7), (8.9), represented in the form

$$G_\alpha(t,x) = \frac{1}{(2\pi)^d} \int_{\mathbb{R}^d} e^{-\frac{t}{2}|\xi|^\alpha} e^{ix\xi} d\xi. \tag{8.10}$$

The question we want to explore is the existence of a random walk approximating the diffusion process governed by equation (8.7). Let $\mathbf{X}$ be an d-dimensional random vector [MS01] which takes values in $h\mathbb{Z}^d$ with the probability mass function $p_k = P(X = hk)$, $k \in \mathbb{Z}^d$. Notice that in the case of simple random walk

$$p_k = \begin{cases} \frac{1}{2d}, & \text{if} \quad |k| = 1, \\ 0, & \text{otherwise.} \end{cases}$$

Further, let the random vectors $\mathbf{X}_1, \mathbf{X}_2, \ldots,$ be an independent and identically distributed random vectors, all having the same probability distribution as $\mathbf{X}$ does. We introduce a spatial grid $\{x_j = jh, j \in \mathbb{Z}^d\}$, with $h > 0$, and a temporal grid $\{t_n = n\tau, n = 0,1,2,\ldots\}$ with a step $\tau > 0$. Consider the sequence of random vectors

$$\mathbf{S}_n = \mathbf{X}_1 + \mathbf{X}_2 + \cdots + \mathbf{X}_n, \quad n = 1,2,\ldots$$

assuming $\mathbf{S}_0 = 0$, for convenience. We interpret $\mathbf{X}_1, \mathbf{X}_2, \ldots,$ as jumps of the particle being at $x = x_0 = \mathbf{0}$ at the starting time $t = t_0 = 0$ and making a jump $\mathbf{X}_n$ from the position $\mathbf{S}_{n-1}$ to the position $\mathbf{S}_n$ at the time instance $t = t_n$. Then the position $\mathbf{S}(t)$, $t > 0$, of the particle at time t is

$$\mathbf{S}(t) = \sum_{1 \le k \le t/\tau} \mathbf{X}_k, \quad t > 0.$$

Due to independence of random vectors $X_1, \ldots, X_n$, the probabilities $p_k = P(X_1 = x_k)$ can be interpreted as transition probabilities from a point $x_j \in h\mathbb{Z}^d$ to a point $x_{j+k} \in h\mathbb{Z}^d$. By this we automatically assume that the particle jumps are isotropic in all directions. They satisfy the following nonnegativity and normalization conditions:

(a) $p_k \ge 0, k \in \mathbf{Z}^d$;
(b) $\sum_{k \in \mathbb{Z}^d} p_k = 1$.

We recall a fact from probability theory that for two random vectors X and Y with probability mass functions p_k and q_k, $k \in \mathbb{Z}^d$, the probability mass function of $X + Y$ is defined as a convolution $p * q$, by the rule

$$(p*q)_k = \sum_{j\in\mathbb{Z}^d} p_j q_{k-j}, \quad k \in \mathbb{Z}^d. \tag{8.11}$$

Also for a random variable X its characteristic function is defined by the formula

$$\hat{p}(\xi) = \sum_{k\in\mathbb{Z}^d} p_k e^{ik\xi}, \quad \xi \in \mathbb{R}^d, \tag{8.12}$$

where $kx = k_1x_1 + \cdots + k_d\xi_d$. We will consider the solution $u(t,x)$ of diffusion equation (8.7) with initial condition (8.9) as a probability density function (with respect to x). Namely, for given time $t > 0$, $u(t,x)$ is the probability of sojourn of a diffusing particle at $x \in \mathbb{R}^d$. For the discrete random walk introduced above we use the notation y_j^n for the (discrete) probability of sojourn (in the time instant t_n) of the wandering particle at the point x_j. Heuristically, we consider y_j^n as an approximation of $h^d u(t_n,x_j) \approx \int_{C_j} u(t_n,x)dx$, the total probability of sojourn inside a cubical cell C_j with the center x_j and side length h.

We also will use the fact that for a continuous function $f(x)$ integrable over $\mathbb{R}^d$, the following convergence of the Riemann sum

$$h^d \sum_{j\in\mathbb{Z}^d} f(jh) \to \int_{\mathbb{R}^d} f(x)dx. \tag{8.13}$$

as $h \to 0$, holds.

Lemma 8.1. *For the probabilities y_j^n the following statements hold:*

1. $y_j^{n+1} = \sum_{k\in\mathbb{Z}^d} p_k y_{j-k}^n, \ j \in \mathbb{Z}^d;$

2. $y_j^n = (\underbrace{p*\cdots*p}_{n\ times})_j.$

Proof. The first statement follows from the recursion $\mathbf{S}_{n+1} = \mathbf{S}_n + \mathbf{X}_n$ and (8.11). The second statement follows from the first one by induction.

Lemma 8.2. *Let $s(\tau), \tau \geq 0$, be a differentiable function, such that*

(a) $s(\tau) \to s_0,\ \tau \to \infty$;
(b) $\tau s'(\tau) \to 0,\ \tau \to \infty$. Then

$$\lim_{\tau\to\infty} \left(1 + \frac{s(\tau)}{\tau}\right)^{\tau} = e^{s_0}. \tag{8.14}$$

Proof. Consider the function

$$g(\tau) = \frac{\ln\left(1 + \frac{s(\tau)}{\tau}\right)}{\frac{1}{\tau}}, \quad \tau > 0.$$

Condition (a) implies that the function $s(\tau)$ is bounded at infinity, and therefore, one can apply the L'Hôpital's rule to find the limit of $g(\tau)$ when $\tau \to \infty$. Then, due to conditions (a) and (b)

$$\lim_{\tau \to \infty} g(\tau) = \frac{[\tau s'(\tau) - s(\tau)]\tau^{-2}}{-\tau^{-2}\left(1 + \frac{s(\tau)}{\tau}\right)} = s_0.$$

The latter implies (8.14).

Theorem 8.1. *Let $\mathbf{X}_1, \mathbf{X}_2, \ldots,$ be an independent and identically distributed random vectors with the probability mass function $p_k = P(\mathbf{X}_j = hk), k \in \mathbb{Z}^d$, of each random vector $\mathbf{X}_j$, $j = 1, \ldots, n$. Assume that*

(a) if $0 < \alpha < 2$, then

$$p_k = \begin{cases} 1 - \mu b(\alpha) \sum_{m \in \mathbf{Z}^d \setminus \{0\}} \frac{1}{|m|^{d+\alpha}}, & \text{if } k = 0; \\ \mu b(\alpha) |k|^{-(d+\alpha)}, & \text{if } k \neq 0, \end{cases} \tag{8.15}$$

with μ satisfying the condition

$$0 < \mu \leq \frac{1}{b(\alpha) \sum_{m \in \mathbf{Z}^d \setminus \{0\}} |m|^{-d-\alpha}},$$

and the space and time steps h and τ being connected by the scaling relation $\tau = \tau(h) = \mu h^{\alpha}$;

(b) if $\alpha = 2$, then

$$p_k = \begin{cases} \frac{1}{2d}, & \text{if } |k| = 1; \\ 0, & \text{if } |k| = 0, \end{cases}$$

with $\tau = \frac{h^2}{d}$.

Then the sequence of random vectors $\mathbf{S}_n = \mathbf{X}_1 + \ldots + \mathbf{X}_n$ converges as $n \to \infty$ in the sense of distributions to the random vector whose probability density is the fundamental solution of the Cauchy problem (8.7), (8.9), i.e., to $G(t,x)$ defined in (8.10).

Proof. Let $0 < \alpha < 2$. We will show that the sequence of random vectors $\mathbf{S}_n$ converges as $n \to \infty$ to the random vector whose probability density function evolving in time is of the form

$$G(t,x) = \frac{1}{(2\pi)^d} \int_{\mathbb{R}^d} e^{-\frac{1}{2}t|\xi|^{\alpha}} e^{ix\xi} d\xi, \quad t > 0, x \in \mathbb{R}^d.$$

It is obvious that the Fourier transform of $G(t,x)$ with respect to the variable x is the function $\hat{G}(t,\xi) = e^{-\frac{1}{2}t|\xi|^{\alpha}}$. Let $\hat{p}_h(\xi)$ be the characteristic function corresponding to the random vector $\mathbf{X}_1 \in h\mathbb{Z}^d$ with the probability mass function p_k defined in (8.15), that is (see (8.12))

$$\hat{p}_h(\xi) = \sum_{k\in\mathbb{Z}^d} p_k e^{ihk\xi}.$$

Due to Lemma 8.1 the characteristic function of $\mathbf{S}_n$ can be represented in the form

$$\hat{y}(t_n,\xi) = \sum_{k\in\mathbb{Z}^d} y_k^n e^{ihk\xi} = \left[\hat{p}_h(\xi)\right]^n.$$

Taking this into account it suffices to show that

$$\left[\hat{p}_h(\xi)\right]^n \to e^{-\frac{t}{2}|\xi|^\alpha}, \quad n\to\infty. \tag{8.16}$$

The latter is equivalent to

$$\lim_{h\to 0} \frac{\ln \hat{p}_h(\xi)}{\tau(h)/2} = -|\xi|^\alpha.$$

where $\tau(h) = \frac{t}{n} = \mu h^\alpha$. Below we show the validity of the limit in (8.16). Using the definition of p_k given in equation (8.15), we have

$$\begin{aligned}\left[\hat{p}_h(\xi)\right]^n &= \left[1-\mu b(\alpha)\sum_{0\neq k\in\mathbb{Z}^d}\frac{1}{|k|^{d+\alpha}} + \mu b(\alpha)\sum_{0\neq k\in\mathbb{Z}^d}\frac{e^{ik\xi h}}{|k|^{d+\alpha}}\right]^n \\ &= \left[1-\mu b(\alpha)\sum_{0\neq k\in\mathbb{Z}^d}\frac{1-e^{ik\xi h}}{|k|^{d+\alpha}}\right]^n. \end{aligned}\tag{8.17}$$

Further, one can easily verify that

$$\sum_{0\neq k\in\mathbb{Z}^d}\frac{1-e^{ik\xi h}}{|k|^{d+\alpha}} = \sum_{0\neq k\in\mathbb{Z}^d}\frac{1-e^{-ik\xi h}}{|k|^{d+\alpha}}.$$

Due to the definition of the symmetric second finite difference of the function $e^{ix\xi}$ at the origin, this implies

$$\begin{aligned}\sum_{0\neq k\in\mathbb{Z}^d}\frac{1-e^{ik\xi h}}{|k|^{d+\alpha}} &= \frac{1}{2}\sum_{0\neq k\in\mathbb{Z}^d}\frac{2-e^{ik\xi h}+e^{-ik\xi h}}{|k|^{d+\alpha}} \\ &= -\frac{1}{2}\sum_{0\neq k\in\mathbb{Z}^d}\frac{(\Delta_{kh}^2 e^{ix\xi})_{|x=0}}{|k|^{d+\alpha}}. \end{aligned}\tag{8.18}$$

Now in equation (8.17) substituting $\mu = \frac{t}{nh^\alpha}$ and using equality (8.18), one obtains

$$\left[\hat{p}_h(\xi)\right]^n = \left[1+\frac{\frac{t}{2}b(\alpha)\sum_{k\in\mathbb{Z}^d\setminus\{0\}}\frac{(\Delta_{kh}^2 e^{ix\xi})_{|x=0}}{|kh|^{d+\alpha}}h^d}{n}\right]^n. \tag{8.19}$$

Further, it follows from Lemma 8.2 that if a sequence $s_n = s(n),\ n = 1,2,\ldots,$ converges to s when $n \to \infty$, and $\tau s'(\tau) \to 0,\ \tau \to \infty$, then

$$\lim_{n\to\infty}(1+\frac{s_n}{n})^n = e^s. \tag{8.20}$$

Due to (8.13) and Corollary 2.11 the expression

$$s_n = s_{n(h)} = \frac{t}{2}b(\alpha)\sum_{k\in\mathbb{Z}^d\setminus\{0\}}\frac{(\Delta_{kh}^2 e^{ix\xi})_{|x=0}}{|kh|^{d+\alpha}}h^d$$

converges to

$$s = \frac{t}{2}b(\alpha)\int_{\mathbb{R}^d}\frac{(\Delta_y^2 e^{ix\xi})_{x=0}}{|y|^{d+\alpha}}dy = \frac{t}{2}(D_0^\alpha e^{ix\xi})_{|x=0} = -\frac{t}{2}|\xi|^\alpha,$$

as $h \to 0$ (or, the same, $n \to \infty$) for all $\alpha \in (0,2)$. Moreover, utilizing the equality $h^\alpha = t/(\mu n)$ the function $s(\tau)$ can be written in the form

$$s(\tau) = \mu b(\alpha)\sum_{k\in\mathbb{Z}^d\setminus\{0\}}\frac{\tau(\cos\frac{k\xi}{\tau}-1)}{|k|^{d+\alpha}}.$$

Exploiting the dominating convergence theorem, it is not hard to see that $\tau s'(\tau) \to 0$, as $\tau \to \infty$. Hence in accordance with (8.20), letting $n \to \infty$ in (8.19), we have

$$\hat{y}(t_n,\xi) = [\hat{p}_h(\xi)]^n \to e^{-\frac{t}{2}|\xi|^\alpha}, \quad \xi \in \mathbb{R}^d.$$

The case $\alpha = 2$, with probabilities p_k given in Part b) of Theorem 8.1, corresponds to the N-dimensional simple random walk and was shown earlier.

Remark 8.1. The constructed random walk can be generalized to the class of stable motions L_t, characteristic functions of which in the one-dimensional case have the form $exp(t\psi(\xi))$, with $\psi(\xi)$ defined in (7.121). In multidimensional case the symbol $\psi(\xi)$ has the form

$$\psi(\xi) = -i(\mathbf{a},\xi) + \int_{|\theta|=1}(-i\xi,\theta)^\alpha M(d\theta), \quad \xi \in \mathbb{R}^d,$$

where $\mathbf{a} \in \mathbb{R}^d$ and $M(d\theta)$ is a probability measure defined on the unit sphere. If $a = 0$ and $M(d\theta) = const\cdot d\theta$, then we get the symmetric case considered above.

8.4 Random walk approximants for mixed Lévy processes

The objective of this section is twofold. The first objective is to describe the driving process of SDEs considered in previous sections, approximating it by a discrete random walk. To achieve this we develop a DODE model in this section and CTRW model in the next section. The second objective is to generalize the random walk construction discussed in Section 8.2 and some random walks models introduced and studied in the papers [GM01, US06].

Definition 8.1. Let ρ be a finite measure with the support $\operatorname{supp}\rho \subseteq (0,2]$. Denote by $\mathbb{SS}$ the class of $(\mathscr{F}_t)$-semimartingales $Z_t, Z_0 = 0$, whose characteristic function is given by

$$\mathbb{E}\left[e^{i\xi Z_t}\right] = \exp\left\{-t\int_0^2 |\xi|^\alpha d\rho(\alpha)\right\}, \quad \xi \in \mathbb{R}^d. \tag{8.21}$$

If $f_{Z_t}(x), x \in \mathbb{R}^d$, is the density function of the process Z_t, then equation (8.21) can be expressed as the Fourier transform $\mathbb{E}\left[e^{i\xi Z_t}\right] = F[f_{Z_t}](\xi)$. The class $\mathbb{SS}$ obviously contains Lévy's $S\alpha S$-processes and all mixtures of their finitely many independent representatives. For the process $Z_t \in \mathbb{SS}$ corresponding to a finite measure ρ, we use the notation $Z_t = X_t^\rho$ to indicate this correspondence.

Suppose X_t^ρ is a stochastic process obtained by mixing of independent Lévy's $S\alpha S$-processes with a mixing measure ρ, $\operatorname{supp}\rho \subset (0,2)$.[3] Then its associated FPK equation has the form

$$\frac{\partial u(t,x)}{\partial t} = \int_0^2 \mathbb{D}_0^\alpha u(t,x) d\rho(\alpha), \quad t > 0, \, x \in \mathbb{R}^d, \tag{8.22}$$

where $\mathbb{D}_0^\alpha$ is given by (8.8). Indeed, it is seen from (8.21) that the Lévy symbol of X_t^ρ equals

$$\Psi(\xi) = -\int_0^2 |\xi|^\alpha d\rho. \tag{8.23}$$

On the other hand, due to the formula $F[D_0^\alpha \varphi](\xi) = -|\xi|^2 F[\varphi](\xi)$ (see (3.82)), the Fourier transform of the right-hand side of (8.22) is

$$F\left[\int_0^2 D_0^\alpha \varphi(x) d\rho(\alpha)\right] = \int_0^2 F[D_0^\alpha \varphi](\xi) d\rho(\alpha) = \left(-\int_0^2 |\xi|^\alpha d\rho\right) F[\varphi](\xi).$$

Hence, the symbol of the pseudo-differential operator on the right-hand side of (8.22) coincides with $\Psi(\xi)$. This means that the FPK equation associated with X_t^ρ is given by equation (8.22). Since $X_0^\rho = 0$ the function $u(t,x)$ in (8.22) satisfies the initial

---

[3] The set $(0,2)$ can be replaced by $(0,2]$, but this requires an additional care (see [US06]).

condition $u(0,x) = \delta_0(x)$. Using the Fourier transform technique, one can verify that the solution of (8.22) satisfying the latter initial condition can be expressed in the form

$$G^\rho(t,x) = \frac{1}{(2\pi)^d} \int_{\mathbb{R}^d} e^{t\Psi(\xi)-ix\xi} d\xi. \tag{8.24}$$

The following theorem provides a random walk approximation of the stochastic process X_t^ρ.

Theorem 8.2. *Let $X_j \in h\mathbb{Z}^d$, $j \geq 1$, be i.i.d. random vectors with the probability mass function*

$$p_k = \mathbb{P}(X_1 = k) = \begin{cases} 1 - 2\tau \sum_{m \neq 0} \frac{Q_m(h)}{|m|^d}, & \text{if } k = 0; \\ 2\tau \frac{Q_k(h)}{|k|^d}, & \text{if } k \neq 0, \end{cases} \tag{8.25}$$

where $\tau > 0$, $h > 0$, and

$$Q_m(h) = \int_0^2 \frac{b(\alpha)d\rho(\alpha)}{(hm)^\alpha}, \quad m \neq 0. \tag{8.26}$$

Assume that

$$\sigma(\tau,h) := 2\tau \sum_{m \neq 0} \frac{Q_m(h)}{|m|^d} \leq 1. \tag{8.27}$$

Then the sequence of random vectors $S_n = X_1 + \ldots + X_n$, converges in law to X_t^ρ as $n \to \infty$.

Proof. In order to construct a random walk relevant to (8.80) we use the approximation (8.13) for the integral on the right-hand side of (8.22), namely

$$\mathbb{D}_0^\alpha u(t,x_j) \approx b(\alpha) \sum_{k \in \mathbb{Z}^d} \frac{u_{j+k}(t) - 2u_j(t) + u_{j-k}(t)}{|k|^{d+\alpha} h^\alpha}, \tag{8.28}$$

and the first order difference ratio

$$\frac{\partial u}{\partial t} \approx \frac{u_j(t_{n+1}) - u_j(t_n)}{\tau}$$

for $\frac{\partial u}{\partial t}$ with the time step $\tau = t/n$, and $u_j(t) = u(x_j,t)$, $x_j \in h\mathbb{Z}^d$. Following notations in Section 8.2, we denote the probability of walker being at x_j at time t_n by $y_j^n = u(t_n,x_j)$, $t_n = n\tau$, $x_j = hj$. Then, taking into account the recursion $\mathbf{S}_{n+1} = \mathbf{S}_n + \mathbf{X}_n$, one has

$$y_j^{n+1} = \sum_{k \in \mathbb{Z}^d} p_k y_{j-k}^n, \; j \in \mathbb{Z}^d, n = 0,1,\ldots. \tag{8.29}$$

with the transition probabilities

$$p_k = \begin{cases} 1 - 2\tau \sum_{m \neq 0} \frac{Q_m(h)}{|m|^d}, & \text{if } k = 0; \\ 2\tau \frac{Q_k(h)}{|k|^d}, & \text{if } k \neq 0, \end{cases} \tag{8.30}$$

where $Q_m(h)$ is defined in (8.26). Assume that the condition (8.27) is fulfilled. Then, obviously, the transition probabilities satisfy the properties:

$$\sum_{k\in\mathbb{Z}^d} p_k = 1, \quad \text{and} \quad p_k \geq 0,\ k\in\mathbb{Z}^d. \tag{8.31}$$

Introduce the function

$$\mathscr{R}(\alpha) = \sum_{k\neq 0} \frac{1}{|k|^{d+\alpha}} = \sum_{m=1}^{\infty} \frac{M_m}{m^{d+\alpha}}, \quad 0<\alpha\leq 2,$$

where $M_m = \sum_{|k|=m} 1$. (In the one-dimensional case $\mathscr{R}(\alpha)$ relates to the Riemann's zeta-function through $\mathscr{R}(\alpha) = 2\zeta(1+\alpha)$.) The inequality (8.27) can be rewritten as

$$\sigma(\tau,h) = 2\tau \int_0^2 \frac{b(\alpha)\mathscr{R}(\alpha)}{h^\alpha} d\rho(\alpha) \leq 1.$$

It follows from the latter inequality that $h\to 0$ yields $\tau\to 0$. This, in turn, yields $n = t/\tau \to \infty$ for any finite t.

In order to prove the theorem we have to show that the sequence of random vectors $\mathbf{S}_n$ tends to the random vector with the density function $G^\rho(t,x)$ (for a fixed t) in (8.24). This means that the discrete function $y_j(t_n)$ tends to $G^\rho(t,x)$ as $n\to\infty$. It is obvious that the Fourier transform of $G^\rho(t,x)$ with respect to the variable x is the function $\widehat{G^\rho}(t,\xi) = e^{t\Psi(\xi)}$. Let $\hat{p}(\xi)$ be the characteristic function corresponding to the discrete function p_k, $k\in\mathbb{Z}^d$, that is

$$\hat{p}(\xi) = \sum_{k\in\mathbb{Z}^d} p_k e^{ik\xi}. \tag{8.32}$$

It follows from the recursion formula (8.29) (which exhibits the convolution) and the well-known fact that convolution goes over in multiplication by the Fourier transform, the characteristic function of $y_j(t_n)$ can be represented in the form

$$\begin{aligned}\hat{y}_j(t_n,\xi) &= \hat{y}_j(t_{n-1},\xi)\hat{p}(\xi) = \cdots = \hat{y}_j(0,\xi)[\hat{p}(\xi)]^n \\ &= [\hat{p}(\xi)]^n, \quad n=1,2,\ldots.\end{aligned}$$

Taking this into account it suffices to show that $[\hat{p}(h\xi)]^n \to e^{t\Psi(\xi)}$, $n\to\infty$. The next step of the proof uses (8.20). We have

$$\begin{aligned}[\hat{p}(h\xi)]^n &= \left(1-\tau\sum_{k\neq 0}\frac{Q_k}{|k|^d}(1-e^{ik\xi h})\right)^n \\ &= \left(1-\tau\sum_{k\neq 0}\frac{1}{|k|^d}\int_0^2 \frac{a(b(\alpha)d\rho(\alpha)}{|k|^\alpha h^\alpha}(1-e^{ik\xi h})\right)^n \\ &= \left(1+\frac{t\int_0^2\{b(\alpha)\sum\frac{(\Delta^2_{kh}e^{ix\xi})_{|x=0}}{|kh|^{d+\alpha}}h^n\}d\rho(\alpha)}{n}\right)^n\end{aligned}$$

In Section 8.2 we have proved that the expression

$$b(\alpha)\sum_{k\in\mathbb{Z}^d}\frac{\left(\Delta_{kh}^2 e^{ix\xi}\right)_{|x=0}}{|kh|^{d+\alpha}}h^d$$

tends to $(D_0^\alpha e^{ix\xi})_{|x=0} = -|\xi|^\alpha$ as $h \to 0$ (or, the same, $n \to \infty$) for all $\alpha \in (0,2)$. Hence

$$s_n = s_{n(h)} = \int_0^2 \left\{ b(\alpha)\sum \frac{\Delta^2 e^{ik\xi h}}{|kh|^{d+\alpha}}h^d \right\} d\rho(\alpha) \to -\int_0^2 |\xi|^\alpha d\rho(\alpha)$$
$$= \Psi(\xi), \quad n \to \infty \; (h \to 0).$$

Moreover, the function $s\tau$ ($s_n = s(n)$) satisfies condition $\tau s'(\tau) \to 0,\ \tau \to \infty$. Thus, in accordance with (8.20) we have

$$[\hat{p}(h\xi)]^n \to e^{t\Psi(\xi)}, \quad n \to \infty,$$

completing the proof.

The random walk related to the multiterm fractional diffusion equation can be derived from Theorem 8.2. Assume that $d\rho(\alpha)$ has the form

$$d\rho(\alpha) = \sum_{m=1}^{M} a_m \delta_{\alpha_j}(\alpha) d\alpha,$$

with $0 < \alpha_1 < \ldots \alpha_M < 2$. So, we again exclude the case $\{2\} \in singsupp \rho$.

Theorem 8.3. *([US06]) Let the transition probabilities $p_k = P(\mathbf{X} = x_k), k \in \mathbb{Z}^d$, of the random vector $\mathbf{X}$ be given as follows:*

$$p_k = \begin{cases} 1 - \sum_{j\neq 0} \frac{1}{|j|^d} \sum_{m=1}^{M} \frac{\mu_m a_m b(\alpha_m)}{|j|^{\alpha_m}}, & if\ k = 0; \\ \frac{1}{|k|^d} \sum_{m=1}^{M} \frac{\mu_m a_m b(\alpha_m)}{|j|^{\alpha_m}}, & if\ k \neq 0, \end{cases}$$

where $\mu_m = \frac{2\tau}{h^{\alpha_m}}, m = 1, \ldots, M$. Assume,

$$\sum_{m=1}^{M} a_m b(\alpha_m) \mathscr{R}(\alpha_m) \mu_m \leq 1.$$

Then the sequence of random vectors $\mathbf{S}_n = h\mathbf{X}_1 + \ldots + h\mathbf{X}_n$, converges as $n \to \infty$ in law to the random vector whose probability density function is the fundamental solution of the multiterm fractional order differential equation

$$\frac{\partial u(t,x)}{\partial t} = \sum_{m=1}^{M} a_m \mathbb{D}_0^{\alpha_m} u(t,x), \quad t > 0,\ x \in \mathbb{R}^d.$$

Remark 8.2. 1. Theorem 8.2 in the particular case of the measure $\rho(d\alpha) = a(\alpha)d\alpha$, where $a(\alpha) \in C[0,2]$, is proved in [US06].
2. The condition $\{2\} \notin singsupp\,\rho$ is required, because the value $\alpha = 2$ is singular in the definition of $\mathbb{D}_0^{\alpha}$ (see (3.75)). The particular case of $\rho(d\alpha) = \delta(\alpha - 2)d\alpha$ reduces equation (8.22) to the classic diffusion equation and as was seen in Section 8.2, the corresponding (simple) random walk converges to Brownian motion.

8.5 Continuous time random walk approximants of time-changed processes

8.5.1 Continuous time random walk. Montroll-Weiss equation

Driving processes of the SDEs associated with time-fractional FPK equations appear to be time-changes of basic processes like Brownian motion, Lévy process, fractional Brownian motion, etc. Donsker's theorem states that Brownian motion is the limit in the weak topology of a scaled sum of a sequence of independent and identically distributed (i.i.d.) random variables $\{X_j\}$, with $X_1 \in L^2(\mathbb{P})$. This fact is important from the approximation point of view since an approximation of the basic driving process B_t yields, under some conditions, an approximation of other processes X_t driven by B_t. Natural approximants of time-changed processes B_W, L_W, etc., where W is the inverse to a stable subordinator, are continuous time random walks (CTRWs). A CTRW is a random walk subordinated to a renewal process. CTRWs are described by two sequences of random variables: one representing the length of the jump, and the other one representing the waiting time between successive jumps. More precisely, let $Y_1, Y_2, \dots, Y_n, \dots,\quad (Y_i \in \mathbb{R}^d)$, be a sequence of i.i.d. random vectors, and let $\tau_1, \tau_2, \dots, \tau_n, \dots,\quad (\tau_i \in \mathbb{R})$ be an i.i.d. sequence of positive real-valued random variables. Then

$$S_n = Y_1 + \cdots + Y_n \tag{8.33}$$

is the position after n jumps, and

$$T_n = \tau_1 + \cdots + \tau_n \tag{8.34}$$

is the time of the nth jump. Assume that $S_0 = 0$ and $T_0 = 0$. The stochastic process

$$X_t = S_{N_t} = \sum_{i=1}^{N_t} Y_i, \tag{8.35}$$

where

$$N_t = \max\{n \geq 0 : T_n \leq t\}, \tag{8.36}$$

is called a *continuous time random walk*. We assume that the random variable τ_1 and random vector Y_1 are independent. This case is referred to as *uncoupled* CTRW. Suppose $\varphi(t,x) = \phi(t)w(x)$, $\tau \geq 0, x \in \mathbb{R}^d$, is the joint density of (τ_1, Y_1), where $\phi(t)$ and $w(x)$ are density functions of τ_1 and Y_1, respectively. Then the probability $p(t,x)$ of being the walker at x at time t satisfies the following integral equation (see, e.g., [MK00, SGM00, GM05])

$$p(t,x) = \delta_0(x)\Phi(t) + \int_0^t \int_{\mathbb{R}^n} \phi(t-\tau)w(x-y)p(\tau,y)dyd\tau, \tag{8.37}$$

called a *master equation*. In this equation $\Phi(t) = 1 - \int_0^t \phi(\tau)d\tau = \int_t^\infty \phi(\tau)d\tau$. One can easily verify that the Laplace transform of $\Phi(t)$ is

$$\tilde{\Phi}(s) = L[\Phi](s) = \frac{1-\tilde{\phi}(s)}{s}, \quad \Re(s) > 0. \tag{8.38}$$

Applying the Laplace transform and then the Fourier transform to both sides of (8.37), one has

$$\hat{\tilde{p}}(s,\xi) = \frac{\tilde{\Phi}(s)}{1-\hat{\tilde{\varphi}}(s,\xi)}, \quad \Re(s) > 0, \xi \in \mathbb{R}^d. \tag{8.39}$$

Equation (8.39) is known as the *Montroll-Weiss equation*, and first was obtained by Montroll and Weiss in [MW65] using a different argumentation, which does not use the master equation. In our case of uncoupled CTRW $\hat{\tilde{\varphi}}(s,\xi) = \tilde{\phi}(s)\hat{w}(\xi)$. Therefore, taking into account (8.38), one can write the Montroll-Weiss equation in the form

$$\hat{\tilde{p}}(s,\xi) = \frac{1-\tilde{\phi}(s)}{s}\frac{1}{1-\tilde{\phi}(s)\hat{w}(\xi)}, \quad \Re(s) > 0, \xi \in \mathbb{R}^d. \tag{8.40}$$

Our aim is to study limit processes of CTRWs. For this purpose, consider τT_n, where $\tau > 0$ is a nonrandom parameter. Due to (8.34), this is equivalent to the change of the sequence τ_n to the sequence $\tau \cdot \tau_n$. The density function of the latter is $\phi_\tau(t) = \tau^{-1}\phi(t/\tau)$ and the corresponding Laplace transform is $\tilde{\phi}_\tau(s) = \tilde{\phi}(\tau s)$. Similarly, with a nonrandom parameter $h > 0$ consider hS_n, which is equivalent to the change of Y_n in (8.33) to hY_n. The density function of the vector hY_n is $w_h(x) = h^{-n}w(x/h)$, and the corresponding characteristic function is $\hat{w}_h(\xi) = \hat{w}(h\xi)$. In this case the CTRW process takes the form $X_t^{\tau,h} = hS_{N_{t/\tau}}$, with the corresponding Montroll-Weiss equation

$$\hat{\tilde{p}}_{\tau,h}(s,\xi) = \frac{1-\tilde{\phi}(\tau s)}{s}\frac{1}{1-\tilde{\phi}(\tau s)\hat{w}(h\xi)}, \quad \Re(s) > 0, \xi \in \mathbb{R}^d. \tag{8.41}$$

Now assume that

$$\tilde{\phi}(s) \sim 1 - \lambda s^{\beta}, \quad s \to 0, \tag{8.42}$$

with some $\lambda > 0$, and

$$\hat{w}(\xi) \sim 1 + A(\xi), \quad |\xi| \to 0, \tag{8.43}$$

where $A(\xi)$ is a continuous radial function taking negative values, and such that $A(0) = 0$. Further, assume that parameters τ and h are connected so that $\tau^{-\beta} A(h\xi) \to \lambda A(\xi)$, as $\tau \to 0$. The validity of the latter limit yields also that h is a function of τ, and $h(\tau) \to 0$ if $\tau \to 0$. Under these assumptions equation (8.41) implies

$$\begin{aligned} \hat{\tilde{p}}_{\tau,h}(s,\xi) &\sim \frac{\lambda \tau^{\beta} s^{\beta-1}}{\lambda \tau^{\beta} s^{\beta} - A(h\xi) + \lambda \tau^{\beta} s^{\beta} A(h\xi)} \\ &= \frac{s^{\beta-1}}{s^{\beta} - \frac{1}{\lambda \tau^{\beta}} A(h\xi) + s^{\beta} A(h\xi)}. \end{aligned} \tag{8.44}$$

Hence, letting $\tau \to 0$, or $h \to 0$, in (8.44), one obtains locally uniform convergence

$$\hat{\tilde{p}}_{\tau,h}(s,\xi) \to \frac{s^{\beta-1}}{s^{\beta} + (-A(\xi))}, \quad \tau \to 0 \; (h \to 0), \tag{8.45}$$

since $A(h\xi) \to 0$ as $h \to 0$. It follows from Proposition 3.7 that the right-hand side of (8.45) is the Laplace transform of $E_{\beta}(t^{\beta} A(\xi))$, which, in turn, is the Fourier transform of the distribution

$$\mathscr{E}(t,x) = E_{\beta}(t^{\beta} A(D)) \delta_0(x), \quad t > 0, x \in \mathbb{R}^d.$$

Due to Theorem 5.2 the latter is a (unique) solution of the Cauchy problem for the fractional order differential equation

$$D_*^{\beta} u(t,x) = A(D) u(t,x), \quad t > 0, \; x \in \mathbb{R}^d, \tag{8.46}$$

$$u(0,x) = \delta_0(x), \quad x \in \mathbb{R}^d. \tag{8.47}$$

Thus, the limiting process of the CTRW under conditions (8.42) and (8.43) is a process associated with the initial value problem (8.46)–(8.47) for the fractional order FPK type equation. Consider an example.

Example 8.1. Let τ_1 be Levy's stable subordinator with the stability index $0 < \beta < 1$. The density function of τ_1 has the asymptotic behavior (see (7.37))

$$\phi(t) \sim \frac{\beta}{\Gamma(1-\beta) t^{1+\beta}}, \quad t \to \infty.$$

It follows that

$$\int_t^{\infty} \phi(u) du \sim \frac{1}{\Gamma(1-\beta) t^{\beta}}, \quad t \to \infty. \tag{8.48}$$

Further, we use the following fact [GM05]: the Laplace transform of a density $\varphi(t)$ satisfying the condition

$$\int_t^\infty \varphi(u)du \sim \frac{C}{\beta t^\beta}, \quad t \to \infty, \, C > 0, \tag{8.49}$$

has the asymptotic behavior (8.42) with

$$\lambda = \frac{C\pi}{\Gamma(1+\beta)\sin(\pi\beta)}.$$

In our case $C = \beta/\Gamma(1-\beta)$. Hence, using the relationship (1.7), one has

$$\lambda = \frac{\beta\pi}{\Gamma(1+\beta)\Gamma(1-\beta)\sin(\pi\beta)} = \frac{\pi}{\Gamma(\beta)\Gamma(1-\beta)\sin(\pi\beta)} = 1.$$

Thus, the density of the stable subordinator τ_1 with the stability index $0 < \beta < 1$ satisfies (8.42) with $\lambda = 1$.

Further, let $0 < \alpha < 2$ and Y_1 be a Lévy's symmetric α-stable distribution, that is $\hat{w}(\xi) = e^{-|\xi|^\alpha} \sim 1 + (-|\xi|^\alpha)$, $|\xi| \to 0$. Hence, the density of Y_1 satisfies (8.43) with $A(\xi) = -|\xi|^\alpha$. This is the symbol of the operator $\mathbb{D}_0^\alpha$ (see Proposition 3.4). Thus, in this particular example the CTRW approximates the process X_t associated with the fractional FPK equation

$$D_*^\beta u(t,x) = \mathbb{D}_0^\alpha u(t,x), \quad t > 0, x \in \mathbb{R}^d.$$

It follows from Theorem 7.10 (the case of one subordinator W_t) that the process X_t is a time-changed process with the time-change process E_t, the inverse to the Lévy's stable subordinator W_t with the stability index β, that is $X_t = Y_{E_t}$, where Y_t is the Lévy's α-stable process.

With appropriate scaling parameters this example can be extended to an arbitrary process $Y_t^\rho \in \mathbb{SS}$, as well. That is, the limiting process is associated with the Cauchy problem

$$D_*^\beta u(t,x) = \Psi(D)u(t,x), \quad t > 0, \, x \in \mathbb{R}^d, \tag{8.50}$$

$$u(0,x) = \delta_0(x), \quad x \in \mathbb{R}^d, \tag{8.51}$$

where $\Psi(D)$ is the pseudo-differential operator with the symbol $\Psi(\xi)$ defined in (8.23). For the sake of clarity, we consider a particular case, namely

$$\rho(d\alpha) = \sum_{j=1}^m a_j \delta_{\alpha_j}(\alpha) d\alpha.$$

Suppose i.i.d. random vectors Y_k in (8.33) have the structure

$$Y_k = \sum_{j=1}^m A_j Y_k^{(j)},$$

where $A_j = a_j^{1/\alpha_j}$ and $Y_k^{(j)}$ are symmetric α_j-symmetric distributions independent for all $k \in \mathbb{N}$. The densities of $Y_k^{(j)}$ are $w_j(x)$ with the corresponding characteristic functions $\hat{w}_j(\xi) = \exp(-|\xi|^{\alpha_j})$. By construction, the characteristic function $\hat{w}(\xi)$ of the random vector Y_k is the product of characteristic functions of $Y_k^{(j)}$, that is

$$\hat{w}(\xi) = \prod_{j=1}^{m} \hat{w}_j(A_j \xi) = \exp\left(- \sum_{j=1}^{m} a_j |\xi|^{\alpha_j} \right) \sim 1 - \sum_{j=1}^{m} a_j |\xi|^{\alpha_j}, \quad |\xi| \to 0.$$

Like the previous case we again rescale τ_k and Y_k introducing nonrandom parameters τ and $\mathbf{h} = (h_1, \dots, h_m)$. Namely, we change τ_k in (8.34) and Y_k in (8.33) respectively by $\tau \cdot \tau_k$ and $Y_k = A_1 h_1 Y_k^{(1)} + \cdots + A_m h_m Y_k^{(m)}$. Then, it follows from the corresponding Montroll-Weiss equation that

$$\begin{aligned} \hat{\tilde{p}}_{\tau,\mathbf{h}}(s,\xi) &\sim \frac{\tau^\beta s^{\beta-1}}{\tau^\beta s^\beta + \sum_{j=1}^m a_j h_j^{\alpha_j} |\xi|^{\alpha_j} - \tau^\beta s^\beta \sum_{j=1}^m a_j h_j^{\alpha_j} |\xi|^{\alpha_j}} \\ &= \frac{s^{\beta-1}}{s^\beta + \sum_{j=1}^m \tau^{-\beta} h_j^{\alpha_j} a_j |\xi|^{\alpha_j} - s^\beta \sum_{j=1}^m a_j h_j^{\alpha_j} |\xi|^{\alpha_j}}. \end{aligned} \tag{8.52}$$

If one selects the multi-scaling parameter $\mathbf{h} \in \mathbb{R}^m$ such that $h_j = \tau^{\beta/\alpha_j}$, $j = 1, \dots, m$, then (8.52) implies the locally uniform convergence

$$\hat{\tilde{p}}_{\tau,\mathbf{h}}(s,\xi) \to \frac{s^{\beta-1}}{s^\beta + \sum_{j=1}^m a_j |\xi|^{\alpha_j}}, \quad \tau \to 0 \ (|\mathbf{h}| \to 0).$$

Hence, in this case the constructed CTRW approximates the process X_t associated with the time-fractional distributed order FPK equation

$$D_*^\beta u(t,x) = \sum_{k=0}^{m} a_k \mathbb{D}_0^{\alpha_k} u(t,x), \quad t > 0, x \in \mathbb{R}^d.$$

8.5.2 Continuous time random walk approximants of time-changed processes

Now we construct a CTRW approximation (called *fully discrete random walk model [GMM02, GV03]*) of the time-changed process associated with the fractional FPK equation (8.50) with initial condition (8.51). Note that the solution $u(t,x)$ is the density function of the time-changed process $X_t = Y_{W_t}$, where Y_t is a driving process and W_t is the inverse to a β-stable subordinator. Since X_t is non-Markovian, an approximating random walk also can not be independent. Therefore, transition probabilities split into two different sets of probabilities:

(a) *non-Markovian transition probabilities,* which express a long non-Markovian memory of past; and

(b) *Markovian transition probabilities,* which express transition from positions at the previous time instant.

Suppose that non-Markovian transition probabilities are given by (see [GMM02])

$$c_\ell = (-1)^{\ell+1}\binom{\beta}{\ell} = \left|\binom{\beta}{\ell}\right|, \quad \ell = 1,\dots,n,$$
$$\gamma_n = \sum_{\ell=0}^{n}(-1)^\ell\binom{\beta}{\ell}, \tag{8.53}$$

and Markovian transition probabilities $\{p_k\}_{k\in\mathbb{Z}^n}$ are given by

$$p_k = \begin{cases} c_1 - 2\tau^\beta Q(h), & \text{if } k = 0; \\ 2\tau^\beta \frac{Q_k(h)}{|k|^d}, & \text{if } k \neq 0, \end{cases} \tag{8.54}$$

where $Q_k(h), k \neq 0$, is defined in (8.26), and $Q(h) = \sum_{k\neq 0} Q_k(h)|k|^{-d}$. Then the probability q_j^{n+1} of the walker being at the site $x_j = jh$, $j \in \mathbb{Z}^d$, at the time instant t_{n+1} is

$$q_j^{n+1} = \gamma_n q_j^0 + \sum_{\ell=1}^{n-1} c_{n-\ell+1} q_j^\ell + \left(c_1 - \tau^\beta Q_0(h)\right) q_j^n + \sum_{k\neq 0} p_k q_{j-k}^n.$$

It follows from Theorems 8.2 and 7.10 that the density function of the process $X_t = Y_{E_t}$ satisfies the equation

$$D_*^\beta u(t,x) = \Psi(D_x)u(t,x), \tag{8.55}$$

where the pseudo-differential operator $\Psi(D_x)$ has the symbol $\Psi(\xi)$ defined in equation (8.23), that is

$$\Psi(D_x)\varphi(x) = \int_0^2 \mathbb{D}_0^\alpha \varphi(x) d\rho,$$

with a finite measure ρ, $\operatorname{supp}\rho \subset (0,2)$, and the initial condition $u(0,x) = \delta_0(x)$. Due to Theorem 5.1, the unique solution to (8.55) with this initial condition is given by (see Example 5.1 and equation (5.15))

$$u(t,x) = E_\beta(\Psi(D)t^\beta)\delta_0(x) = \frac{1}{(2\pi)^n}\int_{\mathbb{R}^n} E_\beta(\Psi(\xi)t^\beta)e^{-ix\xi}d\xi.$$

Therefore, the Fourier-Laplace transform of $u(t,x)$, in accordance with Proposition 3.7, is

$$L[F[u](\xi)](s) = L[E_\beta(\Psi(\xi)t^\beta)](s) = \frac{s^{\beta-1}}{s^\beta + (-\Psi(\xi))}, \quad s > 0, \xi \in \mathbb{R}^d. \tag{8.56}$$

Theorem 8.4. *([Umal15]) Fix $t > 0$ and let $h > 0$, $\tau = t/n$. Let $Y_j \in \mathbb{Z}^d$, $j \geq 1$, be identically distributed random vectors with the non-Markovian and Markovian transition probabilities defined in* (8.53) *and in* (8.54)*, respectively. Assume that*

$$\tau \leq \left(\frac{\beta}{Q(h)}\right)^{\frac{1}{\beta}}. \tag{8.57}$$

Then the sequence of random vectors $S_n = hY_1 + \ldots + hY_n$, converges as $n \to \infty$ in law to $X_t = Y_{E_t}$ whose probability density function is the solution to equation (8.55) with the initial condition $u(0,x) = \delta_0(x)$.

Proof. For the Caputo fractional derivative on the left-hand side of (8.55) we will use the backward Grünwald-Letnikov discretization (3.94):

$$D_*^\beta u_j^n = {}_0\mathscr{D}_t^\beta u_j^n \approx \sum_{m=0}^{n+1} (-1)^m \binom{\beta}{m} \frac{u_j^{n+1-m} - u_j^0}{\tau^\beta} \tag{8.58}$$

where $u_j^n = u(t_n, x_j)$, $n = 0, 1, \ldots$, $j \in \mathbb{Z}^d$, $x_j \in \mathbb{Z}_h^d$, and $t_n = n\tau$, $\tau > 0$. Using notations (8.53) and rearranging terms, equation (8.58) can be expressed in the form

$$D_*^\beta u_j^n \approx \frac{1}{\tau^\beta}\left(u_j^{n+1} - c_1 u_j^n - \sum_{m=2}^{n+1} c_m u_j^{n+1-m} - \gamma_n u_j^0\right). \tag{8.59}$$

For the discretization of the right-hand side of (8.55), due to approximation (8.28), valid for all $0 < \alpha < 2$, one obtains

$$\Psi(D_x) u_j^n \approx \sum_{k \in \mathbb{Z}_h^d} d_k u_{j-k}^n, \quad d_k := \begin{cases} 2\dfrac{Q_k(h)}{|k|^d}, & k \neq 0, \\ -2\displaystyle\sum_{k \neq 0} \frac{Q_k(h)}{|k|^d}, & k = 0, \end{cases} \tag{8.60}$$

where $Q_k(h)$ is defined in (8.26). Setting the discretizations for the time and space-fractional derivatives in (8.58) and (8.60)) equal to each other, we get

$$\frac{1}{\tau^\beta}\left(u_j^{n+1} - c_1 u_j^n - \sum_{m=2}^{n} c_m u_j^{n+1-m} - \gamma_n u_j^0\right) = \sum_{k \in \mathbb{Z}_h^d} d_k u_{j-k}^n. \tag{8.61}$$

Rearranging terms and solving for u_j^{n+1}, the following recursion equation is constructed:

$$u_j^{n+1} = \gamma_n u_j^0 + \sum_{m=2}^{n} c_m u_j^{n+1-m} + \sum_{k\in\mathbb{Z}_h^d} q_k u_{j-k}^n, \tag{8.62}$$

$$q_k = \begin{cases} \tau^\beta d_k = \tau^\beta \int_0^2 \left(\frac{b\alpha}{h^\alpha}\right) \frac{d\rho(\alpha)}{|k|^{d+\alpha}}, & k \neq 0 \\ c_1 - \sum_{k\neq 0} q_k, & k = 0. \end{cases} \tag{8.63}$$

By construction, $u_j^0 = 1$ if $j = 0 = (0,\dots,0)$, and $u_j^0 = 0$ otherwise.

The update u_j^{n+1} in equation (8.62) is determined by Markovian contributions (those values of u at time $t = t_n$) and non-Markovian contributions (those values of u at times $t = \{t_0, t_1, \dots, t_{n-1}\}$). The order of the time fractional derivative β determines the effect that the non-Markovian transition probabilities (γ_n and $c_2, \dots, c_m$) has on u_j^{n+1}. This effect can be measured by sum of all of the transition probabilities in equation (8.62):

$$\left(\gamma_n + \sum_{m=2}^{n} c_m\right) + \sum_{k\in\mathbb{Z}^d} q_k = 1.$$

where

$$\sum_{k\in\mathbb{Z}^d} q_k = (c_1 - q_0) + \sum_{k\neq 0} q_k = c_1 \qquad \text{and} \qquad \gamma_n + \sum_{m=2}^{n} c_m = 1 - c_1.$$

As a result, when $\beta = 1$ one has $c_1 = 1$, $c_2 = \cdots = c_n = \gamma_n = 0$, and hence, equation (8.62) simply reduces to (8.29), with $q_k = p_k$, where p_k, $k \in \mathbb{Z}^d$, are defined by (8.30).

Let $\hat{u}^n(\xi)$ be the characteristic function of the discrete sequence u_j^n for a fixed $n = 0, 1, \dots,$ (see the definition in (8.32)). Then equation (8.62), in terms of characteristic functions, takes the form

$$\hat{u}^{n+1}(\xi) = \gamma_n + \sum_{m=2}^{n} c_m \hat{u}^{n+1-m}(\xi) + \hat{q}(\xi)\hat{u}^n(\xi), \tag{8.64}$$

since $\hat{u}^0(\xi) = 1$. Further, let $\hat{U}_\tau(s,\xi)$ be the discrete Laplace transform of $\hat{u}^{n+1}(\xi)$, namely

$$\hat{U}_\tau(s,\xi) = \tau \sum_{n=0}^{\infty} \hat{u}^{n+1}(\xi) e^{-st_n}, \quad s > 0.$$

Then multiplying both sides of (8.64) by $\tau e^{-n\tau s}$ and summing over the index n, one obtains

$$\begin{aligned}\hat{U}_\tau(s,\xi) &= \gamma_\tau(s) + \tau \sum_{n=0}^{\infty}\left(\sum_{m=2}^{n+1} c_m \hat{u}^{n+1-m}(\xi)\right) e^{-n\tau s} + \hat{q}(\xi)\tau \sum_{n=0}^{\infty} \hat{u}^n(\xi) e^{-sn\tau} \\ &= \gamma_\tau(s) - \tau \sum_{n=0}^{\infty}\left(\sum_{m=1}^{n+1} (-1)^m \binom{\beta}{m} \hat{u}^{n+1-m}(\xi)\right) e^{-n\tau s} \\ &+ \hat{d}(\xi)\tau^{1+\beta} \sum_{n=0}^{\infty} \hat{u}^n(\xi) e^{-sn\tau}, \end{aligned} \tag{8.65}$$

where

$$\gamma_\tau(s) = \tau \sum_{n=0}^{\infty} \gamma_n e^{-sn\tau} = \tau \sum_{n=0}^{\infty} \sum_{m=0}^{n+1} (-1)^m \binom{\beta}{m} e^{-sn\tau}.$$

Changing the order of summation one can show that

$$\gamma_\tau(s) = e^{s\tau}\left(\sum_{n=0}^{\infty} \tau e^{-sn\tau}\right) \sum_{m=0}^{\infty} (-1)^m \binom{\beta}{m} e^{-sm\tau}.$$

Further, in accordance with relation (8.56), in order to prove the theorem we need to show that $\hat{U}_\tau(s,h\xi)$ converges as $h \to 0$ (that implies $\tau \to 0$ too) to

$$L[E_\beta(\Psi(\xi)t^\beta)](s) = \frac{s^{\beta-1}}{s^\beta + (-\Psi(\xi))}, \quad s > 0,\ \xi \in \mathbb{R}^d$$

the Laplace transform of the Mittag-Leffler function $E_\beta(x)$ composed by $\Psi(\xi)t^\beta$. Here $\Psi(\xi) = -\int_0^2 |\xi|^\alpha dp(\alpha)$; see (8.23). Indeed, this convergence implies the convergence $\hat{u}^n(h\xi) \to E_\beta(\Psi(\xi)t^\beta)$, as $n \to \infty$, uniformly for all $\xi \in \mathscr{K}$, where $\mathscr{K}$ is an arbitrary compact in $\mathbb{R}^d$. In turn, the latter convergence is equivalent to the convergence in law of the sequence S_n to the process Y_{W_t}. To show the convergence $\hat{U}_\tau(s,h\xi) \to L[E_\beta(\Psi(\xi)t^\beta)](s)$, we notice that

$$\tau \sum_{n=0}^{\infty} \hat{u}^n(\xi) e^{-sn\tau} = \tau + e^{-s\tau}\hat{U}_\tau(s,\xi),$$

and changing the order of summation

$$\begin{aligned}&\tau \sum_{n=0}^{\infty}\left(\sum_{m=1}^{n+1} (-1)^m \binom{\beta}{m} \hat{u}^{n+1-m}(\xi)\right) e^{-n\tau s} \\ &= -\tau\beta + \left(\tau e^{s\tau} + \hat{U}_\tau(s,\xi)\right)\left(\sum_{n=0}^{\infty} (-1)^n \binom{\beta}{n} e^{-sn\tau} - 1\right). \end{aligned} \tag{8.66}$$

It follows from equations (8.65)–(8.66) that

$$\hat{U}_\tau(s,\xi) = \frac{e^{s\tau} I_\tau(\beta,s)\left(\sum_{n=0}^{\infty} \tau e^{-sn\tau} - \tau\right) + \tau^{1-\beta}(\tau^\beta \hat{d}(\xi) + \beta + e^{s\tau})}{I_\tau(\beta,s) - \hat{d}(\xi)e^{-s\tau}} \tag{8.67}$$

where

$$I_\tau(\beta,s) = \frac{1}{\tau^\beta}\sum_{n=0}^{\infty}(-1)^n \binom{\beta}{n} e^{-sn\tau}.$$

Further, the following limits hold:

$$\lim_{\tau\to 0}\frac{1}{\tau^\beta}\sum_{n=0}^{\infty}(-1)^n\binom{\beta}{n}e^{-sn\tau} = \left({}_{-\infty}\mathscr{D}_t^\beta e^{st}\right)_{|t=0} = s^\beta, \tag{8.68}$$

$$\lim_{\tau\to 0}\left(\sum_{n=0}^{\infty}\tau e^{-sn\tau} - \tau\right) = s^{-1}, \tag{8.69}$$

$$\lim_{h\to 0}\hat{d}(h\xi) = \Psi(\xi), \tag{8.70}$$

$$\lim_{\tau\to 0}\tau^{1-\beta}(\tau^\beta\hat{d}(\xi) + \beta + e^{s\tau}) = 0, \tag{8.71}$$

The relation (8.68) follows from the definition (3.94), the first equality in Example 3.12, and the fact that the Grünwald-Letnikov fractional derivative coincides with Liouville-Weyl derivative in the class of suitable functions (see Remark 3.5). The relations (8.69) and (8.71) can be easily verified by direct calculation. To show the relation (8.70) we have

$$\begin{aligned}
\hat{d}(h\xi) &= -2\sum_{k\neq 0}\frac{Q_k(h)}{|k|^d} + 2\sum_{k\neq 0}\frac{Q_k(h)}{|k|^d}e^{ikh\xi} = 2\sum_{k\neq 0}\frac{Q_k(h)}{|k|^d}(e^{ikh\xi}-1)\\
&= \sum_{k\neq 0}\frac{Q_k(h)}{|k|^d}\left(e^{ikh\xi} - 2 + e^{ikh\xi}\right) = \sum_{k\neq 0}\frac{Q_k(h)}{|k|^d}\left(\Delta_{kh}^2 e^{ix\xi}\right)_{|x=0}\\
&= \int_0^2 b(\alpha)\left(\lim_{N\to\infty}\sum_{|kh|\le N, k\neq 0}\frac{\Delta_{kh}^2 e^{ix\xi}}{|kh|^{d+2}}h^d\right)_{|x=0} d\rho(\alpha).
\end{aligned}$$

Letting $h\to 0$, due to (8.13) and the second formula in Corollary 2.11, we obtain

$$\begin{aligned}
\lim_{h\to 0}\hat{d}(h\xi) &= \int_0^2 b(\alpha)\left(\lim_{N\to\infty}\int_{|y|\le N}\frac{\Delta_y^2 e^{ix\xi}}{|y|^{d+2}}dy\right)_{|x=0} d\rho(\alpha)\\
&= \int_0^2 b(\alpha)\left(\int_{\mathbb{R}^d}\frac{\Delta_y^2 e^{ix\xi}}{|y|^{d+2}}dy\right)_{|x=0} d\rho(\alpha) = \int_0^2 (\mathbb{D}_0^\alpha e^{ix\xi})_{|x=0}d\rho(\alpha)\\
&= -\int_0^2 |\xi|^2 d\rho(\alpha) = \Psi(\xi).
\end{aligned}$$

Now taking into account the relations (8.68)–(8.71) it follows from (8.67) that

$$\lim_{h\to 0} \hat{U}_\tau(s,h\xi) = \frac{s^{\beta-1}}{s^\beta - \Psi(\xi)}, \quad s>0,\ \xi \in \mathbb{R}^d,$$

as desired.

Remark 8.3. 1. The authors of paper [LSAT05] use a different discretization for $D_*^\beta u_j^n$. Namely,

$$D_*^\beta u_j^n \approx \frac{1}{\Gamma(1-\beta)} \sum_{m=0}^{n} \int_{t_n}^{t_{n+1}} \frac{u_j'(t_{n+1}-s)}{s^\beta} ds$$

$$= \frac{1}{\nu\tau^\beta}\left([u_j^{n+1}-u_j^n] + \sum_{m=0}^{n} [(m+1)^{1-\beta} - m^{1-\beta}][u_j^{n+1-m} - u_j^{n-m}], \right) \tag{8.72}$$

where $\nu = \Gamma(2-\beta)$. Setting

$$a(\beta,\tau) = (\nu\tau^\beta)^{-1}, \tag{8.73}$$

$$\gamma_m = (m+1)^{(1-\beta)} - m^{(1-\beta)}, \quad m=0,\dots,n, \tag{8.74}$$

$$c_k = \gamma_{k-1} - \gamma_k, \quad k=1,\dots,n, \tag{8.75}$$

and rearranging terms, equation (8.72) can be expressed in the form

$$D_*^\beta u_j^n \approx a(\beta,\tau)\left(u_j^{n+1} - c_1 u_j^n - \sum_{m=2}^{n} c_m u_j^{n+1-m} - \gamma_n u_j^0 \right). \tag{8.76}$$

The latter unifies both discretizations. Indeed, in the case of the Grünwald-Letnikov time-fractional derivative, $a(\beta,\tau)$, γ_m and c_k are re-defined as the following:

$$a(\beta,\tau) = \frac{1}{\tau^\beta}, \tag{8.77}$$

$$\gamma_m = 1 - \sum_{i=1}^{m} c_i, \tag{8.78}$$

$$c_k = (-1)^k \binom{\beta}{k} = \left(1 - \frac{1+\beta}{k}\right) c_{k-1}, \quad k=1,2,\dots\ (c_0=1). \tag{8.79}$$

Note that for $0<\beta\le 1$, $\gamma_0 = 1$. In addition, for $\beta=1$, $\gamma_m = 0$ for $m=1,\dots,n$, $c_1 = 1$, $c_k = 0$ for $k=2,\dots,n$ and $\nu=1$. As a result, when $\beta=1$, both discretizations reduce to the standard forward-time discretization of $\partial u/\partial t$:

$$D_*^1 u_j^n = \frac{\partial u}{\partial t} \approx \frac{u_j^{n+1}-u_j^n}{\tau}.$$

2. Theorem 8.4 extends to the case when the left-hand side of equation (8.55) is a time distributed fractional order differential operator with a mixing measure μ, whose support satisfies $\operatorname{supp}\mu \subseteq [0,1]$, that is

$$D_\mu u(t,x) = \int_0^1 D_*^\beta u(t,x)d\mu(\beta) = \Psi(D)u(t,x), \quad t > 0, x \in \mathbb{R}^d, \tag{8.80}$$

where $\Psi(D)$ is a pseudo-differential operator with the symbol $\Psi(\xi)$ defined in (8.23). In this case for the left-hand side of (8.80) we again have a discretization of the form (8.76). Namely, we have

$$D_\mu u_j^n \approx a(\tau)\left(u_j^{n+1} - c_1^* u_j^n - \sum_{m=2}^{n} c_m^* u_j^{n+1-m} - \gamma_n^* u_j^0\right),$$

where

$$a(\tau) = \int_0^1 a(\tau,\beta)d\mu(\beta),\ c_k^* = \frac{1}{a(\tau)}\int_0^1 a(\tau,\beta)c_k(\beta)d\mu(\beta),\ k = 1,\ldots,n, \tag{8.81}$$

$$\gamma_n^* = \frac{1}{a(\tau)}\int_0^1 a(\tau,\beta)\gamma_n(\beta)d\mu(\beta),\ n = 1,2,\ldots. \tag{8.82}$$

In equations (8.81) and (8.82) the integrands $a(\tau,\beta)$, $c_k(\beta)$, and $\gamma_n(\beta)$ are defined in (8.77)–(8.79) or (8.73)–(8.75) depending on whether the Grünwald-Letnikov approximation or approximation (8.72) is used for discretization of $D_*^\beta u(t,x)$ in (8.80).
3. We also note that condition (8.57) takes the form

$$\tau \leq \left(\frac{2-2^{1-\beta}}{\Gamma(2-\beta)Q(h)}\right)^{\frac{1}{\beta}}$$

if the non-Markovian probabilities are selected as in (8.73)–(8.75). This condition as well as (8.57) generalize the well-known Lax's stability condition $\tau \leq h^2/2$ arising in the finite-difference method for solution of an initial value problem for the heat equation, which corresponds to the case $\beta = 1$ and $\Psi(D) = \Delta$, the Laplace operator. In this case $Q(h)$ reduces simply to $Q(h) = 2/h^2$.

8.6 Additional notes

1. *Random walk. CTRW.* The random walk problem was first set in a note by Karl Pearson in the journal "Nature" in 1905 [Pea05]. Nowadays it is a mathematical tool broadly used in modeling of various problems arising in science and engineering. For the general theory of random walks and their relation to mathematical problems we refer the reader to books [Spi01, L10]. Some aspects of the modern state of the random walk theory are presented in the following survey papers: various applications of random walk to fractional dynamical processes arising in natural and social sciences in [MK00, MK04], random walk in graphs in [Lov93], applications of

random walk to finance in [SGM00]. Continuous time random walk was introduced by Montroll and Weiss [MW65] in 1965. CTRWs have rich applications in many applied sciences and the literature on CTRWs is still increasing at a rapid rate. See papers [MK00, MK04, SGM00, MS05] and references therein for a discussion of the history of development of the CTRW theory and its connections to fractional differential equations and other relevant fields. There are various approaches to the study of weak CTRW limits, depending on the topology and methods used for the proof of convergence. The methods used include master equations, constructive random walk approximations, and use of abstract continuity theorems. CTRWs also serve as approximate for driving processes of SDEs. Driving processes of the SDEs associated with fractional FPK equations appear to be independent time-changes of basic processes like Brownian motion, Lévy processes, fractional Brownian motions, etc.

2. *Approximation of Brownian motion. Donsker's theorem.*
Let X_k, $k = 1, 2, \ldots$, be a sequence of independent and identically distributed mean zero, variance one random vectors. Consider the sequence of scaled sums

$$S_n(t) = \frac{1}{\sqrt{n}} \sum_{j=1}^{\lfloor nt \rfloor} X_j, \tag{8.83}$$

where $\lfloor r \rfloor$ means the greatest integer not exceeding r. Denote by $D([0,\infty),\mathbb{R}^d)$ the Skorohod space $D([0,\infty),\ \mathbb{R}^d)$ of cádlág processes with the Skorohod topology; see definition in [Bil99].

Theorem 8.5. *(Donsker) The random walk $S_n(t)$ in* (8.83) *converges weakly to d-dimensional standard Brownian motion in the Skorohod space $D([0,\infty),\ \mathbb{R}^d)$.*

If one modifies the path of the nth term making it continuous by linear interpolation of the normalized partial sums, then the same kind of convergence holds in $C([0,\infty),\mathbb{R}^d)$ with the uniform topology.

3. *Approximation of fBW.* In the case of Hurst exponent $H > 1/2$ the weak convergence of scaled sums of random variables to a fractional Brownian motion was studied by Taqqu [Taq75]. He described the class of functions $G(s)$, such that $\frac{1}{d_n}\sum_{j=1}^{\lfloor nt \rfloor} G(X_j)$ with $d_n \sim n^{2H}L(n)$, L is slowly varying, converges weakly to cB_t^H, $c = \mathbb{E}[XG(X)]$. Sottinen [Sot01] proved that the following random walk

$$Z_n(t) = \sum_{j=1}^{\lfloor nt \rfloor} n \int_{\frac{j-1}{n}}^{\frac{j}{n}} z(\lfloor nt \rfloor/n, s)ds \frac{1}{\sqrt{n}} X_j,$$

where $z(t,s) = c_H(H-1/2)s^{1/2-H}\int_s^t u^{H-1/2}(u-s)^{H-1/2}du$, converges weakly to a fractional Brownian motion B_t^H.

4. *Approximation of time-changed stable laws.* The CTRW approximation in different topologies of time-changed stable Lévy processes with the time-change process being the inverse of Lévy's stable subordinator is studied in the paper [MS05]. Suppose that τ_1 belongs to the strict domain of attraction of a stable law with index $\beta \in (0,1)$ and Y_1 belong to the strict domain of attraction of a generalized full operator stable law. Then, under some condition to τ_1 and Y_1, there exists a regularly varying function $B(c)$, $c > 0$, and slowly varying function L, such that $B(c)S_{L^{-1}(ct)}$ converges to A_{E_t}, as $c \to \infty$, in the sense of finite-dimensional distributions. Here S_t is CTRW associated with i.i.d. $\tau_1, \tau_2, \ldots,$, and i.i.d. $Y_1, Y_2, \ldots$; A_t is an operator stable motion, and E_t is the inverse to the stable subordinator W_t with index β. In the paper [BMS04] the convergence of CTRW approximation of time-changed stable Lévy processes in M_1-topology of the Skorokhod space $D([0,\infty),R^d)$, is proved.

5. *Constructive random walk approximations.* Gillis and Weiss [GW70] modeled random walk with jump probabilities $p(r) = p(-r) = Ar^{-(\alpha+1)}$, $0 < \alpha \le 2$, in the 1-D case, and $p(r) = r^{-\beta}$, where $r^2 = r_1^2 + r_2^2$, $1 < \beta \le 2$, in the 2-D case. Here A is a normalizing constant. They found the estimated number of distinct lattice points visited in the course of the random walk. In a series of papers (see [GM99, GM01, GM05]) Gorenflo and Mainardi constructed several classes of

random walk models (Gillis-Weiss, Grünwald-Letnikov, globally binomial, Chechkin-Gonchar, fully discrete, etc.) approximating space- and space-time fractional diffusion processes. For one of these models, called a Gillis-Weiss model, for the case of $0 < \alpha < 2$ they proved the following result (see [GM01]). Let $X_j \in \mathbb{Z}$, $j \geq 1$, be i.i.d. random variables with the probability mass function

$$p_k = \mathbb{P}(X_1 = k) = \begin{cases} 1 - 2\mu b(\alpha)\zeta(\alpha+1), & \text{if } k = 0; \\ \frac{\mu b(\alpha)}{|k|^{\alpha+1}}, & \text{if } k \neq 0, \end{cases}$$

where $\tau > 0$, $h > 0$, $\mu = h^{-\alpha}\tau$, $\zeta(s)$ is Riemann's zeta-function, and $b(\alpha) = \pi^{-1}\Gamma(\alpha + 1)\sin(\alpha\pi/2)$. Assume that $\mu \leq 1/(2b(\alpha)\zeta(\alpha+1))$. Then the sequence of random variables $S_N = hX_1 + \ldots + hX_N$, converges as $N \to \infty$ in law to the process S_t, whose density function (for each fixed t) is $G_t^{(\alpha)}(x) = F^{-1}[e^{-t|\xi|^\alpha}](x)$, where F^{-1} is the inverse Fourier transform.

Another model called fully discrete random walk model is presented in [GV03, GAR04]. In particular, the following statement is obtained. Fix $t > 0$ and let $h > 0$, $\tau = t/n$. Let variables $X_j \in \mathbb{Z}$, $j \geq 1$, be identically distributed random variables whose non-Markovian transition probabilities are defined as in (8.53) and Markovian transition probabilities are defined as

$$p_k = \begin{cases} c_1 - 2\mu, & \text{if } k = 0; \\ \pm\mu, & \text{if } k = \pm 1; \\ 0, & \text{if } |k| \geq 2, \end{cases}$$

where $\mu = \tau^\beta h^{-2}$. Assume that $\mu \leq \beta/2$. Then the sequence of random vectors $S_n = hX_1 + \ldots + hX_n$, converges as $n \to \infty$ in law to $X_t = X_{W_t}$ whose probability density function is the solution to the equation

$$D_*^\beta u(t,x) = \frac{\partial^2 u(t,x)}{\partial x^2}, \quad t > 0,\ x \in \mathbb{R},\ (0 < \beta < 1)$$

satisfying the initial condition $u(0,x) = \delta_0(x)$. A similar result is obtained in [LSAT05] with further analysis of stability and convergence using a numerical approach. The random walk constructed by Abdel-Rehim [A13] approximates in the weak sense the process whose density solves the equation, $D_*^\beta u(t,x) = \mathscr{D}_0^\alpha\, u(t,x)$, $0 < \beta < 1$, $0 < \alpha < 2$, generalizing the above two results. In the paper [US06] a random walk approximating the (multivariate) stable process, whose density function solves the equation (8.22) (in the case $\rho(d\alpha) = a(\alpha)d\alpha$, $a \in C[0,2]$), is constructed. Theorem 8.4 generalizes the results obtained in the papers [GM01, A13, US06].

Chapter 9
Complex ΨDOSS and systems of complex differential equations

9.1 Introduction

In Chapters 4–7 we discussed pseudo-differential equations of integer and fractional orders with ΨDOSS depending on real variables $t \in \mathbb{R}$ and $x \in \mathbb{R}^n$. In this section we will discuss differential and pseudo-differential equations depending on complex variables $t = \tau + i\sigma \in \mathbb{C}$ and $z = x + iy \in \mathbb{C}^n$. Consider two simple examples with the one-dimensional "spatial" variable:

(i) "complex wave" equation, and
(ii) "complex heat" equation.

The first equation is obtained from the wave equation

$$\frac{\partial^2 u(\tau,x)}{\partial \tau^2} = -D^2 u(\tau,x), \quad \tau > 0, x \in \mathbb{R}, \tag{9.1}$$

where $D = -id/dx$, by "complexifying" the variables t and x, that is

$$D_t^2 u(t,z) = D_z^2 u(t,z), \quad t \in \mathbb{C}, z \in \mathbb{C}, \tag{9.2}$$

where $D_t = \frac{\partial}{\partial \tau} + i\frac{\partial}{\partial \sigma}$ and $D_z = \frac{\partial}{\partial x} + i\frac{\partial}{\partial y}$. The solution to (9.1), satisfying the initial conditions $u(0,x) = \varphi(x)$ and $u_t(0,x) = \psi(x)$, was obtained in Section 2.2 in the form

$$u(\tau,x) = [\cos \tau D]\, \varphi(x) + \left[\frac{\sin \tau D}{D}\right] \psi(x). \tag{9.3}$$

Replacing D in (9.3) by D_z one obtains

$$u(t,z) = [\cos t D_z]\, \varphi(z) + \left[\frac{\sin t D_z}{D_z}\right] \psi(z). \tag{9.4}$$

S. Umarov, *Introduction to Fractional and Pseudo-Differential Equations with Singular Symbols*, Developments in Mathematics 41,
DOI 10.1007/978-3-319-20771-1_9

Is $u(t,z)$ in equation (9.4) a solution to complex equation (9.2) satisfying the “initial” conditions $u(0,z) = \varphi(z)$ and $D_t u(0,z) = \psi(z)$? If yes, in what sense the operators $\cos tD_z$ and $\frac{\sin tD_z}{D_z}$ must be understood, and in what spaces these operators act? It is not hard to verify that d’Alembert’s formula in this case takes the form

$$u(t,z) = \frac{\varphi(ze^{it}) + \varphi(ze^{-it})}{2} + \frac{1}{2}\int_{z-t}^{z+t} \psi(\zeta)d\zeta,$$

where the integral is the line integral over a smooth curve connecting points $z-t$ and $z+t$ on the complex plane.

The second equation is obtained from the heat equation

$$\frac{\partial u(\tau,x)}{\partial \tau} = -D^2 u(\tau,x), \quad \tau > 0,\ x \in \mathbb{R}, \tag{9.5}$$

complexifying the variables τ and x, that is

$$D_t u(t,z) = D_z^2 u(t,z), \quad t \in \mathbb{C}, z \in \mathbb{C}. \tag{9.6}$$

Again, replacing D by D_z in the solution representation $u(\tau,x) = \exp(-tD^2)\varphi(x)$ of equation (9.5), satisfying the initial condition $u(0,x) = \varphi(x)$, can we state that

$$u(t,z) = e^{tD_z^2}\varphi(z)$$

solves complex equation (9.6) with the “initial” condition $u(0,z) = \varphi(z)$? If the answer yes, how should we understand the operator $\exp(tD_z^2)$, and in what class of functions it is meaningful?

We note that the complex “wave” equation has a unique solution in the class of analytic functions near $(0,0) \in \mathbb{C}^2$ if φ and ψ are analytic in a neighborhood of $0 \in \mathbb{C}$, while the complex “heat” equation does not possess this property. This is due to the fact that the complex “wave” equation is Kowalevskian (definition is given below), while the complex “heat” equation is not. Hence, in the theory of complex differential and pseudo-differential equations new features appear, making this theory very distinct from its “real” counterpart.

Thus, in this chapter we will discuss the problem of existence and uniqueness of a solution to systems of complex differential and pseudo-differential equations in the complex $(n+1)$-dimensional space. These systems in the general form can be represented as

$$D_t^{p_j} u_j(t,z) + \sum_{k=1}^{N}\sum_{q=0}^{p_k-1} A_{jk}^q(t,z,D_z) D_t^q u_k(t,z) = f_j(t,z), \quad j = 1,\dots,N, \tag{9.7}$$

$$\sum_{k=1}^{N}\sum_{q=0}^{p_k-1} B_{jk}^{mq}(z,D_z) D_t^q u_k(t,z)\Big|_{t=t_0} = \varphi_{jm}(z), \quad m = 0,\dots,p_j-1,\ j = 1,\dots,N, \tag{9.8}$$

where $t \in \mathscr{D}$, a connected domain in $\mathbb{C}$, and $z \in \mathbb{C}^n$. The operators $A^q_{jk}(t,z,D_z)$, $q = 0,\dots,p_j - 1; k, j = 1,\dots,N$, and $B^{mq}_{jk}(D_z), q, m = 0,\dots,p_j - 1; k, j = 1,\dots,N$, are, in general, pseudo-differential operators with analytic symbols (see the definition in Section 9.6) in a domain $G \subset \mathbb{C}^n$, and the functions $f_j(t,z)$, $j = 1,\dots,N$, and $\varphi_{jm}(z)$, $m = 0,\dots,p_j - 1$; $j = 1,\dots,N$, satisfy certain conditions clarified in Section 9.7; $p_j \geq 1$, $j = 1,\dots,N$, are integers. We note that symbols of ΨDO have singularities of finite order at the boundary of G or finite exponential type if $G = \mathbb{C}^n$.

The Cauchy problem is a particular case, corresponding to $B^{mq}_{jk}(z,D_z) = \delta^{mq}_{jk} I$, where I is the identity operator, and

$$\delta^{mq}_{jk} = \begin{cases} 1, & \text{if } q = m, \text{ and } k = j, \\ 0, & \text{otherwise}, \end{cases}$$

is the generalized Kronecker symbol. It is not hard to see that boundary conditions (9.8) can be reduced to the Cauchy conditions

$$D^q_t u_j(t,z)\Big|_{t=t_0} = \psi_{jq}(z), \quad q = 0,\dots,p_j - 1, j = 1,\dots,N, \tag{9.9}$$

where the vector function $\psi_{jq}(z)$ of length $p_1 + \cdots + p_n$ is a solution to the system of pseudo-differential equations

$$\sum_{k=1}^{N} \sum_{q=0}^{p_j - 1} B^{mq}_{jk}(z,D_z)\psi_{kq}(z) = \varphi_{jm}(z), \quad m = 0,\dots,p_j - 1, j = 1,\dots,N. \tag{9.10}$$

Hence, the general boundary value problem (9.7)–(9.8) splits into two problems:

1. the system of pseudo-differential equations (9.10), and
2. the Cauchy problem (9.7), (9.9).

A brief history. We start with a brief history, since it casts light on the question: "Where did the conditions for orders of operators $A^q_{jk}(t,z,D_z)$ and $B^{mq}_{jk}(z,D_z)$ appeared in the theorems of this chapter came from?"

The Cauchy problem, the most important and the most studied amongst boundary value problem (9.7)–(9.8), was always a focus of many classics (d'Alembert, Euler, Fourier, Poisson, Cauchy, Hadamard, Holmgren, Petrovskii, Sobolev, etc.). Mizohata in his book [Miz67] emphasized four problems related to the Cauchy problem:

1. existence of a local solution;
2. uniqueness of a solution;
3. continuous dependence on data;
4. existence of a global solution,

which to some extent reflect the development of the general theory of the Cauchy problem for partial differential equations in the twenties century.

First result on the existence of a local solution was the Cauchy-Kowalevsky theorem (see, e.g., [Miz67, Hor83]). This theorem in the case of differential operators $A^q_{jk}(t,z,D_z)$ of finite order m^q_{jk}, i.e.

$$A^q_{jk}(t,z,D_z) = \sum_{|\alpha|\le m^q_{jk}} a^q_{jk\alpha}(t,z)D^\alpha_z,$$

states that if the coefficients and data of the non-characteristic Cauchy problem are analytic functions, then there exists a unique local solution in the class of analytic functions. An essential contribution to the modern theory of the Cauchy problem was made by Petrovskii [Pet96], Schwartz [Sch51], Hörmander [Hor83], Gårding, Kotake, Leray [LGK67], Mizohata [Miz67, Miz74], Ovsyannikov [Ovs65], Treves [Tre80], Gindikin, Volevich [VG91], Kitagawa [Kit90], etc. The Cauchy problem in the case of infinite order differential operators $A^q_{jk}(D_z)$ (not depending on z, i.e., with constant coefficients) were studied by Korobeynik [K73], Leont'ev [Leo76], Baouendi, Goulaouic [BG76], Dubinskii [Dub84], Napalkov [Nap82], and others.

We note that yet Cauchy and Kowalevsky had known that if the orders m^q_{jk} of operators $A^q_{jk}(z,D_z)$ satisfy the condition $m^q_{jk} \le p_j - q$, then there exists a unique local solution to the Cauchy problem in the class of analytic functions. Kowalevsky [Kow1874] in examples showed that this condition is essential for the analytic solvability, namely, if this condition is not verified then the Cauchy problem may not have a solution in the class of analytic functions. Therefore, systems satisfying this condition are called *Kowalevskian*; see [Miz74]. In the case of one equation (that is $N = 1$)

$$D^m_t u(t,z) = \sum_{k=0}^{m-1} A_k(t,z,D_z)D^k_t u(t,z) + f(t,z),$$

this condition takes the form

$$m_k \le m-k, \quad k = 0,\dots,m-1, \tag{9.11}$$

where m_k is the order of differential operator $A_k(t,z,D_z)$. Obviously, equation (9.2) satisfies condition (9.11), while equation (9.6) does not. In 1974 Mizohata [Miz74] showed that in the case of one equation, condition (9.11) is also necessary for analytic solvability; see also [Kit76].

The sufficient condition for the general system, as was shown by Leray et al. [LGK67] in 1964, is the Leray-Volevich (LV) condition

$$m^q_{jk} \le \mu_j - \mu_k + p_j - q,$$

where $\mu_1,\dots,\mu_N$ are collection of natural numbers (related to the orders of singularities near the boundary). What concerns the necessity of the LV-condition, then as was noted by Dubinskii [Dub90], it depends essentially on the problem setting. Namely, if the singularities of solutions evolve cylindrically, then the necessary condition for existence of a local analytic solution is $m^q_{jk} \le \mu_j - \mu_k$; however, if the singularities evolve along the characteristic cone, then the LV-condition becomes necessary [Dub90].

Apart from analytic theory, well posedness in classes of exponential functions were studied. Tikhonov [Tik35] was the first, who in 1935 indicated the exact exponential growth conditions for uniqueness of a solution of the heat equation. For general parabolic systems the uniqueness and well-posedness classes in terms of exponential classes of functions were studied, in particular, in works [Tac36, GS53, Hay78, K81]. In the above-mentioned references [Dub84, Dub90] Dubinskii showed that the analytic and exponential theories are in a dual relationship.

In the case of real $z \in \mathbb{R}^n$ in the system (9.10) and $p_j = 1, j = 1,\ldots,N$, and $B^{00}_{jk}(z,D_z) = B_{jk}(z,D_z)$ are differential operators of finite order

$$B_{jk}(z,D_z) = \sum_{|\alpha|\leq \nu_{jk}} b^{\alpha}_{jk}(z)D^{\alpha}_z,$$

the elliptic systems were studied by Bernstein [Ber28], Petrovskii [Pet96], Hörmander [Hor83], Douglis and Nirenberg [DN55], Morrey [Mor58], Oleynik and Radkevich [OR73], etc. An important question of analyticity of a solution was always in the focus of many authors; see, e.g., [Ber04, Ber28, MN57, Mor58, OR73, Pet96] and the references therein. Douglis and Nirenberg [DN55] studied elliptic systems under the following conditions for orders of operators $B_{jk}(z,D_z)$:

$$\nu_{jk} \leq \mu_j - \nu_k, \quad k,j = 1,\ldots,N, \tag{9.12}$$

where $\mu_1,\ldots,\mu_N$ and $\nu_1,\ldots,\nu_N$ are some collection of integers. In the modern literature conditions (9.12) are referred to as the Douglis-Nirenberg, or DN conditions. Mizohata [Miz62] and Suzuki [Suz64] found examples of elliptic equations, smooth solutions to which are not analytic. Therefore, finding necessary and sufficient conditions for analytic solvability of systems is a challenging question. See more on the history and other contributions in Section "Additional notes."

In this chapter we will present resent results on necessary and sufficient conditions for analytic and exponential solvability of general boundary value problem (9.7)–(9.8) with pseudo-differential operators $A^q_{jk}(t,z,D_z)$ and $B_{jk}(z,D_z)$ with analytic symbols (Section 9.7). For this purpose we construct an algebra of pseudo-differential operators with meromorphic symbols defined on a complex domain (manyfold) (Sections 9.5 and 9.6).

The main tool of the construction of PsDOs in the real case is the Fourier transform (Chapters 2–7). However, in complex analysis there is no primary analog of the Fourier transform. The existing Borel and Fourier-Laplace transforms, introduced in Section 2.7, do not give desired results. In 1984 Dubinskii [Dub84] introduced a complex Fourier transform of f as an analytic functional, defined as an image of the PsDO with the symbol f on the Dirac delta function (see Section 2.7). This transform inherits many properties of the real Fourier transform and can be easily adapted for the construction of PsDO with analytic symbols defined on a complex manyfold $\Omega \in \mathbb{C}^n$. We note that symbols may have singularities on the boundary of Ω. This construction can be extended to the class of meromorphic symbols, as well [Uma14]. However, in this case the corresponding ΨDOSS become multi-valued (Section 9.5).

We have seen in Section 2.7 that the complex Fourier transform is an extension of Borel's transform to the space of analytic and exponential functionals and the inverse to the Fourier-Laplace transform. The complex Fourier transform in this spirit is adapted to spaces of analytic and exponential functions and functionals, introduced in this chapter and studied in Section 9.4. In contrast to spaces used in [Dub84] (see Section 2.7) we introduce new spaces of analytic and exponential functions and functionals. Under the conditions

$$m_{jk}^q \le \mu_j - \mu_k + p_j - q, \quad \nu_{jk}^{qm} \le \mu_j - \mu_k + q - m,$$
$$q = 0, \dots, p_j - 1, \quad m = 0, \dots, p_k - 1, \quad k, j = 1, \dots, N,$$

to orders of operators A_{ij}^q and B_{jk}^{mq} we show the existence of a unique local solution of (9.7)–(9.8) in the introduced spaces.

9.2 Some Banach spaces of exponential and analytic functions and functionals

Let $\xi \in \mathbb{R}^n$ and α be a multi-index, that is $\alpha = (\alpha_1, \dots, \alpha_n)$, and α_j are non-negative integers. We use notations $|\alpha| = \alpha_1 + \dots + \alpha_n$, $|\xi| = \xi_1 + \dots + \xi_n$, and $\xi^\alpha = \xi_1^{\alpha_1} \dots \xi_n^{\alpha_n}$. Introduce the function

$$G(\xi) = G_{\mu,r}^\alpha(\xi) = \frac{(1+|\xi|)^\mu e^{r|\xi|}}{\xi^\alpha}, \quad \xi \in \mathbb{R}_+^n, \tag{9.13}$$

where $r > 0$ is real and μ is integer fixed numbers and $\mathbb{R}_+^n = \{\xi \in \mathbb{R}^n : \xi_1 > 0, \dots, \xi_n > 0\}$. Obviously, this function is continuous, differentiable, and strictly positive on $\mathbb{R}_+^n$. If one of the components of α is zero, say $\alpha_{j_0} = 0$, then the domain of $G(\xi)$ extends to the hyperplane $\xi_{j_0} = 0$. It follows from the definition of $G(\xi)$ that if $\alpha_j \neq 0, j = 1, \dots, n$, then $G(\xi) \to \infty$ as $\xi \to \partial\mathbb{R}_+^n \cup \{\infty\}$. If $\alpha = 0 = (0, \dots, 0)$, then

$$\inf_{\mathbb{R}_+^n} G(\xi) = \begin{cases} G(0) = 1 & \text{if } \mu \ge -r, \\ e^{-\mu - r}\left(\frac{r}{-\mu}\right)^{-\mu} & \text{if } \mu < -r. \end{cases}$$

Also it is not hard to see that if $\mu > -r$ and $\alpha_{j_0} = 0$ for some $j_0 \in \{1, \dots, n\}$, then the infimum is attained on the hyperplane $\xi_{j_0} = 0$. If all the components of α are not zero, then the following statement on the infimum of $G(\xi)$ holds.

Proposition 9.1. *For each multi-index α, $\alpha_j \neq 0$, $j = 1, \dots, n$, there is a unique infimum of $G(\xi)$ attained at $\xi^* \in \mathbb{R}_+^n$, for which the asymptotic behavior $|\xi^*| \sim O(|\alpha|)$ for large $|\alpha|$ holds. If $\mu = 0$, then $\xi_j^* = \alpha_j / r$, $j = 1, \dots, n$.*

Proof. To prove this proposition we consider the system of equations

$$\frac{\partial G(\xi)}{\partial \xi_j} = 0, \quad j = 1, \dots, n,$$

which reduces to

$$r|\xi|\xi_j + (\mu + r)\xi_j - \alpha_j|\xi| = \alpha_j, \quad j = 1, \dots, n. \tag{9.14}$$

Summing the latter over the indices $j = 1, \dots, n$, we obtain the quadratic equation for $|\xi|$:

$$r|\xi|^2 - (|\alpha| - \mu - r)|\xi| - |\alpha| = 0. \tag{9.15}$$

This equation has one positive and one negative roots if $|\alpha| > 0$. The point ξ^* corresponding to the negative root of (9.15) is out of the domain of $G(\xi)$, and hence, the only stationary point ξ^* delivering the infimum (minimum) corresponds to the positive root of equation (9.15), i.e., $|\xi^*| > 0$. The fact that $\xi^* \in \mathbb{R}^n_+$ follows from equations (9.14):

$$\xi_j^* = \eta \frac{\alpha_j}{r} > 0, \quad j = 1, \dots, n, \tag{9.16}$$

where

$$\eta = \frac{|\xi^*| + 1}{|\xi^*| + 1 + \mu/r}. \tag{9.17}$$

The latter is obviously positive if $\mu \geq 0$. If $\mu < 0$, then using the root representation of equation (9.15), one can see that $r(|\xi^*| + 1) \leq |\alpha| - \mu + r$, which implies $\eta \geq (|\xi^*| + 1)/(|\alpha| + r) > 0$. Further, it follows from (9.16) that if $\mu = 0$, then $\xi_j^* = \alpha_j/r$, and $|\xi^*| = |\alpha|/r$, and if $\mu \neq 0$, then $|\xi^*| \sim O(|\alpha|)$, $|\alpha| \to \infty$. Additionally, equation (9.16) also implies that if $\mu > -r$ and $\alpha_{j_0} = 0$ for some $j = j_0$, then $\xi_{j_0}^* = 0$. Thus, the infimum of $G(\xi)$ in this case is attained on the hyperplane $\xi_{j_0} = 0$.

Define a Banach space $\mathscr{E}_{\mu,r}$ as the set of entire functions $\varphi(z)$ satisfying the inequality

$$|\varphi(z)| \leq C(1 + |z|)^{\mu} e^{r|z|}, \quad z \in \mathbb{C}^n, \tag{9.18}$$

where $C \geq 0$ is a constant. The smallest constant $C = C_\varphi$ in (9.18), that is

$$\|\varphi\|_{\mu,r} = \sup_{z \in \mathbb{C}^n} (1 + |z|)^{-\mu} e^{-r|z|} |\varphi(z)|$$

is a norm in $\mathscr{E}_{\mu,r}$. It follows immediately from (9.18) that if $\nu > \mu$ and/or $s > r$, then the embedding

$$\mathscr{E}_{\mu,r} \subset \mathscr{E}_{\nu,s} \tag{9.19}$$

is continuous.

Let $\mathscr{K}(\alpha) = \mathscr{K}_{\mu,r}(\alpha) = G^{\alpha}_{\mu,r}(\xi_*) = \inf_{\xi \in \mathbb{R}^n_+} G^{\alpha}_{\mu,r}(\xi)$. Due to Proposition 9.1 $\mathscr{K}(\alpha)$ is well defined for all multi-indices α.

Proposition 9.2. *Let $\varphi \in \mathscr{E}_{\mu,r}$. Then*

$$\frac{|D^\alpha \varphi(z)|}{\alpha!} \le \|\varphi\|_{\mu,r} \mathscr{K}_{\mu,r}(\alpha)(1+|z|)^\mu e^{r|z|}, \quad |\alpha| = 0,1,\dots. \tag{9.20}$$

Proof. In accordance with the Cauchy theorem on integral representation, for arbitrary $\xi_j > 0$, $j = 1,\dots,n$, one has

$$D^\alpha \varphi(z) = \frac{\alpha!}{(2\pi i)^n} \int\limits_{|\zeta_1 - z_1| = \xi_1} \cdots \int\limits_{|\zeta_n - z_n| = \xi_n} \frac{\varphi(\zeta)d\zeta}{(\zeta - z)^{\alpha+(1)}}, \tag{9.21}$$

where $\alpha + (1) = (\alpha_1 + 1, \dots, \alpha_n + 1)$. The substitution $\zeta = z + \xi e^{i\theta}$ (i.e., $\zeta_j = z_j + \xi_j e^{i\theta_j}$, $j = 1,\dots,n$), where θ runs over the n-dimensional torus $T^n = \{\theta \in \mathbb{R}^n : 0 \le \theta_j < 2\pi,\ j = 1,\dots,n\}$, reduces (9.21) to

$$\frac{D^\alpha \varphi(z)}{\alpha!} = \frac{1}{(2\pi)^n \xi^\alpha} \int\limits_{T^n} \varphi(z + \xi e^{i\theta}) e^{i(\theta_1 + \cdots + \theta_n)} d\theta, \tag{9.22}$$

Further, multiplying and dividing the integrand in (9.22) by $(1 + |z + \xi e^{i\theta}|)^\mu e^{r|z+\xi e^{i\theta}|}$, and taking into account $\varphi \in \mathscr{E}_{\mu,r}$, one has the following estimate:

$$\frac{|D^\alpha \varphi(z)|}{\alpha!} \le \|\varphi\|_{\mu,r} G(\xi)(1+|z|)^\mu e^{r|z|},$$

where $G(\xi)$ is defined in equation (9.13). Minimizing $G(\xi)$ over all $\xi \in R^n_+$, one obtains (9.20).

Corollary 9.1. *1. Let $\varphi \in \mathscr{E}_{\mu,r}$. Then*

$$\frac{|D^\alpha \varphi(0)|}{\alpha!} \le \mathscr{K}_{\mu,r}(\alpha) \|\varphi\|_{\mu,r}, \quad |\alpha| = 0,1,\dots. \tag{9.23}$$

2. Let $\varphi \in \mathscr{E}_{\mu,r}$. Then

$$\|D^\alpha \varphi\|_{\mu,r} \le \alpha! \mathscr{K}_{\mu,r}(\alpha) \|\varphi\|_{\mu,r}, \quad |\alpha| = 0,1,\dots. \tag{9.24}$$

The converse to the first statement in Corollary 9.1 is also true in the following sense.

Proposition 9.3. *Let an entire function $\varphi(z)$ satisfy the inequalities*

$$\frac{|D^\alpha \varphi(0)|}{\alpha!} \le C \mathscr{K}_{\mu,r}(\alpha), \quad |\alpha| \ge N, \tag{9.25}$$

where $C > 0$ is a constant not depending on α, and N is a nonnegative integer. Then $\varphi \in \mathscr{E}_{\mu,r}$.

Proof. Consider the function

$$\phi(z) = \sum_{\alpha} \frac{|\alpha|^{\mu}(er)^{|\alpha|}}{\alpha^{\alpha}} z^{\alpha}. \tag{9.26}$$

This function belongs to $\mathscr{E}_{\mu,r}$. Indeed, since

$$\frac{1}{e} \limsup_{|\alpha|\to\infty} \left(|\alpha| \sqrt[|\alpha|]{\frac{(er)^{|\alpha|}}{\alpha^{\alpha}}} \right) = r,$$

it follows from the theory of entire functions (see, e.g., [GR09, Hor90]) that the function $\phi(z)$ is an exponential function of type r. The function $\phi(z)$ majorizes $\varphi(z)$. To verify this we recall that $\mathscr{K}_{\mu,r}(\alpha) = G(\xi^*)$, where $\xi^* = \frac{\alpha}{r}\eta$, $\eta = (|\xi^*| + 1)(|\xi^*| + 1 + \mu/r)^{-1}$. If $\mu \geq 0$, then $0 < \eta \leq 1$. If $\mu < 0$, then due to the asymptotics $|\xi^*| = O(|\alpha|)$, for large $|\alpha|$ one obtains from (9.17) that $\eta < 1 + \varepsilon$, where ε is arbitrarily small. Moreover, for large $|\alpha|$ it is easy to see that $\lim_{|\alpha|\to\infty} \eta^{|\alpha|} = e^{-\mu}$. Therefore, there exists an integer N, such that for all $|\alpha| \geq N$ the inequality $\eta^{|\alpha|} \geq e^{-\mu} - \varepsilon$ holds, where $0 < \varepsilon < e^{-\mu}$. Making use of these facts and taking ε small enough, one has

$$\mathscr{K}_{\mu,r}(\alpha) = \frac{(1 + \frac{|\alpha|}{r}\eta)^{\mu} e^{|\alpha|\eta} r^{|\alpha|}}{\alpha^{\alpha}\eta^{|\alpha|}} \leq C \frac{|\alpha|^{\mu}(er)^{|\alpha|}}{\alpha^{\alpha}}, \quad C > 0. \tag{9.27}$$

The latter together with (9.25) implies that series (9.26) for $\phi(z)$ is a majorant for $\varphi(z)$, and hence, $\varphi \in \mathscr{E}_{\mu,r}$ as well.

We denote the space conjugate to the Banach space $\mathscr{E}_{\mu,r}$ by $\mathscr{E}^*_{\mu,r}$. The space $\mathscr{E}^*_{\mu,r}$ is a Banach space with the norm

$$\|h\|^*_{\mu,r} = \sup_{\varphi \neq 0} \frac{|\langle h(z), \varphi(z)\rangle|}{\|\varphi\|_{\mu,r}}, \tag{9.28}$$

where $h \in \mathscr{E}^*_{\mu,r}$, $\varphi \in \mathscr{E}_{\mu,r}$, and the symbol $\langle\cdot,\cdot\rangle$ stands for the duality pair of the spaces $\mathscr{E}^*_{\mu,r}$ and $\mathscr{E}_{\mu,r}$.

Proposition 9.4. *Let $h \in \mathscr{E}^*_{\mu,r}$. Then*

$$\|h\|^*_{\mu,r} = \sum_{|\alpha|=0}^{\infty} \mathscr{K}_{\mu,r}(\alpha) |\langle h, z^{\alpha}\rangle|. \tag{9.29}$$

Proof. Suppose $\varphi_0(z) \neq 0$ is a function in $\mathscr{E}_{\mu,r}$ that delivers sup in equation (9.28), that is

$$\|h\|^*_{\mu,r} = \frac{|\langle h(z), \varphi_0(z)\rangle|}{\|\varphi_0\|_{\mu,r}}.$$

Expanding $\varphi_0(z)$ to Taylor series and using (9.23), we have

$$\|h\|^*_{\mu,r} \leq \sum_{|\alpha|=0}^{\infty} \mathcal{K}_{\mu,r}(\alpha)|\langle h, z^\alpha\rangle|.$$

On the other hand, for an arbitrary $\varphi \in \mathcal{E}_{\mu,r}$,

$$\|h\|^*_{\mu,r} \geq \frac{|\langle h(z), \varphi(z)\rangle|}{\|\varphi\|_{\mu,r}}.$$

We pick the function

$$0 \neq \varphi^*(z) = C_0 \sum_{|\alpha|=0}^{\infty} \mathcal{K}_{\mu r}(\alpha) \frac{\overline{\langle h, z^\alpha\rangle}}{|\langle h, z^\alpha\rangle|} z^\alpha, \tag{9.30}$$

where C_0 is a positive real number. Due to Proposition 9.3, $\varphi^* \in \mathcal{E}_{\mu,r}$. Therefore, $\|\varphi^*\|_{\mu,r} < \infty$. We set $C_0 = \|\varphi^*\|_{\mu,r}$ in (9.30). One can easily see that by definition of φ^*, the expression $\langle h, \varphi^*\rangle$ is a real positive number. Hence, we have

$$\frac{|\langle h, \varphi^*\rangle|}{\|\varphi^*\|_{\mu,r}} \geq \frac{\langle h, \varphi^*\rangle}{\|\varphi^*\|_{\mu,r}} = \sum_{|\alpha|=0}^{\infty} \mathcal{K}_{\mu,r}(\alpha)|\langle h, z^\alpha\rangle|,$$

completing the proof.

Proposition 9.5. *Let $\mu > 0$. Linear combinations of quasi-polynomials $z^\alpha e^{\zeta z}$, where $|\alpha| \leq \mu$ and $|\zeta| \leq r$, form a dense set in $\mathcal{E}_{\mu,r}$.*

Proof. Let h be an arbitrary element in $\mathcal{E}^*_{\mu,r}$. Assume that $\langle h, z^\alpha e^{\zeta z}\rangle = 0$ for all α, $|\alpha \leq \mu|$ and ζ, $|\zeta| = |\zeta_1| + \cdots + |\zeta_n| \leq r$. To prove the statement we have to show that $h = 0$. Let, first, $\mu = 0$, i.e., $\langle h, e^{\zeta z}\rangle = 0$ for all ζ, $|\zeta| \leq r$. Then, in accordance with Proposition 9.4, we have

$$\|h\|^*_{0,r} = \sum_{|\alpha|=0}^{\infty} \mathcal{K}_{0,r}(\alpha)|\langle h, z^\alpha\rangle| = \sum_{|\alpha|=0}^{\infty} \mathcal{K}_{0,r}(\alpha)\left|D^\alpha_\zeta \langle h, e^{\zeta z}\rangle_{|\zeta=0}\right| = 0,$$

which implies $h = 0$. If $\mu > 0$, then obviously, $g_\alpha = \bar{z}^\alpha h \in \mathcal{E}^*_{\mu-|\alpha|,r}$. In particular, when $|\alpha| = \mu$, the functional $g_\alpha \in \mathcal{E}^*_{0,r}$. Therefore,

$$0 = \langle h, z^\alpha e^{\zeta z}\rangle = \langle \bar{z}^\alpha h, e^{\zeta z}\rangle = \langle g_\alpha, e^{\zeta z}\rangle, \forall \zeta : |\zeta| \leq r,$$

implies that $g_\alpha = 0$. Hence, $h = \sum_{|\alpha|\leq\mu} a_\alpha D^\alpha \delta(z)$, where a_α are complex constants and δ is the Dirac delta function. Since, in particular h vanishes at monomials z^β, $|\beta| \leq \mu$, in fact, $a_\alpha = 0$, $|\alpha| \leq \mu$. Thus, $h = 0$.

Remark 9.1. Proposition 9.5 is not valid if either the condition $|\alpha| \leq \mu$ or $|\zeta| \leq r$ is replaced by $|\alpha| < \mu$ or $|\zeta| < r$, respectively. Indeed, assuming $n = 1$ and $\mu = 1$, one can prove this claim showing that the function $\varphi(z) = (1+z)e^{rz}$ cannot be approximated in $\mathscr{E}_{\mu,r}$ by linear combinations of quasi-polynomials $z^k e^{\zeta z}, k = 0,1, |\zeta| < r$, or exponentials $e^{\zeta z}$, $|\zeta| \leq r$. We note also that linear combinations $\{e^{\zeta z}\}$, $|\zeta| \leq r$, form a dense set in $\mathscr{E}_{\mu,r}$ if $\mu \leq 0$ due to embedding (9.19).

Further, introduce a Banach space $\mathscr{O}_{\mu,r}$ of functions analytic on the polydisc $U_r = \{\zeta \in \mathbb{C}^n : |\zeta_j| < r, j = 1,\dots,n\}$, with the norm

$$[\phi]_{\mu,r} = \sum_{|\alpha|=0}^{\infty} \mathscr{K}_{\mu,r}(\alpha)|\phi_\alpha|,$$

where $\phi_\alpha = D^\alpha \phi(0)$, $|\alpha| = 0,1,\dots$. With each function $\phi \in \mathscr{O}_{\mu,r}$ it is associated a differential operator of infinite order defined as

$$\Phi(D)\varphi(z) = \sum_{|\alpha|=0}^{\infty} \frac{\phi_\alpha}{\alpha!} D_z^\alpha \varphi(z). \tag{9.31}$$

Proposition 9.6. *The operator $\Phi(D)$ associated with the function $\phi \in \mathscr{O}_{\mu,r}$ maps continuously the space $\mathscr{E}_{\mu,r}$ into itself. Moreover, for any $\varphi \in \mathscr{E}_{\mu,r}$ the inequality*

$$\|\Phi(D)\varphi(z)\|_{\mu,r} \leq [\phi]_{\mu,r}\|\varphi\|_{\mu,r} \tag{9.32}$$

holds.

Proof. Let $\varphi \in \mathscr{E}_{\mu,r}$. Then using Proposition 9.2, we obtain

$$|\Phi(D)\varphi(z)| \leq \sum_{|\alpha|=0}^{\infty} |\phi_\alpha| \frac{|D_z^\alpha \varphi(z)|}{\alpha!} \leq \|\varphi\|_{\mu,r}(1+|z|)^\mu e^{r|z|} \sum_{|\alpha|=0}^{\infty} \mathscr{K}_{\mu,r}(\alpha)|\phi_\alpha|.$$

This immediately implies inequality (9.32).

By duality, one can define a differential operator of infinite order $\Phi(-D)$ associated with the function $\phi \in \mathscr{O}_{\mu,r}$ in the space $\mathscr{E}^*_{\mu,r}$, as well. Namely, for $h \in \mathscr{E}^*_{\mu,r}$, by definition,

$$\langle \Phi(-D)h, \varphi \rangle = \langle h, \Phi(D)\varphi \rangle, \quad \forall \varphi \in \varphi \in \mathscr{E}_{\mu,r}.$$

Proposition 9.7. *The operator $\Phi(D)$ associated with the function $\phi \in \mathscr{O}_{\mu,r}$ maps continuously the space $\mathscr{E}^*_{\mu,r}$ into itself. Moreover, for $h \in \mathscr{E}^*_{\mu,r}$ the inequality*

$$\|\Phi(-D)h(z)\|^*_{\mu,r} \leq [\phi]_{\mu,r}\|h\|^*_{\mu,r} \tag{9.33}$$

holds.

Proof. Let $\varphi \in \mathscr{E}_{\mu,r}$ be an arbitrary function in $\mathscr{E}_{\mu,r}$. Using (9.32), we obtain

$$\begin{aligned}|\langle \Phi(-D)h(z), \varphi(z)\rangle| &= |\langle h(z), \Phi(D)\varphi(z)\rangle| \leq \|h\|^*_{\mu,r}\|\Phi(D)\varphi\| \\ &\leq [\phi]_{\mu,r}\|h\|^*_{\mu,r}\|\varphi\|_{\mu,r}.\end{aligned}$$

This immediately implies inequality (9.33).

9.3 Complex Fourier transform

Now we define a complex Fourier transform F for functions of the space $\mathscr{O}_{\mu,r}$.

Definition 9.1. The Fourier transform of a function $\phi \in \mathscr{O}_{\mu,r}$ is

$$F[\phi](\zeta) = (2\pi)^n \Phi(-D_\zeta)\delta(\zeta). \tag{9.34}$$

That is the Fourier transform of ϕ is the value of the differential operator of infinite order $(2\pi)^n\Phi(-D)$, associated with $(2\pi)^n\phi$, at the Dirac delta function. It follows from Definition 9.1 and Proposition 9.7 that the mapping

$$F : \mathscr{O}_{\mu,r} \to \mathscr{E}^*_{\mu,r} \tag{9.35}$$

is continuous. Let $\varphi \in \mathscr{E}_{\mu,r}$. Using the definition (9.31) of $\Phi(D)$, for arbitrary $\phi \in \mathscr{O}_{\mu,r}$ we have

$$\begin{aligned}\langle F[\phi](\zeta), \varphi(\zeta)\rangle &= (2\pi)^n\langle \delta(\zeta), \Phi(D)\varphi(\zeta)\rangle = (2\pi)^n \sum_{|\alpha|=0}^{\infty} \frac{\phi_\alpha}{\alpha!}\langle \delta(\zeta), D^\alpha_\zeta \varphi(\zeta)\rangle \\ &= (2\pi)^n \sum_{|\alpha|=0}^{\infty} \frac{1}{\alpha!} D^\alpha_z \phi(0) D^\alpha_\zeta \varphi(0).\end{aligned} \tag{9.36}$$

One can derive useful implications from this representation. Namely, due to estimate (9.23) (see Corollary 9.1) it follows from (9.36) that

$$|\langle F[\phi](\zeta), \varphi(\zeta)\rangle| \leq (2\pi)^n \|\varphi\|_{\mu,r} \sum_{|\alpha|=0}^{\infty} \mathscr{K}_{\mu,r}(\alpha)|\phi_\alpha| = (2\pi)^n[\phi]_{\mu,r}\|\varphi\|_{\mu,r},$$

or

$$\|F[\phi]\|^*_{\mu,r} \leq (2\pi)^n[\phi]_{\mu,r}.$$

Another implication from representation (9.36) is a formula for the inverse Fourier transform F^{-1}. Namely, taking $\varphi(\zeta) = e^{z\zeta}$, where $z\zeta = z_1\zeta_1 + \cdots + z_n\zeta_n$, we have

$$\langle F[\phi](\zeta), e^{z\zeta}\rangle = (2\pi)^n \sum_{|\alpha|=0}^{\infty} \frac{D^\alpha_z \phi(0)}{\alpha!} z^\alpha = (2\pi)^n \phi(z).$$

Rewriting the latter in the form

$$\phi(z) = F^{-1}[F[\phi]](z) = \frac{1}{(2\pi)^n}\langle F[\phi](\zeta), e^{z\zeta}\rangle, \tag{9.37}$$

we can see that *the inverse Fourier transform* coincides with the known Fourier-Laplace transform. This formula implies the following two important formulas:

$$F[D_z^\alpha \phi](\zeta) = \zeta^\alpha F[\phi](\zeta), \tag{9.38}$$

$$F[(-z)^\alpha \phi(z)](\zeta) = D_\zeta^\alpha F[\phi](\zeta). \tag{9.39}$$

Further, differentiating (9.37), we have $D_z^\alpha \phi(0) = \phi_\alpha = (2\pi)^{-n}\langle F[\phi](\zeta), \zeta^\alpha\rangle$. Using this fact and Proposition 9.4, we obtain

$$[\phi]_{\mu,r} = \sum_{|\alpha|=0}^{\infty} |\phi_\alpha| \mathscr{K}_{\mu,r}(\alpha) = \frac{1}{(2\pi)^n}\sum_{|\alpha|=0}^{\infty} \mathscr{K}_{\mu,r}(\alpha)|\langle F[\phi](\zeta), \zeta^\alpha\rangle| = \frac{1}{(2\pi)^n}\|F[\phi]\|_{\mu,r}^*.$$

This equality expresses a complex analog of the Parseval's equality (Theorem 1.3) of the Fourier transform acting in $L_2(\mathbb{R}^n)$. Summarizing, we have proved the following statement.

Theorem 9.1. *The Fourier transform operator* $F : \mathscr{O}_{\mu,r} \to \mathscr{E}_{\mu,r}^*$ *is isometric isomorphism. The inversion formula is given in equation* (9.37).

The representation for the Fourier transform obtained in equation (9.36) is symmetric with respect to $\phi \in \mathscr{O}_{\mu,r}$ and $\varphi \in \mathscr{E}_{\mu,r}$. Therefore, with an appropriate interpretation of the definition of the Fourier transform, similar to Theorem 9.1, one can prove

Theorem 9.2. *The Fourier transform operator* $F : \mathscr{E}_{\mu,r} \to \mathscr{O}_{\mu,r}^*$ *is isometric isomorphism. The inversion formula is again given in equation* (9.37).

Theorems 9.1 and 9.2 imply the following corollary.

Corollary 9.2. *The following commutative diagram holds:*

$$\begin{array}{ccc} \mathscr{E}_{\mu,r} & \stackrel{*}{\longleftrightarrow} & \mathscr{E}_{\mu,r}^* \\ F^{-1}\uparrow\downarrow F & & F^{-1}\downarrow\uparrow F \\ \mathscr{O}_{\mu,r}^* & \stackrel{*}{\longleftrightarrow} & \mathscr{O}_{\mu,r} \end{array}$$

where symbols "$\stackrel{*}{\leftrightarrow}$," "$\stackrel{F^{-1}}{\rightarrow}$," and "$\stackrel{F}{\rightarrow}$" stand for the passage to conjugate, the inverse Fourier transform, and the Fourier transform, respectively.

Now, when the Fourier transform F is defined on $\mathscr{E}_{\mu,r}$ as well, we note that representation (9.36) can also be interpreted as the Parseval equality

$$\langle F[\phi](\zeta), \varphi(\zeta)\rangle = \langle \phi(\zeta), F[\varphi](\zeta)\rangle, \quad \phi \in \mathscr{O}_{\mu,r}, \varphi \in \mathscr{E}_{\mu,r}.$$

Proposition 9.8. *Let $\mu > 0$. A function $f(z)$ analytic on the polydisc U_r belongs to $\mathscr{O}_{-\mu,r}$ if and only if it satisfies the inequality*

$$|f(z)| \leq \frac{M}{(r-|z|)^{\mu-1}}, \quad z \in U_r, \tag{9.40}$$

where $M > 0$ is a constant.

Proof. For the sake of simplicity we show this fact for $n = 1$. Let $f \in \mathscr{O}(U_r)$ satisfy the estimate

$$|f(z)| \leq \frac{M}{(r-|z|)^{\nu}}, \quad z \in U_r,$$

with a positive integer ν. Without loss of generality one can assume that $f(z) = \frac{\phi(z)}{(r-z)^{\mu-1}} + \psi(z)$, where $\phi(z)$ is regular at $z = r$ and a singularity of $\psi(z)$ at $z = r$ is weaker than the first term of the above representation of f. Then one can easily verify that

$$|f_\alpha| = |D_z^\alpha f(0)| \sim \frac{(\alpha+\nu-1)!}{(\nu-1)!}\frac{M}{r^{\alpha+\nu}}, \quad \alpha \to \infty. \tag{9.41}$$

Using inequality (9.27) and the Stirling formula, we have

$$[f]_{-\mu,r} = \sum_{\alpha=0}^{\infty} \mathscr{K}_{-\mu,r}(\alpha)|f_\alpha| \sim C\sum_{\alpha=0}^{\infty} \mathscr{K}_{-\mu,r}(\alpha)\frac{(\alpha+\nu-1)!}{(\nu-1)!r^{\alpha+1}}$$
$$\leq C\sum_{\alpha=0}^{\infty} \frac{\alpha!\alpha^{\nu-1}(er)^\alpha}{\alpha^\mu \alpha^\alpha r^\alpha} \leq C\sum_{\alpha=0}^{\infty} \frac{1}{\alpha^{\mu-\nu+\frac{1}{2}}} < \infty,$$

if $\nu \leq \mu - 1$. Hence, $f \in \mathscr{O}_{-\mu,r}$, if $f \in \mathscr{O}(U_r)$ and satisfies condition (9.40).

Further, if $|f(z)| > \frac{M}{(r-|z|)^\mu}$ near the boundary of the disc $|z| < r$, then using the asymptotic relations

$$\mathscr{K}_{-\mu,r}(\alpha) \sim \frac{(er)^\alpha}{\alpha^\mu \alpha^\alpha}, \quad \alpha! \sim \left(\frac{\alpha}{e}\right)^\alpha \sqrt{2\pi\alpha},$$

when $\alpha \to \infty$, and (9.41), one obtains $[f]_{-\mu,r} = \infty$, that is $f \notin \mathscr{O}_{-\mu,r}$.

Remark 9.2. Dubinskii denoted the class of functions $f \in \mathscr{O}(U_r)$ satisfying estimate (9.40) by $\mathscr{D}_{\mu-1,r}$; see [Dub90]. Proposition 9.8 immediately implies that the space $\mathscr{D}_{\mu-1,r}$, $\mu > 0$, is isomorphic to the space $\mathscr{O}_{-\mu,r}$. Therefore, it follows from Corollary 9.2 that

$$\begin{array}{ccc}
\mathscr{D}_{\mu-1,r} & \overset{*}{\longleftrightarrow} & \mathscr{D}^*_{\mu-1,r} \\
F^{-1}\uparrow\downarrow F & & F^{-1}\downarrow\uparrow F \\
\mathscr{E}^*_{-\mu,r} & \overset{*}{\longleftrightarrow} & \mathscr{E}_{-\mu,r}\,,
\end{array}$$

where $\mathscr{D}^*_{\mu-1,r}$ is the dual space to $\mathscr{D}_{\mu-1,r}$.

9.4 Complex Fourier transform in fiber spaces of analytic and exponential functions and functionals

In this section we introduce fiber spaces of exponential and holomorphic functions and locally convex topological vector spaces and extend the Fourier transform introduced in the previous section to these spaces. Note that these spaces will serve as solution spaces for differential and pseudo-differential equations with complex variables.

Suppose Ω is a connected domain (or connected manifold) in $\mathbb{C}^n$ and let $\zeta \in \Omega$. Let μ and r be a nonnegative and positive real numbers, respectively, such that $r < dist(\zeta, \mathbb{C}^n \setminus \Omega)$. Denote by $\mathscr{E}^{\zeta}_{\mu,r}$ the set of entire functions $\varphi(z) = e^{z\zeta}v(z)$, $v \in \mathscr{E}_{\mu,r}$, where the Banach space $\mathscr{E}_{\mu,r}$ was defined in the previous section. The space $\mathscr{E}^{\zeta}_{\mu,r}$ is also a Banach space with the norm $\|\varphi\|_{\mu,r,\zeta} = \|e^{-z\zeta}\varphi\|_{\mu,r}$.

Further, we introduce a fiber bundle $(E, \Omega, \pi) \equiv E^{\Omega}_{\mu,r}(\mathbb{C}^n)$ with the base Ω and projection

$$\pi : (E, \Omega, \pi) \to \Omega,$$

where $\pi^{-1}(\zeta) = \mathscr{E}^{\zeta}_{\mu,r}$, $\zeta \in \Omega$. It follows from this definition that the fibers $E_{\mu,r,\zeta} = exp(z\zeta)\mathscr{E}^{\zeta}_{\mu,r}$ are Banach spaces with the respective norms $\|\varphi\|_{\mu,r,\zeta} = \|exp(-z\zeta)\varphi\|_{\mu,r}$. It is obvious that $\pi^{-1}(\zeta_1)$ and $\pi^{-1}(\zeta_2)$ are isomorphisms for arbitrary $\zeta_1, \zeta_2 \in \Omega$. The dual space to $E^{\Omega}_{\mu,r}(\mathbb{C}^n)$ is also a fiber bundle $(E^*, -\Omega, \pi_*) \equiv \left(E^{\Omega}_{\mu,r}(\mathbb{C}^n)\right)^*$ with fibers $E^*_{\mu,r,\zeta} \equiv \left(E_{\mu,r,\zeta}\right)^*$, with the base $-\Omega$, and with the projection

$$\pi_*^{-1} : (E^*, -\Omega, \pi_*) \to -\Omega.$$

Schematically the relationship between introduced fiber bundles can be represented as

$$\begin{array}{ccc} E^{\Omega}_{\mu,r}(\mathbb{C}^n) & \overset{*}{\longleftrightarrow} & \left(E^{\Omega}_{\mu,r}(\mathbb{C}^n)\right)^* \\ \pi^{-1} \nwarrow & & \nearrow \pi_*^{-1} \\ & \Omega & \end{array}$$

Since fibers in the above constructions are endowed with norms, one can introduce the structure of convergence. Namely, we say that a sequence $\varphi_m \in E^{\Omega}_{\mu,r}(\mathbb{C}^n)$ converges to $\varphi_0 \in E^{\Omega}_{\mu,r}(\mathbb{C}^n)$, if for arbitrary $\zeta \in \Omega$ we have $\varphi_m \to \varphi_0$ as $m \to \infty$ in $\pi^{-1}(\zeta) = E_{\mu,r,\zeta}$. In the dual space we introduce the weak convergence: a sequence $\Phi_n \in \left(E_{\mu,r,\zeta}\right)^*$ converges weakly to $\Phi_0 \in \left(E_{\mu,r,\zeta}\right)^*$, if for arbitrary $\zeta \in \Omega$ we have $\Phi_m \to \Phi_0$ as $m \to \infty$ in $\pi_*^{-1}(\zeta) = E^*_{\mu,r,\zeta}$.

Similarly, let $(\mathscr{O}, \Omega, \tau) \equiv \mathscr{O}_{\mu,r}(\Omega)$ be a fiber bundle with fibers $\mathscr{O}_{\mu,r,\zeta}$ with the base Ω and the projection $\tau : (\mathscr{O}, \Omega, \tau) \to \Omega$, where $\tau^{-1}(\zeta) = \mathscr{O}_{\mu,r,\zeta}$, $\zeta \in \Omega$. The space $\mathscr{O}_{\mu,r,\zeta}$, $\zeta \in \Omega$, is a Banach space of analytic functions $\phi(z)$ defined on the polydisc with the center at ζ and "poly-radius" $(r,\dots,r)$, i.e., $U_r(\zeta)=\{z\in\mathbb{C}^n : |z_j-\zeta_j|<r,\ j=1,\dots,n\}$, with the norm

$$[\phi]_{\mu,r,\zeta}=\sum_{|\alpha|=0}^{\infty}|\phi_\alpha(\zeta)|\mathscr{K}_{\mu,r}(\alpha),\quad \phi_\alpha(\zeta)=D^\alpha\phi(\zeta).$$

The dual space to $\mathscr{O}_{\mu,r}(\Omega)$ is also a fiber bundle $(\mathscr{O}^*,-\Omega,\pi_*)\equiv\mathscr{O}^*_{\mu,r}(\Omega)$ with fibers $\mathscr{O}^*_{\mu,r,\zeta}\equiv\left(\mathscr{O}_{\mu,r,\zeta}\right)^*$, with the base $-\Omega$, and with the projection

$$\tau_*^{-1}:(\mathscr{O}^*,-\Omega,\pi_*)\to-\Omega.$$

Schematically the relationship between these fiber bundles can be represented as

$$\begin{array}{ccc}\mathscr{O}_{\mu,r}(\Omega) & \overset{*}{\longleftrightarrow} & \mathscr{O}^*_{\mu,r}(\Omega)\\ \tau^{-1}\nwarrow & & \nearrow\tau_*^{-1}\\ & \Omega & \end{array}$$

Since fibers in these constructions are endowed with norms, one can introduce the structure of convergence. Namely, a sequence $h_m\in\mathscr{O}_{\mu,r}(\Omega)$ converges to $h_0\in\mathscr{O}_{\mu,r}(\Omega)$, if for arbitrary $\zeta\in\Omega$ we have $h_m\to h_0$ as $m\to\infty$ in $\tau^{-1}(\zeta)=\mathscr{O}_{\mu,r,\zeta}$. In the dual space we introduce the weak convergence: a sequence $H_n\in\mathscr{O}^*_{\mu,r}(\Omega)$ converges weakly to $f_n\in\mathscr{O}^*_{\mu,r}(\Omega)$, if for arbitrary $\zeta\in\Omega$ we have $H_m\to H_0$ as $m\to\infty$ in $\tau_*^{-1}(\zeta)=\mathscr{O}^*_{\mu,r,\zeta}$.

Finally we introduce the Fourier transform on the spaces $\mathscr{O}_{\mu,r}(\Omega)$ and $E^{\Omega}_{\mu,r}(\mathbb{C}^n)$ as the Fourier transform defined fiberwise, i.e., on fibers $\{\tau^{-1}(\zeta),\zeta\in\Omega\}$ and $\{\pi^{-1}(\zeta),\zeta\in\Omega\}$, respectively. For this purpose we need the following isomorphisms:

$$f_\zeta:\mathscr{O}_{\mu,r,\zeta}\to\mathscr{O}_{\mu,r},\quad f^*_\zeta:\mathscr{O}^*_{\mu,r,\zeta}\to\mathscr{O}^*_{\mu,r},$$
$$g_\zeta:E_{\mu,r,\zeta}\to E_{\mu,r},\quad g^*_\zeta:E^*_{\mu,r,\zeta}\to E^*_{\mu,r}.$$

By definition, the Fourier transform in the fiber space $\mathscr{O}_{\mu,r}(\Omega)$ is the family of mappings $F_{\zeta_0}:\mathscr{O}_{\mu,r,\zeta_0}\to E^*_{\mu,r,\zeta_0}$, $\zeta_0\in\Omega$, defined as

$$F[h](\zeta)=g^{-1}_{\zeta_0}\circ F[f_{\zeta_0}\circ h](\zeta),\quad \zeta_0\in\Omega,$$

where $F[f_{\zeta_0}\circ h]$ is the Fourier transform given in equation (9.34). Similarly, the Fourier transform in the space $E^{\Omega}_{\mu,r}(\mathbb{C}^n)$ is the family of mappings $F_{\zeta_0}:E_{\mu,r,\zeta_0}\to\mathscr{O}^*_{\mu,r,\zeta_0}$, $\zeta_0\in\Omega$, defined as

$$F[\varphi](\zeta)=f^{-1}_{\zeta_0}\circ F[g_{\zeta_0}\circ\varphi](\zeta),\quad \zeta_0\in\Omega.$$

For simplicity the Fourier transform in both fiber spaces $\mathcal{O}_{\mu,r}(\Omega)$ and $E^{\Omega}_{\mu,r}(\mathbb{C}^n)$ will also be denoted by the same letter F. The following statement follows immediately from Corollary 9.2.

Theorem 9.3. *The following commutative diagram holds:*

$$\begin{array}{ccc} E^{\Omega}_{\mu,r}(\mathbb{C}^n) & \overset{*}{\longleftrightarrow} & \left(E^{\Omega}_{\mu,r}(\mathbb{C}^n)\right)^* \\ F^{-1}\uparrow\downarrow F & & F^{-1}\downarrow\uparrow F \\ \mathcal{O}^*_{\mu,r}(\Omega) & \overset{*}{\longleftrightarrow} & \mathcal{O}_{\mu,r}(\Omega) \end{array}$$

Remark 9.3. Dubinskii [Dub84] introduced the space $Exp_{\Omega}(\mathbb{C}^n)$ defined as an inductive limit of Banach spaces composed with the help of $E_{\mu,r,\zeta}$. This space and its dual were used as solution spaces for the Cauchy problem for pseudo-differential equations with holomorphic symbols. The Fourier transform in $Exp_{\Omega}(\mathbb{C}^n)$ is defined by the formula $F[f](\zeta) = (2\pi)^n f(-D_z)\delta_0(\zeta)$, which maps $Exp_{\Omega}(\mathbb{C}^n)$ onto the space $\mathcal{O}^*(\Omega)$ of analytic functionals concentrated on compact sets in Ω. There is a relationship similar to commutative diagram in Theorem 9.3:

$$\begin{array}{ccc} Exp_{\Omega}(\mathbb{C}^n) & \overset{*}{\longleftrightarrow} & Exp^*_{\Omega}(\mathbb{C}^n) \\ F^{-1}\uparrow\downarrow F & & F^{-1}\downarrow\uparrow F \\ \mathcal{O}^*(\Omega) & \overset{*}{\longleftrightarrow} & \mathcal{O}(\Omega). \end{array}$$

For details we refer the reader to [Dub84, Dub90].

9.5 An algebra of matrix-symbols with singularities

In Sections 9.6 and 9.7 we will use linear differential operators of the form

$$L(z,D_z) = \sum_{|\alpha|\le m} a_\alpha(z) D_z^\alpha,$$

with meromorphic coefficients $a_\alpha(z), |\alpha| \le m$, and pseudo-differential operators of the form

$$L(D_z,z) = \sum_{|\alpha|\le m} z^\alpha a_\alpha(D_z),$$

with meromorphic symbols $a_\alpha(z)$ defined in a domain $G \subset \mathbb{C}^n$. Therefore, in this section we will study the symbols of the form

$$a(z,\zeta) = \sum_{|\alpha|\le m} z^\alpha a_\alpha(\zeta).$$

By definition, a symbol of degree m is an ordered collection $a \equiv \{a_\alpha(\zeta)\}_{|\alpha|\leq m}$ of functions a_α from some space X, which is specified below. The class of symbols of degree m is denoted by $S(m,X)$. Note that if m is a degree of the symbol a, then $m+k$ for an arbitrary nonnegative integer k is also a degree. The least degree m for which there is α, $|\alpha| = m$, such that $a_\alpha(\zeta) \neq 0$, but $a_\alpha = 0$ for all $|\alpha| > m$ is called the exact degree of the symbol a, and denoted $\deg(a)$. The identity symbol $j \in S(0,X)$ is the symbol with $a_0(\zeta) \equiv 1$. For the zero-symbol θ the functions $a_\alpha \equiv 0$ for all $|\alpha| \leq m$. We use the following convention: if $\deg(a) < 0$, then we accept $a = 0$ and write $\deg(a) = -\infty$. By this convention $\deg(\theta) = -\infty$, and $\theta \in S(-\infty,X)$.

The sum $a+b$ and product $a \circ b$ of two symbols $a \in S(m_1,X)$ and $b \in S(m_2,X)$ are, respectively, defined by

$$a+b \equiv \{a_\alpha(\zeta)+b_\alpha(\zeta)\}_{|\alpha|\leq \max(m_1,m_2)}, \tag{9.42}$$

and

$$a \circ b \equiv \Big\{ \sum_{\substack{|\gamma|\leq m_1 \\ \gamma \preccurlyeq \alpha}} \sum_{\substack{|\beta|\leq m_2 \\ \beta \succcurlyeq \alpha-\gamma}} \binom{\beta}{\beta+\gamma-\alpha} b_\beta(\zeta) D_\zeta^{\beta+\gamma-\alpha} a_\gamma(\zeta) \Big\}_{|\alpha|\leq m_1+m_2}, \tag{9.43}$$

where $\gamma \preccurlyeq \alpha$ means $\gamma_j \leq \alpha_j$, $j = 1,\ldots,n$, and $\binom{\sigma}{\delta}$ for multi-indices σ and δ means

$$\binom{\sigma}{\delta} = \prod_{j=1}^{n} \frac{\sigma_j!}{\delta_j!(\sigma_j-\delta_j)!}.$$

The formula (9.43) for composition of two symbols follows from the Leibniz rule for pseudo-differential operators. It is easy to see that, in general, $a \circ b \neq b \circ a$. Indeed, for $a = \{a_0(\zeta), a_1(\zeta)\} = \{0,\zeta\}$ with $m_1 = 1$ and $b = \{b_0(\zeta)\} = \{\zeta\}$ with $m_2 = 0$ formula (9.43) implies that

$$a \circ b = \{b_0(\zeta)a_0(\zeta), b_0(\zeta)a_1(\zeta)\} = \{0,\zeta^2\},$$

while

$$b \circ a = \{a_0(\zeta)b_0(\zeta) + a_1(\zeta)b_0'(\zeta), a_1(\zeta)b_0(\zeta)\} = \{\zeta,\zeta^2\}.$$

Hence, the product of two symbols is not a commutative operation. The following properties of symbols immediately follow from the above definitions of the sum and the product of symbols.

Proposition 9.9. *Let $a \in S(m_1,X)$ and $b \in S(m_2,X)$. Then*

1. $\deg(a+b) = \max\{\deg(a),\deg(b)\}$;
2. $\deg(a \circ b) = \deg(a) + \deg(b)$;
3. $a+\theta = \theta+a = a$;
4. $a \circ j = j \circ a = a$.

Let $\Omega \subset C^n$ be an n-dimensional complex domain and $\mathscr{M}(\Omega)$ and $\mathscr{O}(\Omega)$ be sheaf of germs of meromorphic and holomorphic functions, respectively. We assume that $f \in \mathscr{M}(\Omega)$ has a local representation $f(z) = g(z)/h(z)$, where $g,h \in \mathscr{O}(\Omega)$, in a neighborhood of any point of Ω. We denote by P_f and N_f the set of poles $P_f = \{z \in \Omega : h(z) = 0\}$, and the set of zeros (or null-set) $N_f = \{z \in \Omega : g(z) = 0\}$ of the meromorphic function f.

Let M be an $N \times N$-matrix whose entries $m_{i,j}$, $i,j = 1,\dots,N$, are allowed degrees of symbols (i.e., nonnegative integers, or $-\infty$). We denote by $S(M,\mathscr{M}(\Omega))$ the set of $N \times N$ matrix-valued symbols with entries

$$a_{ij}(z,\zeta) = \sum_{|\alpha| \le m_{ij}} z^\alpha a_{ij,\alpha}(\zeta), \quad i,j = 1,\dots,N,$$

where $z \in C^n$, $\zeta \in \Omega$, and $a_{ij,\alpha} \in \mathscr{M}(\Omega)$, $|\alpha| \le m_{ij}$, $i,j = 1,\dots,N$. Introduce the following analytic sets of co-dimension 1:

$$P_j = \bigcup_{i=1}^{N} \left(\bigcup_{|\alpha| \le m_{ij}} P_{a_{\alpha ij}} \right),$$

$$Q_j = (\bigcup_{k=1}^{j} N_{a_{0kk}}) \cup \left(\bigcup_{1 \le k \le j-1} \left(\bigcup_{k+1 \le l \le j} (\bigcup_{|\alpha| \le m_{kl}} P_{a_{\alpha kl}}) \right) \right).$$

Let $A_0(\zeta)$ be *the constant part of* $A(z,\zeta) \in S(M,\mathscr{M}(\Omega))$, i.e., the matrix $A_0(\zeta) = (a_{ij,0}(\zeta))_{i,j=1}^N$ and let $\Delta(\zeta)$ be its determinant $\Delta(\zeta) = \det A_0(\zeta)$. Obviously $\Delta(\zeta)$ is also meromorphic and let the following local representation hold:

$$\Delta(\zeta) = \frac{G(\zeta)}{H(\zeta)}, \quad G,H \in \mathscr{O}(\Omega). \tag{9.44}$$

We call the set $P_A = \{\zeta \in \Omega : H(\zeta = 0)\}$ a *polar* set and the set $N_A = \{\zeta \in \Omega : G(\zeta) = 0\}$ a *null* set of the matrix symbol $A(z,\zeta)$. It follows from general theory of determinants that for the inverse matrix $A_0^{-1}(\zeta)$ one has a local representation $\det(A^{-1}(\zeta)) = H(\zeta)/G(\zeta)$, and therefore, $P_{A^{-1}} = N_A$ and $N_{A^{-1}} = P_A$. Further, we introduce the sets:

$$Z(A) = P_A \cup N_A, \quad Z_{reg}(A) = Z(A) \setminus (P_A \cap N_A), \quad \text{and} \quad Z_{reg,\Omega}(A) = \Omega \cap Z_{reg}(A).$$

It is obvious that these sets are invariant with respect to inversion of the symbol $A(z,\zeta)$.

Since symbols in $S(M,\mathscr{M}(\Omega))$ have entries $a_{ij,\alpha} \in S(m_{i,j},\mathscr{M}(\Omega))$, one can define the addition, composition, and involution operations in $S(M,\mathscr{M}(\Omega))$ using operations introduced in (9.42) and (9.43).

Theorem 9.4. *A symbol $A(z,\zeta) \in S(M,\mathscr{M}(\Omega))$ has the inverse $A^{-1}(z,\zeta) \in S(M,\mathscr{M}(\Omega))$ if and only if there exists a collection of integers $\mu_1,\dots,\mu_N$ such that the inequalities*

$$\deg(a_{ij}) \le \mu_i - \mu_j, \quad i,j = 1,\dots,N, \tag{9.45}$$

hold.

Remark 9.4. Under the condition of this theorem $S(M,\mathcal{M}(\Omega))$ is a noncommutative involutive algebra.

Proof. Sufficiency. Let us first assume that the numbers $\mu_1,\dots,\mu_N$ are strictly ordered in the decreasing order: $\mu_1 > \dots > \mu_N$. Then, due to conditions (9.45), $deg(a_{ij}) < 0$ if $i > j$, and hence $a_{ij} = \theta$. Thus, in this case the symbol $A(z,\zeta)$ is represented in the form:

$$A(z,\zeta) = \begin{bmatrix} a_{11}\ a_{12}\ \dots\ a_{1N} \\ \theta\ \ a_{22}\ \dots\ a_{2N} \\ \dots\ \dots \\ \theta\ \ \theta\ \ \dots\ a_{NN} \end{bmatrix},$$

where $deg(a_{jj}) = 0$, $j = 1,\dots,N$, and $a_{ij} \in S(\mu_i - \mu_j,\mathcal{M}(\Omega))$ if $j > i$. Let b_{ij}, $i,j = 1,\dots,N$, be entries of the inverse symbol $A^{-1}(z,\zeta)$. The requirement $A^{-1}(z,\zeta) \in S(M,\mathcal{M}(\Omega))$ implies $b_{ij} = \theta$ if $i > j$, $deg(a_{jj}) = 0$, $j = 1,\dots,N$, and $b_{ij} \in S(\mu_i - \mu_j,\mathcal{M}(\Omega))$ if $j > i$. This is natural, since the inverse of the right triangular matrix is again a right triangular matrix. The symbols b_{ij} are defined from the system of algebraic equations

$$\begin{cases} a_{jj} \circ b_{jj} = 1, & j = 1,\dots,N, \\ \sum_{k=i}^{j} a_{ik} \circ b_{kj} = 0, & \text{if } i < j, \\ b_{ij} = \theta, & \text{if } i > j. \end{cases} \tag{9.46}$$

These equations define all the components of symbols b_{ij} uniquely. Indeed, it follows from (9.46) immediately that

$$b_{jj} = 1/a_{jj}, \quad j = 1,\dots,N.$$

Setting $j = i+1$, we have

$$a_{ii} \circ b_{ii+1} + a_{ii+1} \circ b_{i+1i+1} = \theta,$$

which implies

$$b_{ii+1} = -\frac{1}{a_{ii}}[a_{ii+1} \circ b_{i+1i+1}], \quad i = 1,\dots,N-1.$$

Similarly, if all the symbols $b_{ii+\ell-1}$, $i = 1,\dots,N-\ell+1$, are found for some $1 \le \ell \le N-2$, then $b_{ii+\ell}$ is defined as

$$b_{ii+\ell} = -\frac{1}{a_{ii}} \sum_{k=i+1}^{i+\ell} a_{ik} \circ b_{ki+\ell}, \quad i = 1,\dots,N-\ell.$$

Now assume that the numbers $\mu_1,\dots,\mu_N$ satisfy the ordering $\mu_1 = \dots = \mu_{k_1} > \mu_{k_1+1} = \dots = \mu_{k_2} > \dots > \mu_{k_p+1} = \dots = \mu_N$. In this case the symbol $A(z,\zeta)$ is represented in the block-matrix form:

$$A(z,\zeta)=\begin{bmatrix} A_{11} & A_{12} & \dots & A_{1p} \\ \Theta & A_{22} & \dots & A_{2p} \\ & \dots & \dots & \\ \Theta & \Theta & \dots & A_{pp} \end{bmatrix},$$

where $k_1+\ldots+k_p=N$, $A_{jj}, j=1,\ldots,p$, are $k_j\times k_j$-matrix-symbols in $S(\Theta, \mathscr{M}(\Omega))$, and $A_{ij}, i<j$, are $k_i\times k_j$-matrix-symbols that belong to $S(M,\mathscr{M}(\Omega))$ with a degree-matrix M, entries of which are positive numbers. The inverse symbol $A^{-1}(z,\zeta)$ is of the structure

$$A^{-1}(z,\zeta)=\begin{bmatrix} B_{11} & B_{12} & \dots & B_{1p} \\ \Theta & B_{22} & \dots & B_{2p} \\ & \dots & \dots & \\ \Theta & \Theta & \dots & B_{pp} \end{bmatrix},$$

where the block B_{ij} belongs to the same class of symbols as the corresponding block A_{ij} does. The blocks B_{ij} are defined from the system of algebraic equations

$$\begin{cases} A_{jj}\circ B_{jj}=1, & j=1,\ldots,p, \\ \sum_{k=i}^{j} A_{ik}\circ B_{kj}=0, & \text{if } i<j, \\ B_{ij}=\Theta, & \text{if } i>j. \end{cases}$$

These equations define all the blocks B_{ij} uniquely. Indeed,

$$B_{jj}=A_{jj}^{-1}, \quad j=1,\ldots,p,$$

and if all the blocks $B_{ii+\ell-1}, i=1,\ldots,p-\ell+1$, are found for some $1\le\ell\le p-2$, then $B_{ii+\ell}$ is defined as

$$B_{ii+\ell}=-A_{ii}^{-1}\sum_{k=i+1}^{i+\ell} A_{ik}\circ B_{ki+\ell}, \quad i=1,\ldots,p-\ell.$$

Finally, if $\mu_1,\ldots,\mu_N$ are arbitrary numbers, then rearranging rows and columns of the matrix-symbol $A(z,\zeta)\in S(M,\mathscr{M}(\Omega))$ we obtain a matrix-symbol $\bar{A}(z,\zeta)$, for which $\bar{\mu}_1,\ldots,\bar{\mu}_N$ are ordered. Indeed, if for indices i and j, $i<j$, of the collection $\mu_1,\ldots,\mu_N$ the relation $\mu_i<\mu_j$ holds, then switching i-th and j-th columns, and then switching i-th and j-th rows of $A(z,\zeta)$, we obtain a collection $\mu_1^{'},\ldots,\mu_N^{'}$, with $\mu_i^{'}=\mu_j>\mu_i=\mu_j^{'}$. These two switchings are equivalent to the multiplication by two matrices C_i and R_i with determinants $\det(C_j)=\det(R_j)=-1$. Performing these operations finitely many times we arrive to the symbol $\bar{A}(z,\zeta)\in S(\bar{M},\mathscr{M}(\Omega))$, where $\bar{M}$ is a matrix of degrees corresponding to the ordered collection $\bar{\mu}_1,\ldots,\bar{\mu}_N$. Hence, the symbol $\bar{A}(z,\zeta)$ is connected with $A(z,\zeta)$ through $\bar{A}(z,\zeta)=CA(z,\zeta)$, where $C=R_kC_k\ldots R_1C_1$ is an invertible $N\times N$ matrix not depending on z and ζ. Therefore, $A^{-1}(z,\zeta)=\bar{A}^{-1}(z,\zeta)C$. As we have seen above in the ordered case the symbol $\bar{A}^{-1}(z,\zeta)$ also belongs to the same class $S(\bar{M},\mathscr{M}(\Omega))$. The multiplication of $\bar{A}^{-1}(z,\zeta)$ by C from the right is equivalent to switching of columns and rows exactly in the reverse order. This implies that $A^{-1}(z,\zeta)\in S(M,\mathscr{M}(\Omega))$.

Necessity. Assume that the inverse symbol $A^{-1}(z,\zeta) = B(z,\zeta) \in S(M,\mathscr{M}(\Omega))$ exists. The relations

$$\sum_{\ell=1}^{N} a_{i\ell} \circ b_{\ell j} = \delta_{ij}, \quad i,j = 1,\dots,N,$$

that indicate that the symbols $A(z,\zeta)$ and $B(z,\zeta)$ are mutually inverse, contain

$$L = \frac{1}{n!} \sum_{i=1}^{N} \sum_{j=1}^{N} \frac{[\max_k (\deg(a_{ik}) + m_{kj} + n)]!}{[\max_k (\deg(a_{ik}) + m_{kj})]!}$$

equations. On the other hand, since each symbol b_{ij} contains $\frac{(m_{ij}+n)!}{m_{ij}!n!}$ components, then the total number of components of $B(z,\zeta)$ is

$$K = \frac{1}{n!} \sum_{i=1}^{N} \sum_{j=1}^{N} \frac{(m_{ij}+n)!}{m_{ij}!}.$$

Due to our assumption on the existence of the inverse symbol, we have $L = K$. This implies

$$\max_k (\deg(a_{ik}) + m_{kj}) = m_{ij}, \quad i,j = 1,\dots,N. \tag{9.47}$$

It follows from (9.47) that the inequalities

$$\deg(a_{ik}) \leq m_{ij} - m_{kj} \tag{9.48}$$

are valid for all $j = 1,\dots,N$. Let μ_i and ν_i are the integer and fractional parts of $(m_{i1} + \dots + m_{iN})/N$, respectively. Then, equation (9.48) can be rewritten in the form

$$\deg(a_{ik}) \leq \mu_i - \mu_k + (\nu_i - \nu_k). \tag{9.49}$$

Finally, since $\deg(a_{ik})$ are integers and $|\nu_i - \nu_k| < 1$, it follows from (9.49) that

$$\deg(a_{ik}) \leq \mu_i - \mu_k, \quad i,k = 1,\dots,N,$$

proving the necessity of the condition (9.45).

Under the additional condition $N_A \cap \Omega = \emptyset$ to $A(z,\zeta)$ the class of symbols $S(M,\mathscr{O}(\Omega))$ becomes an involutive algebra. Namely, the following theorem is valid [Uma91-1]:

Theorem 9.5. *A symbol $A(z,\zeta) \in S(M,\mathscr{O}(\Omega))$ has the inverse $A^{-1}(z,\zeta) \in S(M,\mathscr{O}(\Omega))$ if and only if the following two conditions hold:*

(i) there exists a collection of integers $\mu_1,\dots,\mu_N$ such that the inequalities

$$\deg(a_{ij}) \leq \mu_i - \mu_j, \quad i,j = 1,\dots,N,$$

are fulfilled, and
(ii) $N_A \cap \Omega = \emptyset$.

The proposition below proved in [Vol63] (see also [Miz67]) provides a sufficient condition for the matrix $m_{ij} = \deg(a_{ij})$ to exist a collection μ_k, $k = 1,\ldots,N$, satisfying the condition (9.45).

Proposition 9.10. *Let a matrix M with rational entries m_{ij} (including $-\infty$) satisfy the condition: $m_{ii} = 0$, $i = 1,\ldots,N$, and for any permutation π of the set $\{1,\ldots,N\}$ the inequality $\sum_{i=1}^{N} m_{i,\pi(i)} \leq 0$ holds. Then there exists a collection $\mu_1,\ldots,\mu_N$ of rational numbers satisfying $m_{ij} \leq \mu_j - \mu_i$, $i,j = 1,\ldots,N$.*

Remark 9.5. If entries of M are integers, then in the proposition above the numbers $\mu_1,\ldots,\mu_N$ also can be selected integer. Obviously, if $m_{ij} \leq \mu_j - \mu_i$ for all $i,j = 1,\ldots,N$, then the transposed matrix satisfies $m_{ij}^T \leq \mu_i - \mu_j$ for all $i,j = 1,\ldots,N$.

9.6 Algebras of pseudo-differential operators with complex symbols with singularities

Let a symbol $a = \{a_\alpha(\zeta)\}_{|\alpha|\leq m} \in S(m,X)$, where X is a class of symbols specified below. We define the pseudo-differential operator with the symbol a as

$$Af = \sum_{|\alpha|\leq m} z^\alpha F^{-1}[a_\alpha(\zeta)F[f](\zeta)](z), \tag{9.50}$$

where F is the complex Fourier transform defined in (9.34). The class of pseudo-differential operators with symbols in $S(m,X)$ will be denoted $OPS(m,X)$. We also write $deg(A)$ having in mind the degree of the corresponding symbol. The sum $A+B$ and composition $A \circ B$ of operators $A \in OPS(m_1,X)$ and $B \in OPS(m_2,X)$ are defined as operators with symbols $a+b$ and $a \circ b$, respectively. Hence, $OPS(m,X)$ is an algebra isomorphic to the algebra $S(m,X)$.

Proposition 9.11. *Let $A \in OPS(m, \mathcal{O}_{\mu,r,\zeta_0})$, $\zeta_0 \in \Omega$. Then the mappings*

$$A \equiv A(z,D_z) : \mathcal{E}_{\mu,r,\zeta_0} \to \mathcal{E}_{\mu+m,r,\zeta_0} \quad \text{and} \quad A^* \equiv A(z,-D_z) : \mathcal{E}^*_{\mu+m,r,\zeta_0} \to \mathcal{E}^*_{\mu,r,\zeta_0}$$

are continuous. Moreover, for the norms of the operators A and A^* the estimate

$$\|A\| = \|A^*\| \leq \sum_{|\alpha|\leq m} [a_\alpha]_{\mu,r,\zeta_0} \tag{9.51}$$

holds.

Proof. Since the spaces $\mathcal{E}_{\mu,r,\zeta_0}$ for different $\zeta_0 \in \Omega$ are isomorphic, it suffices to consider the case $\zeta_0 = 0$. Let $\varphi \in \mathcal{E}_{\mu,r}$ be an arbitrary element. It is readily seen that the multiplication operator by a function $\psi(z) \in \mathcal{E}_{\mu_0,r_0}$ is continuous from $\mathcal{E}_{\mu,r}$ to $\mathcal{E}_{\mu+\mu_0,r+r_0}$. In particular, for $z^\alpha \in \mathcal{E}_{|\alpha|,0}$, taking into account Proposition 9.6, one has

$$\|A\varphi\|_{\mu+m,r} \leq \sum_{|\alpha|\leq m} \|z^{\alpha} a_{\alpha}(D)\varphi\|_{\mu+m,r} \leq \sum_{|\alpha|\leq m} \|a_{\alpha}(D)\varphi\|_{\mu,r}$$
$$\leq \|\varphi\|_{\mu,r} \sum_{|\alpha|\leq m} [a_{\alpha}]_{\mu,r}.$$

The second part of the statement now follows by duality.

Introduce the following spaces of direct products with the corresponding direct product topologies:

$$\mathscr{E}_{\bar{\mu},r,\zeta_0} = \bigotimes_{j=1}^{N} \mathscr{E}_{\mu_j,r,\zeta_0}, \qquad \mathscr{E}^*_{\bar{\mu},r,\zeta_0} = \bigotimes_{j=1}^{N} \mathscr{E}^*_{\mu_j,r,\zeta_0},$$
$$\mathscr{O}_{\bar{\mu},r,\zeta_0} = \bigotimes_{j=1}^{N} \mathscr{O}_{\mu_j,r,\zeta_0}, \qquad \mathscr{O}^*_{\bar{\mu},r,\zeta_0} = \bigotimes_{j=1}^{N} \mathscr{O}^*_{\mu_j,r,\zeta_0},$$
$$E^{\Omega}_{\bar{\mu},r}(\mathbb{C}^n) = \bigotimes_{j=1}^{N} E^{\Omega}_{\mu_j,r}(\mathbb{C}^n), \quad \left(E^{\Omega}_{\bar{\mu},r}(\mathbb{C}^n)\right)^* = \bigotimes_{j=1}^{N} \left(E^{\Omega}_{\mu_j,r}(\mathbb{C}^n)\right)^*,$$
$$\mathscr{O}_{\bar{\mu},r}(\Omega) = \bigotimes_{j=1}^{N} \mathscr{O}_{\mu_j,r}(\Omega), \quad \mathscr{O}^*_{\bar{\mu},r}(\Omega) = \bigotimes_{j=1}^{N} \mathscr{O}^*_{\mu_j,r}(\Omega),$$

and

$$\mathscr{M}_{\bar{\mu},r}(\Omega) = \bigotimes_{j=1}^{N} \mathscr{M}_{\mu_j,r}(\Omega),$$

where $\mathscr{M}_{\mu,r}(\Omega) = (\mathscr{M}, \Omega, \pi)$ is a fiber space of meromorphic functions with the base Ω, fibers $\mathscr{M}_{\mu,r,\zeta_0}$, and projection

$$\pi : (\mathscr{M}, \Omega, \pi) \to \Omega,$$

where $\pi^{-1}(\zeta_0) = \mathscr{M}_{\mu,r,\zeta_0}$, $\zeta \in \Omega$. An element of the fiber $\mathscr{M}_{\mu,r,\zeta_0}$ in a neighborhood of the point $\zeta_0 \in \Omega$ has a local representation $m(z) = f(z)/g(z) \in \mathscr{O}_{\mu,r}(\Omega \setminus P_m)$. Hence, one can define a dual space $\mathscr{M}^*_{\bar{\mu},r}(\Omega)$ of meromorphic functionals as well, similar to their analytic and exponential counterparts.

Let $A(z,D_z)$ be a pseudo-differential operator with the matrix-symbol $\mathscr{A}(z,\zeta)$, whose entries $\mathscr{A}_{ij}(z,\zeta) \in S(m_{ij}, \mathscr{O}_{m_j,r,\zeta_0})$, $i,j = 1,\dots,N$. We define the adjoint operator $A^*(z,D_z)$ as a pseudo-differential operator with the matrix-symbol $\mathscr{A}^*(z,\zeta) = \mathscr{A}^T(z,-\zeta)$, that is with entries $\mathscr{A}^*_{ij}(z,\zeta) = \mathscr{A}_{ji}(z,-\zeta) \in S(m_{ji}, \mathscr{O}_{m_j,r,-\zeta_0})$, $i,j = 1,\dots,N$.

Proposition 9.12. *Let $\mathscr{A}(z,\zeta)$ be a matrix-symbol with entries $\mathscr{A}_{ij} \in S(m_{ij}, \mathscr{O}_{m_j,r,\zeta_0})$, and let $m_{ij} \leq \mu_i - \mu_j$ for all $i,j = 1,\dots,N$. Then the mappings*

$$A(z,D_z) : \mathscr{E}_{\bar{\mu},r,\zeta_0} \to \mathscr{E}_{\bar{\mu},r,\zeta_0}; \;\; A^*(z,D_z) : \mathscr{E}^*_{\bar{\mu},r,\zeta_0} \to \mathscr{E}^*_{\bar{\mu},r,\zeta_0}$$

are continuous. Moreover, for the norms of these operators the estimate

$$\|A\| = \|A^*\| \leq \sum_{i=1}^{N} \left(\max_{1 \leq j \leq N} \sum_{|\alpha| \leq m_{ij}} [a_\alpha]_{\mu_j, r, \zeta_0} \right) \tag{9.52}$$

holds.

Proof. Since the operator $A_{ij} \in OPS(m_{ij}, \mathscr{O}_{\mu_j,r,\zeta_0})$, it follows from Proposition 9.11 that it maps the space $\mathscr{E}_{\mu_j,r,\zeta_0}$ continuously onto $\mathscr{E}_{\mu_j+m_{ij},r,\zeta_0}$. The latter is continuously embedded into $\mathscr{E}_{\mu_i,r,\zeta_0}$ due to inequality $\mu_j + m_{ij} \leq \mu_i$ for all $i,j = 1,\ldots,N$. These imply the continuity of the operator $A(z,D_z) : \mathscr{E}_{\bar{\mu},r,\zeta_0} \to \mathscr{E}_{\bar{\mu},r,\zeta_0}$.

To show (9.52) one can use estimate (9.51) for the operator A_{ij} :

$$\|A_{ij}\varphi_j(z)\|_{\mu_i,r,\zeta_0} \leq \sum_{|\alpha| \leq m_{ij}} [a_{ij\alpha}]_{\mu_j,r,\zeta_0} \|\varphi_j\|_{\mu_j,r,\zeta_0},$$

where $\varphi_j \in \mathscr{E}_{\mu_j,r,\zeta_0}$. It follows that

$$\|(A\varphi)_i\|_{\mu_i,r,\zeta_0} \leq \|\varphi\|_{\bar{\mu},r,\zeta_0} \max_{1 \leq j \leq N} \sum_{|\alpha| \leq m_{ij}} [a_{ij\alpha}]_{\mu_j,r,\zeta_0}.$$

Here $(A\varphi)_i$ is the i-th component of the vector-function $A(z,D_z)\varphi(z)$. Summing the latter inequality over all $i = 1,\ldots,N$, one obtains estimate (9.52). The rest of the statement of the theorem follows by duality.

Proposition 9.13. *Let $\mathscr{A}(z,\zeta)$ be a matrix-symbol with entries $\mathscr{A}_{ij} \in S(m_{ij}, \mathscr{O}_{\mu_j,r}(\Omega))$. Suppose that the collection of integers $\{\mu_1,\ldots,\mu_N\}$ such that $m_{ij} \leq \mu_j - \mu_i$ for all $i,j = 1,\ldots,N$. Then the mappings*

$$A(z,D_z) : E^{\Omega}_{\bar{\mu},r}(\mathbb{C}^n) \to E^{\Omega}_{\bar{\mu},r}(\mathbb{C}^n),$$
$$A^*(z,D_z) : \left(E^{\Omega}_{\bar{\mu},r}(\mathbb{C}^n)\right)^* \to \left(E^{\Omega}_{\bar{\mu},r}(\mathbb{C}^n)\right)^*$$

are continuous.

Proof. Follows easily from Proposition 9.12.

Proposition 9.14. *Let $\mathscr{A}(z,\zeta)$ be a matrix-symbol with entries $\mathscr{A}_{ij} \in S(m_{ij}, \mathscr{M}_{\mu_j,r}(\Omega))$. Suppose that the collection of integers $\{\mu_1,\ldots,\mu_N\}$ satisfy inequalities $m_{ij} \leq \mu_j - \mu_i$ for all $i,j = 1,\ldots,N$. Then the pseudo-differential operators corresponding to symbols $\mathscr{A}(z,\zeta)$ and $\mathscr{A}^*(z,\zeta)$ are continuous as mappings*

$$A(z,D_z) : E^{\Omega\setminus P_A}_{\bar{\mu},r}(\mathbb{C}^n) \to E^{\Omega}_{\bar{\mu},r}(\mathbb{C}^n); \tag{9.53}$$
$$A^*(z,D_z) : \left(E^{\Omega}_{\bar{\mu},r}(\mathbb{C}^n)\right)^* \to \left(E^{\Omega\setminus P_A}_{\bar{\mu},r}(\mathbb{C}^n)\right)^*. \tag{9.54}$$

Moreover, the inverse operator $A^{-1}(z,D_z)$ exists and is continuous as a mapping

$$A^{-1}(z,D_z) : E^{\Omega\setminus N_j}_{\bar{\mu},r}(\mathbb{C}^n) \to E^{\Omega}_{\bar{\mu},r}(\mathbb{C}^n). \tag{9.55}$$

Proof. Let $\varphi_j \in \mathscr{E}_{\mu_j,r,\zeta_0}(\mathbb{C}^n)$, $j=1,\dots,N$, where $\zeta_0 \in \Omega \setminus P_j$. In accordance with the definition of $\mathscr{M}_{\mu,r}(\Omega)$ all the functions $a_{ij\alpha}(\zeta)$ in the symbol $\mathscr{A}_{ij}(z,\zeta)$ belong to $\mathscr{O}_{\mu,r,\zeta_0}$ with $r < dist(\zeta_0, \partial(\Omega \setminus P_j))$. Therefore, $\mathscr{A}_{ij}(z,\zeta) \in S(m_{ij}, \mathscr{O}_{\mu_j,r}(\Omega \setminus P_j))$. Now the continuity of mappings (9.53) and (9.55) follow from Proposition 9.13. This fact implies the continuity of the inverse operator $A^{-1}(z,D_z)$ in mapping (9.55) too, since due to Theorem 9.4 the inverse symbol $\mathscr{A}^{-1}(z,\zeta)$ has entries $\mathscr{A}_{ij}^{-1}(z,\zeta) \in S(m_{ij}, \mathscr{M}_{\mu_j,r}(\Omega))$.

Pseudo-differential operators with meromorphic symbols in $S(M,\mathscr{M}(\Omega))$ behave differently. Unlike the previous cases they act in factor-spaces. To formulate the continuity theorem first we study kernels of pseudo-differential operators with meromorphic symbols.

For an operator $A \in OPS(M,\mathscr{M}(\Omega))$ we denote by $\kappa_\pm$ the dimension of the kernel of $A^{\pm 1}$:

$$\kappa_\pm = \kappa_\pm(A,\Omega) = \dim Ker(A^{\pm 1}).$$

The meaning of κ_+ is obvious. If $\kappa_- = m$, then the image of the operator A is a factor space factorized by the m-dimensional space $KerA^{-1}$. Thus, the operator A in this case is multi-valued. The operator $A \in OPS(M,\mathscr{M}(\Omega))$ is single-valued if and only if $\kappa_- = 0$.

Let $\mathscr{P}_k^\pm$, $k=1,\dots,K_\pm$, be connected irreducible components of $P_{\mathscr{A}^{\pm 1}} \cap Z_{reg,\Omega}$ and $L_k^\pm, k=1,\dots,K_\pm$, be their respective orders. Denote by $W_{kl}^\pm(\mathscr{A})$ the span of all linear combinations

$$f_{k,\ell}(z) = F^{-1}[\delta^{(\ell)}(\rho_k^\pm(\zeta))](z), \quad \ell=0,\dots,L_k^\pm,\ k=1,\dots,K_\pm, \tag{9.56}$$

where δ is the Dirac distribution, and $\rho_k^\pm(\zeta)$ are holomorphic functions, locally representing $\mathscr{P}_k^\pm$, that is $\mathscr{P}_k^\pm \equiv \{\zeta : \rho_k^\pm(\zeta) = 0\}$.

Theorem 9.6. *Let $A \in OPS(M,\mathscr{M}(\Omega))$ and there exists a collection of integers $\{\mu_1,\dots,\mu_N\}$ such that $m_{ij} \le \mu_j - \mu_i$ for all $i,j=1,\dots,N$. Then*

$$Ker(A^{\pm 1}) = \bigoplus_{k=1}^{K_\pm} \left(\bigoplus_{\ell=0}^{L_k^\pm - 1} W_{k\ell}^\mp(A) \right).$$

Proof. We will show that $V \in Ker(A^{-1})$ if and only if $V \in Ker(A_0^{-1})$, where A_0 is the constant part of the operator A. Indeed, without loss of generality, one can assume that $\mu_1,\dots,\mu_N$ are ordered, i.e., $\mu_1 = \dots = \mu_{k_1} > \mu_{k_1+1} = \dots = \mu_{k_2} > \dots > \mu_{k_{l-1}+1} = \dots = \mu_{k_l}$, $k_1 + \dots + k_l = N$. Otherwise, with the help of permutations of rows and columns, which correspond to the multiplication of A by a scalar invertible matrices, one gets a desired ordering. Hence, the operator A has the form

$$A = \begin{bmatrix} A_{11} & A_{12} & \dots & A_{1l} \\ \Theta & A_{22} & \dots & A_{2l} \\ & \dots & \dots & \\ \Theta & \Theta & \dots & A_{ll} \end{bmatrix}, \tag{9.57}$$

where $A_{jj} \in OPS(\theta, \mathscr{M}(\Omega))$ form the constant part of the operator A. Due to Theorem 9.4 the inverse matrix A^{-1} also has the same block-matrix structure as (9.57) with entries A^{-1}_{ij} of the same size of A_{ij}. Accordingly, one has $V = (V_1, \ldots, V_l)$, where V_j, $j = 1, \ldots, l$, are vector-functions of length k_j. Let $V \in Ker(A^{-1})$, that is $A^{-1}V = 0$. It follows from matrix structure (9.57) of the inverse operator A^{-1} immediately that $A^{-1}_{ll}V_l = 0$. Further, since

$$A^{-1}_{l-1l-1}V_{l-1} + A^{-1}_{l-1,l}V_l = 0, \text{ and } A^{-1}_{l-1l} = -A^{-1}_{l-1l-1} \circ A_{l-1l} \circ A^{-1}_{ll},$$

which also follows from (9.57), one has

$$A^{-1}_{l-1l-1}V_{l-1} = -A^{-1}_{l-1l-1} \circ A_{l-1l} \circ A^{-1}_{ll}V_l = 0.$$

Consecutively, one obtains $A_{jj}V_j = 0, j = l-2, \ldots, 1$. This implies $Ker(A^{-1}) \subset Ker(A_0^{-1})$. Making use of these formulas on reverse order, we conclude that $Ker(A_0^{-1}) \subset Ker(A^{-1})$. Hence, $Ker(A^{-1}) = Ker(A_0^{-1})$. Therefore, it suffices to consider the equation $A_0^{-1}(D_z)V(z) = 0$. Due to isomorphic property of the Fourier transform, the latter is equivalent to the system of algebraic equations $\mathscr{A}_0^{-1}(\zeta)$ $F[V](\zeta) = 0$ with a parameter $\zeta \in \Omega$. Here $\mathscr{A}_0^{-1}(\zeta)$ is the symbol of A_0^{-1}. It is not hard to see that there exists a matrix $B(\zeta)$, $\det B(\zeta) \neq 0$, such that

$$\mathscr{A}_0^{-1}(\zeta)F[V](\zeta) = B(\zeta)\Big(H(\zeta)F[V](\zeta)\Big) = 0, \tag{9.58}$$

where $H(\zeta) \in \mathscr{O}(\Omega)$ is defined in a local representation of $\det(\mathscr{A}_0)$ given in equation (9.44). Recall a local representation of the meromorphic function $\det(\mathscr{A}_0^{-1}(\zeta)) = H(\zeta)/G(\zeta)$ (see (9.44)). To show (9.58) one can take $B(\zeta) = (H(\zeta))^{-1}\mathscr{A}_0^{-1}(\zeta)$. Then it can be easily verified that $\det(B(\zeta)) = \det(\mathscr{A}_0^{-1}(\zeta))$ $(G(\zeta))^{-1} \neq 0$, $\zeta \in \Omega$. Equation (9.58) means that the problem on description of the kernel of A^{-1} is reduced to equations

$$H(\zeta)F[V_j](\zeta) = 0, \quad j = 1, \ldots, N, \tag{9.59}$$

for each component $F[V_j]$ of the vector-function $F[V](\zeta)$, considered on the space of analytic functionals $\mathscr{O}^*(\Omega)$. Now let $\mathscr{P}_k^-$, $k = 1, \ldots, K_-$, be irreducible components of the analytic set $Z_{reg,\Omega} \cap P_{A^{-1}}$ with orders L_k^-. Then solutions to equation (9.59) have the form $F[V_j](\zeta) = \delta^{(\ell)}(\rho_k(\zeta))$, $\ell = 0, \ldots, L_k^-$, $k = 1, \ldots, K_-$, for each $j = 1, \ldots, N$, where $\rho_k(\zeta)$ locally represents $\mathscr{P}_k^-$. Taking the inverse Fourier transform, one has $V(\zeta) = f_{k,\ell}(z)\mathbf{v} \in Ker(A^{-1})$, where $f_{k,\ell}$ are defined in (9.56), and $\mathbf{v}$ is an arbitrary scalar vector, obtaining the desired result.

Corollary 9.3. *1. Let $\mathscr{A}(z,\zeta) \in S(M, \mathscr{O}(\Omega))$ with a Runge domain Ω and a matrix M, entries of which satisfy $m_{ij} \leq \mu_i - \mu_j$, $i, j = 1, \ldots, N$, for some collection $\mu_1, \ldots, \mu_N$. Then $\kappa_-(A, \Omega) = 0$;*

2. Let $\mathscr{A}(z,\zeta) \in S(M, \mathscr{O}(\Omega))$ with a Runge domain Ω and a matrix M, entries of which satisfy $m_{ij} \leq \mu_i - \mu_j$, $i, j = 1, \ldots, N$, for some collection $\mu_1, \ldots, \mu_N$. If $N_A \cap \Omega = \emptyset$, then $\kappa_+(A, \Omega) = 0$.

It is known [Chi89] that an analytic set in a neighborhood of any regular point represents an analytic submanifold (of co-dimension one in our case). Therefore, for $n \geq 2$ it follows from Theorem 9.6 that $\kappa_{\pm} = \infty$, as long as $\Omega \cap Z_{reg,A^{\pm 1}} \neq \emptyset$, and $\kappa_{\pm} = 0$, otherwise. Hence, if $n \geq 2$ only two possibilities may arise. It is not so in the one-dimensional case.

Theorem 9.7. *Let $n = 1$. Let L_k^+ and L_k^- be orders of poles $\zeta_k^+ \in P_A$, $k = 1\ldots, K_+$, and zeros $\zeta_k^- \in N_A$, $k = 1,\ldots,K_-$, respectively. Then,*

$$\kappa_+(A,\Omega) = \sum_{\zeta_k^- \in \Omega \cap N_A} L_k^- \text{ and } \kappa_-(A,\Omega) = \sum_{\zeta_k^+ \in \Omega \cap P_A} L_k^+. \tag{9.60}$$

Proof. In the one-dimensional case solutions of equation (9.59) are $F[V_j](\zeta) = g_{k,\ell}(\zeta) = \delta^{(\ell)}(\zeta - \zeta_k^+)$, $\ell = 0,\ldots,L_k^+$, $k = 1,\ldots,K_+$. Their Fourier inverses are $V_j(z) = f_{k,\ell}(z) = z^\ell e^{\zeta_k^+ z}$, $\ell = 0,\ldots,L_k^+$, $k = 1,\ldots,K_+$. Obviously, this set of functions is linearly independent. This implies the second formula in (9.60). Since $\kappa_+(A,\Omega) = \kappa_-(A^{-1},\Omega)$, the first formula is also correct.

Theorems 9.6 and 9.7 show that if Ω contains nonempty polar- or null-set of the symbol of a pseudo-differential operator, then the latter has a nontrivial kernel or co-kernel. Therefore, one needs factor-spaces to formulate a continuity statements in this case.

We will use traditional notations: if $\mathscr{X}$ is a generic topological space and K is its subspace, then $\mathscr{X}/K$ denotes the factor-space (with the topology of factor-space) of elements $\phi + \varphi$, where $\phi \in \mathscr{X}$ and $\varphi \in K$. Elements $\Phi = \phi + \varphi$ for all $\varphi \in K$ are considered identical. The conjugate $(\mathscr{X}/K)^*$ to a factor-space $\mathscr{X}/K$ consists of elements $G \in \mathscr{X}^*$ orthogonal to K : $< G, \varphi >= 0, \forall \varphi \in K$. We will denote the conjugate space $\mathscr{X}^*_{K^\perp}$.

Proposition 9.15. *Let $\mathscr{A}(z,\zeta)$ be a matrix-symbol with entries $\mathscr{A}_{ij} \in S(m_{ij}, \mathscr{M}(\Omega))$. Suppose that the collection of integers $\{\mu_1,\ldots,\mu_N\}$ satisfy inequalities $m_{ij} \leq \mu_j - \mu_i$ for all $i,j = 1,\ldots,N$. Then the pseudo-differential operators corresponding to symbols $\mathscr{A}(z,\zeta)$ and $\mathscr{A}^*(z,\zeta)$ are continuous as mappings*

$$A(z,D_z) : E^{\Omega}_{\bar{\mu},r}/Ker(A) \to E^{\Omega}_{\bar{\mu},r}/Ker(A^{-1});$$

$$A^*(z,D_z) : \left(E^{\Omega}_{\bar{\mu},r}\right)^*_{Ker(A)^\perp} \to \left(E^{\Omega}_{\bar{\mu},r}\right)^*_{Ker(A^{-1})^\perp}.$$

Proof. The proof follows from Theorem 9.6 and Proposition 9.14.

Consider the following examples illustrating Theorem 9.6 and 9.7.

Example 9.1. 1. Let a symbol $a \in S(m, \mathscr{O}(\Omega))$, where Ω is an arbitrary Runge domain. Then $\kappa_-(A,\Omega) = 0$, and hence the corresponding operator A is single-valued (uniquely defined).

2. Let $n = 1$ and $0 \in \Omega$. Let the symbol $a(\zeta) = 1/\zeta \in S(0, \mathscr{M}(\Omega))$. Then, $\kappa_-(A,\Omega) = 1$, and the corresponding operator $A(D_z) = D_z^{-1}$ (the primitive) is defined up

to an additive constant. Note that if $0 \notin \Omega$, then D_z^{-1} is uniquely defined and represents the "natural integral" (see [Dub96]):

$$D_z^{-1} f(z) = nat \int f(\zeta)d\zeta, \quad f \in Exp_\Omega(\mathbb{C}).$$

Now, suppose $n = 2$ and $a(\zeta_1, \zeta_2) = 1/\zeta_1$. Assume that Ω is a Runge domain containing $(0,0)$. Then $P_A = \{(\zeta_1, \zeta_2) \in \Omega : \zeta_1 = 0\}$. In this case the corresponding operator $A(D_{z_1}, D_{z_2}) = D_{z_1}^{-1}$ represents the integral with respect to the variable z_1 and is defined up to an arbitrary function of the variable z_2. Hence, $\kappa_-(A, \Omega) = \infty$.

9.7 Systems of pseudo-differential equations with meromorphic symbols

In this section we discuss the existence and uniqueness problems for general boundary value problem (9.7)–(9.8). We first consider a system of pseudo-differential equations

$$B(z, D_z)\Psi(z) = \Phi(z), \tag{9.61}$$

and the Cauchy problem for a system of first order evolution pseudo-differential equations

$$D_t V(t,z) = A(t,z,D_z)V(t,z) + H(t,z), \tag{9.62}$$

$$V(0,z) = V_0(z), \tag{9.63}$$

where $B(z, D_z) \in OPS(M, X)$, $A(t, z, D_z) \in OPS(M_1, X)$ for each fixed t; the space of symbols X, as well as vector-functions (functionals) $\Phi(z)$, $H(t,z)$, and $V_0(z)$ will be specified below.

Theorem 9.8. *Let $B(z, D_z) \in OPS(M, \mathcal{M}(\Omega))$ and assume that there exists a collection $\bar{\mu} = \mu_1, \ldots, \mu_N$, such that the entries of the matrix M satisfy the inequalities $m_{ij} \leq \mu_i - \mu_j$, $i, j = 1, \ldots, N$. Then for any vector-function $\Phi(z) \in E^\Omega_{\bar{\mu},r}/Ker(B^{-1})$ there exists a unique solution $\Psi(z)$ to system (9.61) in the factor-space $E^\Omega_{\bar{\mu},r}/Ker(B)$.*

Proof. Due to Proposition 9.15 the pseudo-differential operator $B = B(z, D_z)$ with the symbol $\mathcal{B}(z, \zeta) \in S(M, \mathcal{M}(\Omega))$ is well defined in the space $E^\Omega_{\bar{\mu},r}/Ker(B)$. In accordance with Theorem 9.4 there exists the inverse symbol $\mathcal{B}^{-1}(z, \zeta) \in S(M, \mathcal{M}(\Omega))$. The corresponding inverse operator $B^{-1} = B^{-1}(z, D_z)$ is well defined in the space $E^\Omega_{\bar{\mu},r}/Ker(B^{-1})$. Let $\Phi(z) \in E^\Omega_{\bar{\mu},r}/Ker(B^{-1})$, i.e., $\Phi(z) = \phi(z) + \varphi(z)$, where $\phi \in E^\Omega_{\bar{\mu},r}$, and $\varphi \in Ker(B^{-1})$. Now one can show that $\Psi(z) = B^{-1}(z, Dz)\Phi(z) + \psi(z)$, for arbitrary $\psi \in Ker(B)$, solves the system (9.61). Indeed,

$$B(z,D_z)\Psi(z) = B(z,D_z)\Big(B^{-1}(z,D_z)\Phi(z)+\psi\Big)$$
$$= \Phi(z) + B(z,D_z)\psi(z) = \Phi(z).$$

Theorem 9.9. *Let $B(z,D_z) \in OPS(M,\mathscr{M}(\Omega))$ and there exists a collection $\bar{\mu} = \mu_1,\dots,\mu_N$, such that the entries of the matrix M satisfy the inequalities $m_{ij} \leq \mu_j - \mu_i, i,j = 1,\dots,N$. Then for any $\Phi(z) \in \left(E^{\Omega}_{\bar{\mu},r}\right)^*_{Ker(B^*)^{\perp}}$ there exists a unique weak solution $\Psi(z)$ to system (9.61) in the space $\left(E^{\Omega}_{\bar{\mu},r}\right)^*_{Ker((B^*)^{-1})^{\perp}}$.*

Proof. Let $\Phi(z) \in \left(E^{\Omega}_{\bar{\mu},r}\right)^*_{Ker(B^*)^{\perp}}$. Then for arbitrary $U \in E^{\Omega}_{\bar{\mu},r}/Ker(B^*)$ one has

$$\langle B(z,D_z)\Psi(z), U(z)\rangle = \langle \Phi(z), U(z)\rangle,$$

or

$$\langle \Psi(z), B^*(z,D_z)U(z)\rangle = \langle \Phi(z), U(z)\rangle.$$

Due to Proposition 9.15 the operator $B^*(z,D_z)$ is continuous from $E^{\Omega}_{\bar{\mu},r}/Ker(B^*)$ to the space $E^{\Omega}_{\bar{\mu},r}/Ker(B^*)^{-1}$. Note that due to Theorem 9.4 there exists the inverse symbol $\mathscr{B}^{-1}(z,\zeta) \in S(M,\mathscr{M}(\Omega))$, and hence, the corresponding inverse operator $(B^*)^{-1}$ exists and well defined in the space $E^{\Omega}_{\bar{\mu},r}/Ker(B^*)^{-1}$. Therefore, if one sets $B^*(z,D_z)U(z) = V(z)$, where $V(z) \in E^{\Omega}_{\bar{\mu},r}/Ker(B^*)^{-1}$, then due to Theorem 9.8 one has $U(z) = (B^*(z,D_z))^{-1}V(z)$. This implies that the functional $\Psi(z)$ defined by

$$\langle \Psi(z), V(z)\rangle = \langle \Phi(z), (B^*(z,D_z))^{-1}V(z)\rangle = \langle \left((B^*(z,D_z))^{-1}\right)^* \Phi(z), V(z)\rangle \quad (9.64)$$

solves system (9.61) in the weak sense. Representation (9.64) also shows that for the inverse the formula $B^{-1}(z,D_z) = \left((B^*(z,D_z))^{-1}\right)^*$ holds, and $\langle \Psi(z), f(z)\rangle = 0$ if $f \in Ker(B^*)^{-1}$.

If one considers the operator $A \in OPS(m,\mathscr{M}(\Omega))$ in the space $E^{\Omega\setminus P_A}_{\bar{\mu},r}$, then it follows from the definition (9.50) of a pseudo-differential operator with a meromorphic symbol, that the polar set of the symbol of A does not intersect with $\Omega \setminus P_A$. This implies that the symbol belongs to $S(m,\mathscr{O}_{\mu,r})$. In this case $Ker(A) = \{0\}$, and therefore, the above theorems take the form:

Theorem 9.10. *Let $B(z,D_z) \in OPS(M,\mathscr{M}(\Omega))$ and assume that there exists a collection $\bar{\mu} = \mu_1,\dots,\mu_N$, such that the entries of the matrix M satisfy the inequalities $m_{ij} \leq \mu_i - \mu_j, i,j = 1,\dots,N$. Then for any vector-function $\Phi(z) \in E^{\Omega\setminus N_B}_{\bar{\mu},r}$ there exists a unique solution $\Psi(z)$ to system (9.61) in the space $E^{\Omega\setminus P_B}_{\bar{\mu},r}$.*

Theorem 9.11. *Let $B(z,D_z) \in OPS(M,\mathscr{M}(\Omega))$ and assume that there exists a collection $\bar{\mu} = \mu_1,\dots,\mu_N$, such that the entries of the matrix M satisfy the inequalities $m_{ij} \leq \mu_j - \mu_i, i,j = 1,\dots,N$. Then for any $\Phi(z) \in \left(E^{\Omega\setminus P_B}_{\bar{\mu},r}\right)^*$ there exists a unique weak solution $\Psi(z)$ to system (9.61) in the space $\left(E^{\Omega\setminus N_B}_{\bar{\mu},r}\right)^*$.*

Remark 9.6. 1. In Theorems 9.10 and 9.11 one can replace the spaces $E_{\bar{\mu},r}^{\Omega\setminus N_B}$, $E_{\bar{\mu},r}^{\Omega\setminus P_B}$, and their conjugates by the spaces $Exp_{\Omega\setminus N_B}(\mathbb{C}^n)$, $Exp_{\Omega\setminus P_B}(\mathbb{C}^n)$ defined in Section 9.4 and their respective conjugates.
2. Similar to the proof of Theorem 9.4 one can show that the conditions $m_{ij} \leq \mu_i - \mu_j$ in the above theorems are also necessary for existence of a solution.

Using formulas (9.38) and (9.39) and the scheme (see Theorem 9.3)

$$\mathscr{O}_{\bar{\mu},r}(\Omega) \underset{F^{-1}}{\overset{F}{\rightleftarrows}} \left(E_{\bar{\mu},r}^{\Omega}\right)^*$$

one can obtain dual results in terms of the Fourier transform. Namely, applying the Fourier transform to equation (9.61), one has

$$B(D_\zeta,\zeta)H(\zeta) = G(\zeta), \quad \zeta \in \Omega \setminus P_B,$$

or, the same

$$\sum_{j=1}^{N} \sum_{|\alpha|\leq m_{ij}} (-1)^\alpha a_{ij\alpha}(\zeta) D_\zeta^\alpha h_j(\zeta) = g_i(\zeta), \quad \zeta \in \Omega \setminus P_B, i = 1,\ldots,N, \qquad (9.65)$$

where $H(\zeta) = (h_1(\zeta),\ldots,h_N(\zeta))^T$ and $G(\zeta) = (g_1(\zeta),\ldots,g_N(\zeta))^T$.

Theorem 9.12. *Let the matrix-symbol $B(z,\zeta) \in S(M,\mathscr{M}(\Omega))$ and there exists a collection $\bar{\mu} = \mu_1,\ldots,\mu_N$, such that the entries of the matrix M satisfy the inequalities $m_{ij} \leq \mu_j - \mu_i, i,j = 1,\ldots,N$. Then for any vector-function $G(\zeta) \in \mathscr{O}_{\bar{\mu},r}(\Omega \setminus P_B)$ there exists a solution $\Psi(z)$ to system (9.65) in the space $\mathscr{O}_{\bar{\mu},r}(\Omega \setminus N_B)$.*

Proof. Consider the system (9.65) in the scale of spaces $\mathscr{O}_{\bar{\mu},r}(\Omega)$. Applying the inverse Fourier transform F^{-1} we have

$$B(z,D_z)F^{-1}[H](z) = F^{-1}[G](z) \qquad (9.66)$$

in the scale of spaces $\left(E_{\bar{\mu},r}^{\Omega}\right)^*$. In accordance with Theorem 9.11, under the condition of our theorem, for any $F^{-1}[G] \in \left(E_{\bar{\mu},r}^{\Omega\setminus P_B}\right)^*$ there is a unique solution $F^{-1}[H] \in \left(E_{\bar{\mu},r}^{\Omega\setminus N_B}\right)^*$ to system (9.66). Now applying the Fourier transform and using isomorphism $F : \left(E_{\bar{\mu},r}^{\Omega\setminus N_B}\right)^* \to \mathscr{O}_{\bar{\mu},r}(\Omega \setminus N_B)$ one obtains the desired result.

Similarly, using the scheme (see Theorem 9.3)

$$\mathscr{O}_{\bar{\mu},r}^*(\Omega) \underset{F}{\overset{F^{-1}}{\rightleftarrows}} E_{\bar{\mu},r}^{\Omega}$$

we can establish the existence of a solution of the system (9.65) in the space of analytic functionals.

Theorem 9.13. *Let $B(z,D_z) \in OPS(M,\mathscr{M}(\Omega))$ and assume that there exists a collection $\bar{\mu} = \mu_1,\dots,\mu_N$, such that the entries of the matrix M satisfy the inequalities $m_{ij} \le \mu_i - \mu_j$, $i,j = 1,\dots,N$. Then for any vector-functional $\Phi(z) \in \mathscr{O}^*_{\bar{\mu},r}(\Omega \setminus N_B)$ there exists a solution $\Psi(z)$ to system (9.65) in the space $\mathscr{O}^*_{\bar{\mu},r}(\Omega \setminus P_B)$.*

Now assume that $\mathscr{D} \subset \mathbb{C}$ is a domain containing t_0 and $\mathscr{X}$ be a topological vector space. Below we use the spaces of the form $\mathscr{O}[\mathscr{D};\mathscr{X}]$, elements $f(t)$ of which for each fixed t belong to $\mathscr{X}$ and analytic in the variable t in the topology of $\mathscr{X}$.

Theorem 9.14. *Let $A = A(t,z,D_z) \in \mathscr{O}\left[\mathscr{D};OPS(M,\mathscr{M}(\Omega))\right]$ and there exists a collection $\bar{\mu} = \mu_1,\dots,\mu_N$, such that the entries of the matrix M satisfy the inequalities $m_{ij} \le \mu_i - \mu_j + 1$, $i,j = 1,\dots,N$. Then there exist numbers $r > 0$ and $\sigma > 0$ such that for any vector-functions $H(t,z) \in \mathscr{O}\left[\mathscr{D};E^{\Omega\setminus P_A}_{\bar{\mu},r+\sigma|t-t_0|}\right]$, and $V_0(z) \in E^{\Omega\setminus P_A}_{\bar{\mu},r}$ a unique solution $V(t,z)$ to the Cauchy problem (9.62)–(9.63) exists in a δ-neighborhood of t_0 and belongs to the factor-space $\mathscr{O}\left[|t-t_0| < \delta;E^{\Omega\setminus P_A}_{\bar{\mu},r+\sigma|t-t_0|}\right]$.*

Proof. The Cauchy problem (9.62)–(9.63) can be written in the equivalent integro-differential form

$$V(t,z) = V_0(z) + \int_{t_0}^{t} A(\tau,z,D_z)V(\tau,z)d\tau + \int_{t_0}^{t} H(\tau,z)d\tau.$$

Consider the operator

$$\mathbf{A}V(t,z) = \int_{t_0}^{t} A(\tau,z,D_z)V(\tau,z)d\tau.$$

For i-th component of this operator one has

$$\begin{aligned}(\mathbf{A}V(t,z))_i &= \sum_{j=1}^{N}\int_{t_0}^{t} A_{ij}(\tau,z,D_z)V_j(\tau,z)d\tau \\ &= \sum_{j=1}^{N}\int_{t_0}^{t}\left(\sum_{|\alpha|\le m_{ij}} z^{\alpha}a_{ij\alpha}(\tau,D_z)V_j(\tau,z)\right)d\tau. \end{aligned} \tag{9.67}$$

In order to prove the theorem it suffices to show the existence of a unique solution for arbitrary fiber of the space $E^{\Omega\setminus P_A}_{\bar{\mu},r+\sigma|t-t_0|}$. Let $\zeta_0 \in \Omega \setminus P_A$ be an arbitrary fixed point, and consider equation (9.67) in the fiber $\mathscr{E}_{\bar{\mu},r+\sigma|t-t_0|,\zeta_0}$. Since ζ_0 is located out of the polar set of the operator $A(t,z,D_z)$, the symbol of this operator belongs to $\mathscr{O}_{\bar{\mu},r,\zeta_0}$. Therefore, making use of Proposition 9.11 and taking into account the evolution of $V(t,z)$ over the scale $\mathscr{O}\left[\mathscr{D};\mathscr{E}_{\bar{\mu},r+\sigma|t-t_0|,\zeta_0}\right]$, one obtains the estimate

$$\begin{aligned}|(\mathbf{A}V(t,z))_i| &\le \sum_{j=1}^{N}\sum_{|\alpha|\le m_{ij}}|z|^{|\alpha|}\int_{0}^{|t-t_0|}|a_{ij\alpha}(\tau,D_z)V_j(\tau,z)||d\tau| \\ &\le I(|t-t_0|)\sum_{j=1}^{N}(1+|z|)^{m_{ij}+\mu_j}\sup_{t\in\mathscr{D}}\|V_j\|_{\mu_j,r+\sigma|t-t_0|,\zeta_0}\sum_{|\alpha|\le m_{ij}}\sup_{t\in\mathscr{D}}[a_{ij\alpha}]_{\mu_j,r,\zeta_0},\end{aligned}$$

where

$$I(|t-t_0|) = \int_0^{|t-t_0|} e^{(r+\sigma|\tau-t_0|)|z|}|d\tau| \leq \frac{|t-t_0|}{\sigma|z|} e^{(r+\sigma|t-t_0|)|z|}.$$

Taking this and the inequality $m_{ij} + \mu_j \leq \mu_i + 1, i, j = 1, \ldots, N$, into account, one has

$$|(\mathbf{A}V(t,z))_i| \leq \frac{|t-t_0|}{\sigma}(1+|z|)^{\mu_i} e^{r+\sigma|t-\tau|} \sum_{j=1}^{N} \sup_{t\in\mathscr{D}} \|V_j(t,z)\|_{\mu_j, r+\sigma|t-t_0|,\zeta_0} \sup_{t\in\mathscr{D}} [\mathscr{A}_{ij}]_{\mu_j,r,\zeta_0}.$$

This implies

$$\|(\mathbf{A}V(t,z))_i\|_{\mu_i r+\sigma|t-t_0|,\zeta_0} \leq \frac{|t-t_0|}{\sigma} \sup_{t\in\mathscr{D}} \|V(t,z)\|_{\bar{\mu},r+\sigma|t-t_0|,\zeta_0} \max_{1\leq j\leq N} \sup_{t\in\mathscr{D}} \|A_{ij}(t,z,D_z)\|.$$

Now summing up by index $i = 1, \ldots, N$, we have

$$\|(\mathbf{A}V(t,z))\|_{\bar{\mu},r+\sigma|t-t_0|,\zeta_0} \leq \frac{|t-t_0|}{\sigma} \sup_{t\in\mathscr{D}} \|A(t,z,D_z)\| \sup_{t\in\mathscr{D}} \|V(t,z)\|_{\bar{\mu},r+\sigma|t-t_0|,\zeta_0}.$$

It follows from this estimate that $\mathbf{A}$ is a contraction operator if the condition

$$|t-t_0| \sup_{t\in\mathscr{D}} \|A(t,z,D_z)\| \leq \sigma$$

holds. Hence, taking $\delta < \sigma / \sup_{t\in\mathscr{D}} \|A(t,z,D_z)\|$ we have that in the δ-neighborhood of t_0 a unique solution to the Cauchy problem (9.62)–(9.63) exists.

Theorem 9.15. *Let $A = A(t,z,D_z) \in \mathscr{O}[\mathscr{D}; M, \mathscr{M}(\Omega)]$ and assume that there exists a collection $\bar{\mu} = \mu_1, \ldots, \mu_N$, such that the entries of the matrix M satisfy the inequalities $m_{ij} \leq \mu_j - \mu_i + 1$, $i, j = 1, \ldots, N$. Then there exist numbers $r > 0$ and $\sigma > 0$ such that for any vector-functionals $H(t,z) \in \mathscr{O}\left[\mathscr{D}; \left(E_{\bar{\mu},r}^{\Omega\setminus P_A}\right)^*\right]$ and $V_0(z) \in \left(E_{\bar{\mu},r}^{\Omega\setminus P_B}\right)^*$ there exists a unique solution $V(t,z)$ to the Cauchy problem* (9.62)–(9.63) *in the space $\mathscr{O}\left[|t-t_0| < \delta; \left(E_{\bar{\mu},r}^{\Omega\setminus P_B}\right)^*\right]$ with some $\delta > 0$.*

Proof. Since the proof follows from Theorem 9.14 by duality, we only briefly sketch its idea. Let $V(t,z) \in\in \mathscr{O}\left[\mathscr{D}; \left(E_{\bar{\mu},r}^{\Omega\setminus P_A}\right)^*\right]$ and $v(t,z) \in \mathscr{O}\left[\mathscr{D}; E_{\bar{\mu},r+\sigma|t-t_0|}^{\Omega\setminus P_A}\right]$. Then the relation

$$D_t\langle V(t,z), v(t,z)\rangle = \langle D_t V(t,z), v(t,z)\rangle + \langle V(t,z), D_t v(t,z)\rangle$$

implies

$$\langle V(t,z), v(t,z)\rangle = \langle V(t_0,z), v(t_0,z)\rangle + \int_{t_0}^{t} \langle D_s V(s,z), v(s,z)\rangle ds + \int_{t_0}^{t} \langle V(s,z), D_s v(s,z)\rangle ds.$$

The latter due to equation (9.62) and the initial condition in (9.63) takes the form

$$\begin{aligned}\langle V(t,z),v(t,z)\rangle &= \langle \Phi(z),v(t_0,z)\rangle + \int_{t_0}^{t} \langle A(s,z,D_z)V(s,z)+H(s,z),v(s,z)\rangle ds \\ &+ \int_{t_0}^{t} \langle V(s,z),D_s v(s,z)\rangle ds \\ &= \langle \Phi(z),v(0,z)\rangle + \int_{t_0}^{t} \langle V(s,z),D_s v(s,z)+A^*(s,z,D_z)v(s,z)\rangle ds \\ &+ \int_{t_0}^{t} \langle H(s,z),v(s,z)\rangle ds. \end{aligned} \tag{9.68}$$

Since the latter is valid for arbitrary $v(t,z)$, it is also valid for $v(t,\tau,z)$, which solves the Cauchy problem

$$D_\tau v(t,\tau,z) + A^*(\tau,z,D_z)v(\tau,z) = 0, \quad t_0 < \tau < t, \tag{9.69}$$

$$v(t,\tau,z)_{|\tau=t} = v(t,z). \tag{9.70}$$

For the symbol of the adjoint operator $A^*(t,z,D_z)$ the order-matrix m^*_{ij} satisfies the inequality $m^*_{ij} \le \mu_i - \mu_j + 1, i,j = 1,\dots,N$. Therefore, in accordance with Theorem 9.14 the Cauchy problem (9.69)–(9.70) has a unique solution in the space $\mathscr{O}\left[|t-t_0| < \delta; E^{\Omega\setminus P_A}_{\bar{\mu},r+\sigma|t-t_0|}\right]$ for any fixed $v(t,z)$, if $|t-t_0| < \delta$, where $\delta > 0$ small enough. Substituting $v(t,z)$ in equation (9.68) by $v(t,\tau,z)$, we have

$$\langle V(t,z),v(t,z)\rangle = \langle \Phi(z),v(t,t_0,z)\rangle + \int_{t_0}^{t} \langle H(s,z),v(t,\tau,z)\rangle d\tau, \quad |t-t_0| < \delta. \tag{9.71}$$

The functional $V(t,z)$ defined by (9.71) is a unique solution to the Cauchy problem (9.62)–(9.63). It can be readily seen that $V(t,z) \in \mathscr{O}\left[|t-t_0| < \delta; \left(E^{\Omega\setminus N_B}_{\bar{\mu},r}\right)^*\right]$, and hence is a desired solution.

Now consider general boundary value problems for the first order systems

$$D_t V(t,z) = A(t,z,D_z)V(t,z) + H(t,z), \tag{9.72}$$

$$B(z,D_z)V(t,z)_{|t=0} = \Phi(z), \tag{9.73}$$

This problem can be reduced to the equivalent Cauchy problem for system (9.72) with the initial condition

$$V(0,z) = \Psi(z),$$

where $\Psi(z)$ is a solution to the system of pseudo-differential equations

$$B(z,D_z)\Psi(z) = \Phi(z).$$

Combining the above proved Theorems 9.8 and 9.14 (in the dual case Theorems 9.9 and 9.15) one can prove the following statements.

Theorem 9.16. *Let operators* $A = A(t,z,D_z) \in \mathcal{O}\big[\mathcal{D}; OPS(M,\mathcal{M}(\Omega))\big]$ *and* $B(z,D_z) \in OPS(\mathcal{N},\mathcal{M}(\Omega))$. *Suppose there exists a collection* $\bar{\mu} = \mu_1,\dots,\mu_N$, *such that*

i) the entries of the matrix M *satisfy the inequalities* $m_{ij} \le \mu_i - \mu_j + 1, i,j = 1,\dots,N$;
ii) the entries of the matrix $\mathcal{N}$ *satisfy the inequalities* $n_{ij} \le \mu_i - \mu_j, i, j = 1,\dots,N$.

Then there exist numbers $r > 0$ *and* $\sigma > 0$ *such that for any vector-functions* $H(t,z) \in \mathcal{O}\left[\mathcal{D}; E^{\Omega\setminus P_A}_{\bar{\mu},r+\sigma|t-t_0|}\right]$, *and* $\Phi(z) \in E^{\Omega}_{\bar{\mu},r}/Ker(B^{-1})$ *a unique solution* $V(t,z)$ *to the Cauchy problem* (9.72)–(9.73) *exists in a* δ*-neighborhood of* t_0 *and belongs to the space* $\mathcal{O}\left[|t-t_0| < \delta; E^{\Omega\setminus P_A}_{\bar{\mu},r+\sigma|t-t_0|}\right]$. *Moreover, the kernel of this problem is isomorphic to the kernel of the operator* $B(z,D_z)$.

Theorem 9.17. *Let operators* $A = A(t,z,D_z) \in \mathcal{O}\big[\mathcal{D}; OPS(M,\mathcal{M}(\Omega))\big]$ *and* $B(z,D_z) \in OPS(\mathcal{N},\mathcal{M}(\Omega))$. *Suppose there exists a collection* $\bar{\mu} = \mu_1,\dots,\mu_N$, *such that*

i) the entries of the matrix M *satisfy the inequalities* $m_{ij} \le \mu_j - \mu_i + 1, i, j = 1,\dots,N$;
ii) the entries of the matrix $\mathcal{N}$ *satisfy the inequalities* $n_{ij} \le \mu_j - \mu_i, i,j = 1,\dots,N$.

Then there exist numbers $r > 0$ *and* $\sigma > 0$ *such that for any vector-functionals* $H(t,z) \in \mathcal{O}\left[\mathcal{D}; \left(E^{\Omega\setminus P_A}_{\bar{\mu},r+\sigma|t-t_0|}\right)^*\right]$, *and* $\Phi(z) \in \left(E^{\Omega\setminus P_A}_{\bar{\mu},r}\right)^*_{Ker(B^*)^\perp}$ *a unique solution* $V(t,z)$ *to the Cauchy problem* (9.72)–(9.73) *exists in a* δ*-neighborhood of* t_0 *and belongs to the space* $\mathcal{O}\left[|t-t_0| < \delta; \left(E^{\Omega\setminus P_A}_{\bar{\mu},r+\sigma|t-t_0|}\right)^*_{Ker((B^*)^{-1})^\perp}\right]$.

As an example of application of these theorems consider the following boundary value problem for a pseudo-differential equation of higher order

$$D_t^m u(t,z) + \sum_{k=0}^{m-1} A_k(t,z,D_z) D_t^k u(t,z) = h(t,z), \quad t \in \mathcal{D}, z \in \mathbb{C}^n, \tag{9.74}$$

$$\sum_{j=0}^{m-1} B_{ij}(z,D_z) D_t^j u(t,z)\Big|_{t=0} = \varphi_i(z), \quad z \in \mathbb{C}^n, i = 0,\dots,m-1, \tag{9.75}$$

where $A_k(t,z,D_z)$, $k = 0,\dots,m-1$, are pseudo-differential operators with symbols

$$\mathcal{A}_k(t,z,\zeta) = \sum_{|\alpha|\le m_k} a_{k\alpha}(t,\zeta) z^\alpha.$$

This problem is equivalent to the following system:

$$D_t v(t,z) + \tilde{A}(t,z,D_z) v(t,z) = H(t,z),$$
$$B(z,D_z) v(t,z)\Big|_{t=0} = \phi(z),$$

where the vector-functions $v(t,z) = (u(t,z),\dots,u_t^{(m-1)}(t,z))^T$, $H(t,z) = (0,\dots,h(z))^T$, $\phi(z) = (\varphi_0(z),\dots,\varphi_{m-1}(z))^T$, and the operator $\tilde{A}(t,z,D_z)$ has the matrix-symbol with entries

$$\tilde{A}_{ij}(t,z,\zeta) = \begin{cases} 1, & \text{if } j = i+1, i = 0,\dots,m-2, \\ A_j(t,z,\zeta), & \text{if } i = m,\ j = 0\dots,m-1, \\ \theta, & \text{otherwise.} \end{cases}$$

In the matrix form

$$\tilde{A}(t,z,\zeta) = \begin{bmatrix} \theta & 1 & \dots & \theta & \theta \\ \theta & \theta & \dots & \theta & \theta \\ & \dots & \dots & & \\ \theta & \theta & \dots & \theta & 1 \\ A_0 & A_1 & \dots & A_{m-2} & A_{m-1} \end{bmatrix},$$

Applying Theorem 9.16 one has $\mu_j = j, j = 0,\dots,m-1$. Therefore, boundary value problem (9.74)–(9.75) have a local solution in the scale $E^{\Omega\setminus P_A}_{\bar{\mu},r+\sigma|t-t_0|}$, $\bar{\mu} = (0,\dots,m-1)$, if the polynomial degrees m_k and m_{ij} of symbols $\mathscr{A}_k(t,z,\zeta)$ and $\mathscr{B}_{ij}(z,\zeta)$, satisfy, respectively, the following inequalities:

$$m_k \le m-k, \quad k = 0,\dots,m-1,$$

and

$$m_{ij} \le i-j, \quad i,j = 0,\dots,m-1.$$

9.8 Reduction to a system of first order

The general system of pseudo-differential equations (9.7) can be reduced to a system of first order of the form (9.72). Boundary condition (9.8) in this process also changes to the form (9.73). We prove the following statement:

Lemma 9.1. *Let a vector-function $u(t,x) = (u_1(t,x),\dots,u_N(t,x))$ solve the general problem* (9.7)–(9.8)*. Then the vector-function*

$$V(t,x) = (u_1(t,x),\dots,D_t^{p_1-1}u_1(t,x),\dots,u_N(t,x),\dots,D_t^{p_N-1}u_N(t,x))$$

of length $p_1+\dots+p_N$ solves a problem of the form (9.72)–(9.73)*, with vector-functions $H(t,z)$ and $\Phi(z)$*

$$H(t,z) = (\mathbf{h}_1(t,z),\dots,\mathbf{h}_N(t,z)),\ \mathbf{h}_j(t,z) = (0,0,\dots,f_j(t,z)), \tag{9.76}$$

$$\Phi(z) = (\phi_1(z),\dots,\phi_N(z)),\ \phi_j(z) = (\varphi_{j0}(z),\dots,\varphi_{jp_j-1}), \tag{9.77}$$

where $\mathbf{h}_j(t,z)$ *is a vector of length* p_j *with only nonzero* p_j*-th component* $f_j(t,z)$*; and the matrix-operators* $A(t,z,D_z) = \mathbf{A}_{ij}(t,z,D_z)$ *and* $B(z,D_z) = \mathbf{B}_{ij(z,D_z)}$, $i,j = 1,\dots,N$, *are block-matrices with respective blocks of sizes* $p_i \times p_j$ *:*

$$\mathbf{A}_{ij}(t,z,D_z) = \begin{cases} \begin{bmatrix} \theta & 1 & \dots & \theta & \theta \\ \theta & \theta & \dots & \theta & \theta \\ & \dots & \dots & . & \\ \theta & \theta & \dots & \theta & 1 \\ A_{jj}^0 & A_{jj}^1 & \dots & A_{jj}^{p_j-2} & A_{jj}^{p_j-1} \end{bmatrix}, & \text{if } i = j, \\ \begin{bmatrix} \theta & \theta & \dots & \theta & \theta \\ \theta & \theta & \dots & \theta & \theta \\ & \dots & \dots & & \\ \theta & \theta & \dots & \theta & \theta \\ A_{ij}^0 & A_{ij}^1 & \dots & A_{ij}^{p_j-2} & A_{ij}^{p_j-1} \end{bmatrix}, & \text{if } i \neq j. \end{cases} \tag{9.78}$$

and

$$\mathbf{B}_{i,j}(z,D_z) = \begin{bmatrix} B_{ij}^{00} & B_{ij}^{01} & \dots & B_{ij}^{0p_j-1} \\ B_{ij}^{10} & B_{ij}^{11} & \dots & B_{ij}^{1p_j-1} \\ & \dots & \dots & \\ B_{ij}^{p_i-10} & B_{ij}^{p_i-11} & \dots & B_{ij}^{p_i-1p_j-1} \end{bmatrix}. \tag{9.79}$$

Proof. In accordance with the definition of the vector-function $V(t,z)$, it can be represented in the form $V(t,z) = (\mathbf{v}_1(t,z),\dots,\mathbf{v}_N(t,z))$, where

$$\mathbf{v}_1(t,z) = (v_1,\dots,v_{p_1}) \equiv \left(u_1(t,z),\dots,D_t^{p_1-1}u_1(t,z)\right),$$
$$\mathbf{v}_2(t,z) = (v_{p_1+1},\dots,v_{p_1+p_2}) \equiv \left(u_2(t,z),\dots,D_t^{p_2-1}u_2(t,z)\right),$$
$$\dots$$
$$\mathbf{v}_N(t,z) = (v_{p_1+\dots+p_{N-1}+1},\dots,v_{p_1+\dots+p_N}) \equiv \left(u_N(t,z),\dots,D_t^{p_N-1}u_N(t,z)\right).$$

This together with equation (9.7) implies that

$$D_t v_1(t,z) = v_2(t,z),$$
$$D_t v_2(t,z) = v_3(t,z),$$
$$\dots$$
$$D_t v_{p_1-1}(t,z) = v_{p_1}(t,z),$$
$$D_t v_{p_1}(t,z) = \sum_{j=1}^{N} \left[A_{1j}^0 v_{p_1+\dots+p_{j-1}+1}(t,z) + \dots + A_{1j}^{p_j-1} v_{p_1+\dots+p_j}(t,z)\right] + f_1(t,z),$$
$$\dots\dots\dots\dots$$

$$D_t v_{p_1+\cdots+p_{N-1}+1}(t,z) = v_{p_1+\cdots+p_{N-1}+2}(t,z),$$
$$D_t v_{p_1+\cdots+p_{N-1}+2}(t,z) = v_{p_1+\cdots+p_{N-1}+3}(t,z),$$
$$\cdots$$
$$D_t v_{p_1+\cdots+p_N-1}(t,z) = v_{p_1+\cdots+p_N}(t,z),$$
$$D_t v_{p_1+\cdots+p_N}(t,z) = \sum_{j=1}^{N}\left[A^0_{Nj} v_{p_1+\cdots+p_{j-1}+1}(t,z) + \cdots + A^{p_j-1}_{Nj} v_{p_1+\cdots+p_j}(t,z)\right] + f_N(t,z).$$

These equations show that the vector-function $V(t,z)$ satisfies equation (9.72) with the operator $A(t,z,D_z)$ in (9.78) and $H(t,z)$ in (9.76). Similarly, one can show that $V(t,z)$ satisfies boundary conditions (9.73) with the operator $B(z,D_z)$ in (9.79) and vector-function $\Phi(z)$ in (9.77).

9.9 Existence theorems for general boundary value problems

Theorem 9.18. *Let operators* $A \equiv \{A^q_{jk}(t,z,D_z)\} \in \mathscr{O}\left[\mathscr{D}; OPS(M^q, \mathscr{M}(\Omega))\right]$ *and* $B \equiv \{B^{mq}_{jk}(z,D_z)\} \in OPS(\mathscr{N}^{mq}, \mathscr{M}(\Omega))$. *Suppose there exists a collection* $\bar{\mu} = \mu_1, \ldots, \mu_N$, *such that*

i) the entries of the matrices $M^q, q = 0, \ldots, p_j - 1$, *satisfy the inequalities*

$$m^q_{jk} \le \mu_j - \mu_k + p_j - q, \quad j,k = 1,\ldots,N,\ q = 0,\ldots,p_j - 1;$$

ii) the entries of the matrix $\mathscr{N}^{mq}$, $m = 0, \ldots, p_j - 1, q = 0, \ldots, p_k - 1$, *satisfy the inequalities*

$$n^{mq}_{jk} \le \mu_j - \mu_k + m - q, \quad j,k = 1,\ldots,N,\ m = 0,\ldots,p_j - 1,\ q = 0,\ldots,p_k - 1.$$

Then there exist numbers $r > 0$ *and* $\sigma > 0$ *such that for any vector-functions* $H(t,z) \in \mathscr{O}\left[\mathscr{D}; E^{\Omega\setminus P_A}_{\bar{\mu},r+\sigma|t-t_0|}\right]$, *and* $\Phi(z) \in E^{\Omega\setminus P_A}_{\bar{\mu},r}/Ker(B^{-1})$ *a solution* $V(t,z)$ *to boundary value problem (9.7)–(9.8) exists in a* δ*-neighborhood of* t_0 *and belongs to the space* $\mathscr{O}\left[|t-t_0| < \delta; E^{\Omega\setminus P_A}_{\bar{\mu},r+\sigma|t-t_0|}\right]$. *Moreover, the kernel of this problem is isomorphic to the kernel of the operator B.*

Proof. Applying Lemma 9.1 we can reduce problem (9.7)–(9.8) to the first order system of the form (9.72)–(9.73). Now the proof follows immediately due to Theorem 9.16.

The theorem below follows from the previous by duality.

Theorem 9.19. *Let operators* $A \equiv \{A^q_{jk}(t,z,D_z)\} \in \mathscr{O}\left[\mathscr{D}; OPS(M^q, \mathscr{M}(\Omega))\right]$ *and* $B \equiv \{B^{mq}_{jk}(z,D_z)\} \in OPS(\mathscr{N}^{mq}, \mathscr{M}(\Omega))$. *Suppose there exists a collection* $\bar{\mu} = \mu_1, \ldots, \mu_N$, *such that*

i) the entries of the matrices $M^q, q = 0,\dots,p_j-1$, satisfy the inequalities

$$m^q_{jk} \le \mu_k - \mu_j + p_j - q, \quad j,k = 1,\dots,N,\ q = 0,\dots,p_j-1;$$

ii) the entries of the matrix $\mathcal{N}^{mq}$, $m = 0,\dots,p_j-1, q = 0,\dots,p_k-1$, satisfy the inequalities

$$n^{mq}_{jk} \le \mu_k - \mu_j + m - q, \quad j,k = 1,\dots,N,\ m = 0,\dots,p_j-1,\ q = 0,\dots,p_k-1.$$

Then there exist numbers $r > 0$ and $\sigma > 0$ such that for any vector-functionals $H(t,z) \in \mathcal{O}\left[\mathcal{D}; \left(E^{\Omega\setminus P_A}_{\bar{\mu},r+\sigma|t-t_0|}\right)^*\right]$, and $\Phi(z) \in \left(E^{\Omega\setminus P_A}_{\bar{\mu},r}\right)^*_{Ker(B^*)^\perp}$ a solution $V(t,z)$ to the Cauchy problem (9.7)–(9.8) exists in a δ-neighborhood of t_0 and belongs to the space $\mathcal{O}\left[|t-t_0| < \delta; \left(E^{\Omega\setminus P_A}_{\bar{\mu},r+\sigma|t-t_0|}\right)^*_{Ker((B^*)^{-1})^\perp}\right]$.

Finally, using the duality relations between exponential and analytic functions and functionals through the Fourier transform established in Theorem 9.3, we can prove the existence results for general boundary value problems for systems of differential equations of the form

$$D_t^{p_j} U_j(t,\zeta) + \sum_{k=1}^{N} \sum_{q=0}^{p_k-1} A^q_{jk}(t,D_\zeta,\zeta) D_t^q U_k(t,\zeta) = G_j(t,\zeta), \tag{9.80}$$

$$t \in \mathcal{D},\ \zeta \in \Omega,\ j = 1,\dots,N,$$

$$\sum_{k=1}^{N} \sum_{q=0}^{p_k-1} B^{mq}_{jk}(D_\zeta,\zeta) D_t^q U_k(t,\zeta)\Big|_{t=t_0} = \Psi_{jm}(\zeta), \tag{9.81}$$

$$\zeta \in \Omega,\ m = 0,\dots,p_j-1,\ j = 1,\dots,N,$$

where $\mathcal{D} \subset \mathbb{C}$ is a connected domain containing t_0; $\Omega \subset \mathbb{C}^n$ does not contain polar sets P_A and P_B associated with operators A^q_{jk} and B^{mq}_{jk}, whose symbols are

$$\mathcal{A}^q_{jk}(t,z,\zeta) = \sum_{|\alpha| \le m^q_{jk}} a_{jk\alpha}(t,\zeta) z^\alpha \in \mathcal{O}[\mathcal{D}; S(M^q, \mathcal{M}(\Omega))], \tag{9.82}$$

$$\mathcal{B}^{mq}_{jk}(z,\zeta) = \sum_{|\beta| \le n^{mq}_{jk}} b_{jk\beta}(\zeta) z^\beta \in S(\mathcal{N}^{qm}, \mathcal{M}(\Omega)), \tag{9.83}$$

$$q = 0,\dots,p_j-1,\ m = 0,\dots,p_k-1,\ j,k = 1,\dots,N.$$

Due to formulas (9.38) and (9.39), applying the Fourier transform, one can reduce boundary value problem (9.80)–(9.81) to the problem of the form (9.7)–(9.8). Hence, by duality, Theorems 9.18 and 9.19 imply the following statements.

Theorem 9.20. *Let the symbols of differential operators $A \equiv \{A^q_{jk}(t,D_\zeta,\zeta)\}$ and $B \equiv \{B^{mq}_{jk}(D_\zeta,\zeta)\}$ satisfy conditions (9.82) and (9.83), respectively. Suppose there exists a collection $\bar{\mu} = \mu_1,\dots,\mu_N$, such that*

i) the entries of the matrices $M^q, q = 0,\dots,p_j - 1$, *satisfy the inequalities*

$$m_{jk}^q \le \mu_k - \mu_j + p_j - q,\ j,k = 1,\dots,N,\ q = 0,\dots,p_j - 1;$$

ii) the entries of the matrix $\mathscr{N}^{mq}$, $m = 0,\dots,p_j - 1, q = 0,\dots,p_k - 1$, *satisfy the inequalities*

$$n_{jk}^{mq} \le \mu_k - \mu_j + q - m, \quad j,k = 1,\dots,N,\ m = 0,\dots,p_j - 1,\ q = 0,\dots,p_k - 1.$$

Then there exist numbers $r > 0$ *and* $\sigma > 0$ *such that for any vector-functions* $H(t,z) \in \mathscr{O}\left[\mathscr{D}; \mathscr{O}_{\bar{\mu},r+\sigma|t-t_0|}(\Omega)\right]$, *and* $\Phi(z) \in \mathscr{O}_{\bar{\mu},r}(\Omega \setminus N_B)$ *a solution* $U(t,z)$ *to boundary value problem* (9.80)–(9.81) *exists in a* δ*-neighborhood of* t_0, *where* $\delta < r/\sigma$, *and belongs to the space* $\mathscr{O}\left[|t-t_0| < \delta; \mathscr{O}_{\bar{\mu},r+\sigma|t-t_0|}(\Omega)\right]$.

Theorem 9.21. *Let the symbols of differential operators* $A \equiv \{A_{jk}^q(t, D_\zeta, \zeta)\}$ *and* $B \equiv \{B_{jk}^{mq}(D_\zeta, \zeta)\}$ *satisfy conditions* (9.82) *and* (9.83), *respectively. Suppose there exists a collection* $\bar{\mu} = \mu_1,\dots,\mu_N$, *such that*

i) the entries of the matrices $M^q, q = 0,\dots,p_j - 1$, *satisfy the inequalities*

$$m_{jk}^q \le \mu_j - \mu_k + p_j - q,\ j,k = 1,\dots,N,\ q = 0,\dots,p_j - 1;$$

ii) the entries of the matrix $\mathscr{N}^{mq}$, $m = 0,\dots,p_j - 1, q = 0,\dots,p_k - 1$, *satisfy the inequalities*

$$n_{jk}^{mq} \le \mu_j - \mu_k + m - q, \quad j,k = 1,\dots,N,\ m = 0,\dots,p_j - 1,\ q = 0,\dots,p_k - 1.$$

Then there exist numbers $r > 0$ *and* $\sigma > 0$ *such that for any vector-functions* $H(t,z) \in \mathscr{O}\left[\mathscr{D}; \mathscr{O}^*_{\bar{\mu},r+\sigma|t-t_0|}(\Omega)\right]$, *and* $\Phi(z) \in \mathscr{O}^*_{\bar{\mu},r}(\Omega)$ *a solution* $U(t,z)$ *to boundary value problem* (9.80)–(9.81) *exists in a* δ*-neighborhood of* t_0, *where* $\delta < r/\sigma$, *and belongs to the space* $\mathscr{O}\left[|t-t_0| < \delta; \mathscr{O}^*_{\bar{\mu},r+\sigma|t-t_0|}(\Omega \setminus N_B)\right]$.

9.10 Additional notes

1. *The Cauchy problem.* The Cauchy problem has a long and rich history. We refer the reader to survey papers [Miz67, VG91, S88, Dub90] on the history and modern state of this theory. In the general form the Cauchy problem was first posed by Augustin Louis Cauchy and the existence of a unique local solution of this problem was proved in his paper [Cau42] in 1842. Sophie von Kowalevsky[1] was not aware of Cauchy's result and reproved [Kow1874] this theorem in 1875. The theorem was later named the Cauchy-Kowalevsky theorem. Kowalevsky showed the importance of the condition $m_k \le m - k, k = 0,\dots,m-1$, for existence of an analytic solution of the Cauchy problem for equation (9.74) in the following example:

[1] Under this name she published her paper [Kow1874]. Her original Russian full name is Sofia Vasilyevna Kovalevskaya.

$$D_t u(t,z) = D_z^2 u(t,z), \quad |t| < 1, |z| \leq 1,$$
$$u(0,z) = \varphi(z), \quad |z| < 1,$$

with analeptic function $\varphi(z)$ in the unit disc $|z| < 1$. The solution of this problem has the representation

$$u(t,z) = e^{tD_z^2}\varphi(z) = \sum_{n=0}^{\infty} \frac{D_z^{2n}\varphi(z)}{n!} t^n.$$

Now taking $\varphi(z) = (1-z)^{-1}$, one can see that $D_z^n \varphi(z) = n!(1-z)^{-n-1}$, one obtains a power series

$$u(t,z) = \sum_{n=0}^{\infty} \frac{(2n)!}{n!} \frac{t^n}{(1-z)^{n+1}},$$

divergent for all t and z in any neighborhood of the origin (except $t = 0$).

2. *On necessary conditions for existence of a solution.* Mizohata [Miz74] (see also [Kit76]) showed that the condition $m_k \leq m-k$, $k = 0,\dots,m-1$, is necessary for the existence of an analytic solution of the Cauchy problem for equation (9.74). More precisely, he proved the following statement.

Theorem 9.22. *(Mizohata [Miz74]) In order that the Cauchy-Kowalevsky theorem for the Cauchy problem for equation* (9.74) *hold at the origin, it is necessary that*

$$m_k \leq m-k, k = 0,\dots,m-1, \tag{9.84}$$

Let $p = \max_{k,\alpha}\{|\alpha|/(k+n(k,\alpha))\}$, where $n(k,\alpha) = \min\{\mu : a_{k,\alpha}^{\mu} \not\equiv 0\}$, and $a_{k,\alpha}^{\mu(x)}$ are coefficients of the operator $A_k(t,z,D_z) = \sum_{\alpha,\mu} a_{k,\alpha} t^{\mu} D_z^{\alpha}$. Mizohata showed that $p \leq 1$, which is equivalent to condition (9.84). Kitagawa [Kit90] introduced weights p_k and p_v by

$$p_* = \max_{k,\alpha}\{|\alpha|/(k+n(k,\alpha)), |\alpha| \leq k\} \text{ and } p^* = \max_{k,\alpha}\{|\alpha|/(k+n(k,\alpha)), |\alpha| > k\},$$

and proved that in order that the Cauchy-Kowalevsky theorem for the Cauchy problem for equation (9.74) hold at the origin, it is necessary that $p^* < p_*$. The latter again implies condition(9.84).

Leray-Volevich's (LV) condition (9.1), that is

$$m_{kj}^q \leq \mu_k - \mu_j + p_k - q, \quad k,j = 1,\dots,N,$$

first appeared in Volevich [Vol63] in 1963, and in the context of the Cauchy problem for systems of differential equations in Gårding-Kotake-Leray [LGK67], in 1964. Mizohata [Miz74] called systems satisfying LV conditions (9.1) Kowalevskian in the sense of Volevich. The case $\mu_k = k$ was used by Leray in 1953 [Ler53]. Usual Kowalevskian systems correspond to the case $\mu_k = 0, k = 1,\dots,N$.

3. *Infinite order differential operators.* Differential operators of infinite order obviously do not satisfy LV conditions, and therefore, the corresponding system with such operators are not Kowalevskian in the sense of Volevich. The Cauchy problem for equations and systems with differential operators of infinite order was studied by Korobeynik [K73], Leont'ev [Leo76], Baouendi and Goulaouic [BG76], Dubinskii [Dub84], Napalkov [Nap82], and others. The related theory of analytic pseudo-differential operators is in the focus of many researchers; see survey paper [S88] on results up to 1988, and in works [Dub96, Ren10] on its current state. The analytic solutions of differential equations with the real time variable and complex spatial variables are studied in [Gal08] in model cases.
4. *On uniqueness of a solution.* Holmgren [Hol01] in 1901 showed that the Cauchy problem for equations with analytic coefficients, but not necessarily analytic data, cannot have more than

one solution. However, if coefficients of the equation are C^{∞} functions, then the Cauchy problem may not have a unique solution. Namely, Plis [Pl54] in 1954 constructed an example of fourth order equation with C^{∞}-coefficients for which the uniqueness does not hold. Later other examples were constructed; see [Met93]. Calderon [Cal58] in 1958 proved the uniqueness theorem, which played a key role for further development of the Cauchy theory. Later, other variations or weaker versions of uniqueness conditions were found. In particular, the uniqueness of a solution to the Cauchy problem for differential equations with partially holomorphic coefficients is obtained in works [Hor83, Uch04] and for systems of such equations in [Tam06].

References

[A13] Abdel-Rehim, E.A.: Explicit approximation solutions and proof of convergence of the space-time fractional advection dispersion equations. Appl. Math., **4,** 1427–1440 (2013)

[A26] Abel, N.H.: Solution of a mechanical problem. (Translated from the German) In: D. E. Smith (ed) A Source Book in Mathematics, Dover Publications, New York, 656–662 (1959)

[AS64] Abramowitz, M., Stegun, I.: Handbook of Mathematical Functions with Formulas, Graphs, and Mathematical Tables. Dover Publications, New York (1972)

[ADN69] Agmon, S., Douglis, A., Nirenberg, L.: Estimates near the boundary for solutions of elliptic partial differential equations satisfying general boundary conditions. Comm. Pure and Appl. Math., **12,** 623–727 (1969)

[Agr04] Agraval, Om.: Analytical solution for stochastic response of a fractionally damped beam. J. Vibration and Acoustics, **126**, 561–566 (2004)

[AP89] Alimov, Sh.A., Ashurov, R.R., Pulatov A.K.: Multiple series and Fourier integrals, Commutative harmonic analysis - 4. Springer-Ferlag, 1–97 (1992)

[AS09] Almeida, A., Samko, S.: Fractional and hypersingular operators in variable exponent spaces on metric measure spaces. Meditter. J. Math., **6,** 215–232 (2009)

[AUS06] Andries, E., Umarov, S.R., Steinberg, St.: Monte Carlo random walk simulations based on distributed order differential equations with applications to cell biology. Frac. Calc. Appl. Anal., **9** (4), 351–369 (2006)

[AB92] Antipko, I.I., Borok, V.M.: The Neumann boundary value problem in an infinite layer. J. Sov. Math. **58**, 541–547 (1992) (Translation from: Theoret. Funktion. Anal. Prilozh.**53**, 71–78 (1990))

[App09] Applebaum, D.: Lévy Processes and Stochastic Calculus. Cambridge University Press (2009)

[Ata02] Atakhodzhaev, M.A.: Ill-posed internal boundary value problems for the biharmonic equation. VSP, Utrecht (2002)

[BRV05] Bacchelli, B., Bozzini, M., Rabut, C., Varas, M.: Decomposition and reconstruction of multidimensional signals using polyharmonic pre-wavelets. Applied and Computational Harmonic Analysis, **18,** 282–299 (2005)

[BT00] Bagley, R.L., Torvic P.J.: On the existence of the order domain and the solution of distributed order equations I, II. Int. J. Appl. Math. **2**, 865–882, 965–987 (2000)

[Bal60] Balakrishnan, A.V.: Fractional powers of closed operators and the semigroups generated by them. Pacific J. Math. **10** (2), 419–437 (1960)

S. Umarov, *Introduction to Fractional and Pseudo-Differential Equations with Singular Symbols*, Developments in Mathematics 41,
DOI 10.1007/978-3-319-20771-1

[BG76] Baouendi M.S., Goulaouic, C. Cauchy problem for analytic pseudo-differential operators. Comm. in Part. Diff. Equat., **1** (2), 135–189 (1976)

[BMR01] Barndorff-Nielsen, O.E., Mikosch, T., Resnick S. (eds): Lévy processes: Theory and applications. Birkhäuser (2001)

[BC07] Baudoin, F., Coutin, L.: Operators associated with a stochastic differential equation driven by fractional Brownian motions. Stoch. Process. Appl. **117** (5), 550–574 (2007)

[Baz98] Bazhlekova E.: The abstract Cauchy problem for the fractional evolution equation. Frac. Calc. Appl. Anal., **1,** 255–270 (1998)

[Baz01] Bazhlekova, E.: Fractional evolution equations in Banach spaces. Dissertation, Technische Universiteit Eindhoven, 117 pp (2001)

[BMS04] Becker-Kern, P., Meerschaert, M.M., Scheffler, H.-P.: Limit theorems for coupled continuous time random walks. Ann. Probab. **32** (1B), 730–756 (2004)

[Ben03] Bender, C.: An Itô formula for generalized functionals of a fractional Brownian motion with arbitrary Hurst parameter. Stoch. Process. Appl., **104** (1), 81–106 (2003)

[BGR90] Bensoussan A., Glowinski, R., Rascanu, R.: Approximations of Zakai equation by the splitting up method. SIAM Journal on Control and Optimization, **28** (6), 1420–1431 (1990)

[Ber96] Bertoin, J.: Lévy processes. Cambridge University Press (1996)

[BJS64] Bers, L., John, F., Schechter M.: Partial Differential Equations. Interscience Publishers, New York - London - Sydney (1964)

[BL76] Bergh, J., Löfström, J.: Interpolation Spaces: An Introduction. Springer (1976)

[Ber04] Bernstein, S.N.: Sur la natur analytique des solutions des équations aux dérivées partielles du second ordre. Math. Ann. **59**, 20–76 (1904)

[Ber28] Bernstein, S.N.: Demonstration du theoreme de M. Hilbert sur la nature analytique des solutions des equations du type elliptique sans l'emploi des series normales. Math. Ztschr. **28**, 330–348 (1928)

[BIN75] Besov, O.B., Il'in, V.P., Nikolskii, S.M.: Integral Representation of Functions and Embedding Theorems I, II. Willey, New York (1979)

[BHOZ08] Biagini, F., Hu, Y., Oksendal, B., Zhang, T.: Stochastic calculus for fractional Brownian motion and applications. Springer (2008)

[Bil99] Billingsley, P.: Convergence of Probability Measures. 2nd ed. Wiley-Interscience publication (1999)

[BD64] Björken, J., Drel, S.: Relativistic Quantum Theory, V I. McGraw Hill Book Co., New York (1964)

[Bo55] Bochner, S.: Harmonic Analysis and the Theory of Probability. California Monographs in Mathematical Science, University of California Press, Berkeley (1955)

[Bor69] Borok, B.M.: Uniqueness classes for the solution of a boundary problem with an infinite layer for systems of linear partial differential equations with constant coefficients. Mat. Sb., **79** (121): 2(6), 293–304 (1969)

[Bor71] Borok, V.M.: Correctly solvable boundary-value problems in an infinite layer for systems of linear partial differential equations. Math. USSR. Izv. **5**, 193–210 (1971)

[BL81] Brezis, H., Lions, J.L.: Nonlinear Partial Differential Equations and Their Applications. Chapman & Hall (1981)

[BK96] Budhiraja, A., Kallianpur, G.: Approximation to the solutions of Zakai equations using multiple Wiener and Stratonovich expansions. Stochastics, **56,** 271–315 (1966)

[Cal58] Calderon, A.P.: Uniqueness in the Cauchy problem for partial differential equations. Amer. J. of Math., **80,** 16–35 (1958)

[CV71] Calderon, A.P., Vaillancourt, R.: On the boundedness of pseudo-differential operators. J. Math. Soc. Japan, **23** (2) 374–378 (1971)

[CZ58] Calderon, A.P., Zygmund, A. Uniqueness in the Cauchy problem for partial differential equations. Amer. J. Math. **58,** 16–36 (1980)

[Cap67] Caputo M.: Linear models of dissipation whose Q is almost frequency independent, II. Geophys. J. R. Astr. Soc., **13**, 529–539 (1967)

[Cap69] Caputo M.: Elasticitá e Dissipazione, Zanichelli, Bologna (1969)

[Cap95] Caputo, M.: Mean fractional order derivatives. Differential equations and filters. Annals Univ. Ferrara - Sez. VII - SC. Mat., **16**, 73–84 (1995)

[Cap01] Caputo, M.: Distributed order differential equations modeling dielectric induction and diffusion. Fract. Calc. Appl. Anal., **4**, 421–442 (2001)

[Cau42] Cauchy, A.-L.: Mémoire sur l'emploi du calcul des limites dans l'intégration des équations aux dérivées partielles. Comptes rendus **15**, 44–59 (1842)

[CKS03] Chechkin, A.V., Klafter, J., Sokolov I.M.: Fractional Fokker-Planck equation for ultraslow diffusion. EPL, **63** (3), 326–334 (2003)

[CGSG03] Chechkin, A.V., Gorenflo, R., Sokolov, I.M., Gonchar V.Yu.: Distributed order time fractional diffusion equation. Fract. Calc. Appl. Anal., **6**, 259–279 (2003)

[CGS05] Chechkin, A.V., Gorenflo, R., Sokolov I.M.: Fractional diffusion in inhomogeneous media. J. Physics. A: Math. Gen., **38**, 679–684 (2005)

[CGKS8] Chechkin, A.V., Gonchar, V.Yu., Gorenflo, R., Korabel, N., Sokolov, I.M.: Generalized fractional diffusion equations for accelerating subdiffusion and truncated Lévy flights. Phys. Rev. E, **78**, 021111(13) (2008)

[CSK11] Chechkin, A.V., Sokolov, I.M., Klafter, J.: Natural and modified forms of distributed order fractional diffusion equations. J. Klafter, S.C. Lim, R. Metzler (eds): Fractional Dynamics: Recent Advances. Singapore: World Scientific, Ch. 5, 107–127 (2011)

[Chi89] Chirka, E.M.: Complex analytic sets. Kluwer Academic Publishers (1989)

[CW77] Cordes, H.O., Williams, D.A.: An algebra of pseudo-differential operators with nonsmooth symbol. Pacific J. Math., **78** (2), 278–290 (1978)

[Co65] Courrége, Ph.: Sur la forme intégro-différentielle des opérateurs de C_k^∞ dans C satisfaisant au principe du maximum. Sém. Théorie du Potentiel, Exposé 2 (1965/66)

[DPZ02] Da Prato, G., Zabczyk, J.: Second order partial differential equations in Hilbert spaces. Cambridge University Press (2002)

[Da87] Daum, F.E.: Solution of the Zakai equation by separation of variables. IEEE Transactions on automatic control. AC-32 (10), 941–943 (1987)

[DU98] Decreusefond, L., Üstünel, A.S.: Stochastic analysis of the fractional Brownian motion. Potential Analysis. **10** (2), 177–214 (1998)

[DF01] Diethelm, K., Ford, N.J.: Numerical solution methods for distributed order time fractional diffusion equation. Fract. Calc. Appl. Anal., **4**, 531–542 (2001)

[DF09] Diethelm, K., Ford, N.J.: Numerical analysis for distributed-order differential equations. J. Comp. Appl. Math. **225** (1), 96–104 (2009)

[DP65] Ditkin, V.A., Prudnikov A.P.: Integral Transforms and Operational Calculus. Oxford, Pergamon (1965)

[Djr66] Djrbashian M.M.: Integral Transforms and Representations of Functions in the Complex Plane. Nauka, Moscow (1966) (in Russian)

[Don52] Donsker, M.D.: Justification and extension of Doob's heuristic approach to the Kolmogorov-Smirnov theorems. Ann. Math. Stat., **23**, 277–281 (1952)

[DN55] Douglis, A., Nirenberg, L.: Interior estimates for elliptic systems of partial differential equations. Comm. Pure. & Appl. Math. **8**, 503–538 (1955)

[Dub81] Dubinskii, Yu. A.: On a method of solving partial differential equations. Sov. Math. Dokl. **23**, 583–587 (1981)

[Dub82] Dubinskii, Yu.A.: The algebra of pseudo-differential operators with analytic symbols and its applications to mathematical physics. Russ. Math. Surv., **37**, 109–153 (1982)

[Dub84] Dubinskii, Yu.A.: Fourier transformation of analytic functions. The complex Fourier method. Dokl. Akad. Nauk SSSR., **275** (3), 533–536 (1984) (in Russian)

[Dub90] Dubinskii, Yu. A.: The Cauchy problem and pseudo-differential operators in the complex domain. Russ. Math. Surv., **45** (2), 95–128 (1990)

[Dub91] Dubinskii, Yu.A.: Analytic Pseudo-differential Operators and Their Applications. Kluwer Academic Publishers, Dordrecht (1991)

[Dub96] Dubinskii Yu. A.: Cauchy problem in complex domains. Moscow (1996) (In Russian)

[Du33] Duhamel, J.M.C.: Mémoire sur la méthode générale relative au mouvement de la chaleur dans les corps solides plongés dans les milieux dont la température varie avec le temps. J. Ec. Polyt. Paris **14**, Cah. 22, 20 (1833)

[DS88] Dunford, N., Schwartz, J.T.: Linear Operators III: Spectral Operators. Wiley-Interscience. NY-London-Sydney-Toronto (1988)

[Ede75] Edenhofer, E.: Integraldarstellung einer m-polyharmonischen Funktion, deren Funktionswerte und erste $m-1$ Normalableitungen auf einer Hypersphäre gegeben sind Math. Nachr. **68**, 105–113 (1975)

[Edi97] Edidin, M.: Lipid microdomains in cell surface membranes. Curr. Opin. Struct. Biol., **7**, 528–532 (1997)

[Ego67] Egorov, Yu.V.: Hypoelliptic pseudodifferential operators, Tr. Mosk. Mat. Obs., **16**, 99–108 (1967)

[ES95] El-Sayed, A.M. Fractional order evolution equations. J. of Frac. Calc., **7**, 89–100 (1995)

[EK04] Eidelman, S.D., Kochubei, A.N.: Cauchy problem for fractional diffusion equations. Journal of Differential Equations, **199**, 211–255 (2004)

[EN99] Engel, K.-J., Nagel, R.: One-parameter Semigroups for Linear Evolution Equations. Springer (1999)

[Fef71] Fefferman, Ch.: The multiplier theorem for the ball. The Annals of Mathematics, **94** (2), 330–336 (1971)

[Fel52] Feller, W.: On a generalization of Marcel Riesz potentials and the semi-groups generated by them. Meddelanden Lunds Universitets Matematiska Seminarium (Comm. Sém. Mathém. Université de Lund), Tome suppl. dédié a M. Riesz. Lund, 73–81 (1952)

[Fel68] Feller, W. An Introduction to Probability Theory and Its Applications, 3rd ed. John Wiley and Sons. NY-London-Sydney (1968)

[FJG08] Feng, B., Ji, D., Ge, W.: Positive solutions for a class of boundary value problem with integral boundary conditions in Banach spaces. J. Comput. Appl. Math., **222**, 351–363 (2008)

[FGT12] Fugére, J., Gaboury, S., Tremblay, R.: Leibniz rules and integral analogues for fractional derivatives via a new transformation formula. Bulletin of Math. Anal. Appl. **4** (2), 72–82 (2012)

[FKK72] Fujisaki M., Kallianpur, G., Kunita, H.: Stochastic differential equations for the non-linear filtering problem. Osaka J. of Mathematics **9**(1), 19–40 (1972)

[Fuj90] Fujita, Y.: Integrodifferential equation which interpolates the heat and the wave equations. Osaka J. Math. **27**, 309–321, 797–804 (1990)

[FOT94] Fukushima, M., Oshima, Y., Takeda, M.: Dirichlet forms and symmetric Markov processes. De Gruyter Studies in Mathematics, 19, Walter de Gruyter Verlag, Berlin-New-York (1994)

[Gal08] Gal, C.G., Gal, S.G., Goldstein, G.A.: Evolution equations with real time variable and complex spatial variables. Compl. Var. Elliptic Equ. 53, 753–774; Higher-order heat and Laplace-type equations with real time variable and complex spatial variable. 55, 357–373; Wave and telegraph equations with real time variable and complex spatial variables. 57, 91–109 (2008–2010–2012)

[GR09] Ganning, R., Rossi, H.: Analytic Functions of Several Complex Variables. AMS Chelsea Publishing (2009)

[Går98] Gårding, L.: Hyperbolic equations in the twentieth century. Seminaires et Congres, **3**, 37–68 (1998)

[LGK67] Gårding, L., Leray, J., Kotake, T.: Probleme de Cauchy. Moscou, Mir (1967) (in Russian)

[GS53] Gel'fand I.M., Shilov, G.E.: Fourier transforms of rapidly growing functions and questions of uniqueness of the solution of Cauchy's problem. Usp. Mat. Nauk. **8** (6), 3–54 (1953)

[Ger48] Gerasimov, A.: A generalization of linear laws of deformation and its applications to problems of internal friction, Prikl. Matem. i Mekh. **12** (3), 251–260 (1948) (in Russian)

[GW94] Ghosh, R.N., Webb, W.W.: Automated detection and tracking of individual and clustered cell surface low density lipoprotein receptor molecules. Biophys. J., **66**, 1301–1318 (1994)

[GT83] Gilbarg, D., Trudinger, N.S.: Elliptic partial differential equations of second order. Springer-Verlag, Berlin (1977)

[GW70] Gillis, J.E., Weiss, G.H.: Expected number of distinct sites visited by a random walk with an infinite variance. J. Math. Phys. **11**, 1307–1312 (1970)

[VG91] Gindikin S.G., Volevich, L.R.: The Cauchy problem. In Partial Differential Equations III. Egorov, Yu.V., Shubin M.A. (eds), Springer-Verlag, Berlin, 1–87 (1991)

[GN95] Glöckle, W.G., Nonnenmacher, T.F.: A fractional calculus approach to self-similar protein dynamics. Biophys J. **68** (1), 46–53 (1995)

[GK54] Gnedenko, B.V., Kolmogorov, A.N.: Limit Distributions for Sums of Independent Random Variables. Addison-Wesley (1954)

[Goo10] Goodrich, C.S.: Existence of a positive solution to a class of fractional differential equations. Appl. Math. Lett. **23**, 1050–1055 (2010)

[G84] Gorbachuk, V.I., Gorbachuk M.L.: Boundary Value Problems for Operator Differential Equations. Kluwer Academic Publishers. Netherlands (1990)

[Gor97] Gorenflo, R.: Fractional calculus: some numerical methods. In Carpinteri, A., Mainardi, F. (eds): Fractals and Fractional Calculus in Continuum Mechanics. Springer Verlag, Wien and New York 277–290 (1997)

[GAR04] Gorenflo, R., Abdel-Rehim, E.: Convergence of the Grünwald-Letnikov scheme for time-fractional diffusion. J. Comp. and Appl. Math. **205** 871–881 (2007)

[GM97] Gorenflo, R., Mainardi, F.: Fractional calculus: integral and differential equations of fractional order. In Carpinteri, A., Mainardi, F. (eds): Fractals and Fractional Calculus in Continuum Mechanics. Springer Verlag, Wien and New York 223–276 (1997)

[GM98-1] Gorenflo, R., Mainardi, F.: Random walk models for space-fractional diffusion processes. Fract. Calc. Appl. Anal., **1**, 167–191 (1998)

[GM98-2] Gorenflo, R., Mainardi, F.: Fractional calculus and stable probability distributions. Archives of Mechanics, **50**, 377–388 (1998)

[GM99] Gorenflo, R., Mainardi, F.: Approximation of Lévy-Feller diffusion by random walk. ZAA, **18** (2) 231–246 (1999)

[GM01] Gorenflo, R., Mainardi, F.: Random walk models approximating symmetric space-fractional diffusion processes. In Elschner, Gohberg and Silbermann (eds): Problems in Mathematical Physics (Siegfried Prössdorf Memorial Volume). Birkhäuser Verlag, Boston-Basel-Berlin, 120–145 (2001)

[GM05] Gorenflo, R., Mainardi, F.: Simply and multiply scaled diffusion limits for continuous time random walks. J. of Physics. Conference Series, **7**, 1–16 (2005)

[GV91] Gorenflo, R., Vessella, S.: Abel Integral Equations: Analysis and Applications. Lecture Notes in Mathematics, **1461**, Springer Verlag, Berlin (1991)

[GV03] Gorenflo, R., Vivoli, A,: Fully discrete random walks for space-time fractional diffusion equations. Signal Processing, **83**, 2411–2420 (2003)

[GLU00] Gorenflo, R., Luchko, Yu., Umarov, S.R.: The Cauchy and multi-point partial pseudo-differential equations of fractional order. Fract. Calc. Appl. Anal., **3** (3), 249–275 (2000)

[GLU00a] Gorenflo, R., Luchko, Yu., Umarov, S.R.: On boundary value problems for pseudo-differential equations with boundary operators of fractional order. Fract. Calc. Appl. Anal., **3** (4), 454–468 (2000)

[GMM02] Gorenflo, R., Mainardi, F., Moretti, D., Pagnini, G., Paradisi, P.: Discrete random walk models for space-time fractional diffusion, Chemical Physics, **284**, 521–541 (2002)

[GK14] Gorenflo, R., Kilbas, A., Mainardi, F., Rogozin, S.V.; Mittag-Leffler Functions, Related Topics and Applications. Springer Monographs in Mathematics. Springer (2014)

[Gra10] Graf, U.: Introduction to Hyperfunctions and Their Integral Transformations. Birkhäuser, Basel (2010)

[GK10] Guezane-Lakoud, A., Kelaiaia, S.: Solvability of a three-point nonlinear boundary value problem. Electron. J. Differ. Equat. **139**, 1–9 (2010)

[Had02] Hadamard, J.: Sur les problémes aux dérivées partielles et leur signification physique. Princeton University Bulletin, 49–52 (1902)

[HU11] Hahn, M.G., Umarov, S.R.: Fractional Fokker-Plank-Kolmogorov type equations and their associated stochastic differential equations. Frac. Calc. Appl. Anal., **14** (1), 56–79 (2011)

[HKU10] Hahn, M.G., Kobayashi, K., Umarov, S.R.: SDEs driven by a time-changed Lévy process and their associated time-fractional order pseudo-differential equations. J. Theoret. Prob., **25** (1), 262–279 (2012)

[HKU11] Hahn, M.G., Kobayashi, K., Umarov, S.R.: Fokker-Planck-Kolmogorov equations associated with time-changed fractional Brownian motion. Proceed. Amer. Math. Soc., **139** (2), 691–705 (2011)

[HKRU11] Hahn, M.G., Kobayashi, K., Ryvkina, J., Umarov, S.R.: On time-changed Gaussian processes and their associated Fokker-Planck-Kolmogorov equations. Electron. Commun. Probab. **16**, 150–164 (2011)

[HK07] Haußmann, W., Kounchev, O.: On polyharmonic interpolation. J. Math. Anal. Appl. **331** (2), 840–849 (2007)

[HK93] Hayman, W.K., Korenblum, B.: Representation and uniqueness theorems for polyharmonic functions. J. d'Analyse Mathématique, **60**, 113–133 (1993)

[Hay78] Hayne, R.M.:. Uniqueness in the Cauchy problem for parabolic equations. Trans. Am. Math. Soc., **241**, 373–399 (1978)

[HMS11] Haubold, H.J., Mathai, A.M., Saxena, R.K.: Mittag-Leffler functions and their applications. J. Appl. Math., 51 pp (2011)

[Hil00] Hilfer R. (ed): Applications Of Fractional Calculus In Physics. World Scientific (2000)

[Hoh00] Hoh, W.: Pseudo-differential operators with negative definite symbols of variable order. Rev. Mat. Iberoam, **16** (2), 219–241 (2000)

[Hol01] Holmgren, E.: Über Systeme von linearen partiellen Differentialgleichungen. Öfversigt af Kongl. Vetenskaps-Academien Förhandlinger, **58**, 91–103 (1901)

[Hor61] Hörmander, L.: Hypoelliptic differential operators. Annales de l'lnstitut Fourier. **11**, 477–492 (1961)

[Hor65] Hörmander, L.: Pseudo-differential operators. Comm. Pure Appl. Math. **18**, 501–517 (1965)

[Hor67] Hörmander, L.: Hypoelliptic second order differential equations. Acta Math. **119** (3–4), 147–171 (1967)

[Hor68] Hörmander, L.: Pseudo-differential operators and hypoelliptic equations. Proc. Symp. Pure. Math. **10**, 131–183 (1969)

[Hor83] Hörmander, L.: The Analysis of Linear Partial Differential Operators, I - IV. Springer-Verlag, Berlin-Heidelberg-New-York (1983)

[Hor90] Hörmander, L.: An introduction to complex analysis in several variables. North-Holland (1990)

[Ibr14] Ibrahim, R.W.: Solutions to systems of arbitrary-order differential equations in complex domains. Electronic Journal of Differential Equations, **46**, 2014, 1–13 (2014)

[IW81] Ikeda, N., Watanabe, Sh.: Stochastic Differential Equations and Diffusion Processes. Amsterdam-Oxford-New York, North-Holland Publishing Co. (1981)

[IX00] Ito, K., Xiong, K.: Gaussian filters for nonlinear filtering problems. IEEE Transactions on Automatic. Control. 1, **45** (5), 910–927 (2000)

[Jac01] Jacob, N.: Pseudo-differential Operators and Markov Processes. Vol. I. Fourier Analysis and Semigroups Vol. II. Generators and Their Potential Theory, Vol. III. Markov Processes and Applications. Imperial College Press, London (2001, 2002, 2005)

[JL93] Jacob, N., Leopold, H.-G.: Pseudo differential operators with variable order of differentiation generating Feller semigroups. Integr. Equat. Oper. Th., **17**, 544–553 (1993)

[JSc02] Jacob, J., Schilling, R.L.:. Lévy-type processes and pseudo-differential operators. In Barndorff-Nielsen, O., Mikosch, T., Resnick S. (eds.), Levy Processes: Theory and Applications. Boston, Bikhäser, 139–168 (2001)

[Jac79] Jacod, J.: Calcul Stochastique et Problèmes de Martingales. Lecture Notes in Mathematics, **714**, Springer, Berlin (1979)

[JCP12] Jiao, Zh., Chen, Y.-Q., Podlubny, I.: Distributed-Order Dynamic Systems. Springer, London (2012)

[KO14] Kadem, A., Kirane, M., Kirk, C.M., Olmstead, W.E.: Blowing-up solutions to systems of fractional differential and integral equations with exponential non-linearities. IMA Journal of Applied Mathematics, **79**, 1077–1088 (2014)

[KB61] Kalman, R.E., Bucy, R.C.: New results in linear filtering and prediction theory. Journal of basic engineering, **83**, 95–108 (1961)

[Kat12] Katsikadelis, J.T.: Numerical solution of distributed order fractional differential equations. J. Comp. Phys. **259**, 11–22 (2012)

[K81] Kamynin, L.I., Khimchenko, B.N.: Tikhonov-Petrovskii problem for second-order parabolic equations. Siberian Mathematical Journal, **22** (5), 709–734 (1981)

[Kan72] Kaneko, A.: Representation of hyperfunctions by measures and some of its applications. J. Fac. Sci. Univ. Tokyo, Sect. IA Math. **19**, 321–352 (1972)

[KSh91] Karatzas, I., Shreve, S.E.: Brownian Motion and Stochastic Calculus. Springer (1991)

[KAN10] Kazemipour, S.A., Ansari, A., Neyrameh, A.: Explicit solution of space-time fractional Klein-Gordon equation of distributed order via the Fox H-functions. M. East J. Sci. Res. **6** (6), 647–656 (2010)

[KJ11] Kehue, L., Jigen, P.: Fractional abstract Cauchy problems. Integr. Equ. Oper. Theory, **70**, 333–361 (2011)

[KMSL13] Keyantuo, V., Miana, P.J., Sánches-Lajusticia, L.: Sharp extensions for convoluted solutions of abstract Cauchy problems. Integr. Equat. Oper. Theory, **77**, 211–241 (2013)

[KST06] Kilbas, A.A., Srivastava, H.M., Trujillo, J.J.: Theory And Applications of Fractional Differential Equations. Elsevier (2006)

[Kir94] Kiryakova, V.: Generalized Fractional Calculus and Applications. Longman Sci. & Techn., Harlow and J. Wiley & Sons, New York (1994)

[Kit76] Kitagawa, K.: A remark on a necessary condition of the Cauchy-Kowalevsky. Publ. RIMS, Kyoto Univ. **11**, 523–534 (1976)

[Kit90] Kitagawa, K.: Sur le téorème de Cauchy-Kowalevski. J. Math. Kyoto Univ. **30** (1), 1–32 (1990)

[KM67] Klass, D.L., Martinek, T.W.: Electroviscous fluids. I. Rheological properties. J. Appl. Phys., **38** (1), 67–74 (1967)

[Kob11] Kobayashi, K. Stochastic calculus for a time-changed semimartingale and the associated stochastic differential equations. J. Theoret. Prob., **24** (3), 789–820 (2011)

[Koc89] Kochubei, A.: Parabolic pseudo-differential equations, hypersingular integrals and Markov processes. Math. USSR, Izvestija **33**, 233–259 (1989)

[Koc08] Kochubey, A. Distributed order calculus and equations of ultraslow diffusion. J. Math. Anal. and Appl. **340** (1), 252–281 (2008)

[KN65] Kohn, J.J., Nirenberg, L.: An algebra of pseudo-differential operators. Comm. Pure Appl. Math. **18**, 269–305 (1965)

[Kol26] Kolmogorov, A.N.: Une série de Fourier-Lebesgue divergente partout. C. R. Acad. Sci. Paris **183** , 1327–1328 (1926)

[Kol40] Kolmogorov, A.N.: Wienersche Spiralen und einige andere interessante Kurven im Hilbertschen Raum. Dokl. Acad. Sci. URSS, **26** 115–118 (1940)

[Kom66] Komatsu, H.: Fractional powers of operators. Pacific J. Math. 19(2), 285–346 (1966)

[Kon67] Kondrat'ev, V.A.: Boundary value problems for elliptic equations with conical points. Trudy Mosk. Mat. Obsh., 209–292 (1967)

[K73] Korobeinik, Yu.F., Kubrak, V.K.: The existence of particular solutions of a differential equation of infinite order with prescribed growth. Izv. Vys. Matematika. **9** (136), 36–45 (1973) (in Russian)

[Kos93] Kostin, V.A.: The Cauchy problem for an abstract differential equation with fractional derivatives. Russ. Dokl. Math. **46**, 316–319 (1993)

[Kow1874] Kowalevsky, S.: Zür theorie der partiellen differentialgleichungen. Journal für die reine und angevandte mathematik, 1–32 (1874)

[Ku90] Kunita, H.: Stochastic Flows and Stochastic Differential Equations. Cambridge University Press (1990)

[Kus67] Kushner, H.J.: Dynamical equations for optimal nonlinear filtering. J. Diff. Eq. **3**, 179–190 (1967)

[L10] Lawler, G., Limic, V.: Random walk: A Modern Introduction. Cambridge University Press (2010)

[Leb01] Lebesgue, H.: Sur une généralisation de l'intégrale définie. Comptes Rend. l'Acad. Sci., **132**, 1025–1028 (1901)

[Leo76] Leont'ev, A.F.: Series of Exponents, Nauka, Moscow (1976) (in Russian)

[Ler53] Leray, J.: Hyperbolic differential equations. Princeton (1953)

[Let68] Letnikov, A.V.: The theory of differentiation of arbitrary power. Mat. Sbornik **3**, 1–68 (1968) (in Russian)

[LS70] Levitan, B.M., Sargsjan, I.S.: Introduction to Spectral Theory: Selfadjoint Ordinary Differential Equations. Transl. Math. Mon. 39 (1975)

[LW12] Liang, J.-R., Wang, J., Lŭ, L.-J., Hui, G., Qiu, W.-Y., Ren, F.-Y.: Fractional Fokker-Planck equation and Black-Scholes formula in composite-diffusive regime. J. Stat. Phys., **146**, 205–216 (2012)

[LSAT05] Liu, F., Shen, S., Anh, V., Turner, I.: Analysis of a discrete non-Markovian random walk approximation for the time fractional diffusion equation. ANZIAM J., **46**, 488–504 (2005)

[Lim06] Lim S. C.: Fractional derivative quantum fields at positive temperature. Physica A: Statistical Mechanics and its Applications 363, 269–281 (2006)

[LS02] Lipster, R.Sh., Shiryaev, A.N.: Statistics of Random Processes, I, II. Springer, New-York (2002)

[Liz63] Lizorkin P.I.: Generalized Liouville differentiation and functional spaces $L^r_p(E_n)$. Embedding theorems. Mat. Sb. **60**, 325–353 (1963) (in Russian)

[Liz67] Lizorkin, P.I.: Multipliers of Fourier integrals in the spaces L_p. Proc. Steklov Inst. Math. **89**, 269–290 (1967)

[Liz69] Lizorkin, P.I.: Generalized Liouville differentiation and the method of multipliers in the theory of embeddings of classes of differentiable functions. Proc. Steklov Inst. Math. **105**, 105–202 (1969)

[Lop53] Lopatinskiĭ, Ya.B.: On a method of reducing boundary problems for a system of differential equations of elliptic type to regular integral equations. Ukrain. Mat. Zb. **5**, 123–151 (1953)

[LH02] Lorenzo, C.F., Hartley T.T.: Variable order and distributed order fractional operators. Nonlinear Dynamics **29**, 57–98 (2002)

[LMR97] Lototsky, S., Mikulevicius, R., Rozovskii, R.: Nonlinear filtering revisited: a spectral approach. SIAM J. Control Optimization. **35**, 435–461 (1997)

[Lov93] Lovász, L.: Random walks on graphs: a survey. Bolyai Society Math. Studies, Combinatorics **2**, 1–46 (1993)

[LQR12] Lv, L., Qiu, W., Ren, F.: Fractional Fokker-Planck equation with space and time dependent drift and diffusion. J. Stat. Phys. **149**, 619–628 (2012)

[Mag09] Magdziarz, M.: Black-Scholes formula in subdiffusive regime. J. Stat. Phys. **136**, 553–564 (2009)

[MGZ14] Magdziarz, M., Gajda, J., Zorawik, T.: Comment on fractional Fokker-Planck equation with space and time dependent drift and diffusion. J. Stat. Phys. **154**, 1241–1250 (2014)

[Mag06] Magin R.: Fractional Calculus in Bioengineering. Begell House Publishers Inc. (2006)

[Mai96] Mainardi, F.: Fractional relaxation-oscillation and fractional diffusion-wave phenomena. Chaos, Solitons and Fractals. **7** (9), 1461–1477 (1996)

[MPG99] Mainardi, F., Paradisi, P., Gorenflo, R.: Probability distributions generated by fractional diffusion equations. Fracalmo Center publ., 46 pp. (1999) (available online: http://arxiv.org/pdf/0704.0320v1.pdf)

[MPG07] Mainardi, F., Mura, A., Pagnini, G., Gorenflo, R.: Sub-diffusion equations of fractional order and their fundamental solutions. In "Mathematical methods in engineering", Springer, 23–55 (2007)

[Mai10] Mainardi, F.: Fractional calculus and waves in linear viscoelasticity: An introduction to mathematical models. Imperial College Press (2010)

[MLP01] Mainardi, F., Luchko, Yu., Pagnini, G.: The fundamental solution of the space-time fractional diffusion equation. Frac. Calc. Appl. Anal., **4** (2), 153–192 (2001)

[MVN68] Mandelbrot, B.B., Van Ness, J. W.: Fractional Brownian motions, fractional noises and applications. SIAM Rev. **10**, 422–437 (1968)

[McC96] McCulloch, J.: Financial applications of stable distributions. In Statistical Methods in Finance: Handbook of Statistics **14**, Madfala, G., Rao, C.R. (eds). Elsevier, Amsterdam, 393–425 (1996)

[MBB01] Meerschaert, M.M., Benson, D., Bäumer, B.: Operator Lévy motion and multiscaling anomalous diffusion. Phys. Rev. E **63**, 021112–021117 (2001)

[MS01] Meerschaert, M.M., Scheffler, H.-P.: Limit Distributions for Sums of Independent Random Vectors. Heavy Tails in Theory and Practice. John Wiley and Sons, Inc. (2001)

[MS05] Meerschaert, M.M., Scheffler, H.-P.: Limit theorems for continuous time random walks with slowly varying waiting times. Stat. Probabil. Lett. **71** (1), 15–22 (2005)

[MS06] Meerschaert, M.M., Scheffler, H.-P.: Stochastic model for ultraslow diffusion. Stochastic professes and their applications, **116** (9), 1215–1235 (2006)

[MNX09] Meerschaert, M.M., Nane, E., Xiao, Y.: Correlated continuous time random walks. Stat. Probabil. Lett. **79**, 1194–1202 (2009)

[Met93] Métivier, G.: Counterexamples to Hölmgren's uniqueness for analytic non linear Cauchy problems. Invent. Math., **112**, 217–222 (1993)

[MK00] Metzler, R,. Klafter, J.: The random walk's guide to anomalous diffusion: a fractional dynamics approach. Physics Reports, **339**, 1–77 (2000)

[MK04] Metzler, R., Klafter, J.: The restaurant in the end of random walk. Physics A: Mathematical and General. **37** (31), 161–208 (2004)

[MBK99] Metzler, R., Barkai, E., Klafter, J.: Anomalous diffusion and relaxation close to thermal equilibrium: a fractional Fokker-Planck equation approach. Phys. Rev. Lett. **82**, 3563–3567 (1999)

[MN14] Mijena, J.B., Nane, N.: Space-time fractional stochastic partial differential equations. (2014) (available online: http://arxiv.org/pdf/1409.7366v1.pdf)

[MR93] Miller, K.C., Ross, B.: An Introduction to the Fractional Calculus and Fractional Differential Equations. John Wiley and Sons, Inc., New York (1993)

[Mih56] Mikhlin, S.G.: On the multipliers of Fourier integrals, Dokl. Akad. Nauk. **109**, 701–703 (1956)(in Russian)

[Miz62] Mizohata, S.: Solutiones nulles et solutiones non analytiques. J. Math. Kioto. Univ. **2** (1), 271–302 (1962)

[Miz67] Mizohata, S.: The Theory of Partial Differential Equations. Cambridge University Press (1979)

[Miz74] Mizohata, S.: On Kowalewskian systems. Russ. Math. Surv. **29** (2), 223–235 (1974) doi:10.1070/RM1974v029n02ABEH003837

[Miz74] Mizohata, S.: On Cauchy-Kowalevski's Theorem: A Necessary Condition. Publ. RIMS, Kyoto Univ. **10**, 509–519 (1974)

[MW65] Montroll, E.W., Weiss, G.H.: Random walk on Lattices. II. J. Math. Phys. 6 (2), 167–181 (1965)

[Mor58] Morrey, C.: On the analyticity of the solutions of analytic nonlinear elliptic systems of partial differential equations. Amer. J. Math. **80** (1), 198–277 (1958)

[MN57] Morrey, C., Nirenberg, L.: On the analyticity of the solutions of linear elliptic systems of partial differential equations. Comm. Pure Appl. Math. **10**, 271–290 (1957)

[Nag77] Nagase, M.: The L^p-boundedness of pseudo-differential operators with non-regular symbols. Comm. Partial Differential Equations, **2**, 1045–1061 (1977)

[Nai67] Naimark, M.A.: Linear differential operators, 1,2, F. Ungar (1967) (Translated from Russian)

[Nak75] Nakhushev, A.M.: A mixed problem for degenerate elliptic equations. Differ. Equations, **11**, 152–155 (1975)

[Nap82] Napalkov, V.V.: Convolution equations in multidimensional spaces. Mathematical Notes of the Acad. Sci. USSR, **25** (5), 393 pp., Springer (1979)

[N12] Napalkov, V.V., Nuyatov, A.A.: The multipoint de la Vallée-Poussin problem for a convolution operator. Sbornik: Mathematics, **203** (2), 224–233 (2012)

[Nat86] Natterer, F.: The mathematics of computerized tomography. Chichester, UK, Wiley (1986)

[Naz97] Nazarova, M. Kh.: On weak well-posedness of certain boundary value problems generated by a singular Bessel's operator. Uzb. Math. J. **3**, 63–70 (1997) (in Russian)

[Nik77] Nikolskii, S.M.: Approximation of Functions of Several Variables and Embedding Theorems. Springer-Ferlag (1975)

[Nig86] Nigmatullin, R.R.: The realization of generalized transfer equation in a medium with fractal geometry. Phys. Stat. Sol. B **133**, 425–430 (1986)

[Nua06] Nualart, D.: The Malliavin calculus and related topics, 2nd ed. Springer (2006)

[OR73] Oleynik, O.A., Radkevich, E.V.: On the analyticity of solutions of linear partial differential equations. Mathematics of the USSR-Sbornik, **19** (4), 581–596 (1973)

[OS74] Oldham, K., Spanier, J.: The Fractional Calculus: Theory and Application of Differentiation and Integration to Arbitrary Order. Acad. Press, Dover Publications, New York - London (1974)

[Osl72] Osler, T.J.: A further extension of the Leibniz rule to fractional derivatives and its relation to Parseval's formula. SIAM J. Math. Anal. **3**, 1–16 (1972)

[Ovs65] Ovsyannikov, L.V.: A singular operator in a scale of Banach space. Dokl. Akad. Nauk SSSR, **163**, 819–822 (1965) (English transl. Soviet Math. Dokl., **6** (1965))

[PR92] Päivärinta, L., Rempel, S.: Corner singularities of solutions to $\Delta^{\pm 1/2}u = f$ in two dimensions. Asymptotic analysis, **5**, 429–460 (1992)

[Pea05] Pearson, K.: On the problem of random walk. Nature, **72, 294 (1905)**

[Pee76] Peetre, J.: New thoughts on Besov spaces. Duke Univ. Math. Series, Duke Univ., Durham (1976)

[Pet96] Petrovskii, I.G.: Selected works: Systems of partial differential equations and algebraic geometry, Part I. Gordon and Breach Publishers (1996)

[Pla86] Plamenevskii, B.: Algebras of Pseudo-differential Operators. Kluwer Academic Publishers (1989)

[Pl54] Plis, A.: The problem of uniqueness for the solution of a system of partial differential equations. Bull. Acad. Polon. Sci. **2**, 55–57 (1954)

[Pod99] Podlubny, I.: Fractional Differential Equations. Mathematics in Science and Engineering, V 198. Academic Press, San Diego, Boston (1999)

[Pok68] Pokornyi, Yu.V.: On estimates for the Green's function for a multipoint boundary problem. Mat. Zametki, **4** (5), 533–540 (1968)

[Pro91] Protter, P.: Stochastic Integration and Differential Equations. Springer-Verlag, Berlin-New York (1991)

[Pta84] Ptashnik, B.I.: Ill-Posed Boundary Value Problems for Partial Differential Equations. Kiev (1984) (in Russian)

[Pul99] Pulkina, L.S.: A non-local problem with integral conditions for hyperbolic equations. Electronic Journal of Differential Equations, **45**, 1–6 (1999)

[Rad82] Radyno, Ya.V.: Linear equations and bornology. BSU, Minsk (1982) (in Russian)

[RC10] Ramirez, L.E.S., Coimbra, C.F.M.: On the selection and meaning of variable order operators for dynamic modeling. Intern. J. Diff. Equ., 16 pp. (2010)

[RS80] Reed, M., Simon, B.: Methods of Modern Mathematical Physics I. Functional Analysis. Academic Press (1980)

[RSc82] Rempel, S., Schulze, B.-W.: Index Theory of Elliptic Boundary Problems. Akademie-Verlag, Berlin (1982)

[Ren08] Render, H.: Real Bargmann spaces, Fischer decompositions and Sets of uniqueness for polyharmonic functions. Duke Math. J., **142** (2), 313–352 (2008)

[Ren10] Render, H.: Goursat and Dirichlet problems for holomorphic partial differential equations. Comput. Meth. Funct. Theory. **10**, 519–554 (2010)

[Ri10] Riesz, F.: Untersuchungen über Systeme integrierbarer Funktionen. Mathematische Annalen **69** (4), 449–497 (1910)

[R64] Robertson, A.P., Robertson, W.: Topological Vector Spaces. Cambridge University Press (1964)

[Ros75] Ross, B. (ed): Proceedings of the first international conference "Fractional Calculus and Its Applications. University of New Haven, June 1974", Springer, Berlin-Heidelberg-New-York (1975)

[RSh10] Rossikhin, Yu.A., Shitikova, M.V.: Application of fractional calculus for dynamic problems of solid mechanics: Novel trends and recent results. Applied Mechanics Reviews, **63**, 52 pp. (2010)

[Roz90] Rozovskii, B.L., Stochastic Evolution Systems. Linear Theory and Applications to Nonlinear Filtering. Kluwer Academic Publishers, Dordrecht (1990)

[Rub96] Rubin, B.: Fractional Integrals and Potentials. Pitman Monographs and Surveys in Pure and Applied Math., **82**, Longman (1996)

[Sam77] Samko, S.G.: Fundamental functions vanishing on a given set and division by functions. Mathematical notes, **21** (5), 379–386 (1977)

[Sam80] Samko, S.G.: Generalized Riesz potentials and hypersingular integrals with homogeneous characteristics; their symbols and inversion. Proceedings of Steklov Inst. Math., **2**, 173–243 (1983)

[Sam82] Samko, S.G.: Denseness of Lizorkin-type spaces Φ_V in L_p. Mat. Notes, **31**, 432–437 (1982)

[Sam83] Samarov, K.L.: Solution of the Cauchy problem for the Schrödinger equation for a relativistic free particle. Soviet Phys. Docl., **271** (2), 334–337 (1983)

[Sam95] Samko, S.G.: Fractional integration and differentiation of variable order. Analysis Mathematica, **21**, 213–236 (1995)

[Sam95] Samko, S.G.: Denseness of the spaces Φ_V of Lizorkin type in the mixed in $L_p(\mathbb{R}^n)$-spaces. Stud. Math. **113**, 199–210 (1995)

[SR93] Samko, S.G., Ross, B.: Integration and differentiation to a variable fractional order. Integral Transforms and Special Functions, **1** (4), 277–300 (1993)

[SKM87] Samko, S.G., Kilbas, A.A., Marichev, O.I.: Fractional Integrals and Derivatives: Theory and Applications. Gordon and Breach Science Publishers, New York and London (1993)

[ST94] Samorodnitsky, G., Taqqu, M.S. Stable Non-Gaussian Random Processes. Stochastic Models with Infinite Variance. Chapman & Hall, New York (1994)

[Sat99] Sato, K-i.: Lévy Processes and Infinitely Divisible Distributions. Cambridge University Press (1999)

[Sat59] Sato, M.: Theory of Hyperfunctions, I. Journal of the Faculty of Science, University of Tokyo. Sect. 1, Mathematics, astronomy, physics, chemistry, **8** (1), 139–193 (1959)

[Sat60] Sato, M.: Theory of Hyperfunctions, II. Journal of the Faculty of Science, University of Tokyo. Sect. 1, Mathematics, astronomy, physics, chemistry, **8** (2), 387–437 (1960)

[SKK73] Sato, M., Kawai, T., Kashiwara, M.: Microfunctions and pseudo-differential equations. Lecture Notes in Mathematics, **287**, Springer-Verlag, Berlin-Heidelberg-New York, 265–529 (1973)

[Sax01] Saxton, M.J.: Anomalous Subdiffusion in Fluorescence Photobleaching Recovery: A Monte Carlo Study. Biophys. J., **81**(4), 2226–2240 (2001)

[SJ97] Saxton, M.J., Jacobson, K.: Single-particle tracking: applications to membrane dynamics. Ann. Rev. Biophys. Biomol. Struct., **26**, 373–399 (1997)

[Say06] Saydamatov, E.M.: Well-posedness of the Cauchy problem for inhomogeneous time-fractional pseudo-differential equations, Frac. Calc. Appl. Anal., **9**(1), 1–16 (2006)

[Say07] Saydamatov, E.M.: Well-posedness of general nonlocal nonhomogeneous boundary value problems for pseudo-differential equations with partial derivatives. Siberian Advances in Mathematics, **17** (3), 213–226 (2007)

[SGM00] Scalas, E., Gorenflo, R., Mainardi, F.: Fractional calculus and continuous-time finance. Physica A **284**, 376–384 (2000)

[SW66] Schaefer, H.H., Wolff, M.P.: Topological vector spaces, 2nd ed. Springer (1999)

[SLDYL01] Schertzer, D., Larchevêque, M., Duan, J., Yanovsky, V.V., Lovejoy, S.: Fractional Fokker-Planck equation for nonlinear stochastic differential equations driven by non-Gaussian Lévy stable noises. J. Math. Phys., **42**(1), 200–212 (2001)

[ScS01] Schmitt, F.G., Seuront, L.: Multifractal random walk in copepod behavior. Physica A, **301**, 375–396 (2001)

[Sch90] Schneider, W.R.: Fractional diffusion. Lect. Notes Phys. **355**, Heidelberg, Springer, 276–286 (1990)

[SW89] Schneider, W.R., Wyss, W.: Fractional diffusion and wave equations. Journal of Mathematical Physics, **30**, 134–144 (1989)

[Sch51] Schwartz, L.: Théorie des distributions I, II. Hermann, Paris (1951)

[Sha53] Shapiro, Z.Ya.: On general boundary value problems of elliptic type. Isz. Akad. Nauk, Math. Ser. **17**, 539–562 (1953)

[Shu78] Shubin, M.A.: Pseudodifferential Operators and Spectral Theory. Springer (2001)

[Sit05] Situ, R: Theory of stochastic differential equations with jumps and applications. Springer (2005)

[Sob35] Sobolev, S.L.: Le probléme de Cauchy dans l'espace des fonctionelles (Russian and French). Dokl. Akad. Nauk SSSR (Comptes Rend. l'Acad. Sci. URSS) **3** (8), **7** (67), 291–294 (1935)

[Sob36] Sobolev, S.L.: Méthode nouvelle á resoudre le probléme de Cauchy pour les équations linéaires hyperboliques normales. Mat. Sb. **1** (43), 39–72 (1936)

[Sob38] Sobolev, S.L.: On a theorem in functional analysis. Mat. Sb. **4** (46), 471–497 (1938)(Russian) (Engl. transl.: Amer. Math. Soc. Transl. **34** (2), 39–68 (1963))

[Sob50] Sobolev, S.L.: Some applications of functional analysis in mathematical physics (Russian). 1st ed.: Leningr. Goz. Univ., Leningrad (1950) 3rd enlarged ed.: Izd. Nauka, Moskva (1988) (Engl. transl.: Amer. Math. Soc., Providence, R. I. (1963))

[Sob74] Sobolev, S.L.: Introduction to the theory of cubature formulas. Nauka, Moscow (1974)(in Russian) (Engl. transl: Cubature formulas and modern analysis: An introduction. Gordon and Breach (1992))

[SK06] Sokolov, I.M., Klafter, J.: Field-induced dispersion in subdiffusion. Phys. Rev. Lett. **97**, 140602 (2006)

[SCK04] Sokolov, I.M., Chechkin, A.V., Klafter, J.: Distributed-order fractional kinetics. Acta Physica Polonica (2004)

[Sot01] Sottinen, T.: Fractional Brownian motion, random walks and binary market models. Finance and Stochastics. **5**, 343–355 (2001)

[SCK05] Soon, C.M., Coimbra, C.F.M., Kobayashi, M.H.: The variable viscoelasticity oscillator. Ann. Phys. (Leipzig) **14**, 378–389 (2005)

[Spi01] Spitzer, F.: Principles of Random Walk. Springer (2001)

[Ste58] Stein, E.M.: Localization and summability of multiple Fourier series. Acta Math., **100**, 93–147 (1958)

[Ste70] Stein, E.M.: Singular Integrals and Differentiability Properties of Functions. Princeton University Press (1970)

[S88] Sternin, B.Y., Shatalov, V.E.: Differential equations on complex manifolds and Maslov's canonical operators. Sov. Math. Survs. 43, 99–124 (1988)

[Sto13] Stojanović, M.: Well-Posedness of diffusion-wave problem with arbitrary finite number of time fractional derivatives in Sobolev spaces H^s. Frac. Calc. Appl. Anal. **13** (1), 21–22 (2010)

[SCWC11] Sun, H.G., Chen, W., Wei, H., Chen, Y.Q.: A comparative study of constant-order and variable-order fractional models in characterizing memory property of systems. Eur. Phys. J. Special Topics **193**, 185–192 (2011)

[Sut86] Sutradhar, B.C.: On the characteristic function of multivariate Student t-distribution. The Canadian Journal of Statistics (La Revue Canadienne de Statistique), **14** (4), 329–337 (1986)

[Suz64] Suzuki, H.: Analytic hypoelliptic differential operators of first order in two independent variables. J. Math. Society Japan. **16** (4), 367–374 (1964)

[Tac36] Täcklind, S.: Sur les classes quasianalitiques des solutions des l'equations aux derivées partielles du type parabolique. Nord. Acta. Regial. Sociatis schientiarum Uppsaliensis. Ser. 4, **10** (3), 3–55 (1936)

[Tai91] Taira, K.: Boundary Value Problems and Markov Processes. Lecture notes in Mathematics, 1499. Springer-Verlag, Berlin-Heidelberg- New York-Tokyo (1991)

[Tam06] Tamura, M.: On uniqueness in the Cauchy problem for systems for systems with partial analytic coefficients. Osaka J. Math. **43**, 751–769 (2006)

[Taq75] Taqqu, M.S.: Weak convergence to fractional Brownian motion and to the Rosenblatt process. Z. Wahrsch. Verw. Gebiete **31**, 287–302 (1975)

[Tat14] Tatar, S.: Existence and uniqueness in an inverse source problem for a one-dimensional time-fractional diffusion equation. (available online: http://person.zirve.edu.tr/statar/s12.pdf)

[Tay81] Taylor, M.: Pseudo differential operators. Prinston University Press (1981)

[Tik35] Tikhonov, A.N.: Uniqueness theorems for heat equations, Mat. Sb., **42** (2), 199–216 (1935)

[TS66] Tikhonov, A.N., Samarskij, A.A.: Equations of Mathematical Physics. Pergamon-Press, New York (1963)

[Ton28] Tonelli, L.: Su un problema di Abel.] Math. Ann. **99**, 183–199 (1928)

[Tre80] Treves, F.: Introduction to Pseudo-Differential and Fourier Integral Operators. Plenum Publishing Co., New York (1980)

[Tri77] Triebel, H.: Interpolation Theory, Function Spaces, Differential Operators. Birkhäuser, Basel (1977)

[Tri83] Triebel, H.: Theory of Function Spaces. Leipzig, Birkhäuser Verlag, Basel-Boston-Stuttgart (1983)

[Tsa09] Tsallis, C.: Introduction to Nonextensive Statistical Mechanics: Approaching a Complex World. Springer, New York (2009)

[Tsk94] Tskhovrebadze, G.D.: On a multipoint boundary value problem for linear ordinary differential equations with singularities. Archivum Mathematicum, **30** (3), 171–206 (1994)

[TU94] Turmetov, B. Kh., Umarov, S.R.: On a boundary value problem for an equation with the fractional derivative. Russ. Acad. Sci., Dokl., Math., **48**, 579–582 (1994)

[Uma86] Umarov, S.R.: Boundary value problems for differential operator and pseudo-differential equations. Izv. Acad. Sci., RU, **4**, 38–42 (1986) (in Russian)

[Uma88] Umarov, S.R.: A non-triviality criterion of spaces of infinitely differentiable vectors of an operator with empty spectrum. Docl. Ac. Sci. RU., **1**, 11–13 (1988)

[Uma91-1] Umarov, S.R.: Algebra of pseudo-differential operators with variable analytic symbols and well-posedness of corresponding equations. Differential Equations. **27** (6), 1056–1063 (1991)

[Uma91-2] Umarov, S.R.: On well-posedness of systems of pseudo-differential equations with variable analytic symbols. Russ. Ac. Sci., Dokl., Math., **318** (4), 835–839 (1991)

[Uma92] Umarov, S.R.: On well-posedness of boundary value problems for pseudo-differential equations with analytic symbols. Russ. Acad. Sci., Dokl., Math. **45**, 229–233 (1992)

[Uma94] Umarov, S.R.: On some boundary value problems for elliptic equations with a boundary operator of fractional order. Russ. Acad. Sci., Dokl., Math. **48**, 655–658 (1994)

[Uma97] Umarov, S.R.: Nonlocal boundary value problems for pseudo-differential and differential operator equations I. Differ. Equations, **33**, 831–840 (1997)

[Uma98] Umarov, S.R.: Nonlocal boundary value problems for pseudo-differential and differential operator equations II. Differ. Equations, **34**, 374–381 (1988)

[Uma12] Umarov, S.R.: On fractional Duhamel's principle and its applications. J. Differential Equations **252** (10), 5217–5234 (2012)

[Uma14] Umarov, S.R.: Pseudo-differential operators with meromorphic symbols and systems of complex differential equations. Complex variables and elliptic equations: An International Journal. **60** (6), 829–863 (2015) DOI: 10.1080/17476933.2014.979812

[Uma15] Umarov, S.R.: Continuous time random walk models associated with distributed order diffusion equations. Frac. Calc. Appl. Anal. **18** (3), 821–837 (2015)

[UG05-1] Umarov, S.R., Gorenflo, R.: On multi-dimensional symmetric random walk models approximating fractional diffusion processes. Frac. Calc. Appl. Anal., **8**, 73–88 (2005)

[UG05-2] Umarov, S.R., Gorenflo, R. The Cauchy and multipoint problem for distributed order fractional differential equations. ZAA, **24**, 449–466 (2005)

[US06] Umarov, S.G., Steinberg, St. Random walk models associated with distributed fractional order differential equations. IMS Lecture Notes - Monograph Series. High Dimensional Probability, **51**, 117–127 (2006)

[US06] Umarov, S.R., Saydamatov, E.M.: A fractional analog of the Duhamel principle. Frac. Calc. Appl. Anal, **9** (1), 57–70 (2006)

[US07] Umarov, S.R., Saydamatov E.M.: A generalization of the Duhamel principle for fractional order differential equations. Doklady Mathematics, 75 (1), 94–96 (2007)

[US09] Umarov, S.R., Steinberg, St.: Variable order differential equations with piecewise constant order-function and diffusion with changing modes. ZAA, **28** (4), 131–150 (2009)

[UDN14] Umarov, S.R., Daum, F., Nelson, K.: Fractional generalizations of filtering problems and their associated fractional Zakai equations. Frac. Calc. and Appl. Anal., **17** (3), 745–764 (2014)

[UZ99] Uchaykin, V.V., Zolotarev, V.M.: Chance and Stability. Stable Distributions and their Applications. VSP, Utrecht (1999)

[Uch04] Uchida, M.: Hörmander form and uniqueness for the Cauchy problem. Advances in Mathematics, **189**, 237–245 (2004)

[VC11] Valerio, D., Sa-da Costa, J.: Variable-order fractional derivatives and their numerical approximations I - real orders. Signal processing, **91** (3), 470–483 (2011)

[Van89] Van, T. D.: On the pseudo-differential operators with real analytic symbol and their applications. J. Fac. Sci. Univ. Tokyo, IA, Math. **36**, 803–825 (1989)

[VH94] Van, T.D., Haó, D.N.: Differential Operators of Infinite Order With Real Arguments and Their Applications. Singapore, World Scientific (1994)

[Vla79] Vladimirov, V.S.: Generalized Functions in Mathematical Physics. Mir Publishers, Moscow (1979)

[Vol63] Volevich, L.R.: A problem in linear programming stemming from differential equations. Uspekhi Mat. Nauk, **18**, 3 (111), 155–162 (1963)

[Wie28] Wiener, N.: Differential space. Journal of Mathematical Physics **2**, 131–174 (1923)

[Wid46] Widder, D.V.: The Laplace transform. Princeton University Press (1946)

[WZh14] Wei, T., Zhang, Z.Q.: Stable numerical solution to a Cauchy problem for a time fractional diffusion equation. Engineering analysis with boundary elements, **40**, 128–137 (2014)

[WEKN04] Weiss, M., Elsner, M., Kartberg, F., Nilsson, T.: Anomalous subdiffusion Is a measure for cytoplasmic crowding in living cells. Biophysical Journal, **87**, 3518–3524 (2004)

[Wei12] Weisz, F.: Summability of multidimensional trigonometric Fourier series. (2012) Available at http://arxiv.org/abs/1206.1789

[Won99] M.M. Wong, An Introduction to Pseudo-differential Operators, 2nd ed. World Scientific, Singapore (1999)

[Wis86] Wyss, W.: The fractional diffusion equation. J. Math. Phys. **27**, 2782–2785 (1986)

[Zak69] Zakai, M.: On the optimal filtering of diffusion processes. Z. Wahrsch. Verw. Gebiete. **11** (3), 230–243 (1969)

[Zas02] Zaslavsky, G.: Chaos, fractional kinetics, and anomalous transport. Physics Reports, **371**, 461–580 (2002)

[ZhX11] Zhang, Y., Xiang Xu.: Inverse source problem for a fractional diffusion equation. Inverse problems, **27**, 035010, 12 pp. (2011)

[Zh14] Zhang, H.-E.: Multiple positive solutions of nonlinear BVPs for differential systems involving integral conditions. Boundary Value Problems, **6**, 13 pp. (2014)

[Zyg45] Zygmund, A.: Smooth functions. Duke Math. J. **12** (1), 47–76 (1945)

Index

S. Umarov, *Introduction to Fractional and Pseudo-Differential Equations with Singular Symbols*, Developments in Mathematics 41,
DOI 10.1007/978-3-319-20771-1

Printed in the United States
By Bookmasters